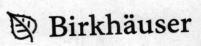

Daniel Alpay

An Advanced
Complex Analysis
Problem Book

Topological Vector Spaces, Functional Analysis, and Hilbert Spaces of Analytic Functions

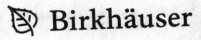 Birkhäuser

Daniel Alpay
Ben-Gurion University of the Negev
Beer Sheva, Israel

ISBN 978-3-319-16058-0   ISBN 978-3-319-16059-7   (eBook)
DOI 10.1007/978-3-319-16059-7

Library of Congress Control Number: 2015956341

Mathematics Subject Classification (2010): 97I80, 46C07, 46E20, 46E22, 47B32

Springer Cham Heidelberg New York Dordrecht London

Printed on acid-free paper

Springer International Publishing AG Switzerland is part of Springer Science+Business Media
(www.birkhauser-science.com)

It is a pleasure to dedicate this work to my son Gabriel. It is also a pleasure to thank the Earl Katz family for endowing the chair (Earl Katz Family Chair in Algebraic System Theory), which supported this research.

Contents

I   Analytic Functions   **9**

1   Algebraic Prerequisites   **11**
   1.1   Sets   11
   1.2   Groups   15
   1.3   Vector spaces   19
   1.4   Fields   23
   1.5   Matrices   24
   1.6   Positive matrices   28
   1.7   Rational functions   34
   1.8   Solutions of the exercises   42

2   Analytic Functions   **61**
   2.1   Some warm-up exercises   61
   2.2   Some important theorems to recall   75
   2.3   Cauchy's formula and Laurent expansion   77
   2.4   Star-shaped and simply connected sets   82
   2.5   Various   84
   2.6   Harmonic and subharmonic functions   85
   2.7   Meromorphic functions in $\mathbb{C}$ 89
   2.8   Solutions of the exercises   90

II   Topology and Functional Analysis   **119**

3   Topological Spaces   **121**
   3.1   Topological spaces   122
   3.2   Compact sets   127
   3.3   Connected sets   131
   3.4   Continuous maps   132
   3.5   Products   136
   3.6   Closed, open and proper maps   137
   3.7   Quotient topology   140

3.8   Manifolds and surfaces 142
3.9   Metric spaces 148
3.10 Solutions of the exercises 157

4   Normed Spaces   **199**
4.1   Normed Banach and Hilbert spaces 199
4.2   Operators on normed spaces 207
4.3   Unbounded operators 222
4.4   Indefinite inner product spaces 225
4.5   Solutions of the exercises 227

5   Locally Convex Topological Vector Spaces   **249**
5.1   Topological vector spaces 249
5.2   Countably normed spaces and Fréchet spaces 253
5.3   Operators in countably normed spaces 257
5.4   Dual of a Fréchet space 259
5.5   Positive operators 262
5.6   Topology of the space of analytic functions 264
5.7   Normal families 267
5.8   The dual of the space of analytic functions 268
5.9   Solutions of the exercises 268

6   Some Functional Analysis   **285**
6.1   Fourier transform 285
6.2   Stieltjes integral 298
6.3   Density results 302
6.4   Solutions of the exercises 306

III   Hilbert Spaces of Analytic Functions   **329**

7   Reproducing Kernel Hilbert Spaces   **331**
7.1   Positive definite kernels 332
7.2   Examples of positive definite functions and kernels 338
7.3   Conditionally negative functions 347
7.4   Vector-valued functions 350
7.5   Reproducing kernel Hilbert spaces 352
7.6   Linear operators in reproducing kernel Hilbert spaces . . . 361
7.7   Finite-dimensional reproducing kernel spaces 367
7.8   Solutions of the exercises 376

Contents

8 Hardy Spaces **405**
  8.1 Reproducing kernel Hilbert spaces of analytic functions 405
  8.2 The Hardy space of the open unit disk 406
  8.3 Some operator theory in $\mathbf{H_2}(\mathbb{D})$ 409
  8.4 Composition operators 412
  8.5 Cuntz relations 413
  8.6 The Hardy space of the open upper half-plane 415
  8.7 The fractional Hardy space $\mathcal{H}_2^\nu$ of the open
       upper half-plane 420
  8.8 Solutions of the exercises 421

9 de Branges–Rovnyak Spaces **443**
  9.1 de Branges–Rovnyak spaces 443
  9.2 Interpolation 448
  9.3 Carathéodory's theorem 449
  9.4 A few words on Schur analysis 450
  9.5 Solutions of the exercises 453

10 Bergman Spaces **461**
  10.1 The Bergman space of analytic functions analytic in $\mathbb{D}$ 461
  10.2 The Bergman space of analytic functions analytic in an ellipse ... 463
  10.3 The Bergman space of the annulus 464
  10.4 The Bergman spaces of polyanalytic functions 464
  10.5 Solutions of the exercises 466

11 Fock Spaces **475**
  11.1 The Bargmann–Fock–Segal spaces of analytic functions 475
  11.2 The Bargmann–Fock–Segal spaces of
        polyanalytic functions 478
  11.3 Solutions of the exercises 479

Bibliography **487**

Index **509**

Name Index **518**

Notation Index **521**

Prologue

Prologue

Quite often, at the undergraduate level, topology is taught in parallel with complex analysis, and functional analysis is taught the following year. Then, interactions of complex analysis with topology and functional analysis are minimal. More precisely:

On the topology side, the Riemann sphere $\widehat{\mathbb{C}}$ is introduced, usually by adding the point at infinity in an *ad hoc* way, $\widehat{\mathbb{C}} = \mathbb{C} \cup \{\infty\}$. It may allow us to define the notion of simply connected set, at least in a pictorial way, using stereographic projection: A subset of the complex plane is simply connected if the complement of its image under stereographic projection is connected in $\widehat{\mathbb{C}}$ (this definition does not require the set to be open, but see the discussion [91, p. 40] in Burckel's book for a proviso). The Riemann sphere is not seen as the one point compactification of $\mathbb{C}$, let alone as a compact manifold. Another example is uniform convergence on compact sets, which may be dealt with by using Morera's theorem, but no hint is given that it corresponds to convergence in a metric space. The student may also meet as an exercise the fact that a meromorphic function on the Riemann sphere is a rational function, but no similar characterization of functions meromorphic on the torus will in general be given, nor will elliptic functions (or, automorphic functions, for that matter) be discussed. Although all the needed tools are at hand, the condition for conformal equivalence of two annuli seems a bit beyond the scope of a first course. Finally, a proof of Riemann's theorem is difficult to present without precise compactness arguments.

On the functional analysis side, analytic functions are really considered as individuals, and not as elements of some underlying space. The student does not meet Banach spaces, or even Hilbert spaces, of analytic functions. And if she/he meets a Fréchet space, let alone a nuclear space, of analytic functions (for instance the space of entire functions; see [164, 165] and [251, pp. 105–106]) it would be analogous to Monsieur Jourdain's discovery that he had been speaking prose for 40 years without realizing it [238].

The aim of this book is to illustrate, at the beginning graduate level, some of the connections between functional analysis and the theory of functions of one complex variable. Our road-map when writing the present book was to help a beginning graduate student to get familiar with various aspects of complex variables and functional analysis (such as the ones alluded to above) not always present in the curriculum. *We use as a bridge the notion of reproducing kernel Hilbert space.* The book contains exercises of three kinds, aimed at a beginning graduate student: On analysis (in a broad sense, essentially functional analysis), on analytic functions, and at the interface of these two topics. There is also some review material, for instance elementary results on topological spaces.

Connections of complex analysis with other fields are numerous. On two completely different avenues, operator theory on the one hand and the theory of stochastic processes on the other hand give two examples of a wide spectrum of applications. This is why we tried to present, or sometimes merely hint at, relationships to other fields, such as stochastic processes, covariance functions, Gaussian measures. We elaborate on these examples in the second part of the introduction.

The book consists of three parts. In the first part, called *Algebra and analytic functions*, we review various elementary but useful results in linear algebra, rational functions, and related topics. We also present exercises and complements in analytic functions.

In the second part, called *Topology and functional analysis*, we give exercises in topological spaces, manifolds and metric spaces, (and in particular countably normed spaces). We also review some facts on indefinite inner product spaces, and in particular Pontryagin spaces. Operator theory in these spaces has been a source of definitions of new class of analytic functions, some of which are presented in the book. We also discuss the Stieltjes integral. We assume known elementary measure theory, and in particular the Lebesgue integral. We refer the reader to Chapter 15 of our previous book [CABP]. Although some complex analysis does appear in this part, the focus is on topology and functional analysis (at an elementary level). Section 6.3 on density results may be of independent interest.

In the third part, entitled *Spaces of analytic functions*, we first discuss positive definite kernels and the associated reproducing kernel Hilbert spaces. We then focus on the case of Hilbert spaces of analytic functions, and important examples of these, such as the Bargmann–Segal–Fock space, the Bergman space and the

Prologue

Hardy space. Each of these spaces has a geometric description and an analytic one, and we discuss these in detail.

Positive definite kernels

We here briefly explain where positive definite functions play a key role. The definition of these functions is recalled in Section 7.1. We here start from an equivalent definition (see Exercise 7.1.7). Let Ω denote some set. The function

$$K(z,w) \; : \; \Omega \times \Omega \; \longrightarrow \; \mathbb{C}$$

is positive definite if it can be written as (7.1.3), that is:

$$K(z,w) = \langle h_w, h_z \rangle_{\mathscr{H}},$$

where $\mathscr{H}$ is a Hilbert space, with inner product $\langle \cdot, \cdot \rangle_{\mathscr{H}}$, and where $z \mapsto h_z$ is a $\mathscr{H}$-valued function defined on Ω. For instance the functions

$$K(z,w) = \frac{1}{1 - z\overline{w}}, \tag{0.0.1}$$

$$K(z,w) = \frac{1}{(1 - z\overline{w})^2}, \tag{0.0.2}$$

where z, w belong to the open unit disk $\mathbb{D}$, the functions

$$K(z,w) = \frac{1}{(z + \overline{w})^{1+\nu}}, \quad \nu \in (-1, \infty), \tag{0.0.3}$$

where z, w run through the open right half-plane $\mathbb{C}_r$, the function

$$K(a,b) = e^{\langle a,b \rangle_{\ell_2}}, \quad a, b \in \ell_2,$$

where $\ell_2 \, (= \ell_2(\mathbb{N}_0))$ denotes the space of square summable sequences of complex numbers indexed by $\mathbb{N}_0 = \{0, 1, 2, \dots\}$, and the function

$$K(t,s) = \min(t,s), \quad t, s \in \mathbb{R}_+, \tag{0.0.4}$$

(where $\mathbb{R}_+ = [0, \infty)$) are positive definite in their respective domain of definition. Sometimes classical formulas intermingle with reproducing kernel formulas in a maybe unexpected way. For instance Taylor's formula with remainder applied to functions f with m continuous derivatives in $[0,1]$ and such that

$$f(0) = \cdots = f^{(m-1)}(0) = 0,$$

that is:

$$f(a) = f(0) + af'(0) + a^2 \frac{f^{(2)}(0)}{2!} + a^{m-1} \frac{f^{(m-1)}(0)}{(m-1)!} + \cdots +$$

$$+ \int_0^a \frac{(a-x)^{m-1}}{(m-1)!} f^{(m)}(x) dx$$

$$= \int_0^a \frac{(a-x)^{m-1}}{(m-1)!} f^{(m)}(x) dx, \quad a \in [0,1],$$

is such an example when this space of functions is endowed with an appropriate inner product. See Exercise 7.5.6, where it is shown that the reproducing kernel of this space is

$$K_m(x, y) = \int_0^1 G(u, x)G(u, y)du,$$

with

$$G(t, x) = \begin{cases} \frac{(x-t)^{m-1}}{(m-1)!}, & t \leq x, \\ 0, & \text{else.} \end{cases}$$

These different examples are the subject of various exercises in the book, but the reader might want to find already proofs that they are instances of positive definite functions.

Associated to a positive definite function K on Ω is a Hilbert space $\mathcal{H}(K)$ of functions defined on Ω, which contains the function $z \mapsto K(z, w)$ for every $w \in \Omega$ and uniquely defined by the reproducing kernel property:

$$\langle f(\cdot), K(\cdot, w)\rangle_{\mathcal{H}(K)} = f(w),$$

for all $f \in \mathcal{H}(K)$ and all $w \in \Omega$.

For the six examples above the associated reproducing kernel Hilbert space is respectively the Hardy space of the unit disk, the Bergman space of the unit disk [70], the fractional ν-Hardy space, the Fock space (or, more precisely, the symmetric Fock space associated with $\mathbb{C}$; see [60, 61]), and Sobolev spaces (see Remark 6.1.26 for the definition).

The associated reproducing kernel Hilbert space often provides a convenient structure to solve underlying problems. For instance, shift-invariant subspaces of the space ℓ_2 have a nice representation when the problem is transposed in the Hardy space. This is Beurling's theorem; see Exercise 8.3.9 for a finite-dimensional version of this result. In the case of the Bergman space, problems in complex variables are considered, in the case of the Fock space, problems from quantum physics, and problems from stochastic processes for the Sobolev space. The function $K(t, s) = \min(t, s)$ (with $t, s \in \mathbb{R}_+$) is the covariance function of the Brownian motion. In fact, any positive definite function is a covariance function of an underlying stochastic process: One can always choose $\mathcal{H}$ in (7.1.3) to be of the form

$$\mathcal{H} = \mathbf{L}_2(E, \mathcal{B}, P),$$

where $P(E) = 1$, (that is, $(E, \mathcal{B}, P)$ is a probability space).

Remark 0.0.1. Quite often, the reproducing kernel Hilbert space associated with a given positive definite function has a number of equivalent characterizations, each stressing a particular aspect of the space. We mention in particular:

Prologue

1. The reproducing kernel itself.

2. An analytic characterization, in terms of the coefficients of an underlying orthonormal basis.

3. A geometric characterization, in terms of an underlying measure.

4. Last, but not least, a transform-like characterization, in terms of an appropriate transform (for instance, the Bargmann–Fock–Segal transform for the Fock space).

See Remark 11.1.2 for a related discussion.

As illustrated by the kernel (0.0.3), it is not always easy to verify that a given function is positive definite. Another example is given by the function

$$\frac{1}{2}\left(|t|^{2H} + |s|^{2H} - |t-s|^{2H}\right),\tag{0.0.5}$$

where $0 < H < 1$, which is positive definite on $\mathbb{R}$. It is the covariance function of the fractional Brownian motion (the case $H = \frac{1}{2}$ corresponds to the Brownian motion). The positivity of (0.0.5) is the topic of Exercise 7.2.9.

Other examples of functions which are not readily seen to be positive definite arise from the theory of polyanalytic functions in one variable; see [57, pp. 169–170], where one can find in particular the following examples: The functions

$$K_N(z,w) = \frac{N}{(1-z\overline{w})^{2N}} \sum_{k=0}^{N-1} (-1)^k \binom{N}{k+1}\binom{N+k}{N} |1-z\overline{w}|^{2(N-1-k)} |z-w|^{2k}\tag{0.0.6}$$

and

$$F_N(z,w) = e^{z\overline{w}} \sum_{k=0}^{N-1} (-1)^k \binom{N}{k+1}\frac{1}{k!}|z-w|^{2k}\tag{0.0.7}$$

are positive for $N = 1, 2, \ldots$, respectively in the open unit disk and in the whole complex plane. Note that $K_1(z,w)$ is equal to (0.0.2).

An apparently unrelated example is as follows: Now Ω is the group S_n of permutations of $\{1, 2, \ldots, n\}$. We denote by i the parity of the signature σ (see Definition 7.2.2). Then for $q \in [-1, 1]$ the function

$$q^{i(\sigma_1^{-1}\sigma_2)}$$

is positive definite in S_n. This example plays an important role in the theory of q-Fock spaces.

Some notations. An index of notations appears at the end of the book. We also mention the following:

- We set $\mathbb{N}_0 = \{0\} \cup \mathbb{N} = \{0, 1, 2, \ldots\}$.

- Given a Hilbert space $\mathcal{H}$, we will remove the dependence on $\mathcal{H}$ in the inner product and associated norm when these are understood from the context. For instance we often will write $\langle \cdot, \cdot \rangle$ rather that $\langle \cdot, \cdot \rangle_{\mathcal{H}}$. A similar remark holds for Banach spaces norms and dualities between topological vector spaces.

- When considering a Lebesgue space $\mathbf{L}_2(\mathbb{R}, \mathcal{B}, d\mu)$ (where $\mathcal{B}$ denote the underlying sigma-algebra) we often remove the dependence on the latter, and write $\mathbf{L}_2(\mathbb{R}, d\mu)$. When $d\mu$ is the Lebesgue measure dx we often write $\mathbf{L}_2(\mathbb{R})$ rather than $\mathbf{L}_2(\mathbb{R}, \mathcal{B}, dx)$. We write sometimes $\langle \cdot, \cdot \rangle_{d\mu}$ or $\langle \cdot, \cdot \rangle_2$ when $d\mu$ is the Lebesgue measure for the inner product in $\mathbf{L}_2(\mathbb{R}, \mathcal{B}, d\mu)$.

- Still related to inner products, we will sometimes put the variable inside the inner product, and for instance write

$$\langle f(z), g(z) \rangle \tag{0.0.8}$$

  rather than $\langle f, g \rangle$ or $\langle f(\cdot), g(\cdot) \rangle$.

- References to our previous book [10] are given as [CAPB].

Prerequisites. The student will need as prerequisites:

(a) Elementary theory of function of a complex variable (we do not recall in detail facts appearing in [CAPB]).
(b) Elementary measure theory and integration (some facts beyond what is recalled in Chapter 15 of [CAPB] are needed and will be mentioned in the text).

Final remarks

In this book, the exercises related to analytic functions are only in the one variable setting. Exercises left without solutions are labeled as *Question* or *Problem*, depending on the assumed difficulty. Sometimes *Question* also refers to an exercise given in [CABP], and whose solution is not repeated here. In a book in preparation we consider settings which go beyond one variable, including in particular some aspects of quaternionic analysis and function theory on a compact real Riemann surface.

Although most, if not all, of the material is classical and has a long history, we tried to always give the specific sources we used, or which inspired us, for a given exercise. For instance, quite a number of exercises has been taken from, or inspired from the reading of, Narasimhan's book on compact Riemann surfaces [242] and Bergman's book on the kernel function [70]. We mention in particular exercises 3.6.5, and 5.3.5 for [242] and exercises 2.2.8, 2.6.7 for [70]. The work [248] is a very friendly introduction to Fock spaces, and has been the direct inspiration of a number of exercises and questions, such as Question 7.2.3. Some of the material has been taken from a chapter [39] from an unpublished project on the history of reproducing kernels.

We also note that some exercises (for instance the construction of the tensor product; see Exercise 1.3.5) are in fact classical results which the student will have seen in class.

> In the pages following this prologue we single out a number of exercises which we think could be of special interest to the reader.

Acknowledgment. Quite a number of exercises were taken from, or inspired by, collaborations. I wish to thank in particular my student Guy Salomon (see for instance Exercises 5.5.2, 5.4.6 and 5.4.8), Professor Mamadou Mboup (see for instance Question 6.1.30) and Professor Izchak Lewkowicz (see for instance Exercise 2.1.8). It is a pleasure to thank Prof. John Snygg for a discussion on Fourier rotation operators in Tartu, Estonia, in August 2014; see Exercise 6.1.33. It is a pleasure to thank Doctor Haim Attia and Doctor Yossi Peretz for a very careful reading of various versions of this book.

Special thanks are due to Professor Palle Jorgensen. Throughout our collaboration, I came to learn a wide range of new mathematical topics, some of which have been translated into exercises; see for instance Exercises 1.6.2 and 7.5.11.

Last but not least, it is a pleasure to thank Doctor Thomas Hempfling from Springer Basel for his help and encouragement over the years for this book and quite a number of other projects.

A choice of exercises. The book contains 323 exercises (besides questions or problems given without answers). We here give a selection of exercises which could be of special interest to the reader.

1. Exercise 1.6.14 deals with positive Toeplitz matrices, but has connections with stochastic processes, maximum entropy and the theory of functions analytic and with a positive real part in the open unit disk.

2. Exercise 1.7.8 exhibits a univalence property for a certain family of rational functions. The arguments are elementary.

3. Exercise 2.1.1 deals with Kaluza's theorem, which characterizes a class of functions f analytic in the open unit disk, with value 1 at the origin, and such that the Taylor coefficients of index greater than or equal to 1 of $1/f$ are negative. These functions relate to an important class of reproducing kernels, called complete Nevanlinna–Pick kernels.

4. Exercise 3.6.8 presents an open mapping which is not topologically equivalent to an analytic function.

5. Exercise 3.8.27 presents the result on the conformal equivalence of two annuli. Only elementary tools are used.

6. Exercise 4.2.18 deals with lower semi-continuous functions, a notion not always so well known, but which plays an important role in the theory of topological vector spaces.

7. In Exercises 5.4.6 and 5.4.8 we discuss a maybe not so well-known aspect of the space of tempered distributions and of a related space of distributions. Behind the lines lies a new class of topological algebras.

8. Interesting Fourier transforms can be found in Exercises 6.1.5 and 6.1.9.

9. An important family of conditionally negative functions is presented in Exercise 7.3.3.

10. Exercise 8.5.2 leads to interesting links with the Cuntz relations, and new kind of interpolation problems.

11. Connections between operator theory and reproducing kernel spaces of a specified form are considered in Exercise 9.1.6.

12. Finally, Exercises 10.4.4 and 11.2.1 contain interesting examples of reproducing kernels in the setting of polyanalytic functions.

Part I

Analytic Functions

Chapter 1

Algebraic Prerequisites

We survey some material of essentially algebraic flavor. We mention in particular exercises on the theory of rational functions and important results and definitions, such as the Hahn–Banach theorem and the tensor algebra.

1.1 Sets

We begin with some elementary results on functions of sets. Given a set X, we denote by $\mathcal{P}(X)$ the set of all subsets of X. For $A \subset X$ the complement of A is

$$X \setminus A = \{x \in X,\ x \notin A\}.$$

More generally, for $A, B \subset X$,

$$A \setminus B = \{x \in X,\ x \in A \text{ and } x \notin B\}.$$

It is useful to introduce the characteristic function 1_A of the set A:

Definition 1.1.1. Let X be some set. The characteristic function 1_A of the set $A \subset X$ is defined by:

$$1_A(x) = \begin{cases} 1, & \text{if } x \in A, \\ 0, & \text{if } x \notin A. \end{cases}$$

Clearly,

$$1_{A \cap B} = 1_A 1_B \quad \text{and} \quad 1_{A \cup B} = 1_A + 1_B - 1_A 1_B.$$

Question 1.1.2. *Prove that*

$$1_{X \setminus A} = 1 - 1_A,$$
$$1_{A \setminus B} = 1_A - 1_{A \cap B},$$
$$1_{A \setminus (A \setminus B)} = 1_{A \cap B} \quad (\textit{and so, } A \setminus (A \setminus B) = A \cap B).$$

The characteristic function of a set appears in a number of places in the book; see for instance Exercise 7.2.1.

We recall a definition:

Definition 1.1.3. Let X and Y be two sets (no topologies are assumed on X or Y), and let f be a function from X into Y. Let $B \in \mathcal{P}(Y)$. The set $f^{-1}(B) \in \mathcal{P}(X)$ defined by

$$f^{-1}(B) = \{x \in X, \; f(x) \in B\}$$

is called the inverse image of B under f.

Example 1.1.4. *In the notation of the previous definition:*

(i) *Show that:*

$$f^{-1}(\cup_{i \in I} B_i) = \cup_{i \in I} f^{-1}(B_i), \tag{1.1.1}$$
$$f^{-1}(\cap_{i \in I} B_i) = \cap_{i \in I} f^{-1}(B_i),$$
$$f^{-1}(Y \setminus B) = X \setminus f^{-1}(B), \tag{1.1.2}$$

where I is any set of indices and where B and the B_i are subsets of Y.

(ii) *Let now $(A_i)_{i \in I}$ be a family of sets of X, and $A \subset X$. Show that:*

$$f(\cup_{i \in I} A_i) = \cup_{i \in I} f(A_i), \tag{1.1.3}$$
$$f(\cap_{i \in I} A_i) \subset \cap_{i \in I} f(A_i), \tag{1.1.4}$$
$$f(X) \setminus f(A) \subset f(X \setminus A).$$

Show by counterexamples that equality will not hold in general in the above inclusions.

Exercise 1.1.5. *Let f be a map from X into Y. Show that*

$$f(f^{-1}(U)) \subset U, \quad \forall U \subset Y, \tag{1.1.5}$$
$$f^{-1}(f(V)) \supset V, \quad \forall V \subset X. \tag{1.1.6}$$

Show by examples that the inclusions may be strict.

More precisely, we have the following question, taken from the book [92] of Calvo, Doyen, Calvo and Boschet. See [92, Exercice 1.2, p. 8].

Question 1.1.6. *Let f be a map from X into Y. Show that*

$$f(f^{-1}(U)) = U, \quad \forall U \subset Y, \quad \Longleftrightarrow \quad f \text{ is onto.} \tag{1.1.7}$$
$$f^{-1}(f(V)) = V, \quad \forall V \subset X, \quad \Longleftrightarrow \quad f \text{ is one-to-one.} \tag{1.1.8}$$

We note that part of the preceding results are the opening exercises in Ramis' classical book of exercises [255]. They are used in numerous places in the present book. For instance, (1.1.2) and (1.1.5) are utilized in the proof of Exercise 4.2.39.

We also note that (1.1.1) is used in the solution of Question 3.5.2, and that we employ (1.1.3) in the proof of Exercise 4.2.7 to show that a certain map is open.

A relation in a set X is a subset R of $X \times X$. We will denote by $x \sim y$ the fact that the pair (x, y) belongs to R.

Definition 1.1.7. An equivalence relation is a relation for which the following conditions hold for all $x, y, z \in X$:

(1) $x \sim x$,   (reflexivity)

(2) If $x \sim y$, then $y \sim x$,   (symmetry)

(3) If $x \sim y$ and $y \sim z$, then $x \sim z$.   (transitivity)

The equivalence class of $x \in X$ is the set of all elements equivalent to x. An equivalence relation $\sim$ divides the set X into disjoint equivalence classes. We will denote by $X/\sim$ the set of equivalence classes, and call it the *quotient space*. The map q which to $x \in X$ associates its equivalence class is called the *canonical projection* from X onto $X/\sim$. Topological properties of q are very important (for instance, being continuous, closed, open) and will be considered in Section 3.7, which is devoted to the quotient topology.

Definition 1.1.8. Let R be an equivalence relation on a set X. A subset U of X is called saturated if

$$U = q^{-1}(q(U)) \tag{1.1.9}$$
$$= \{x \in X \, ; \, q(x) \in q(U)\} .$$

In other words, if $x \in U$ then U also contains the whole equivalence class of x.

Exercise 1.1.9. *In the notation of Definition 1.1.8 assume U saturated. Then, $F = X \setminus U$ is also saturated.*

Equation (1.1.9) makes sense in fact for any function f on the set X. The associated equivalence relation will then be

$$x \sim y \quad \Longleftrightarrow \quad f(x) = f(y),$$

and we mention the following question (see [92, Exercice 1.8, pp. 13–14]). We give the proof of the second item after the statement of the question.

Question 1.1.10. *Let $\mathscr{S}$ denote the family of saturated sets of the function f. Then:*

(1) *Show that if $U \in \mathscr{S}$ so does $f^{-1}(f(U))$.*
(2) *Show that $\mathscr{S}$ is closed under arbitrary union and intersection.*
(3) *Let $U \in \mathscr{S}$ and $V \subset X$ such that $U \cap V = \emptyset$. Show that U and $f^{-1}(f(V))$ do not intersect.*
(4) *If $V \subset U$ with $U, V \in \mathscr{S}$. Show that $V \setminus U \in \mathscr{S}$.*

To prove (2) for the union, we write :

$$f^{-1}\left(f(\cup_{i\in I}A_i)\right) = f^{-1}\left(\cup_{i\in I}f(A_i)\right) \quad \text{(by (1.1.3))}$$
$$= \cup_{i\in I}f^{-1}(f(A_i)) \quad \text{(by (1.1.1))}$$
$$= \cup_{i\in I}A_i,$$

if each of the set A_i is saturated.

Similarly, for the intersection and a family of sets $(A_i)_{i\in I}$ of $\mathscr{S}$, we have:

$$\cap_{i\in I}A_i \subset f^{-1}\left(f(\cap_{i\in I}A_i)\right) \quad \text{(by (1.1.6))}$$
$$\subset f^{-1}\left(\cap_{i\in I}f(A_i)\right) \quad \text{(by (1.1.4))}$$
$$= \cap_{i\in I}f^{-1}(f(A_i)) \quad \text{(by (1.1.1))}$$
$$= \cap_{i\in I}A_i.$$

The notion of equivalence relation permeates all fields of mathematics. For instance:

Question 1.1.11. *Let $\mathcal{V}$ be a vector space and $\mathcal{M} \subset \mathcal{V}$ be a vector subspace of $\mathcal{V}$. The condition*

$$x \sim y \quad \Longleftrightarrow \quad x - y \in \mathcal{M} \tag{1.1.10}$$

is an equivalence relation.

Another equivalence relation of importance is the following: Two open subsets of $\mathbb{C}$ will be called conformally equivalent if there is a one-to-one and onto analytic map from one subset to the other. Riemann's mapping theorem answers this question for simply connected open sets. See for instance [95, Chapitre VI]. We mention at this point that the topological structure of the vector space of functions analytic in a given open set, and a compactness argument, play a key role in the proof of this theorem.

We conclude this elementary section related to sets with a remark pertaining to the axiom of choice.

Remark 1.1.12. An important class of spaces considered in this book consists of reproducing kernel Hilbert spaces. These are (see Section 7.5) Hilbert spaces of functions (say, defined on a set Ω) in which the point evaluations $f \mapsto f(w)$ are continuous for all $w \in \Omega$. This raises the question (already addressed to in Aronszajn's 1944 paper [51, Remarque II, p. 138 and pp. 146–147], in the paper [117], and in Halmos' book [169, Problem 29, p. 32]; see also [40]) to build a Hilbert space of functions which is not a reproducing kernel Hilbert space. This is possible using the axiom of choice (see Exercise 7.5.14). On the other hand, under the set of axioms **ZF** to which is added the axiom of dependent choice **DC** and a hypothesis called **BP** (Baire property), it was shown in [322] that all everywhere defined linear operators in a Hilbert space are continuous.

1.2 Groups

L'étude des sous groupes discrets de **G** *est inséparable de celle des surfaces de Riemann.*

G. de Rham, Sur les polygones générateurs de groupes fuchsiens, [106, p. 61].

In the above exergue, **G** denotes the group of invertible matrices of $\mathrm{GL}(2, \mathbb{C})$ with determinant equal to 1, that is, the special linear group (or modular group) $\mathrm{SL}(2, \mathbb{C})$. The associated linear fractional transformations

$$t_M(z) = \frac{az + b}{cz + d}, \quad \text{where} \quad \begin{pmatrix} a & b \\ c & d \end{pmatrix} \in \mathrm{SL}(2, \mathbb{C}),$$

describe the group of automorphisms of the Riemann sphere (see Exercise 3.1.10 for the latter if need be). Remark that to define t_M one can always suppose that $\det M = 1$ (as is done by dividing all the elements of M by a squareroot of $\det M$). In this section we review some definitions and present exercises related to groups, essentially having in view subgroups of $\mathrm{SL}(2, \mathbb{C})$, which leave invariant either the open unit disk (or the open upper half-plane), or the complex plane. At this stage of the book topology does not appear, and we do not give the definition of a discrete group.

Note 1.2.1. In the discussion we mention various notions (such as quotient space, lattice, elliptic function, Riemann surface) some of which will be considered at a later stage of the book. These notions are used as motivation, but the exercises can be done without referring to them.

First recall the following definition: Let G be a group (which we denote multiplicatively) and let H be a subgroup of G. One defines an equivalence relation (see Definition 1.1.7) via:

$$x \sim_r y \iff xy^{-1} \in H.$$

The right index of H is the cardinality of the set of equivalence classes. The definition of left index is defined similarly with the equivalence relation

$$x \sim_\ell y \iff y^{-1}x \in H,$$

See for instance [214, p. 12].

Exercise 1.2.2. Let $r \in \{2, 3, \ldots\}$.

(1) *Show that the set M_r of elements $M = \begin{pmatrix} a & b \\ c & d \end{pmatrix} \in \mathrm{SL}(2, \mathbb{Z})$ such that*

$$M \equiv \begin{pmatrix} 1 & 0 \\ 0 & 1 \end{pmatrix} \pmod{r}$$

is a subgroup of $\mathrm{SL}(2, \mathbb{Z})$.

(2) *Show that the set of elements* $M = \begin{pmatrix} a & b \\ c & d \end{pmatrix} \in \mathrm{SL}(2, \mathbb{Z})$ *such that c is divisible by r is a subgroup of* $\mathrm{SL}(2, \mathbb{Z})$.

(3) *Show that the natural homomorphism*

$$\mathrm{SL}(2, \mathbb{Z}) \xrightarrow{\tau} \mathrm{SL}(\mathbb{Z}/r\mathbb{Z})$$

is a group homomorphism and is onto.

(4) *Define a group isomorphism identifying*

$$\mathrm{SL}(2, \mathbb{Z})/\mathrm{M_r} \quad and \quad \mathrm{SL}(\mathbb{Z}/r\mathbb{Z}).$$

See, e.g., [219, p. 256] for the case of Euclidean rings for the claims in the previous exercise. The group in item (1) is called the homogeneous principal congruence subgroup of level n (see for instance [249, IV.1]). Its index is equal to

$$i(r) = r^3 \prod_{\substack{p \text{ prime} \\ \text{divisor of } r}} \left(1 - \frac{1}{p^2}\right). \tag{1.2.1}$$

The index of the group in item (2) is equal to

$$i(r) = r \prod_{\substack{p \text{ prime} \\ \text{divisor of } r}} \left(1 + \frac{1}{p}\right). \tag{1.2.2}$$

See [249, IV.1].

The group of translations along the elements of a lattice is an important example of subgroup of $\mathrm{SL}(2, \mathbb{Z})$. It appears, albeit in implicit form, in [CAPB] in the discussion of elliptic functions and of the theta function.

Exercise 1.2.3. *Let* ω_1 *and* ω_2 *be two nonzero complex numbers linearly independent over the real numbers (that is,* $\mathrm{Im}\, \frac{\omega_1}{\omega_2} \neq 0$*).*

(1) *Show that the maps*

$$T_{m,n}(z) = z + m\omega_1 + n\omega_2, \quad m, n \in \mathbb{Z}$$

form a subgroup G of automorphisms, which leaves invariant the complex plane.

(2) *Find the element of* $\mathrm{SL}(2, \mathbb{C})$ *corresponding to* $T_{m,n}$.

Elliptic functions are meromorphic functions f invariant under the above group:

$$f(T_{m,n}(z)) = f(z), \quad \forall n, m \in \mathbb{Z}.$$

Furthermore, the quotient space $\mathbb{C}/G$ is a Riemann surface (a torus), uniquely determined by the corresponding elliptic functions. It is interesting to check if

1.2. Groups

similar results hold when the group G is replaced by another subgroup of $SL(2, \mathbb{C})$. This leads to the theory of automorphic functions.

The following exercise corresponds to the group $SL(2, \mathbb{Z})$, that is, to the group of 2×2 matrices with entries in $\mathbb{Z}$ and determinant 1.

Exercise 1.2.4. *Let* $M = \begin{pmatrix} a & b \\ c & d \end{pmatrix} \in SL(2, \mathbb{Z}).$

(1) *Show that the map*

$$(n, m) \mapsto (an + mc, bn + md)$$

is one-to-one from $\mathbb{Z}^2 \setminus \{(0,0)\}$ *onto itself.*

(2) *Let* $k \geq 3 \in \mathbb{N}$. *Show that the series*

$$g_k(z) = \sum_{(n,m) \in \mathbb{Z}^2 \setminus \{(0,0)\}} \frac{1}{(nz + m)^k} \tag{1.2.3}$$

converges in $\mathbb{C} \setminus \mathbb{R}$.

(3) *Show that*

$$g_k(t_M(z)) = (cz + d)^k g_k(z). \tag{1.2.4}$$

(4) *Let* $k_1, k_2 \in \mathbb{N} \setminus \{1, 2\}$ *and let* $f = \dfrac{g_{k_1} g_{k_2}}{g_{k_1 + k_2}}$. *Show that*

$$f(t_M(z)) = f(z), \quad \forall M \in SL(2, \mathbb{Z}). \tag{1.2.5}$$

Functions which are meromorphic in the open upper half-plane and satisfy (1.2.5) are called automorphic with respect to the group $SL(2, \mathbb{Z})$ of invertible 2×2 matrices with integer entries and determinant equal to 1.

More generally, for appropriate subgroups $G \subset SL(2, \mathbb{Z})$ (called circle groups in Siegel's book [289]) and given $r = p/q$ a rational function with no poles in the open upper half-plane $\mathbb{C}_+$ and bounded there in modulus, the series

$$f(z) = \sum_{M \in G} (cz + d)^{-N} r(t_M(z)) \tag{1.2.6}$$

converges in $\mathbb{C}_+$ for N large enough. Such series are called *Theta Fuchsian series*, see [135, p. 45], or *Poincaré Theta series* or *Poincaré series*. When r is a constant, these are called *Poincaré series of special type* (see [289, p. 55]). As earlier, the matrix M is written as

$$M = \begin{pmatrix} a & b \\ c & d \end{pmatrix}.$$

Exercise 1.2.5. *With* f *given by* (1.2.6), *show that:*

$$f(t_{M_0}(z)) = (c_0 z + d_0)^N f(z), \tag{1.2.7}$$

where $M_0 = \begin{pmatrix} a_0 & b_0 \\ c_0 & d_0 \end{pmatrix} \in G.$

Hint: Considering the partial fraction decomposition of r (see [CAPB, p. 329]), it is enough to look at functions r which are either constant or of the form

$$\frac{1}{(z+a)^m},\tag{1.2.8}$$

where $a \in \mathbb{C}_+$ and $m \geq 1$.

Exercise 1.2.10 below, taken from [95, pp. 185–186], plays an important role in the characterization of the automorphisms of the open unit disk (that is, of the analytic functions which are one-to-one from $\mathbb{D}$ onto itself). Recall that these are exactly the Blaschke factors (corresponding to the disk) multiplied by unitary constant, that is functions of the form $c \cdot b_a(z)$, where $|c| = 1$ and

$$b_a(z) = \frac{z-a}{1-\overline{a}z},\tag{1.2.9}$$

with $a \in \mathbb{D}$. We first need a notation and a number of definitions. Let X be a set. We denote by $\mathrm{Aut}\,(X)$ the group of automorphisms of X (one-to-one mappings from X onto itself).

Definition 1.2.6 (see, e.g., [218, p. 17]). Let G be a group and X be some set. A group automorphism, also called group action, is a map

$$a : G \mapsto \mathrm{Aut}\,(X)$$

satisfying

$$a(gh) = a(g)a(h), \quad \forall g, h \in G, \qquad \text{and} \qquad a(e) = I,$$

where e denotes the neutral element of G and I the identity automorphism.

In particular it holds that

$$a(g^{-1}) = a(g)^{-1}, \quad \forall g \in G.\tag{1.2.10}$$

Definition 1.2.7. The group $\mathcal{G} \subset \mathrm{Aut}\,(X)$ is said to act transitively on the set X if every two points of X can be joined by an element of $\mathcal{G}$:

$$\forall x, y \in X, \ \exists\, a \in \mathcal{G} \text{ such that } y = a(x).\tag{1.2.11}$$

The automorphism a in (1.2.11) is of course not unique. But one has the following result, taken from the paper [13] by Bolotnikov, Rodman and the author. See also Question 8.5.4.

Question 1.2.8.

(1) *Characterize all maps $z \mapsto \frac{az+b}{cz+d}$ which send $\mathbb{D}$ onto itself and are involutive.*
(2) *Given any two different points w_1 and w_2 in the open unit disk, show that there is a map in (1) and sending w_1 to w_2. Is this map unique?*

Definition 1.2.9. The isotropy group of a point $w_0 \in X$ is by definition the group of elements of $\mathrm{Aut}\,(X)$ which leave invariant w_0.

Exercise 1.2.10 (see [95, pp. 185–186]). *Let $\mathcal{G} \subset \mathrm{Aut}\,(X)$ be a group acting transitively on $\mathrm{Aut}\,(X)$, and assume that there exists $w_0 \in X$ such that the isotropy group of w_0 is in $\mathcal{G}$. Then $\mathcal{G} = \mathrm{Aut}\,(X)$.*

Hint: Let $g \in \mathrm{Aut}\,(X)$ and let $w_0 = g(t_0)$. There is an element $h \in \mathcal{G}$ such that $h(w_0) = t_0$.

For another exercise involving groups, together with topology, see Exercise 3.8.7.

1.3 Vector spaces

We discuss in this preliminary section a number of topics related to vector spaces (over the real or the complex numbers), with no topology involved, and which play a key role in the sequel. First we recall that every vector space has a basis, called a Hamel basis. It will not be countable when the space is infinite dimensional, and its existence relies then on the axiom of choice. It follows from the existence of a Hamel basis that every vector subspace, say $\mathcal{A}$, of a vector space $\mathcal{V}$, has an algebraic supplement, meaning a vector subspace $\mathcal{B}$ such that

$$\mathcal{A} \cap \mathcal{B} = \{0\} \quad \text{and} \quad \mathcal{A} + \mathcal{B} = \mathcal{V}.$$

We will use the notation

$$\mathcal{A} \dotplus \mathcal{B} = \mathcal{V}. \tag{1.3.1}$$

In the present book, Hamel basis are used in particular to prove the existence of everywhere defined unbounded linear operators in a Hilbert space, and the existence of a Hilbert space of functions which is not a reproducing kernel Hilbert space. See Exercise 7.5.14.

We begin with the Hahn–Banach theorem.

Theorem 1.3.1 (Hahn–Banach). *Let $\mathcal{V}$ be a real vector space and let p be a real-valued function which satisfies*

$$\begin{aligned} p(v+w) &\le p(v) + p(w), \quad &&\forall v, w \in \mathcal{V}, \\ p(tv) &= tp(v), \quad &&\forall v \in \mathcal{V} \text{ and } \forall t > 0. \end{aligned} \tag{1.3.2}$$

Furthermore, let φ be a real-valued linear functional on a linear subspace $\mathcal{W}$ of $\mathcal{V}$ such that

$$\varphi(w) \le p(w), \quad \forall w \in \mathcal{W}.$$

Then φ has an extension Φ to $\mathcal{V}$ which is bounded by p:

$$\Phi(v) \le p(v), \quad \forall v \in \mathcal{V}.$$

Note that the function p will be a semi-norm when in the second condition t is real (and not only nonnegative) and one requires $p(tv) = |t|p(v)$. See Definition 4.1.1 below. We also remark that the Hahn–Banach theorem can also be proved when conditions (1.3.2) are replaced by the condition

$$p(tv + (1-t)w) \leq tp(v) + (1-t)p(w), \tag{1.3.3}$$

where $t \in [0,1]$ and $v, w \in \mathcal{V}$. See [256, p. 75].

There are also geometric versions of the Hahn–Banach theorem, stating when two convex sets satisfying various auxiliary hypothesis are separated by a closed hyperplane (see Remark 4.1.16 for the latter). See for instance [90, §I.2, pp. 4–7]. We will not write them here explicitly, but mention that one uses such a version of the Hahn–Banach theorem in the study of the Schur–Agler classes (see [3, Proof of Theorem 11.5, p. 169], and (9.1.8) for the definition of these functions).

An important consequence of the Hahn–Banach theorem is the *Basic separation theorem*. See [120, Theorem 12, p. 412]. In the statement we need the following definition (see [120, Definition 6, p. 410]).

Definition 1.3.2. An internal point to a convex set C is a point $c \in C$ such that for every $v \in \mathcal{V}$ there exists $\epsilon > 0$ such that all vectors of the form $c + zv \in C$ for all $|z| \leq \epsilon$.

When $\mathcal{V}$ is a metric space and C has a non-empty interior, any point of the interior $\overset{\circ}{C}$ is an internal point. See (3.1.3) for the definition of the interior of a set.

Theorem 1.3.3. *Let C_1 and C_2 be disjoint convex subsets of the vector space $\mathcal{V}$. Assume that C_1 has an internal point. Then there exists a nonzero linear functional which separates C_1 and C_2.*

This result is used in particular in Exercise 4.1.21 to prove that, in a Hilbert space, a closed and convex set is weakly closed.

Another important consequence of the Hahn–Banach theorem is that the dual of a locally convex topological vector space contains nonzero elements. See Question 5.4.2 below.

Continuous linear forms play a key role in the theory of topological vector spaces. At this stage of the book, no topology has been defined, and the following question, taken from Dieudonné's paper on duality in topological vector spaces, see [115, Théorème 1, p. 109] (see also [183, Lemma, p. 186]), deals with general linear forms.

Question 1.3.4 (see [183, Lemma, p. 186], [183, p. 186]). *Let $\lambda_1, \ldots, \lambda_n$ be linear forms on the complex vector space $\mathcal{V}$. Let λ be a linear form on $\mathcal{V}$ such that*

$\cap_{u=1}^{n} \ker \lambda_u \subset \ker \lambda$. Then there exist complex numbers $c_1, \ldots, c_n$ such that

$$\lambda = \sum_{u=1}^{n} c_u \lambda_u.$$

We now turn to tensor products of vector spaces. In the setting of Hilbert spaces, and with appropriate completion, they play an important role in the characterization of the reproducing kernel Hilbert space associated with a product of positive definite kernels and in the construction of the various Fock spaces. Here we only discuss the algebraic setting. See the introduction of [307, Part III] for a discussion on the various completions.

Theorem 1.3.5. *Let $\mathcal{V}_1$ and $\mathcal{V}_2$ be two vector spaces (over the complex or the real numbers).*

(1) *There is a vector space $\mathcal{V}_3$ and a bilinear map i from $\mathcal{V}_1 \times \mathcal{V}_2$ into $\mathcal{V}_3$ with the following property: For every vector space $\mathcal{V}_4$ and every bilinear map f from the Cartesian product $\mathcal{V}_1 \times \mathcal{V}_2$ into $\mathcal{V}_4$ there exists a uniquely defined linear map g from $\mathcal{V}_3$ into $\mathcal{V}_4$ such that*

$$f = g \circ i.$$

(2) *The space $\mathcal{V}_3$ is unique up to an isomorphism of vector spaces.*

The space $\mathcal{V}_3$ is called the *tensor product* and is denoted $\mathcal{V}_1 \otimes \mathcal{V}_2$. We write

$$i(v_1, v_2) = v_1 \otimes v_2, \quad \text{for } v_1 \in \mathcal{V}_1 \text{ and } v_2 \in \mathcal{V}_2. \tag{1.3.4}$$

We refer to [214, Chapter XVI] and to [247, Chapter VI] for the main properties of tensor products. More generally, one defines the tensor product of n vector spaces, with multilinear maps replacing bilinear maps.

Exercise 1.3.6 (see, for instance, [214, Proposition 1.1, p. 604]). *Show that the tensor product of vector spaces is associative: Given $\mathcal{V}_1, \mathcal{V}_2$ and $\mathcal{V}_3$ three vector spaces, the map defined by*

$$(v_1 \otimes v_2) \otimes v_3 \;\mapsto\; v_1 \otimes (v_2 \otimes v_3) \tag{1.3.5}$$

is an isomorphism from $(\mathcal{V}_1 \otimes \mathcal{V}_2) \otimes \mathcal{V}_3$ onto $\mathcal{V}_1 \otimes (\mathcal{V}_2 \otimes \mathcal{V}_3)$, with the inverse map

$$v_1 \otimes (v_2 \otimes v_3) \;\mapsto\; (v_1 \otimes v_2) \otimes v_3. \tag{1.3.6}$$

Hint: The proof follows the arguments from [214, Proposition 1.1, p. 604]. Given $v_1 \in \mathcal{V}_1$ (resp. $v_3 \in \mathcal{V}_3$) define bilinear maps

$$(v_2, v_3) \;\mapsto\; (v_1 \otimes v_2) \otimes v_3, \quad \text{and} \quad (v_1, v_2) \;\mapsto\; v_1 \otimes (v_2 \otimes v_3)$$

from $\mathcal{V}_2 \times \mathcal{V}_3$ onto $(\mathcal{V}_1 \otimes \mathcal{V}_2) \otimes \mathcal{V}_3$ and $\mathcal{V}_1 \times \mathcal{V}_2$ onto $\mathcal{V}_1 \otimes (\mathcal{V}_2 \otimes \mathcal{V}_3)$ respectively, and use the definition of the tensor product.

In view of the associativity the following definitions and exercises make sense.

Definition 1.3.7. Let $\mathcal{V}$ be a vector space. The direct sum

$$\mathbb{C} \oplus \left(\oplus_{n=1}^{\infty} \mathcal{V}^{\otimes n} \right)$$

is called the tensor algebra associated with $\mathcal{V}$, and will be denoted by $\mathscr{T}(\mathcal{V})$.

The term *tensor algebra* is justified in the following exercise:

Exercise 1.3.8. *Show that the map*

$$(f, g) \; \mapsto \; f \otimes g, \quad f \in \mathcal{V}^{\otimes n} \text{ and } g \in \mathcal{V}^{\otimes m}, \quad n, m \in \mathbb{N}_0,$$

makes $\mathscr{T}(\mathcal{V})$ into an algebra.

Given two vector spaces $\mathcal{V}$ and $\mathcal{W}$ and a linear operator T from $\mathcal{V}$ into $\mathcal{W}$ one defines an operator $\Gamma(T)$ from $\mathscr{T}(\mathcal{V})$ into $\mathscr{T}(\mathcal{W})$ by the formula

$$\Gamma(T)(f_1 \otimes \cdots \otimes f_n) = (Tf_1) \otimes \cdots \otimes (Tf_n). \tag{1.3.7}$$

Remark 1.3.9. When one considers Hilbert spaces, each of the spaces $\mathcal{V}^{\otimes n}$ has a Hilbert space structure, and the closure of the tensor algebra is then the full Fock space associated with $\mathcal{V}$. When one considers symmetric tensor products, one has the symmetric (or bosonic) Fock space associated with $\mathcal{V}$. In the setting of Hilbert spaces, $\Gamma(T)$ is called the second quantization of T.

Remark 1.3.10. The symmetric Fock space and the second quantization are considered in Exercise 1.5.12 in the finite-dimensional case.

Exercise 1.3.11. *Show that (1.3.7) indeed defines a linear operator.*

We conclude this section with an important concept, which plays a role in various places in the book:

Definition 1.3.12. Let $\mathcal{V}$ be a vector space on the complex number. A Hermitian form is a map

$$[\cdot, \cdot] \; : \; \mathcal{V} \times \mathcal{V} \longrightarrow \mathbb{C}$$

with the following properties:

(1) It is Hermitian, in the sense that

$$[v, w] = \overline{[w, v]}, \quad \forall v, w \in \mathcal{V}.$$

(2) It is linear in the first variable:

$$[c_1 v_1 + c_2 v_2, w] = c_1[v_1, w] + c_2[v_2, w], \quad \forall c_1, c_2 \in \mathbb{C} \text{ and } \forall v_1, v_2, w \in \mathcal{V}.$$

In the next question one finds a useful identity, called the polarization identity (see [CAPB, (14.3.3), p. 484]).

Question 1.3.13. *Let $\mathcal{V}$ denote a vector space endowed with a Hermitian form $[\cdot, \cdot]$. Let A denote a linear operator from $\mathcal{V}$ into $\mathcal{V}$. Prove the polarization identity:*

$$[Au, v] = \frac{1}{4} \sum_{k=0}^{3} i^k [A(u + i^k v), u + i^k v]. \qquad (1.3.8)$$

Remark 1.3.14. Polarization identities for multilinear maps appeared already in 1934 in a paper [229] of Mazur and Orlicz; see the historical note in the paper [305] of Erik Thomas, where a simpler proof of the results of [229] can be found.

1.4 Fields

Rational functions in n variables with complex coefficients form a field, which we will denote by $\mathbb{C}(X_1, \ldots, X_n)$. When $n = 1$, and setting $X_1 = X$, the field $\mathbb{C}(X)$ can also be viewed as the field of meromorphic functions on the Riemann sphere. Meromorphic functions in $\mathbb{C}$ which are biperiodic also form a field. They can be viewed as the field of meromorphic functions on a torus (see Exercise 3.8.30 for the latter). They are an example of an algebraic function field, and provide the motivation for this section We here review the main relevant definitions. It is first convenient to review some preliminary notions.

Definition 1.4.1. The field C_1 is a finite degree extension of the field C_2 if C_1 is a finite-dimensional vector space over C_2 (say, of dimension N). The number N is the degree of the extension.

Exercise 1.4.2 (see, e.g., [214, Proposition 1.1, p. 224]). *Let C_1 be a finite degree extension of C_2. Then every element of C_1 is algebraic over C_2, that is, it is the solution of a polynomial equation with coefficients in C_2.*

Definition 1.4.3. The field C_1 is a finitely generated extension (or, an extension of finite type) of the field C_2 if there exist $z_1, \ldots, z_M \in C_1$ such that $C_1 = C_2(z_1, \ldots, z_M)$. The number M is the type of the extension.

As an illustration, let C_1 be the field of biperiodic meromorphic functions (with a given lattice of periods). Then there exist two elements f and g of C_1 such that every element h in C_1 can be written as

$$h(z) = r_1(f(z)) + g(z) r_2(f(z)),$$

where r_1 and r_2 are rational functions (in fact, one can choose $f = \wp$ and $g = \wp'$, where $\wp$ is the Weierstrass function defined from the lattice).

Exercise 1.4.4 (see, e.g., [214, Proposition 1.5, p. 226], [225, Proposition 1.7, p. 55]). *Show that a finite degree extension is finitely generated, and show that the converse need not hold.*

In the case of an algebraic extension one has:

Exercise 1.4.5 (see, e.g., [225, Proposition 1.8, p. 55]). *Let $C_1 = C_2(z_1, \ldots, z_M)$ be a finitely generated extension of C_2 and assume that the z_j are algebraic over C_2. Then, C_1 is a finite degree extension, and $C_1 = C_2[z_1, \ldots, z_N]$.*

In view of the previous result, the term "finite" is not ambiguous in the following definition:

Definition 1.4.6. The field C_1 is an algebraic function field (over $\mathbb{C}$) if there exist $z_1, \ldots, z_N$, which are transcendental and such that C_1 is a finite algebraic extension of the field of rational functions $\mathbb{C}(z_1, \ldots, z_N)$.

The elements $z_1, \ldots, z_N$ are called a basis of transcendence if they are algebraically independent. The number N is then called the transcendence degree. This number is well defined in view of:

Theorem 1.4.7 (see, e.g., [225, Théorème 4.5, p. 64]). *Let C_1 be an algebraic function field. Two basis of transcendence have the same number of elements.*

1.5 Matrices

Exercise 1.5.1. *Let $A, B \in \mathbb{C}^{n \times n}$. Show that the largest subspace $\mathcal{M}$ invariant under A and B and on which A and B coincide is given by*

$$\mathcal{N} = \bigcap_{u=1}^{\infty} \ker(A^u - B^u). \tag{1.5.1}$$

We note that (1.5.1) can be rewritten as

$$\mathcal{N} = \bigcap_{\lambda \in \rho(A) \cap \rho(B)} \ker\left((\lambda I_n - A)^{-1} - (\lambda I_n - B)^{-1}\right),$$

where $\rho(A)$ denotes the resolvent set of A. This formulation has the advantage of still making sense for unbounded operators. See [85, 82] for some applications in the setting of unbounded self-adjoint operators.

Recall that a matrix $M \in \mathbb{C}^{n \times n}$ is called semi-simple if $\mathbb{C}^n$ has a basis made of eigenvectors of M. The spectral theorem implies in particular that Hermitian matrices are semi-simple. The following result is useful in the theory of linear algebraic groups. See [296, Lemma 2.4.2, p. 44]. Here we give it as a preparation to Question 1.5.3.

Question 1.5.2. *Let $M_1, \ldots, M_N$ be a family of pairwise commuting $n \times n$ Hermitian matrices with complex entries. Show that they can be simultaneously diagonalized.*

When $N = 1$ the next question is just a rewriting of the spectral theorem for a Hermitian matrix. The result is a direct consequence of Exercise 1.5.2, and is used for instance in the study of Hecke operators. See [249, pp. III-13 and III-14].

1.5. Matrices

Question 1.5.3. Let $M_1, \ldots, M_N$ be a family of commuting $n \times n$ Hermitian matrices with complex entries. Show that $\mathbb{C}^n$ has a basis made of common eigenvectors of $M_1, \ldots, M_N$.

More generally, in the non-hermitian case, we have:

Exercise 1.5.4. Let $M_1, M_2, \ldots,$ be a (not necessarily finite) family of commuting $n \times n$ matrices.

(1) Show that they have a common invariant subspace.
(2) Show that they have a common eigenvector.
(3) Show that they can be simultaneously triangularized.

In the statement of the next theorem, C is a smooth Jordan curve. See Definition 3.4.15 for a Jordan curve.

Exercise 1.5.5. Let $A \in \mathbb{C}^{n \times n}$ and let C denote a smooth simple closed curve which does not intersect the spectrum of A. Show that

$$P = \frac{1}{2\pi i} \int_C (z I_n - A)^{-1} dz \tag{1.5.2}$$

is a projection operator.

Hint: Use the Jordan form of A to compute the integral.

The operator P in (1.5.2) is called the Riesz projection corresponding to the spectrum of A inside C. The above result still holds in the case of operators in Banach spaces. See Exercise 4.2.13. Then the Jordan form is not available, and the proof requires a different approach, as follows. Recall first that a Jordan curve divides the plane into three connected parts: the curve itself, its interior and its exterior (the latter is unbounded). In view of Exercise 4.2.13 we make the following remark: Consider another Jordan curve, say D, such that the interior of D is inside the interior of C but such that the closed set determined by C and D does not contain elements in the spectrum of A. Then:

$$P = \frac{1}{2\pi i} \int_D (z I_n - A)^{-1} dz.$$

Taking into account the formulas

$$\frac{1}{2\pi i} \int_C \frac{dz}{z - w} = 1 \text{ if } w \in D,$$
$$\frac{1}{2\pi i} \int_D \frac{dw}{w - z} = 0 \text{ if } w \in C, \tag{1.5.3}$$

one can get another proof that P is a projection by considering the integral

$$\int_C \int_D (z I_n - A)^{-1} (w I_n - A)^{-1} dz dw$$

without resorting to the Jordan form of A. This is a way to consider the case where A acts in a Hilbert space, or, more generally in a Banach space. See for instance [155, Proof of Lemma 2.1, pp. 9–10].

The Riesz projection is used in particular in the present book in Exercise 1.7.6 to obtain explicit formulas for the Fourier coefficients of a rational function with no poles on the unit circle (such a function belongs to the Wiener algebra, as is easily seen from its partial fraction expansion).

We now consider an important class of matrices, which appear in various places in the sequel (see for instance Exercise 4.2.46). First a definition: A matrix $J \in \mathbb{C}^{m \times m}$ which satisfies

$$J = J^* = J^{-1} \tag{1.5.4}$$

is called a signature matrix.

Definition 1.5.6. Let $J \in \mathbb{C}^{m \times m}$ be a signature matrix. The matrix $M \in \mathbb{C}^{m \times m}$ is called J-unitary if

$$MJM^* = J. \tag{1.5.5}$$

Question 1.5.7. *Let $J \in \mathbb{C}^{m \times m}$ be a signature matrix. Show that the J-unitary matrices form a multiplicative group.*

Equation (1.5.5) means that M is unitary with respect to the (in general indefinite metric) defined by J. See Section 4.4 for the latter. The following exercise, left here as a question, already appears in Godement's algebra book; see [154, Exercice 51, p. 652]. A proof can be found in Dym's CBMS' lecture notes [122, Theorem 1.2, p. 17].

Question 1.5.8. *Assume that*

$$J = \begin{pmatrix} I_p & 0 \\ 0 & -I_q \end{pmatrix}. \tag{1.5.6}$$

Show that $M \in \mathbb{C}^{(p+q) \times (p+q)}$ is J-unitary if and only if it can be written as

$$M = H(K) \begin{pmatrix} U & 0 \\ 0 & V \end{pmatrix}, \tag{1.5.7}$$

where $U \in \mathbb{C}^{p \times p}$ and $V \in \mathbb{C}^{q \times q}$ are unitary matrices, where $K \in \mathbb{C}^{p \times q}$ is strictly contractive and $H(K)$ denotes the Halmos extension of K:

$$H(K) = \begin{pmatrix} (I_p - KK^*)^{-1/2} & K(I_q - K^*K)^{-1/2} \\ K^*(I_p - KK^*)^{-1/2} & (I_q - K^*K)^{-1/2} \end{pmatrix}.$$

Question 1.5.9 (see also [122, Theorem 1.3, p. 18]). *Let J be the signature matrix given by (1.5.6), and let $c \in \mathbb{C}^{p+q}$ be such that $c^* Jc = 0$. Let $a \in \mathbb{R}$. Show that the matrix*

$$M = I_{p+q} + iacc^* J \tag{1.5.8}$$

is J-unitary and find its representation (1.5.7).

1.5. Matrices

Problem 1.5.10. *When $q = 0$ or $p = 0$, no nonzero vector c satisfies $c^* Jc = 0$. On the other hand, one can ask what is the structure of the multiplicative group generated by the matrices of the form* (1.5.8).

Matrix equations will play an important role in the sequel; see (7.7.14) and (7.7.31). We recall the following result:

Theorem 1.5.11. *Let $A, B, C, D \in \mathbb{C}^{n \times n}$. The matrix equation*

$$AXB - CXD = E$$

has a unique solution for every $E \in \mathbb{C}^{n \times n}$ if and only if

$$ab - cd \neq 0,$$

for all $a \in \sigma(A)$, $b \in \sigma(B)$, $c \in \sigma(C)$, and $d \in \sigma(D)$.

We conclude this section with the study of the symmetric Fock space and second quantization in the case of finite-dimensional spaces. We recall a notation: For $A \in \mathbb{C}^{n \times m}$ and $B \in \mathbb{C}^{p \times q}$ we denote (see also the proof of Exercise 7.4.3 and [CAPB, Exercise 14.3.3, p. 485]) by $A \otimes B$ the Hadamard product (also called Schur product, or tensor product) of the matrices A and B, that is:

$$A \otimes B = \begin{pmatrix} a_{11}B & a_{12}B & \cdots & a_{1m}B \\ a_{21}B & a_{22}B & \cdots & a_{2m}B \\ \vdots & & & \vdots \\ a_{n1}B & a_{n2}B & \cdots & a_{nm}B \end{pmatrix}. \tag{1.5.9}$$

Exercise 1.5.12. *In the exercise, M, N, P and Q denote elements of $\mathbb{N}$. Let $e_1, \ldots, e_N$ (resp. $f_1, \ldots, f_M$) be the canonical basis of $\mathbb{C}^N$ (resp. $\mathbb{C}^M$).*

(1) *Show that the map*

$$\tau(e_i, f_j) = e_i \otimes f_j$$

allows to define $\mathbb{C}^{NM}$ as the tensor product $\mathbb{C}^N \otimes \mathbb{C}^M$.

(2) *Let $A \in \mathbb{C}^{P \times N}$ and $B \in \mathbb{C}^{Q \times M}$. Show that the matrix $A \otimes B$ (see (1.5.9)) defines the tensor product of A and B, meaning that*

$$(A \otimes B)(f \otimes g) = Af \otimes Bg.$$

(3) *Show that*

$$\|(A \otimes B)(f \otimes g)\| = \|Af\| \cdot \|Bg\|, \tag{1.5.10}$$

where the norms are the Euclidean norms.

(4) *Let $M = N = P = Q$. Define $\Gamma(A)$ via*

$$\Gamma(A)(f_n)_{n=0,\ldots} = (A^{\otimes n} f_n)_{n=0,\ldots},$$

where $f_0 \in \mathbb{C}$ and $f_n \in \mathbb{C}^{Nn}$ for $n \geq 1$. Then $\Gamma(A)$ defines an operator from the tensor algebra $\mathscr{T}(\mathbb{C}^N)$ into itself, and

$$\|\Gamma(A)\| \leq \frac{1}{1 - \|A\|} \tag{1.5.11}$$

when $\|A\| < 1$.

Remark 1.5.13. The condition $\|A\| < 1$ is in fact too strong. It is enough to assume that $\|A\| \leq 1$ to ensure that $\Gamma(A)$ is a bounded operator. See for instance [193, Theorem 4.5].

As a consequence of (1.5.10) we have:

Question 1.5.14. *Let A and B be unitary (resp. isometric) (resp. coisometric) matrices. Then $A \otimes B$ is unitary (resp. isometric) (resp. coisometric).*

The case of positive matrices was considered in [CABP, Exercise 14.3.3, p. 485]. See the following section for the definition of, and exercises on, these matrices.

1.6 Positive matrices and finite-dimensional Hilbert spaces

Positivity plays an important role in this book, and in this section we present exercises relative to positive matrices and associated finite-dimensional Hilbert spaces (the notion of positive matrix is reviewed in [CAPB, Definition 14.3.1, pp. 484–485], and the notion of inner product p. 480 there).

Definition 1.6.1. A positive (or non-negative) matrix $M \in \mathbb{C}^{n \times n}$ is defined by the condition

$$c^* M c \geq 0, \quad \forall c \in \mathbb{C}^n. \tag{1.6.1}$$

If the inequality is strict for $c \neq 0$ we say that M is strictly positive.

We recall, see, e.g., [CAPB, p. 484], that condition (1.6.1) implies in particular that $M = M^*$. This is checked using the polarization identity (1.3.8).

Positive matrices are particular cases of positive operators from a Hilbert space into itself (see (4.2.5)), and more generally, from a topological vector space into its anti-dual. See Definition 5.5.1 for the latter.

The following example of a positive matrix will appear another two times in the sequel. First, since (1.6.2) is an instance of a positive definite function (see Exercise 7.2.4), and next in the study of the Fock space of polyanalytic functions; see Exercise 11.2.1. The example itself is amplified in the joint work [33] with Palle Jorgensen.

Exercise 1.6.2. *Let $N \in \mathbb{N}$. Show that the $(N+1) \times (N+1)$ matrix M_N with (m,n) entry equal to*

$$\binom{m+n}{n}, \quad m,n = 0, \ldots, N, \tag{1.6.2}$$

is strictly positive.

Hint: Use the Chu–Vandermonde formula

$$\binom{m+n}{n} = \sum_{\ell=0}^{m \wedge n} \binom{n}{\ell} \binom{m}{\ell} \tag{1.6.3}$$

to obtain a factorization $M_N = L_N L_N^*$, where L_N is the lower triangular matrix (with nonzero diagonal elements) with (n, ℓ) entry $(n \geq \ell)$ equal to

$$(L_N)_{n,\ell} = \binom{n}{\ell},$$

that is:

$$L_N = \begin{pmatrix} 1 & 0 & 0 & \cdots & & 0 \\ 1 & 1 & 0 & 0 & \cdots & 0 \\ 1 & 2 & 1 & 0 & \cdots & 0 \\ \vdots & & & \ddots & & \\ \vdots & & & & & \\ \binom{N}{0} & \binom{N}{1} & & & & \binom{N}{N} \end{pmatrix}.$$

The inverse of L_N can be seen to be the lower triangular matrix with (n, ℓ) entry $(n \geq \ell)$ equal to

$$(L_N^{-1})_{n,\ell} = (-1)^{n+\ell} \binom{n}{\ell}, \tag{1.6.4}$$

from which M_N^{-1} can be computed.

Question 1.6.3 (see [33, Lemma 2.1]).

(1) *Prove that, for $m \leq n$ and $j \in \{m, \ldots, n\}$,*

$$\binom{n}{j} \binom{j}{m} = \binom{n}{m} \binom{n-m}{j-m}.$$

(2) *Prove (1.6.4).*

Having in view Remark 11.3.3 on the reproducing kernel of the Fock space of polyanalytic functions, we now ask:

Exercise 1.6.4. *Show that the first column of M_N^{-1} is given by (11.3.5), that is:*

$$\begin{pmatrix} (-1)^0 \dbinom{N+1}{0+1} \\ \vdots \\ (-1)^k \dbinom{N+1}{k+1} \\ \vdots \\ (-1)^N \dbinom{N+1}{N+1} \end{pmatrix}.$$

In the previous exercise, setting $N = 2$ leads to

$$M_2 = \begin{pmatrix} 1 & 1 & 1 \\ 1 & 2 & 3 \\ 1 & 3 & 6 \end{pmatrix} \quad \text{and first column of the inverse equal to} \quad \begin{pmatrix} 3 \\ -3 \\ 1 \end{pmatrix}.$$

Yet another example is given by (see also [CAPB, Exercise 3.1.11, p. 91]):

Example 1.6.5. *Let $N \in \mathbb{N}_0$. The matrix*

$$\begin{pmatrix} \frac{1}{1} & \frac{1}{2} & \cdots & \frac{1}{1+N} \\ \frac{1}{2} & \frac{1}{3} & \cdots & \frac{1}{2+N} \\ & & & \\ \frac{1}{N+1} & \frac{1}{N+2} & \cdots & \frac{1}{2N+1} \end{pmatrix} \tag{1.6.5}$$

is strictly positive.

The matrix (1.6.5) is an example of a Hankel matrix, that is of a matrix constant on the anti-diagonals. The fact that it is strictly positive is differed to Exercise 4.2.34.

More generally:

Example 1.6.6. *Let $\nu \in (-1, \infty)$ and let $N \in \mathbb{N}_0$. Show that the Hankel matrix*

$$\begin{pmatrix} \frac{1}{1^{\nu+1}} & \frac{1}{2^{\nu+1}} & \cdots & \frac{1}{(1+N)^{\nu+1}} \\ \frac{1}{2^{\nu+1}} & \frac{1}{3^{\nu+1}} & \cdots & \frac{1}{(2+N)^{\nu+1}} \\ & & & \\ \frac{1}{(N+1)^{\nu+1}} & \frac{1}{(N+2)^{\nu+1}} & \cdots & \frac{1}{(2N+1)^{\nu+1}} \end{pmatrix} \tag{1.6.6}$$

is strictly positive.

The proof that (1.6.6) is strictly positive will be given in Exercise 8.7.3. The reader may want to consider the following question. Note that the case $\mu \in \mathbb{N}$ is easily considered using the fact that the pointwise product of two positive matrices is positive. This is Schur's lemma. See [20, Theorem 2.7, p. 23] for a result for strictly positive matrices.

Question 1.6.7. *Let $\mu > 0$. Let $(a_{jk})_{j,k \in 1,\dots,N} \in \mathbb{C}^{N \times N}$ be a positive (resp. strictly positive) matrix. When is the matrix*

$$(a_{jk}^\mu)_{j,k \in 1,\dots,N}$$

positive (resp. strictly positive)?

The above examples illustrate the fact that it is not always easy to see that a Hermitian matrix is positive. As yet another illustration of this fact we give the following example, which can be checked using Exercise 7.2.15.

Example 1.6.8. *Let $N \in \mathbb{N}$. Show that the $(N+1) \times (N+1)$ matrix with (m,n) entry equal to $\frac{1}{1+(m-n)^2}$, $n, m = 0, \dots, N$, is strictly positive.*

Remark 1.6.9. Part of the results presented here still hold in the infinite-dimensional case.

We denote by

$$\langle u, v \rangle_{\mathbb{C}^n} = v^* u$$

the standard inner product in $\mathbb{C}^n$. Furthermore, the range of a matrix (say A) is denoted by $\operatorname{ran} A$.

Exercise 1.6.10. *Let $A \in \mathbb{C}^{n \times n}$ be a positive matrix. Show that the space $\operatorname{ran} A$ endowed with the form*

$$\langle Au, Av \rangle_A = \langle Au, v \rangle_{\mathbb{C}^n} \tag{1.6.7}$$

is a finite-dimensional Hilbert space.

Hint: To show that $\operatorname{ran} A$ is nondegenerate it is useful to recall the following fact: if $\langle \cdot, \cdot \rangle$ is a nonnegative (but possibly degenerate) Hermitian form on a (not necessarily finite-dimensional) vector space $\mathcal{V}$, the Cauchy–Schwarz inequality holds in $(\mathcal{V}, \langle \cdot, \cdot \rangle)$:

$$|\langle u, v \rangle|^2 \leq \langle u, u \rangle \cdot \langle v, v \rangle, \quad \forall u, v \in \mathcal{V}.$$

One can prove Exercise 1.6.11 below very easily using the following fact: A positive matrix $M \in \mathbb{C}^{n \times n}$ can be written as

$$M = \sum_{j=1}^{r} a_j u_j u_j^*,$$

where r is the rank of M, the $a_j > 0$ and $u_1, \dots, u_r$ is an orthonormal basis of $\operatorname{ran} M$. The proof we suggest in the hint is much more involved, but has some advantages. In particular it extends with appropriate modifications to the infinite-dimensional case. See Exercise 4.2.41 and Theorem 7.5.22.

Exercise 1.6.11. *Let $M_1, M_2 \in \mathbb{C}^{n \times n}$ be two positive matrices. Show that*

$$\mathrm{ran}\,(M_1 + M_2) = (\mathrm{ran}\,M_1) + (\mathrm{ran}\,M_2).$$

More precisely, endow the various ranges with the inner product (and corresponding norm) defined in Exercise 1.6.10 and show that for every $u \in \mathbb{C}^n$ there exists a pair $(a, b) \in (\mathrm{ran}\,M_1) \times (\mathrm{ran}\,M_2)$ such that

$$(M_1 + M_2)u = a + b, \tag{1.6.8}$$

and that, for any such pair, one has

$$\|(M_1 + M_2)u\|^2_{M_1 + M_2} \leq \|a\|^2_{M_1} + \|b\|^2_{M_2}. \tag{1.6.9}$$

Show that there is a unique decomposition (1.6.8) for which equality holds in (1.6.9).

Hint: Show that the linear space R

$$R = \{((M_1 + M_2)u, (M_1 u, M_2 u)),\ u \in \mathbb{C}^n\}$$
$$\subset (\mathrm{ran}\,(M_1 + M_2)) \times ((\mathrm{ran}\,M_1) \times (\mathrm{ran}\,M_2))$$

is the graph of an isometry, when the cartesian product $(\mathrm{ran}\,M_1) \times (\mathrm{ran}\,M_2)$ is endowed with the norm defined by

$$\|(f_1, f_2)\|^2_{M_1, M_2} = \|f_1\|^2_{M_1} + \|f_2\|^2_{M_2}. \tag{1.6.10}$$

Compute the adjoint of this isometry to conclude.

Exercise 1.6.12. *Let $M = \begin{pmatrix} A & B \\ B^* & D \end{pmatrix} \in \mathbb{C}^{(r+s) \times (r+s)}$ be a Hermitian matrix, with $A \in \mathbb{C}^{r \times r}$. Assume that $M \geq 0$. Show that*

$$\ker D \subset \mathrm{ran}\,B,$$
$$\mathrm{ran}\,B \subset \mathrm{ran}\,A. \tag{1.6.11}$$

Hint: When D is invertible use the Schur complement formula

$$\begin{pmatrix} A & B \\ C & D \end{pmatrix} = \begin{pmatrix} I_r & BD^{-1} \\ 0 & I_s \end{pmatrix} \begin{pmatrix} A - BD^{-1}C & 0 \\ 0 & D \end{pmatrix} \begin{pmatrix} I_r & 0 \\ D^{-1}C & I_s \end{pmatrix}, \tag{1.6.12}$$

where $(A, B, C, D) \in \mathbb{C}^{r \times r} \times \mathbb{C}^{r \times s} \times \mathbb{C}^{s \times r} \times \mathbb{C}^{s \times s}$.

Definition 1.6.13. The matrix $A - BD^{-1}C$ is called the Schur complement of A in $\begin{pmatrix} A & B \\ C & D \end{pmatrix}$. It plays an important role in various applications. See for instance Exercise 1.7.7.

Matrices with constant diagonals are called Toeplitz matrices. We already met an example of such a matrix in Example 1.6.8.

Exercise 1.6.14. Let $r_{-N}, \ldots, r_0, \ldots, r_N$ be complex numbers such that the matrix $T_N \in \mathbb{C}^{(N+1) \times (N+1)}$ with (n, m) entry equal to r_{m-n} is positive (note that this forces in particular $\overline{r_{-j}} = r_j$ for $j = 0, 1, \ldots, N$).

(1) Assume $T_N > 0$. Find all $r_{N+1} \in \mathbb{C}$ such that the matrix

$$
T_{N+1} = \begin{pmatrix} & & & r_{N+1} \\ & T_N & & r_N \\ & & & \vdots \\ & & & r_1 \\ \overline{r_{N+1}} & \overline{r_N} & \cdots & \overline{r_1} & r_0 \end{pmatrix} \geq 0.
$$

(2) Still with $T_N > 0$, assume T_{N+1} singular. Show that there is a uniquely defined number r_{N+2} such that the corresponding Toeplitz matrix T_{N+2} is positive.

Hints:

(1) Apply formula (1.6.12) to T_{N+1}. Apply it also to T_N to compute T_N^{-1}.

(2) Apply Exercise 1.6.12.

Remark 1.6.15. The problem set in the preceding exercise is called the covariance extension problem, and has numerous consequences related to prediction of stochastic processes, maximum entropy, and the Carathéodory–Fejér interpolation problem. See for instance van den Bos' paper [313].

For a solution to the following question see [CAPB, p. 162]. See also Exercise 2.1.36. Note that the function defined in (1.6.14) is an example of a positive definite function (as defined in Section 7.1).

Question 1.6.16. Let $(r_n)_{n \in \mathbb{Z}}$ be a sequence of complex numbers such that all the Toeplitz matrices T_N (defined in Exercise 1.6.14) are non-negative.

(1) Show that the power series

$$
\varphi(z) = r_0 + 2 \sum_{n=1}^{\infty} z^n r_{-n} \tag{1.6.13}
$$

converges in the open unit disk.

(2) Show that

$$
\frac{\varphi(z) + \overline{\varphi(w)}}{2(1 - z\overline{w})} = \sum_{n,m=0}^{\infty} z^n \overline{w}^m r_{m-n}, \quad z, w \in \mathbb{D}. \tag{1.6.14}
$$

Hint: Note that (1.6.14) is equivalent to

$$
\sum_{u \in \mathbb{Z}} z^u r_{-u} = \sum_{n,m=0}^{\infty} r_{m-n} (z^n \overline{w}^m - z^{n+1} \overline{w}^{m+1}).
$$

Remark 1.6.17. The function (1.6.14) is an example of a positive definite kernel (see Definition 7.1.1).

Question 1.6.18. *Let $h_1, h_2, \ldots$ be a bounded sequence of real numbers such that all the Hankel matrices*

$$H_N = \begin{pmatrix} h_1 & h_2 & \cdots & h_N \\ h_2 & h_3 & \cdots & h_{N+1} \\ & & & \\ h_N & h_{N+1} & \cdots & h_{2N-1} \end{pmatrix}, \quad N = 1, 2, \ldots$$

are nonnegative.

(1) *Compute the sum*

$$\sum_{n,m=0}^{\infty} \frac{h_{n+m+1}}{z^n \overline{w}^m},$$

and show that the function

$$h(z) = -\sum_{n=1}^{\infty} \frac{h_n}{z^n}$$

has a positive imaginary part in the open upper half-plane.

(2) *Consider the case where the sequence $(h_n)_{n \in \mathbb{N}}$ is not assumed bounded (but the Hankel matrices are still nonnegative).*

Hint: Compute

$$\sum_{n,m=0}^{\infty} \frac{h_{n+m+1}}{z^n \overline{w}^m} = \sum_{u=1}^{\infty} h_u \left(\sum_{\substack{n+m=u-1 \\ n,m \in \mathbb{N}_0}} \frac{1}{z^n \overline{w}^m} \right),$$

and compare with $z\overline{w} \frac{h(z) - \overline{h(w)}}{z - \overline{w}}$. This argument shows only that $\operatorname{Im} h(z) \geq 0$ near the origin in $\mathbb{C}_+$. To get the conclusion to the whole of $\mathbb{C}_+$ one needs an analytic extension argument using the positive definite kernel $\frac{h(z) - \overline{h(w)}}{z - \overline{w}}$.

1.7 Rational functions

In [CAPB] we reviewed briefly the notion of realization of a rational function, and we refer to that book (see [CAPB, Exercise 7.5.3, p. 329]) for a proof of the following result:

Theorem 1.7.1. *A rational matrix-valued function R analytic at infinity can be written as*

$$R(z) = D + C(zI_N - A)^{-1}B, \tag{1.7.1}$$

where A, B, C and D are matrices of appropriate sizes.

A representation of the form (1.7.1) is called a realization of the function R. One can also consider realizations centered at the origin, of the form

$$R(z) = D + zC(I_N - zA)^{-1}B. \tag{1.7.2}$$

We now outline a proof of (1.7.2) based on a different method than the one presented in [CAPB] (there for rational functions analytic at infinity). We denote by R_a the resolvent operator

$$(R_a f)(z) = \begin{cases} \dfrac{f(z) - f(a)}{z - a}, & z \neq a, \\ f'(a), & z = a, \end{cases} \tag{1.7.3}$$

where f is rational without a pole at a (or, more generally, f is analytic in a neighborhood of the point a).

Question 1.7.2. *Show that the operators R_a satisfy the resolvent equation*

$$R_a - R_b = (a - b)R_a R_b, \tag{1.7.4}$$

(for functions analytic in neighborhoods of a and b), and that

$$((I + aR_a)f)(z) = ((I - aR_0)^{-1}f)(z) = \frac{zf(z) - af(a)}{z - a}. \tag{1.7.5}$$

Question 1.7.3. *Let M be a $\mathbb{C}^{m \times n}$-valued rational function analytic at the origin.*
(1) *Show that the linear span $\mathcal{M}(M)$ of the functions $R_a M c$, where a varies in the set of points of analyticity of M (which we will denote by $\Omega(M)$) and c varies in $\mathbb{C}^n$, is finite dimensional.*
(2) *Show that M admits a realization of the form (1.7.2) with*

$$\begin{pmatrix} A & B \\ C & D \end{pmatrix} : \mathcal{M}(M) \times \mathbb{C}^n \longrightarrow \mathcal{M}(M) \times \mathbb{C}^m$$

defined by

$$\begin{aligned} Af &= R_0 f, \\ Bc &= R_0 M c, \\ Cf &= f(0), \\ Dc &= M(0)c. \end{aligned} \tag{1.7.6}$$

Hint: In the proof use is made of the fact that the operators R_a satisfy the resolvent identity (1.7.4).

The realization (1.7.6) is the celebrated backward shift realization; see for instance Fuhrmann's books [142, 143].

Realizations of the form (1.7.2) play an important role in the theory of Schur functions (functions analytic and contractive in the open unit disk). See the book

[86] of L. de Branges and J. Rovnyak, where the backward shift realization associated with a Schur function, say s, is defined and shown to be co-isometric. The space $\mathcal{M}(s)$ is then replaced by the reproducing kernel Hilbert space with reproducing kernel (see (7.7.17))

$$k_s(z, w) = \frac{1 - s(z)\overline{s(w)}}{1 - z\overline{w}}.$$

Realizations of the form (1.7.2) have far-reaching generalizations to various situations in several complex variables, see for instance the book [3] of J. Agler and J. McCarthy, and in noncommutative function theory. See for instance the paper of Paul Muhly and Baruch Solel [239] for the latter.

Question 1.7.4.

(1) *Compute the Laurent expansion at ∞ of the function (1.7.1) and the Taylor expansion at the origin of the function (1.7.2).*

(2) *Specialize Questions 1.6.16 and 1.6.18 when the coefficients r_n and h_n are of the form*

$$r_n = CU^nC^* \quad and \quad h_n = CA^{n-1}B,$$

where A, B, C, U are matrices of appropriate sizes, with U unitary and A, B, C with real entries in the case of the numbers $h_1, \ldots$.

Realizations (1.7.1) and (1.7.2) are called minimal if N is minimal. The number N is then called the McMillan degree (or, for short, the degree) of the function. The definition of the degree can be given for any rational function, regardless of analyticity at the origin or at infinity; see [64, Chapter IV, p. 77]. We refer also to [64] for the following discussion on factorization. The product $R(z) = R_1(z)R_2(z)$ of two $\mathbb{C}^{n \times n}$-valued rational functions is called minimal if

$$\deg R = \deg R_1 + \deg R_2. \tag{1.7.7}$$

Conversely, given a $\mathbb{C}^{n \times n}$-valued rational function R, a factorization $R = R_1R_2$ of R into a product of two $\mathbb{C}^{n \times n}$-valued rational functions is called minimal if (1.7.7) is in force. A given $\mathbb{C}^{n \times n}$-valued rational function may lack any (nontrivial) factorizations into a product of two $\mathbb{C}^{n \times n}$-valued rational functions. The description of all minimal factorizations of a given rational function analytic and invertible at infinity is given in [63], [64, Theorem 4.8, p. 90].

Examples of realizations centered at the origin and at infinity are given in Exercises 7.7.3 and 7.7.17, where two important classes of rational matrix-valued functions appear (namely functions which take unitary values with respect to a possibly indefinite metric on the real line or the unit circle). These exercises illustrate an important fact: It is of interest to relate the properties of the matrix

$$\begin{pmatrix} A & B \\ C & D \end{pmatrix}$$

with the properties of $R(z)$.

1.7. Rational functions

In preparation to Exercise 1.7.6 we introduce:

Definition 1.7.5. The space $\mathcal{W}$ of functions of the form

$$f(t) = \sum_{n \in \mathbb{Z}} a_n e^{int},$$

with $(a_n)_{n \in \mathbb{Z}}$ such that

$$\sum_{n \in \mathbb{Z}} |a_n| < \infty$$

is called the Wiener algebra.

We leave to the student to check that it is an algebra (in fact a commutative Banach algebra), when endowed with the pointwise product. A key result is the Wiener–Lévy theorem, which states that an element of $\mathcal{W}$ is invertible in $\mathcal{W}$ if and only if it does not vanish on the unit circle (in other words, invertibility in the algebra is equivalent to pointwise invertibility). It is instructive to look at the original proof of Wiener [321], although the result is a simple consequence of the theory of commutative Banach algebra; see [265].

Exercise 1.7.6. *Show that a rational function R with no poles on the unit circle and analytic at the origin belongs to the Wiener algebra.*

Hint: Two ways are possible. One is to use the partial fraction expansion and compute the Laurent series at the origin. In fact, because of the algebra structure, it is enough to show that the functions

$$\frac{1}{z - a}, \quad a \notin \mathbb{T},$$

belong to the Wiener algebra. This approach does not require the function to be analytic at the origin. The second way is more involved, but gives more precise results. It consists in considering a minimal realization of R and consider the Riesz projection corresponding to the part of the spectrum of A in the open unit disk. One gets then formulas for the coefficients of the Laurent series in terms of the realization. Such formulas are also available when the function has a pole at the origin by considering more general type of realizations for the rational function.

The computations in the following exercises are not very difficult, but they are conducive to study various connections between rational (and more generally analytic) functions and operator models and perturbation theory. See for instance [30, 31] for some applications. See also Problem 1.7.12.

Exercise 1.7.7. *Let R be a $\mathbb{C}^{n \times n}$-valued rational function analytic and invertible at infinity, with minimal realization (1.7.1). Show that*

$$\det R(z) = (\det D) \cdot \frac{\det(zI_N - A^\times)}{\det(zI_N - A)}, \tag{1.7.8}$$

$$\operatorname{Tr} R^{-1}(z) R'(z) = \operatorname{Tr} \left((zI_N - A^\times)^{-1} - (zI_N - A)^{-1} \right), \tag{1.7.9}$$

where

$$A^\times = A - BD^{-1}C. \tag{1.7.10}$$

One recognizes in (1.7.10) the Schur complement (see Definition 1.6.13).

Replacing R by R^{-1} in these formulas we obtain

$$\det R^{-1}(z) = (\det D^{-1}) \cdot \frac{\det(zI_N - A)}{\det(zI_N - A^\times)}, \tag{1.7.11}$$

$$\begin{aligned} \operatorname{Tr}\,(R^{-1})^{-1}(z)(R^{-1})'(z) &= \operatorname{Tr}\,(R(z) \cdot (-R^{-1}(z)R'(z))R^{-1}(z)) \\ &= \operatorname{Tr}\,(-R'(z)R^{-1}(z)) \\ &= \operatorname{Tr}\,\left((zI_N - A)^{-1} - (zI_N - A^\times)^{-1}\right). \end{aligned} \tag{1.7.12}$$

In the above computation we made use of the formula

$$(R^{-1}(z))' = -R^{-1}(z)R'(z)R^{-1}(z), \tag{1.7.13}$$

valid for a rational matrix-valued function R at the points z where $\det R(z) \neq 0$. The formula is valid, as its easy proof will show, more generally for analytic matrix-valued (or even operator-valued) functions, or for differentiable functions of a real variable.

Even in the complex-valued (as opposed to the matrix-valued) case, rational functions, an apparently easy concept, can be the object of surprising phenomenona, as illustrated by the following result of F. Reza.

Exercise 1.7.8 (see [259, Theorem 1, p. 231]).

(1) *Let φ be a function of the form*

$$\varphi(z) = \sum_{n=1}^{N} \frac{c_n}{z + a_n}, \tag{1.7.14}$$

where $N \in \mathbb{N}$ and $0 \le a_1 < a_2 < \cdots < a_N$ and the numbers $c_1, \ldots, c_N$ are all strictly positive. Show that φ is one-to-one in the open half-plane $\operatorname{Re} z > -a_1$.

(2) *Same question for the function*

$$\psi(z) = \sum_{n=1}^{N} \frac{zc_n}{z + a_n}, \tag{1.7.15}$$

where we now assume $a_1 \neq 0$.

Hint: Take two different points z_1 and z_2 in the assumed set such that $\varphi(z_1) = \varphi(z_2)$ and compute the real and imaginary parts of

$$\frac{\varphi(z_1) - \varphi(z_2)}{z_2 - z_1}.$$

Then, distinguish the cases $(\operatorname{Im} z_1)(\operatorname{Im} z_2) > 0$ and $(\operatorname{Im} z_1)(\operatorname{Im} z_2) \le 0$.

Functions of the form (1.7.14) have two important properties:

(a) They have a real positive part in the open right half-plane $\mathbb{C}_r$.
(b) They are real, that is take real values on the real axis, i.e.,

$$\varphi(\bar{z}) = \overline{\varphi(z)}.$$

Rational functions satisfying (a) and (b) are called *positive real functions* and are transfer functions of RLC networks, that is of electrical networks composed of resistances (R), inductors (L), and condensators (C). This is an important instance of a connection between the property of an analytic function (here a rational function which is positive real) and a corresponding engineering concept (an RLC network). Such parallels go way beyond the rational case and functions of one variable. See for instance [9] for some related discussions.

Reza's theorem expresses that RC and RL systems admit an univalent (or schlicht, that is, one-to-one) behavior in a half-plane which includes $\mathbb{C}_r$. See also [187].

Question 1.7.9. *Find the structure of a general rational positive real function.*

Hint: Use the partial fraction expansion of the function.

Exercise 1.7.10.

(1) *Give an example of a (non-constant) rational function with a positive real part in $\mathbb{C}_r$, and which is not one-to-one there.*
(2) *The functions (1.7.14) and (1.7.15) are real (that is, map the part of the real axis where they are defined into the real axis). Is it true that a rational positive real function is always univalent in $\mathbb{C}_r$?*

Hint: Take $N > 1$ and

$$b(z) = \prod_{n=1}^{N} \frac{z - w_n}{z + \overline{w_n}} \tag{1.7.16}$$

where the points $w_1, \ldots, w_N \in \mathbb{C}_r$.

More generally, functions analytic in the open unit disk or in a half-plane and with a real positive part there, play an important role in analysis.

The problem of constructing a rational function with given pole and zero structures is trivial in the scalar case. In the matrix-valued case it is much more involved. As a sample problem, let us mention:

Problem 1.7.11. *Given $\xi_1, \ldots, \xi_N \in \mathbb{C}^p$, $\eta_1, \ldots, \eta_M \in \mathbb{C}^q$, and (not necessarily distinct) complex numbers $z_1, \ldots, z_N, w_1, \ldots, w_M$, find all $\mathbb{C}^{p \times q}$-valued rational functions R such that*

$$\xi_i^* R(z_i) = 0, \quad i = 1, \ldots, N \text{ and}$$
$$R(w_j)\eta_j = 0, \quad j = 1, \ldots, M.$$

Such problems are of special interest when additional structure is imposed on the function R, for instance being J-inner (see Definition 7.7.9 for the latter). The problems are then closely related to classical interpolation problems; see for instance [58].

We conclude this section with some remarks. The material and exercises given here are of course classical. Still, they are a source of new problems, for instance when considering counterparts of the notion of realization in the setting of several complex variables. Various forms of realizations are used. We mention in particular

$$R(z_1, \ldots, z_m) = D + C(I_N - r(z)A)^{-1} r(z)B,$$

where, for instance,

$$r(z) = \begin{cases} \begin{pmatrix} z_1 I_N & z_2 I_N & \cdots & z_m I_N \end{pmatrix}, \\ \text{or} \\ \text{diag}\,(z_1 I_{n_1}, z_2 I_{n_2}, \ldots, z_m I_{n_m}), \quad \text{where } N = \sum_{j=1}^m n_j, \end{cases}$$

and where A, B, C are matrices of appropriate sizes.

The first case is called a Fornasini–Marchesini realization, and the second case is called a Givone–Roesser realization. See [137] and [151] respectively.

These aspects will be considered in the planned sequel to the present book. Even in the setting of one complex variable interesting problems occur. For instance:

Problem 1.7.12. *Given R analytic and invertible at infinity with minimal realization (1.7.1), recall that a minimal realization of R^{-1} is given by*

$$R^{-1}(z) = D^{-1} - D^{-1}C(zI_N - A^\times)^{-1} BD^{-1},$$

where $A^\times = A - BD^{-1}C$ (see (1.7.10)). Study the transformation

$$A \mapsto A^\times = A - BD^{-1}C$$

in terms of perturbation theory (see [85] for a related question).

Remark 1.7.13. Note that $(A^\times)^\times = A$.

Problem 1.7.14. *Given $n \in \mathbb{N}$ and given a scalar rational function $r_0(z)$, study the multiplicative group of $\mathbb{C}^{n \times n}$-valued functions $R(z)$, with non-identically vanishing determinant and such that*

$$\det R(z) = r_0(z)^m$$

for some $m \in \mathbb{Z}$ (which of course depends on R). In particular study the minimal factorizations of elements in this group.

1.7. Rational functions

Question 1.7.15. *For $a \in \mathbb{C}^n$ and $w_0 \in \mathbb{C} \setminus \mathbb{T}$ define*

$$B_{a,w_0}(z) = \left(I_n - \frac{aa^*}{a^*a} + \frac{z - w_0}{1 - z\overline{w_0}} \frac{aa^*}{a^*a} \right). \tag{1.7.17}$$

Show that

$$\frac{I_n - B_{a,w_0}(z)B_{a,w_0}(w)^*}{1 - z\overline{w}} = \frac{1 - |w_0|^2}{(1 - z\overline{w_0})(1 - \overline{w}w_0)} aa^*.$$

Definition 1.7.16. The function B_{a,w_0} is called an elementary Blaschke–Potapov factor when $w_0 \in \mathbb{D}$ (see also (7.7.18) and (7.7.19)). The form

$$\widetilde{B}_{a,w_0}(z) = \left(I_n - \frac{aa^*}{a^*a} + \frac{1 - \overline{w_0}}{1 - w_0} \frac{z - w_0}{1 - z\overline{w_0}} \frac{aa^*}{a^*a} \right), \tag{1.7.18}$$

which insures that $\widetilde{B}_{a,w_0}(1) = I_n$ is also used, in particular in the case of infinite products.

Remarks 1.7.17.

(a) We note that (possibly infinite product) of factors of the form (1.7.17) (or (1.7.18) for infinite products) with $w_0 \in \mathbb{D}$ are the matrix counterpart of Blaschke products.

(b) The functions B_{a,w_0} and $\widetilde{B}_{a,w_0}$ (and finite products of such) are special instances, with $J = I_n$, of rational functions J-unitary on the unit circle. See Exercise 7.7.17.

(c) We note the interpolation property

$$B_{a,w_0}(w_0)a = \widetilde{B}_{a,w_0}(w_0)a = 0,$$

which hints at connections with Problem 1.7.11.

(d) In fact any $\mathbb{C}^{n \times n}$-valued rational function which takes unitary values on the unit circle is a finite product of terms of the form B_{a,w_0} (or $\widetilde{B}_{a,w_0}$) with $w_0 \in \mathbb{C} \setminus \mathbb{T}$. See [28]. This is a special instance of a result of Kreĭn and Langer. See [208] and Remark 7.7.6.

(e) The functions (1.7.17) and (1.7.18) differ by a right or left multiplicative matrix, namely it holds that

$$\left(I_n - \frac{aa^*}{a^*a} + \frac{z - w_0}{1 - z\overline{w_0}} \frac{aa^*}{a^*a} \right)$$

$$= \left(I_n - \frac{aa^*}{a^*a} + \frac{1 - \overline{w_0}}{1 - w_0} \frac{z - w_0}{1 - z\overline{w_0}} \frac{aa^*}{a^*a} \right) \left(I_n - \frac{aa^*}{a^*a} + \frac{1 - w_0}{1 - \overline{w_0}} \frac{aa^*}{a^*a} \right)$$

$$= \left(I_n - \frac{aa^*}{a^*a} + \frac{1 - \overline{w_0}}{1 - w_0} \frac{aa^*}{a^*a} \right) \left(I_n - \frac{aa^*}{a^*a} + \frac{1 - \overline{w_0}}{1 - w_0} \frac{z - w_0}{1 - z\overline{w_0}} \frac{aa^*}{a^*a} \right).$$

1.8   Solutions of the exercises

Solution of Exercise 1.1.4:

(i) We have:

$$x \in f^{-1}(\cup_{i \in I} B_i) \iff f(x) \in \cup_{i \in I} B_i$$
$$\iff x \in \cup_{i \in I} f^{-1}(B_i),$$

and

$$x \in f^{-1}(\cap_{i \in I} B_i) \iff f(x) \in \cap_{i \in I} B_i$$
$$\iff x \in \cap_{i \in I} f^{-1}(B_i).$$

Furthermore,

$$x \in X \setminus f^{-1}(Y \setminus B) \iff f(x) \notin Y \setminus B$$
$$\iff f(x) \in B$$
$$\iff x \in f^{-1}(B),$$

and so

$$X \setminus f^{-1}(Y \setminus B) = f^{-1}(B),$$

and hence

$$f^{-1}(Y \setminus B) = X \setminus f^{-1}(B).$$

(ii) We now have:

$$y \in f(\cup_{i \in I} A_i) \iff \exists x \in \cup_{i \in I} A_i \; : \; y = f(x)$$
$$\iff \exists i \in I \text{ and } \exists x \in A_i \; : \; y = f(x)$$
$$\iff \exists i \in I \; : \; y \in f(A_i)$$
$$\iff y \in \cup_{i \in I} f(A_i).$$

The case of intersection of sets does not go as smoothly. Indeed,

$$y \in f(\cap_{i \in I} A_i) \iff \exists x \in \cap_{i \in I} A_i \; : \; y = f(x)$$
$$\iff \exists x \in X, \; \forall i \in I \; : \; x \in A_i \text{ and } y = f(x)$$
$$\implies y \in \cap_{i \in I} f(A_i).$$

To see that equality will not hold in general, take $I = \{1, 2\}$, $X = \mathbb{R}$,

$$A_1 = (-\infty, 0], \quad \text{and} \quad A_2 = [0, \infty),$$

and

$$f(x) = \begin{cases} 0, & \text{if } x = 0, \\ 1, & \text{if } x \neq 0. \end{cases} \tag{1.8.1}$$

Then $f(A_1) = f(A_2) = \{0, 1\}$ while $f(A_1 \cap A_2) = \{0\}$.

Finally, let $y \in f(X)$. We have:

$$y \in f(X) \setminus f(A) \iff \forall x \in A,\ y \neq f(x)$$
$$\implies y \in f(X \setminus A).$$

The fact that inequality may hold is illustrated by the example (1.8.1) and $A = A_1$. Then,

$$f(X) \setminus f(A) = \emptyset \quad \text{while} \quad f(X \setminus A) = \{1\}.$$

A constant function defined on a set of at least two elements will of course provide an easier example. $\qquad\square$

Solution of Exercise 1.1.5: Let $U \subset Y$. Then,

$$f(f^{-1}(U)) = \{\ f(x) \text{ with } x \in X \text{ and } f(x) \in U\ \} \subset U.$$

A trivial counterexample for the equality case is given by a constant function $f : X \longrightarrow Y$, say $f(x) \equiv y_0$ for some $y_0 \in Y$ (with Y having at least two elements). Then

$$f(f^{-1}(U)) = \begin{cases} \{y_0\} & \text{if } y_0 \in U, \\ \emptyset & \text{if } y_0 \notin U. \end{cases}$$

Similarly,

$$f^{-1}(f(V)) = \{\ x \in X \text{ with } f(x) \in f(V)\ \} \supset V.$$

Here too, a counterexample for the equality case is obtained by taking a constant function, say $f(x) \equiv y_0$ for some $y_0 \in Y$. Then, for $V \neq \emptyset$,

$$f^{-1}(f(V)) = X,$$

which is different from V, unless $V = X$. $\qquad\square$

Solution of Exercise 1.1.9: Assume that $q^{-1}(q(F)) \supsetneq F$. Then there exists $x \in U \cap q^{-1}(q(F))$, and so there is $y \in F$ such that $q(x) = q(y)$. Hence $q^{-1}(q(F))$ contains the element $y \in F$ equivalent to $x \in U$. This is not possible since U is saturated so that $y \in U$, and $U \cap F = \emptyset$. $\qquad\square$

Solution of Exercise 1.2.2:

(1) is left to the reader.

(2) Let

$$M_1 = \begin{pmatrix} a_1 & b_1 \\ c_1 & d_1 \end{pmatrix} \quad \text{and} \quad M_2 = \begin{pmatrix} a_2 & b_2 \\ c_2 & d_2 \end{pmatrix}$$

be two matrices in G_r. Then the $(2,1)$ entry of the product $M_1 M_2$ is equal to

$$c_1 a_2 + d_1 c_2,$$

and is divisible by r since both c_1 and c_2 are divisible by r.

(3) (see for instance [219, p. IV.2] and [219, p. 255]) We leave to the reader to check that τ is a group homomorphism, that is:

$$\tau(M_1 M_2) = \tau(M_1)\tau(M_2), \quad \forall M_1, M_2 \in \mathrm{SL}(2, \mathbb{Z}).$$

To prove that τ is onto, let $\overset{\circ}{M} \in \mathrm{SL}(\mathbb{Z}/r\mathbb{Z})$, and let $\begin{pmatrix} a & b \\ c & d \end{pmatrix}$ be a representative of $\overset{\circ}{M}$. Thus

$$ad - bc = 1 + kr \tag{1.8.2}$$

for some $k \in \mathbb{Z}$. This equation implies in particular that c, d and r are relatively prime together. We first show that there is $\ell \in \mathbb{N}$ such that c and $d + r\ell$ are prime together. To that purpose we partition the set $\mathcal{C}$ of prime divisors of c as $\mathcal{C}_1 \cup \mathcal{C}_2$, where

$$\mathcal{C}_1 = \{p \text{ prime}; p \mid d\}, \quad \text{and} \quad \mathcal{C}_2 = \{p \text{ prime}; p \nmid d\}.$$

Then

$$c \wedge (d + r\ell) = 1, \quad \text{with} \quad \ell = \prod_{p \in \mathcal{C}_2} p. \tag{1.8.3}$$

Indeed, let m which divides both c and $d + r\ell$, and let q be a prime divisor of m with power γ_q. Then there exist integers t and s such that

$$c = q^{\gamma_q} t \quad \text{and} \quad d + r\ell = q^{\gamma_q} s.$$

This is impossible. Indeed:

If $q \in \mathcal{C}_1$: Then q does not divide r, since c, d and r are together prime. But $r\ell = q^{\gamma_q} s - d$ is divisible by q (since q divides d) while $r\ell$ is not divisible by d by construction of ℓ.

If $q \in \mathcal{C}_2$: Then $q \mid \ell$ and so $r\ell - q^{\gamma_q} s$ is divisible by q while d is not.

We now can finish the proof that τ is onto. Take ℓ such that $c \wedge (d + r\ell) = 1$, and $u, v \in \mathbb{Z}$ such that

$$u(d + r\ell) - vc = -(k + ar),$$

where k is as in (1.8.2). This is possible since there exist integers u_0 and v_0 such that $u_0(d + r\ell) - v_0 c = 1$. Then,

$$\begin{aligned}
(a + ru)(d + r\ell) - (b + rv)c &= ad - bc + ar\ell + r(u(d + r\ell) - vc) \\
&= 1 + kr + ar\ell - r(k + a\ell) \\
&= 1.
\end{aligned}$$

1.8. Solutions of the exercises

Thus

$$\begin{pmatrix} a + ru & b + rv \\ c & d + r\ell \end{pmatrix}$$

belongs to $\mathrm{SL}(2, \mathbb{Z})$. It is a representative of $\overset{\circ}{M}$, and so τ is onto.

(4) The map τ induces a group homomorphism from $\mathrm{SL}(2, \mathbb{Z})/M_r$ onto $\mathrm{SL}(\mathbb{Z}/r\mathbb{Z})$, which is onto since $\ker \tau = M_r$. □

Solution of Exercise 1.2.3: We just mention that $T_{m,n}$ corresponds to the matrix

$$\begin{pmatrix} 1 & m\omega_1 + n\omega_2 \\ 0 & 1 \end{pmatrix}.$$ □

Solution of Exercise 1.2.4:

(1) The result is clear from the formula

$$\begin{pmatrix} an + mc & bn + md \end{pmatrix} = \begin{pmatrix} n & m \end{pmatrix} \begin{pmatrix} a & b \\ c & d \end{pmatrix}$$

since $ad - bc = \pm 1$.

(2) This follows from the fact that the family

$$\left(\frac{1}{(n^2 + m^2)^{3/2}} \right)_{(n,m) \in \mathbb{Z}^2 \setminus \{(0,0)\}}$$

is summable; see, e.g., [CAPB, pp. 132–133].

(3) We have

$$\frac{1}{(nt_M(z) + m)^k} = \frac{(cz + d)^k}{(n(az + b) + m(cz + d))^k}$$
$$= \frac{(cz + d)^k}{((na + mc)z + nb + md)^k}.$$

So

$$g_k(t_M(z)) = \sum_{(n,m) \in \mathbb{Z}^2 \setminus \{(0,0)\}} \frac{(cz + d)^k}{((na + mc)z + nb + md)^k}$$

$$= (cz + d)^k \left(\sum_{(n,m) \in \mathbb{Z}^2 \setminus \{(0,0)\}} \frac{1}{((na + mc)z + nb + md)^k} \right)$$

$$= (cz + d)^k \left(\sum_{(n,m) \in \mathbb{Z}^2 \setminus \{(0,0)\}} \frac{1}{(nz + m)^k} \right)$$

$$= (cz + d)^k g_k(z),$$

where we have used item (1) to go from the second to the third line. □

Solution of Exercise 1.2.5: The map $M \mapsto MM_0$ is one-to-one from G onto itself. Furthermore we recall that

$$t_{MM_0}(z) = t_M(t_{M_0}(z)).$$

See for instance [CAPB, (2.3.5), p. 67]. But

$$MM_0 = \begin{pmatrix} aa_0 + bc_0 & ab_0 + bd_0 \\ ca_0 + dc_0 & cb_0 + dd_0 \end{pmatrix} = \begin{pmatrix} a_1 & b_1 \\ c_1 & d_1 \end{pmatrix}$$

and so

$$(cT_{M_0}(z) + d)^{-N} = \left(\frac{(ca_0 + dc_0)z + (cb_0 + dd_0)}{c_0 z + d_0} \right)^{-N} = \frac{(c_0 z + d_0)^N}{(c_1 z + d_1)^N}.$$

So

$$f(T_{M_0}(z)) = \sum_{M \in \mathrm{SL}(2,\mathbb{Z})} (cT_{M_0}(z) + d)^{-N} f(T_M(T_{M_0}(z)))$$

$$= \sum_{M \in \mathrm{SL}(2,\mathbb{Z})} \frac{(c_0 z + d_0)^N}{(c_1 z + d_1)^N} f(T_{MM_0}(z))$$

$$= (c_0 z + d_0)^N \sum_{M \in \mathrm{SL}(2,\mathbb{Z})} \frac{1}{(c_1 z + d_1)^N} f(T_{MM_0}(z))$$

$$= (c_0 z + d_0)^N f(z). \qquad \square$$

Solution of Exercise 1.2.10: We follow [95, p. 185]. Let $g \in \mathrm{Aut}\, X$. Since $\mathcal{G}$ acts transitively on X there is $h \in \mathcal{G}$ which sends w_0 to $g(w_0)$, that is, such that

$$h(w_0) = g(w_0).$$

Thus $h^{-1}g$ is in the group of isotropy of w_0 and by hypothesis belongs to $\mathcal{G}$. Thus $g = h \circ (h^{-1} \circ g) \in \mathcal{G}$. $\qquad \square$

Solution of Exercise 1.3.6: The bilinear map

$$(v_2, v_3) \mapsto (v_1 \otimes v_2) \otimes v_3$$

factorizes to the (uniquely defined) linear map

$$v_2 \otimes v_3 \mapsto (v_1 \otimes v_2) \otimes v_3$$

from $\mathcal{V}_2 \otimes \mathcal{V}_3$ into $(\mathcal{V}_1 \otimes \mathcal{V}_2) \otimes \mathcal{V}_3$. It follows that

$$(v_1, v_2 \otimes v_3) \mapsto (v_1 \otimes v_2) \otimes v_3$$

1.8. Solutions of the exercises

defines a bilinear map from $\mathcal{V}_1 \times (\mathcal{V}_2 \otimes \mathcal{V}_3)$ into $(\mathcal{V}_1 \otimes \mathcal{V}_2) \otimes \mathcal{V}_3$, which factorizes to the linear map (1.3.6). The map (1.3.5) is defined in a similar way, and we have

$$i \circ j(v_1 \otimes (v_2 \otimes v_3)) = i((v_1 \otimes v_2) \otimes v_3) = v_1 \otimes (v_2 \otimes v_3),$$

$\square$

and similarly for $j \circ i$.

Solution of Exercise 1.3.8: This follows from the associativity of the tensor product.
$\square$

Solution of Exercise 1.3.11: The map

$$\widehat{T_n}(f_1, \ldots, f_n) = (Tf_1) \otimes \cdots \otimes (Tf_n)$$

is multilinear from $\mathcal{V}^n$ into $\mathcal{W}^{\otimes n}$. By Theorem 1.3.5 it factorizes to a linear map
from $\mathcal{V}^{\otimes n}$ into $\mathcal{W}^{\otimes n}$.
$\square$

Solution of Exercise 1.4.2: (see, e.g., [214, p. 224]) Let N be the dimension of C_1 as a C_2 vector space, and take $c \in C_1$. The elements $1, c, c^2, \ldots, c^N$ are linearly dependent over C_2 and so there exist $a_0, a_1, \ldots, a_N \in C_2$, not all equal to 0, and such that

$$a_0 + a_1 c + a_2 c^2 + \cdots + a_N c^N = 0.$$

Let N_0 be the smallest index j such that $a_j \neq 0$. If $N_0 = N$ we have $c = 0$, which is algebraic. If $N_0 < N$ we have

$$a_{N_0} + a_{N_0+1} c + \cdots + a_N c^{N-N_0} = 0,$$

$\square$

and so c is algebraic.

Solution of Exercise 1.4.4: Let $v_1, \ldots, v_N \in C_1$ be a basis of C_1 as a C_2-vector space. In particular, every power v_1^n can be written as a linear combination of $v_1, \ldots, v_N$. Thus, $C_1 \subset C_2[v_1, \ldots, v_N]$. Since

$$C_2[v_1, \ldots, v_N] \subset C_2(v_1, \ldots, v_N) \subset C_1,$$

the result follows.

That the converse is not true is seen by taking the field of rational functions
in one variable.
$\square$

Solution of Exercise 1.4.5: Following for instance [225] we proceed by induction and consider first the case $N = 1$. In the arguments it is well to recall that the quotient of a commutative ring by a maximal ideal is a field.

Case $N = 1$: The space

$$I_{z_1} = \{f \in C_2[X] \; ; f(z_1) = 0\}$$

is a maximal ideal, and its quotient field is $C_2[z_1]$. This last space is a finite-dimensional vector space over C_2 since z_1 is algebraic over C_2. This allows us to conclude since

$$C_1 = C_2(z_1) = C_2[z_1].$$

Induction: The induction follows easily since

$$(C_2[z_1, \ldots, z_{N-1}])[z_N] = C_2[z_1, \ldots, z_N].$$
□

Solution of Exercise 1.5.1: Let $\mathcal{N}$ be the required space, and let $\mathcal{M}$ be a subspace both invariant under A and B and on which they coincide. Then,

$$\mathcal{M} \subset \ker(A - B).$$

Since $A\mathcal{M} \subset \mathcal{M}$ we have, for $m \in \mathcal{M}$,

$$A(Am) = B(Am),$$

and so, since $Am = Bm$,

$$A(Am) = B(Bm),$$

and so $m \in \ker(A^2 - B^2)$. Thus the required space is included in

$$\cap_{u=1}^{\infty} \ker(A^u - B^u).$$

This space is invariant under A and B, and $A = B$ on it. Thus

$$\mathcal{N} = \cap_{u=1}^{\infty} \ker(A^u - B^u).$$
□

Remark 1.8.1. Since $U \mapsto \cap_{u=1}^{U} \ker(A^u - B^u)$ is a decreasing sequence of linear subspaces in $\mathbb{C}^n$ there exists an integer m such that

$$\mathcal{N} = \cap_{u=1}^{m} \ker(A^u - B^u).$$

Solution of Exercise 1.5.4:

(1) We can assume without loss of generality that at least one matrix, say M_1, is not a multiple of the identity. Let λ be an eigenvalue of M_1. We have

$$\{0\} \subsetneq \ker(M_1 - \lambda I_n) \subsetneq \mathbb{C}^n,$$

and the space $\mathcal{M}_1 = \ker(M_1 - \lambda I_n)$ is invariant under all the elements of the family.

1.8. Solutions of the exercises

(2) The claim is proved if the space $\mathcal{M}_1$ has dimension 1. Otherwise consider the restrictions of $M_1, M_2 \ldots,$ to $\mathcal{M}_1$ and reiterate the argument in (1). After a finite number of iterations one gets to the result.

(3) Write $\mathbb{C}^n = \mathcal{M}_1 \dotplus \mathcal{N}_1$, where $\dotplus$ denotes a direct sum, and let

$$M_j = \begin{pmatrix} a_j & m_j \\ 0 & B_j \end{pmatrix}, \quad j = 1, 2, \ldots$$

be the matrix representation of M_j with respect to this decomposition, with $B_j \in \mathbb{C}^{(n-1)\times(n-1)}$. The assumed commutativity of the matrices M_j implies that the matrices $B_1, B_2, \ldots$ pairwise commute. We can reiterate the arguments in (2) to this family. After a finite number of iterations one obtains a basis with respect to which $M_1, M_2, \ldots$ are simultaneously triangularized. $\qquad\square$

Solution of Exercise 1.5.5: Let $A = M J_A M^{-1}$, where $M \in \mathbb{C}^{n\times n}$ and where J_A is the Jordan form of A. Thus, J_A is a block diagonal matrix with blocks being elementary Jordan cells of the form

$$J(\lambda) = \begin{pmatrix} \lambda & 1 & 0 & \cdots & & 0 \\ 0 & \lambda & 1 & 0 & \cdots & \\ 0 & 0 & \lambda & 1 & 0 & \cdots \\ \vdots & & & & \vdots & \\ 0 & 0 & \cdots & & \lambda & 1 \\ 0 & 0 & \cdots & & & \lambda \end{pmatrix} = \lambda I_{n_\lambda} + N \in \mathbb{C}^{n_\lambda \times n_\lambda}, \qquad (1.8.4)$$

with

$$N = \begin{pmatrix} 0 & 1 & 0 & \cdots & & 0 \\ 0 & 0 & 1 & 0 & \cdots & \\ 0 & 0 & 0 & 1 & 0 & \cdots \\ \vdots & & & & \vdots & \\ 0 & 0 & \cdots & & 0 & 1 \\ 0 & 0 & \cdots & & & 0 \end{pmatrix},$$

and where $\lambda \notin C$ and where $n_\lambda \in \mathbb{N}$ (of course there may be various cells, of possibly different sizes, with the same λ; furthermore we set $N = 0$ if $n_\lambda = 1$). For a given cell we have

$$(zI_{n_\lambda} - J(\lambda))^{-1} = ((z - \lambda)I_{n_\lambda} - N)^{-1}$$

$$= \frac{1}{z - \lambda}\left(I_{n_\lambda} - \frac{N}{z - \lambda}\right)^{-1}$$

$$= \sum_{k=0}^{n_\lambda - 1} \frac{N^k}{(z - \lambda)^{k+1}},$$

since $N^{n_\lambda} = 0$.

Since C is a simple contour, Cauchy's theorem gives

$$\frac{1}{2\pi i} \int_C (zI_{n_\lambda} - J(\lambda))^{-1} dz = \begin{cases} I_{n_\lambda}, & \text{if } \lambda \text{ lies inside } C, \\ 0, & \text{if } \lambda \text{ lies outside } C. \end{cases}$$

It follows that $P_{J_A} = \frac{1}{2\pi i} \int_C (zI_n - J_A)^{-1} dz$ is a projection, and so is $M P_{J_A} M^{-1}$.

$\square$

Solution of Exercise 1.5.12:

(1) Let u be a bilinear map from $\mathbb{C}^N \times \mathbb{C}^M$ into a finite-dimensional vector space, say $\mathbb{C}^R$. In view of the bilinearity, the formula

$$v(e_i \otimes f_j) = u(e_i, f_j)$$

defines in a unique way a linear operator from $\mathbb{C}^{NM}$ into $\mathbb{C}^R$.

(2) We denote by $e_1, \ldots, f_1, \ldots, g_1, \ldots,$ and $h_1, \ldots,$ the canonical basis of $\mathbb{C}^N$, $\mathbb{C}^M$, $\mathbb{C}^P$ and $\mathbb{C}^Q$ respectively. Let $e = \sum_{i=1}^N \alpha_i e_i \in \mathbb{C}^N$ and $f = \sum_{j=1}^M \beta_j f_j \in \mathbb{C}^M$. We have (with $Ae_i = \sum_{k=1}^P a_{k,i} g_k$ and $Bf_j = \sum_{\ell=1}^Q b_{\ell,j} h_\ell$)

$$(A \otimes B)(e \otimes f) = \sum_{i=1}^N \sum_{j=1}^M \alpha_i \beta_j (Ae_i \otimes Bf_j)$$

$$= \sum_{i=1}^N \sum_{j=1}^M \sum_{k=1}^P \sum_{\ell=1}^Q \alpha_i \beta_j a_{k,i} b_{\ell,j} g_k \otimes h_\ell$$

$$= \sum_{k=1}^P \sum_{\ell=1}^Q \left(\sum_{i=1}^N a_{k,i} \alpha_i \right) \left(\sum_{j=1}^M b_{\ell,j} \beta_j \right) g_k \otimes h_\ell$$

$$= \left(\sum_{k=1}^P (Ae)_k g_k \right) \otimes \left(\sum_{\ell=1}^Q (Bf)_\ell h_\ell \right)$$

$$= Ae \otimes Bf.$$

(3) In view of the previous item, we have

$$(A \otimes B)(e \otimes f) = \sum_{k=1}^P \sum_{\ell=1}^Q \left(\sum_{i=1}^N a_{k,i} \alpha_i \right) \left(\sum_{j=1}^M b_{\ell,j} \beta_j \right) g_k \otimes h_\ell$$

and so:

$$\|(A \otimes B)(e \otimes f)\|^2 = \sum_{k=1}^P \sum_{\ell=1}^Q \left| \sum_{i=1}^N a_{k,i} \alpha_i \right|^2 \cdot \left| \sum_{j=1}^M b_{\ell,j} \beta_j \right|^2 = \|Ae\|^2 \cdot \|Bf\|^2.$$

(4) The item follows from (3) and its proof will be omitted.

$\square$

1.8. Solutions of the exercises

Solution of Exercise 1.6.2: Formula (1.6.3) leads to $M_N = L_N L_N^*$, with

$$
L_N = \begin{pmatrix}
1 & 0 & 0 & 0 & 0 & 0 & \cdots & 0 \\
1 & 1 & 0 & 0 & 0 & 0 & \cdots & 0 \\
1 & 2 & 1 & 0 & 0 & 0 & \cdots & 0 \\
1 & 3 & 3 & 1 & 0 & 0 & \cdots & 0 \\
1 & 4 & 6 & 4 & 1 & 0 & \cdots & 0 \\
\vdots & \vdots & & & & & & \\
1 & n & \binom{n}{2} & \cdots & \binom{n}{2} & n & 1 & 0 \\
1 & n+1 & \binom{n+1}{2} & \cdots & & \binom{n+1}{2} & n+1 & 1
\end{pmatrix}.
$$
$\square$

Solution of Exercise 1.6.4: We need to show that

$$
\sum_{k=0}^{N} (-1)^k \binom{N+1}{k+1} \binom{j+k}{j} = \delta_{0,j}, \quad j = 0, 1, \ldots, N. \tag{1.8.5}
$$

When $j = 0$ this equation reduces to

$$
\sum_{k=0}^{N} (-1)^k \binom{N+1}{k+1} = 1,
$$

which holds since

$$
0 = (-1+1)^{N+1} = \sum_{k=0}^{N+1} (-1)^k \binom{N+1}{k} = 1 + \sum_{k=0}^{N} (-1)^{k+1} \binom{N+1}{k+1}.
$$

When $j = 1$ we need to show that

$$
\sum_{k=0}^{N} (-1)^k \binom{N+1}{k+1} (k+1) = 0, \tag{1.8.6}
$$

which holds since

$$
\sum_{k=0}^{N} (-1)^k \binom{N+1}{k+1} (k+1) = (N+1) \sum_{k=0}^{N} (-1)^k \binom{N}{k} = (1-1)^N = 0.
$$

For the convenience of the reader we also explicit the case $j = 2$. The identity to prove is now (after multiplication by 2):

$$
\sum_{k=0}^{N} (-1)^k \binom{N+1}{k+1} (k+2)(k+1) = 0.
$$

In view of the case $j = 1$ we need to check:

$$\sum_{k=0}^{N}(-1)^k \binom{N+1}{k+1} k(k+1) = 0,$$

or, equivalently

$$-N(N+1) \sum_{k=1}^{N}(-1)^{k-1} \binom{N-1}{k-1} = 0,$$

which clearly holds.

To consider the cases $j = 2, \ldots, N$, we will proceed by induction, and first note that the functions of k:

$$1, k+1, (k+1)(k+2), \ldots, \underbrace{(k+1)(k+2)\cdots(k+j-1)}_{\text{product of } (j-1) \text{ terms}}$$

and

$$\varphi(k) = \underbrace{(k+1)k(k-1)\cdots(k-j+2)}_{\text{product of } j \text{ terms}} \tag{1.8.7}$$

form a basis of the vector space of polynomials (in the variable k) of degree less or equal to j. Thus there exist numbers $a_{uj}, u = 1, \ldots, j-1$ such that

$$\underbrace{(j+k)\cdots(k+2)(k+1)}_{\text{product of } j \text{ terms}} = \left(\sum_{u=0}^{j-1} a_{uj}(k+1)\cdots(k+u)\right)$$

$$+ a_{jj}(k+1)k(k-1)\cdots(k-j+2),$$

since the polynomial $(j+k)\cdots(k+2)(k+1)$ has no constant term. In view of this equality and assuming the induction to hold for $u = 1, \ldots, j-1$, to prove (1.8.5) amounts to proving

$$\sum_{k=0}^{N}(-1)^k \binom{N+1}{k+1}(k+1)\cdots(k-j+2) = 0,$$

that is, in fact

$$\sum_{k=j-1}^{N}(-1)^k \binom{N+1}{k+1}(k+1)\cdots(k-j+2) = 0,$$

since φ in (1.8.7) vanishes for $k = 0, 1, \ldots, j-2$. But this is clear since, for $k \geq j-1$ we have

$$\binom{N+1}{k+1}(k+1)\cdots(k-j+2) = (N+1)\cdots(N-j+2)\binom{N-j+1}{k-j+1},$$

1.8. Solutions of the exercises

and

$$\sum_{k=j-1}^{N} (-1)^k \binom{N-j+1}{k-j+1} = (-1)^{j-1} \sum_{u=0}^{N-j+1} (-1)^u \binom{N-j+1}{u}$$
$$= (-1)^{j-1} \cdot (1-1)^{N-j+1} = 0. \qquad \square$$

Solution of Exercise 1.6.10: When A is not invertible a given element f in the range of A is written in a non-unique way as $f = Au$, and we first need to check that (1.6.7) is well defined. So let $f = Au = Au'$ and $g = Av = Av'$ be two elements in ran A, with $u, u', v, v' \in \mathbb{C}^n$. We have

$$\begin{aligned}
\langle Au, Av \rangle_A &= \langle Au, v \rangle_{\mathbb{C}^n} \\
&= \langle Au', v \rangle_{\mathbb{C}^n} \\
&= \langle u', Av \rangle_{\mathbb{C}^n} \\
&= \langle u', Av' \rangle_{\mathbb{C}^n} = \langle Au', Av' \rangle_A,
\end{aligned}$$

and so $\langle \cdot, \cdot \rangle_A$ is well defined. The fact that it is linear in the first variable follows from the definition of the inner product in $\mathbb{C}^n$. That it is Hermitian follows from the fact that $A = A^*$. Since $A \geq 0$ we have in fact

$$\langle Au, Au \rangle_A \geq 0, \quad \forall u \in \mathbb{C}^n. \tag{1.8.8}$$

To end the proof, we check that $\langle \cdot, \cdot \rangle_A$ is nondegenerate. Inequality (1.8.8) implies that the Cauchy–Schwarz inequality holds in the (possibly degenerate) space $(\text{ran } A, \langle \cdot, \cdot \rangle_A)$. See the hint after the statement of the exercise. Let $u \in \mathbb{C}^n$ be such that $\langle Au, Au \rangle_A = 0$. By the Cauchy–Schwarz inequality

$$|\langle Au, Av \rangle_A|^2 \leq \langle Au, Au \rangle_A \langle Av, Av \rangle_A = 0,$$

and so $\langle Au, Av \rangle_A = 0$ for all $v \in \mathbb{C}^n$. This implies that $Au = 0$ since $\langle Au, Av \rangle_A = \langle Au, v \rangle_{\mathbb{C}^n}$. $\qquad \square$

Solution of Exercise 1.6.11: We follow the hint. Let

$$F = (M_1 + M_2)u \quad \text{and} \quad G = (M_1 u, M_2 u), \quad u \in \mathbb{C}^n.$$

Then

$$\|F\|_{M_1+M_2}^2 = u^*(M_1 + M_2)u = u^* M_1 u + u^* M_2 u = \|G\|_{M_1, M_2}^2. \tag{1.8.9}$$

The space R is a linear subspace of the Cartesian product

$$(\text{ran}\,(M_1 + M_2)) \times ((\text{ran } M_1) \times (\text{ran } M_2)),$$

and contains no elements of the form $(0, G)$ with $G \neq (0, 0)$ in view of (1.8.9).

Thus we can define an isometric linear operator T by

$$TF = G \; : \; \mathrm{ran}\,(M_1 + M_2) \quad \longrightarrow \quad (\mathrm{ran}\,M_1) \times (\mathrm{ran}\,M_2).$$

Furthermore, for $(M_1 v_1, M_2 v_2) \in (\mathrm{ran}\,M_1) \times (\mathrm{ran}\,M_2)$, let $v \in \mathbb{C}^n$ be such that

$$T^*(M_1 v_1, M_2 v_2) = (M_1 + M_2)v. \tag{1.8.10}$$

For $u \in \mathbb{C}^n$ we have

$$\begin{aligned}
u^*(M_1 + M_2)v &= \langle (M_1 + M_2)v, (M_1 + M_2)u \rangle_{M_1 + M_2} \\
&= \langle T^*(M_1 v_1, M_2 v_2), (M_1 + M_2)u \rangle_{M_1 + M_2} \\
&= \langle (M_1 v_1, M_2 v_2), (M_1 u, M_2 u) \rangle \\
&= u^*(M_1 v_1 + M_2 v_2),
\end{aligned}$$

and so

$$T^*(M_1 v_1, M_2 v_2) = M_1 v_1 + M_2 v_2. \tag{1.8.11}$$

This ends the proof since $\ker T = 0$ and so

$$\mathrm{ran}\,(M_1 + M_2) = \mathrm{ran}\,T^* \dotplus \ker T = \mathrm{ran}\,T^*, \tag{1.8.12}$$

where $\dotplus$ denotes an orthogonal and direct sum.

Let now $v \in \mathbb{C}^n$ such that (1.8.10) holds and write:

$$(M_1 v_1, M_2 v_2) = (M_1 v, M_2 v) + (M_1 v_1 - M_1 v, M_2 v_2 - M_2 v).$$

In view of (1.8.10) and (1.8.11) we have $(M_1 v_1 - M_1 v, M_2 v_2 - M_2 v) \in \ker T^*$. Since $(M_1 v, M_2 v) \in \mathrm{ran}\,T$, and using the orthogonal sum

$$(\mathrm{ran}\,M_1) \times (\mathrm{ran}\,M_2) = \mathrm{ran}\,T \dotplus \ker T^*,$$

we have

$$\begin{aligned}
\|(M_1 v_1, M_2 v_2)\|^2_{M_1, M_2} \\
= \|(M_1 v, M_2 v)\|^2_{M_1, M_2} + \|(M_1 v_1 - M_1 v, M_2 v_2 - M_2 v)\|^2_{M_1, M_2} \\
= \|(M_1 + M_2)v\|^2_{M_1 + M_2} + \|(M_1 v_1 - M_1 v, M_2 v_2 - M_2 v)\|^2_{M_1, M_2},
\end{aligned}$$

and so

$$\|M_1 v_1\|^2_{M_1} + \|M_2 v_2\|^2_{M_2} \geq \|(M_1 + M_2)v\|^2_{M_1 + M_2}, \tag{1.8.13}$$

with equality if and only if $M_1 v_1 = M_1 v$ and $M_2 v_2 = M_2 v$. $\qquad\square$

Remarks 1.8.2.

(1) R is a linear relation; see Definition 4.2.49. The proof uses the finite-dimensional version of the following fact: A densely defined contractive linear relation between Hilbert spaces extends to the graph of an everywhere defined contraction. See Exercise 4.2.50.

(2) See formula (4.2.15) in Exercise 4.2.41 for a formula for the minimal decomposition which gives equality in (1.8.13).

1.8. Solutions of the exercises

Solution of Exercise 1.6.12: Let $v_0 \in \mathbb{C}^s$ be in the kernel of D. For every $u \in \mathbb{C}^r$ and $z \in \mathbb{C}$ we have

$$u^* A u + z u^* B v_0 + \bar{z} v_0^* B^* u \geq 0.$$

The choice $z = t \in \mathbb{R}$ gives

$$u^* A u + t(u^* B v_0 + v_0^* B^* u) \geq 0, \quad \forall t \in \mathbb{R},$$

and so $u^* B v_0 + v_0^* B^* u = 0$. Similarly the choice $z = it$ leads to

$$u^* A u + t(i(u^* B v_0 - v_0^* B^* u)) \geq 0, \quad \forall t \in \mathbb{R},$$

and so $u^* B v_0 - v_0^* B^* u = 0$. It follows that $u^* B v_0 = 0$ for all $u \in \mathbb{C}^r$ and so $B v_0 = 0$, that is $v_0 \in \ker B_0$.

Still with $M \geq 0$ we have

$$\begin{pmatrix} A & B \\ B^* & D + \epsilon I_s \end{pmatrix} \geq 0, \quad \forall \epsilon > 0.$$

Formula (1.6.12) leads to

$$A - B(D + \epsilon I_s)^{-1} B^* \geq 0.$$

Exercise 1.6.10 gives then $\operatorname{ran} B(D + \epsilon I_s)^{-1} B^* \subset \operatorname{ran} A$. To conclude we verify that $\operatorname{ran} B(D + \epsilon I_s)^{-1} B^* = \operatorname{ran} B$. Note first that (since $D + \epsilon I_s > 0$)

$$u^* B(D + \epsilon I_s)^{-1} B^* u = 0 \iff Bu = 0,$$

and hence

$$\ker B(D + \epsilon I_s)^{-1} B^* = \ker B^*. \tag{1.8.14}$$

But we have the direct and orthogonal sums

$$\mathbb{C}^s = \operatorname{ran} B \dotplus \ker B^* = \operatorname{ran} B(D + \epsilon I_s)^{-1} B^* \dotplus \ker B(D + \epsilon I_s)^{-1} B^*,$$

$\square$

and (1.8.14) implies $\operatorname{ran} B(D + \epsilon I_s)^{-1} B^* = \operatorname{ran} B$.

Solution of Exercise 1.6.14:

(1) Write

$$T_{N+1} = \begin{pmatrix} T_N & b_{N+1} \\ b_{N+1}^* & r_0 \end{pmatrix},$$

with

$$b_{N+1} = \begin{pmatrix} r_{N+1} \\ r_N \\ \vdots \\ r_1 \end{pmatrix} = \begin{pmatrix} r_{N+1} \\ b_N \end{pmatrix}. \tag{1.8.15}$$

Formula (1.6.12) implies that $T_{N+1} \geq 0$ if and only if $r_0 - b_{N+1}^* T_N^{-1} b_{N+1} \geq 0$.
Write now

$$T_N = \begin{pmatrix} r_0 & a_N \\ a_N^* & T_{N-1} \end{pmatrix} = \begin{pmatrix} 1 & a_N T_{N-1}^{-1} \\ 0 & 1 \end{pmatrix} \begin{pmatrix} r_0 - a_N T_{N-1}^{-1} a_N^* & 0 \\ 0 & T_{N-1} \end{pmatrix} \begin{pmatrix} 1 & 0 \\ T_{N-1}^{-1} a_N^* & 1 \end{pmatrix},$$

with $a_N = (r_1 \quad \cdots \quad r_N)$. We have

$$r_0 - b_{N+1}^* T_N^{-1} b_{N+1} = r_0 - (\overline{r_{N+1}} \quad b_N^*) \begin{pmatrix} 1 & 0 \\ -T_{N-1}^{-1} a_N^* & 1 \end{pmatrix}$$

$$\times \begin{pmatrix} (r_0 - a_N T_{N-1}^{-1} a_N^*)^{-1} & 0 \\ 0 & T_{N-1}^{-1} \end{pmatrix}$$

$$\times \begin{pmatrix} 1 & -a_N T_{N-1}^{-1} \\ 0 & 1 \end{pmatrix} \begin{pmatrix} r_{N+1} \\ b_N \end{pmatrix}$$

$$= r_0 - \frac{|\overline{r_{N+1}} - b_N^* T_{N-1}^{-1} a_N^*|^2}{r_0 - a_N T_{N-1}^{-1} a_N^*} - b_N^* T_{N-1}^{-1} b_N.$$

It follows that the set of complex numbers r_{N+1} such that $T_{N+1} \geq 0$ is the closed disk with center $a_N T_{N-1}^{-1} b_N$ and radius $\sqrt{(r_0 - b_N^* T_{N-1}^{-1} b_N)(r_0 - a_N T_{N-1}^{-1} a_N^*)}$. Furthermore, T_{N+1} is invertible if and only if r_{N+1} is in the interior of this disk.

(2) We fix r_{N+1} such that T_{N+1} is positive but singular. We show that there is a uniquely defined number r_{N+2} such that the corresponding Toeplitz matrix $T_{N+2} \geq 0$. By Exercise 1.6.12 we have that a necessary condition is

$$\begin{pmatrix} r_{N+2} \\ r_{N+1} \\ \vdots \\ r_1 \end{pmatrix} \in \operatorname{ran} T_{N+1}.$$

Equivalently, there should exist $(u, v) \in \mathbb{C} \times \mathbb{C}^{N+1}$ such that

$$\begin{pmatrix} r_{N+2} \\ r_{N+1} \\ \vdots \\ r_1 \end{pmatrix} = \begin{pmatrix} r_0 & a_{N+1} \\ a_{N+1}^* & T_N \end{pmatrix} \begin{pmatrix} u \\ v \end{pmatrix}. \tag{1.8.16}$$

Since T_{N+1} is singular we have by formula (1.6.12) that

$$r_0 = a_{N+1} T_N^{-1} a_{N+1}^* = b_{N+1}^* T_N^{-1} b_{N+1}, \tag{1.8.17}$$

1.8. Solutions of the exercises

and (1.8.16) can be rewritten as

$$
\begin{pmatrix} r_{N+2} \\ r_{N+1} \\ \vdots \\ r_1 \end{pmatrix} = \begin{pmatrix} 1 & a_{N+1}T_N^{-1} \\ 0 & 1 \end{pmatrix} \begin{pmatrix} 0 & 0 \\ 0 & T_N \end{pmatrix} \begin{pmatrix} 1 & 0 \\ T_N^{-1}a_{N+1}^* & 1 \end{pmatrix} \begin{pmatrix} u \\ v \end{pmatrix}
$$

$$
= \begin{pmatrix} 1 & a_{N+1}T_N^{-1} \\ 0 & 1 \end{pmatrix} \begin{pmatrix} 0 \\ a_{N+1}^*u + T_N v \end{pmatrix}
$$

$$
= \begin{pmatrix} a_{N+1}T_N^{-1}(a_{N+1}^*u + T_N v) \\ (a_{N+1}^*u + T_N v) \end{pmatrix},
$$

and thus (recall (1.8.15)) we have $r_{N+2} = a_{N+1}T_N^{-1}b_{N+1}$. We show that this (uniquely defined) value of r_{N+1} indeed leads to a positive T_{N+2}. This is done by verifying that, with this value of r_{N+2},

$$
T_{N+2} = \begin{pmatrix} 1 & a_{N+1}T_N^{-1} & 0 \\ 0 & I_{N+1} & 0 \\ 0 & 0 & 1 \end{pmatrix} \begin{pmatrix} 0 & 0 & 0 \\ 0 & T_N & b_{N+1} \\ 0 & b_{N+1}^* & b_{N+1}^*T_N^{-1}b_{N+1} \end{pmatrix} \begin{pmatrix} 1 & a_{N+1}T_N^{-1} & 0 \\ 0 & I_{N+1} & 0 \\ 0 & 0 & 1 \end{pmatrix}^* .
$$

We leave this easy computation (done using in particular (1.8.17)) to the reader. $\qquad\square$

Solution of Exercise 1.7.6: Let

$$
R(z) = D + zC(I_N - zA)^{-1}B
$$

be a minimal realization of R centered at the origin. Then, $\sigma(A) \cap \mathbb{T} = \emptyset$, where $\sigma(A)$ denotes the spectrum of A, that is the set of eigenvalues of A. Let P be the Riesz projection corresponding to $\sigma(A) \cap \mathbb{D}$. We have

$$
P = \frac{1}{2\pi i} \int_{|z|=1} (zI_N - A)^{-1}dz.
$$

We have for $z \in \mathbb{T}$

$$
R(z) = D + zC(I_N - zA)^{-1}B
$$
$$
= D + zC(I_N - zPA - z(I_N - P)A)^{-1}B
$$
$$
= D + zCP(I_N - zPA)^{-1}B + zC(I_N - P)(I_N - z(I_N - P)A)^{-1}B
$$
$$
= -\underbrace{\sum_{n=1}^{\infty} z^{-n}C(I_N - P)A^{-n-1}B}_{\text{negative Fourier coefficients}} + \underbrace{D - C(I_N - P)((I_N - P)A)^{-1}B}_{\text{zeroth Fourier coefficient}}
$$

$$
+ \underbrace{\sum_{n=1}^{\infty} z^n CA^{n-1}PB}_{\text{positive Fourier coefficients}} ,
$$

and by definition of P,

$$\sum_{n=1}^{\infty} |CA^{n-1}PB| < \infty \quad \text{and} \quad \sum_{n=1}^{\infty} |CA^{-n-1}(I_N - P)B| < \infty,$$

Note that A is *not assumed invertible*. What is meant by $A^{-n-1}(I_N - P)$ is the inverse of the restriction of A to the range of $I_N - P$. $\square$

In the preceding exercise the computations are done for a complex-valued R, but are the same for a matrix-valued function. Then, the absolute values need to be replaced by matrix norms. See [157, pp. 35–37], [155, 156].

Solution of Exercise 1.7.7: The function R is analytic at infinity, and so the limit $R(\infty) = \lim_{z \to \infty} R(z)$ exists entry-wise and is a finite matrix, equal to D. Since $R(\infty)$ is assumed invertible we have

$$\begin{aligned}
\det R(z) &= (\det D)(\det(I_n + D^{-1}C(zI_N - A)^{-1}B)) \\
&= (\det D)(\det(I_N + BD^{-1}C(zI_N - A)^{-1})) \\
&= (\det D)(\det((zI_N - A + BD^{-1}C)(zI_N - A)^{-1})) \\
&= (\det D) \cdot \frac{\det(zI_N - A^{\times})}{\det(zI_N - A)},
\end{aligned}$$

where we have used the well-known formula

$$\det(I_p + UV) = \det(I_q + VU),$$

where $U \in \mathbb{C}^{p \times q}$ and $V \in \mathbb{C}^{q \times p}$. See [CAPB, Proposition 7.7.3, p. 362] if need be. This proves (1.7.8). To prove (1.7.9) we follow the arguments of [30, §6]. We have

$$R'(z) = -C(zI_N - A)^{-2}B,$$

and so

$$\begin{aligned}
R^{-1}(z)R'(z) &= \left(D^{-1} - D^{-1}C(zI_N - A^{\times})^{-1}BD^{-1}\right)\left(-C(zI_N - A)^{-2}B\right) \\
&= -D^{-1}C(zI_N - A)^{-2}B + D^{-1}C(zI_N - A^{\times})^{-1}BD^{-1}C(zI_N - A)^{-2}B \\
&= -D^{-1}C(zI_N - A)^{-2}B + D^{-1}C(zI_N - A^{\times})^{-1}(A - A^{\times})(zI_N - A)^{-2}B \\
&= -D^{-1}C(zI_N - A)^{-2}B + \\
&\quad + D^{-1}C(zI_N - A^{\times})^{-1}(A - zI_N + zI_N - A^{\times})(zI_N - A)^{-2}B \\
&= -D^{-1}C(zI_N - A^{\times})^{-1}(zI_N - A)^{-1}B.
\end{aligned}$$

Hence

$$\begin{aligned}
\operatorname{Tr} R(z)^{-1}R'(z) &= \operatorname{Tr}\left(-D^{-1}C(zI_N - A^{\times})^{-1}(zI_N - A)^{-1}B\right) \\
&= \operatorname{Tr}(zI_N - A)^{-1}(-BD^{-1}C)(zI_N - A^{\times})^{-1} \\
&= \operatorname{Tr}(zI_N - A)^{-1}(A^{\times} - A)(zI_N - A^{\times})^{-1} \\
&= \operatorname{Tr}(zI_N - A)^{-1}\left(A^{\times} - zI_N + zI_N - A\right)(zI_N - A^{\times})^{-1},
\end{aligned}$$

1.8. Solutions of the exercises

and thus:

$$\operatorname{Tr} R(z)^{-1} R'(z) = \operatorname{Tr} \left((z I_N - A^\times)^{-1} - (z I_N - A)^{-1} \right). \tag{1.8.18}$$

$\square$

Solution of Exercise 1.7.8:

(1) Let $z_1 = x_1 + iy_1$ and $z_2 = x_2 + iy_2$ be two different points with real part strictly bigger than $-a_1$. Then both $\varphi(z_1)$ and $\varphi(z_2)$ are defined, and we have:

$$
\begin{aligned}
\frac{\varphi(z_1) - \varphi(z_2)}{z_2 - z_1} &= \sum_{n=1}^{N} \frac{c_n}{(z_1 + a_n)(z_2 + a_n)} \\
&= \sum_{n=1}^{N} c_n \frac{(\overline{z_1} + a_n)(\overline{z_2} + a_n)}{(|z_1 + a_n|^2)(|z_2 + a_n|^2)} \\
&= \sum_{n=1}^{N} c_n \frac{(x_1 + a_n - iy_1)(x_2 + a_n - iy_2)}{(|z_1 + a_n|^2)(|z_2 + a_n|^2)} \\
&= \sum_{n=1}^{N} c_n \frac{(x_1 + a_n)(x_2 + a_n) - y_1 y_2}{(|z_1 + a_n|^2)(|z_2 + a_n|^2)} - \\
&\quad - i \sum_{n=1}^{N} c_n \frac{(x_1 + a_n)y_2 + (x_2 + a_n)y_1}{(|z_1 + a_n|^2)(|z_2 + a_n|^2)}.
\end{aligned}
$$

Assume that $\varphi(z_1) = \varphi(z_2)$. We obtain (see [259, equations (11) and (12)])

$$
\begin{aligned}
\sum_{n=1}^{N} c_n \frac{(\operatorname{Re} z_1 + a_n)(\operatorname{Re} z_2 + a_n) - (\operatorname{Im} z_1)(\operatorname{Im} z_2)}{(|z_1 + a_n|^2)(|z_2 + a_n|^2)} &= 0, \\
\sum_{n=1}^{N} c_n \frac{(\operatorname{Re} z_1 + a_n)(\operatorname{Im} z_2) + (\operatorname{Re} z_2 + a_n)(\operatorname{Im} z_1)}{(|z_1 + a_n|^2)(|z_2 + a_n|^2)} &= 0.
\end{aligned}
\tag{1.8.19}
$$

Assume now $(\operatorname{Im} z_1)(\operatorname{Im} z_2) \le 0$ (and in particular it may be that $\operatorname{Im} z_1 = \operatorname{Im} z_2 = 0$). The first equality in (1.8.19) is then impossible since, by hypothesis

$$(\operatorname{Re} z_1 + a_n)(\operatorname{Re} z_2 + a_n) > 0, \quad n = 1, \dots, N.$$

Suppose now $(\operatorname{Im} z_1)(\operatorname{Im} z_2) > 0$. So both $\operatorname{Im} z_1$ and $\operatorname{Im} z_2$ are strictly positive or strictly negative, and it is the second equation in (1.8.19) which cannot hold. So $\varphi(z_1) \ne \varphi(z_2)$, and φ is univalent in the open half-plane $\operatorname{Re} z > -a_1$.

(2) With z_1 and z_2 as above we now have

$$\frac{\psi(z_1) - \psi(z_2)}{z_1 - z_2} = \sum_{n=1}^{N} c_n a_n \frac{1}{(z_1 + a_n)(z_2 + a_n)},$$

and the same argument as in (1) holds, since $a_1 \ne 0$.

$\square$

Remark 1.8.3. The fact that φ and ψ take real values on the real line is essential in the proof.

Solution of Exercise 1.7.10:

(1) The function b is contractive in the right half-plane since (see, e.g., [CAPB, (1.1.51), p. 21]),

$$1 - \left|\frac{z - w_n}{z + \overline{w_n}}\right|^2 = \frac{4(\operatorname{Re} z)(\operatorname{Re} w_n)}{|z + \overline{w_n}|^2}.$$

Thus the function $\varphi(z) = \frac{1 - b(z)}{1 + b(z)}$ has a real positive part in $\mathbb{C}_r$. It takes the value 1 at the points $w_1, \ldots, w_N$ and so is not univalent in $\mathbb{C}_r$ when $N > 1$.

(2) Taking the points w_n to be real in the product (1.7.16) leads to a real φ which is not univalent.

$\square$

Chapter 2

Analytic Functions

In this second chapter we gather a number of exercises on complex variables. Most of them are used in the sequel, applied to the study of spaces of analytic functions.

2.1 Some warm-up exercises

We here present a number of exercises, some elementary, some a bit less. We begin with a result due to Kaluza, see [198], and appearing in Hardy's book on divergent series; see [171, Theorem 22, p. 68]. In the general index of Hardy's book, the result is referred as Kaluza's theorem. It is used there to study summation methods called Nörlund means (Hardy mentions that Voronoi introduced them earlier and quotes a 1902 paper, republished in 1932 in [318]). Nörlund means are summation methods defined as follows (see Chapter 4 in [171]): Given a sequence $(p_n)_{n \in \mathbb{N}_0}$ of positive numbers with $p_0 > 0$ and given a possibly divergent series $\sum_{n=0}^{\infty} a_n$ of complex numbers, one sets $s_n = \sum_{k=0}^{n} a_k$ and

$$S_n = \frac{\sum_{u=0}^{n} p_u s_{n-u}}{\sum_{u=0}^{n} p_u}. \tag{2.1.1}$$

If $\lim_{n \to \infty} S_n$ exists, it is called the Nörlund sum of the series $\sum_{n=0}^{\infty} a_n$. The reader will have recognized the Cesàro mean of the sequences of partial sums $(s_0, s_1, \ldots)$ when $p_n \equiv 1$. The result in the exercise is used to compare the series summable with respect to two sequences $p_0, p_1, \ldots$ and $q_0, q_1, \ldots$, under a supplementary hypothesis, namely that both $(p_n)_{n \in \mathbb{N}_0}$ and $(q_n)_{n \in \mathbb{N}_0}$ satisfy (2.1.6). See [171, Theorem 23, p. 69].

It is all the more interesting that the result presented in the exercise is used to prove that certain positive definite kernels (such as the reproducing kernel of the Dirichlet space) are complete Nevanlinna–Pick kernels. See Remark 7.6.17 for more information on these kernels.

Exercise 2.1.1 (Kaluza's theorem; see [198], [171, Theorem 22, p. 68], [3, p. 90]). *Let f be analytic in the open unit disk, with power series expansion at the origin* $f(z) = 1 + \sum_{n=1}^{\infty} f_n z^n$. *Let*

$$\frac{1}{f(z)} = 1 + \sum_{n=1}^{\infty} g_n z^n \qquad (2.1.2)$$

be the power series expansion of $1/f$ at the origin.

(1) *Prove that*

$$f_2 g_1 = f_1(g_1 f_1 + g_2), \qquad (2.1.3)$$

and that, for $n \geq 2$,

$$f_{n+1}(g_1 f_{n-1} + \cdots + g_{n-1} f_1 + g_n) = f_n(g_1 f_n + \cdots + g_n f_1 + g_{n+1}). \quad (2.1.4)$$

(2) *We now assume that all the coefficients f_n are different from 0 and set (with $f_0 = 1$)*

$$c_{m,n} = \frac{f_{n+1} f_{n-m}}{f_n} - f_{n-m+1}, \quad m = 1, \ldots, n.$$

Show that

$$g_{n+1} = c_{1,n} g_1 + c_{2,n} g_2 + \cdots + c_{n,n} g_n, \quad n = 1, 2, \ldots \qquad (2.1.5)$$

(3) *We now assume that $f_n > 0$ for $n = 1, 2, \ldots$ and that, moreover,*

$$\frac{f_{n+1}}{f_n} \geq \frac{f_n}{f_{n-1}}, \quad n = 1, 2, \ldots \qquad (2.1.6)$$

Show that

$$g_1 \leq 0 \implies g_n \leq 0, \quad n = 2, 3, \ldots$$

(4) *Show that*

$$\sum_{n=0}^{\infty} (-g_n) < \infty.$$

(5) *What happens in the cases $f_n \equiv 1$ and $f_n = \frac{1}{n+1}$?*

Remark 2.1.2. Condition (2.1.6) appears in Kaluza's 1928 paper [198]. Lamperti gave in 1958 in [211] a necessary and sufficient condition for the coefficients g_n to be nonnegative. We will call the corresponding sequence $(f_n)_{n \in \mathbb{N}}$ a Kaluza sequence, or (following [62]), a sequence with the Kaluza sign property. These questions relate to the theory of renewal sequences and of infinitely divisible renewal processes, and we refer to Kendall's paper [201] for a discussion. We mention now right after this remark two results which are direct consequences of Lamperti's criteria (see [211, Theorem 2 and Theorem 3]), and which the reader may try to prove directly. More general results, and some historical discussions, can be found in the paper [62] of Baricz, Vesti, and Vuorinen (see in particular [62, Theorem 2.11, p. 11]), which also contains examples of Kaluza's sequences in the setting of Gaussian hypergeometric series.

Question 2.1.3 (see [211, Theorem 2 and Theorem 3]). *In the notation of Exercise 2.1.1 assume that the two sequences $(f_n^{(1)})$ and $(f_n^{(2)})$, $n = 0, 1, 2, \ldots$ (with $f_0^{(1)} = f_0^{(2)} = 1$) have the Kaluza sign property. Show that each of the sequences*

$$h_n = f_n^{(1)} f_n^{(2)} \quad and \quad k_n = \sum_{u=0}^{n} \binom{n}{u} f_u^{(1)} f_{n-u}^{(2)}, \quad n = 0, 1, \ldots$$

has also this property.

Question 2.1.4.

(1) *Show that the power series expansion of* tan *at the origin is (see [128, 11.26 (1), p. 132], [271, (2.17–22), p. 88])*

$$\tan z = \sum_{n=1}^{\infty} (-1)^n \frac{2^{2n}(1 - 2^{2n})}{(2n)!} b_{2n} z^{2n-1} \tag{2.1.7}$$

where the b_{2n} are the Bernoulli numbers (see [CAPB, p. 323]),[1] defined from

$$\frac{z}{e^z - 1} = \sum_{n=0}^{\infty} \frac{b_n}{n!} z^n.$$

(2) *Show that the sequence $\left((-1)^n \frac{2^{2n}(1-2^{2n})}{(2n)!} b_{2n} \right)_{n \in \mathbb{N}_0}$ is a Kaluza sequence.*

The next exercise is an easy consequence of Green's theorem, and has applications in the theory of the Bergman space of polyanalytic functions. See Section 10.4 (and in particular Remark 10.4.7) for the latter. The exercise itself is based on [57, p. 169] and [207]. In the statement of the exercise, recall that for a complex-valued function $h(z) = u(x, y) + iv(x, y)$ with continuous partial derivatives one defines

$$\frac{\partial h}{\partial \overline{z}} = \frac{1}{2}\left(\frac{\partial h}{\partial x} + i\frac{\partial h}{\partial y} \right) \quad and \quad \frac{\partial h}{\partial z} = \frac{1}{2}\left(\frac{\partial h}{\partial x} - i\frac{\partial h}{\partial y} \right).$$

Furthermore, it holds that

$$\frac{\partial hg}{\partial \overline{z}} = \frac{\partial h}{\partial \overline{z}} g + h\frac{\partial g}{\partial \overline{z}} \tag{2.1.8}$$

if g is another such function. See [CABP, pp. 152–153] for some exercises involving these operators.

Exercise 2.1.5.

(1) *Let f, g be of class C^1 in a neighborhood of the closed unit disk. Show that (with $\overline{g}(z) \overset{\text{def.}}{=} \overline{g(z)}$)*

$$\iint_{\mathbb{D}} f(z) \frac{\partial \overline{g}}{\partial \overline{z}}(z) dx dy = \frac{1}{2i} \int_{\mathbb{T}} f(z)\overline{g(z)} dz - \iint_{\mathbb{D}} \frac{\partial f}{\partial \overline{z}}(z)\overline{g(z)} dx dy \tag{2.1.9}$$

---

[1]Note the unfortunate misprint in the formula above Exercise 7.2.18 in [CAPB]; the formula should be $\sum_{n=1}^{\infty} \frac{1}{n^{2p}} = \frac{\pi^{2p}}{(-1)^{p-1} b_{2p}}$. Note that (2-17-22) in [271] also contains a misprint.

and

$$\iint_{\mathbb{D}} f(z) \frac{\partial \overline{g}}{\partial z}(z) dx dy = -\frac{1}{2i} \int_{\mathbb{T}} f(z) \overline{g(z)} d\overline{z} - \iint_{\mathbb{D}} \frac{\partial f}{\partial z}(z) \overline{g(z)} dx dy. \quad (2.1.10)$$

(2) *Let for* $j, k \in \mathbb{N}_0$,

$$e_{k,j}(z) = \sqrt{\frac{k+j+1}{\pi}} \frac{1}{(k+j)!} \frac{\partial^{k+j}}{\partial z^k \partial \overline{z}^j} (|z|^2 - 1)^{k+j}. \quad (2.1.11)$$

Show that

$$\iint_{\mathbb{D}} e_{k_1,j_1}(z) \overline{e_{k_2,j_2}(z)} dx dy = \delta_{k_1,k_2} \delta_{j_1,j_2}. \quad (2.1.12)$$

Hint: The first item is a trivial consequence of Green's formula

$$\int_{\Gamma} P(x,y) dx + Q(x,y) dy = \iint_{R} \left(\frac{\partial Q}{\partial x}(x,y) - \frac{\partial P}{\partial y}(x,y) \right) dx dy, \quad (2.1.13)$$

where Γ is a smooth Jordan curve with interior R and where P and Q are smooth in a neighborhood of $R \cup \Gamma$.

We note that (2.1.9) and (2.1.10) still hold in a multiply connected domain (say Ω), with smooth boundary (say Γ). To prove the corresponding formulas it suffices to use this version of Green's theorem for such domains. The choice $f(z) = 1$ and $g(z) = z$ in (2.1.9) leads then to a formula for the area of the given domain:

$$\iint_{\Omega} dx dy = \frac{1}{2\pi} \int_{\Gamma} \overline{z} dz.$$

As a particular and trivial case one has:

Exercise 2.1.6. *Let* $R \in (1, \infty)$ *and let* A_R *denote the annulus*

$$A_R = \{z \in \mathbb{C} \, ; 1 < |z| < R\}. \quad (2.1.14)$$

Using the analog of (2.1.9) for the annulus show that the area of A_R *is equal to* $\pi(R^2 - 1)$.

We now present an easy application of Cauchy's formula. This exercise is used for instance in [102] in the topological analysis proof that a function differentiable in an open set admits a development in power series at every point. The exercise is also used in the proof of Question 3.2.11 where a result of M. Zorn on power series in two variables is presented.

Exercise 2.1.7. *Let* f *be analytic in* $|z| < R$ *with power series*

$$f(z) = \sum_{n=0}^{\infty} f_n z^n, \quad |z| < R.$$

2.1. Some warm-up exercises

Let $r \in (0, R)$ and let $K \geq \max_{|\zeta|=r} |f(\zeta)|$. Show that, for every number z of modulus less than or equal to r and every $m \in \mathbb{N}_0$,

$$|f_m z^m| \leq K. \tag{2.1.15}$$

Hint: Consider first the case of polynomials, and then approximate f by finite sums $\sum_{n=0}^{N} f_n z^n$. See the proof of Lemma 4 in [102].

We now turn to a simple exercise pertaining to functions analytic and with a positive real part in the open unit disk. These functions play an important role in various fields in mathematics and signal processing. We denote the class of such functions by $\mathcal{C}$.

Exercise 2.1.8. *Let f and g belong to $\mathcal{C}$. Show that $h = \dfrac{f}{1 + fg}$ also belongs to $\mathcal{C}$.*

Remark 2.1.9. Part of the problem is to show that $1 + fg$ does not vanish for $z \in \mathbb{D}$.

The following exercise is used in the proof of Abel's theorem on elliptic functions (see for instance [141, Theorem V.6.1, p. 301]).

Exercise 2.1.10. *Let f be a meromorphic periodic function with period $T \in \mathbb{C} \setminus \{0\}$. Let $z_0 \in \mathbb{C}$ be such that f is analytic in a neighborhood of the interval $[z_0, z_0 + T]$, and does not vanish on this interval. Show that*

$$\frac{T}{2\pi i} \int_{[z_0, z_0+T]} \frac{f'(z)}{f(z)} dz \in \mathbb{Z}.$$

Hint: Recall the definition of the winding number of a closed curve around a point. See [CAPB, p. 195].

Still related to periodic functions, we propose the following classical exercise (see for instance [271, p. 235] for the first item):

Exercise 2.1.11.

(1) *Let f be an entire periodic function. Show that ∞ is an essential singularity of f.*
(2) *Let f be a periodic function, analytic in the plane, with isolated singularities. Show that ∞ is not an isolated singularity.*
(3) *What type of singular point is ∞ for the function* tan.

Remarks 2.1.12. In (1) the point ∞ is in particular an isolated singularity. In some books one can read that infinity is an essential singularity in the second case.

In the case of entire functions, the following exercise is [CAPB, Exercise 7.6.1, p. 332 and solution p. 363].

Exercise 2.1.13. *Let f be a function analytic in the open upper half-plane $\mathbb{C}_+$, and assume that $f(z+1) = f(z)$. Show that there is a function g analytic in the pointed open unit disk such that*

$$g(e^{2\pi i z}) = f(z), \quad z \in \mathbb{C}_+. \tag{2.1.16}$$

Hint: Define for $\zeta \in \mathbb{D} \setminus (-1,0]$ a function g by

$$g(\zeta) = f\left(\frac{\ln \zeta}{2\pi i}\right). \tag{2.1.17}$$

The following exercise is related to the Fock space, where (2.1.18) can be rewritten as

$$(\operatorname{Re} M_z)\, u = 2\lambda u.$$

See Exercise 11.1.8. It is closely related to Hermite functions and Hermite polynomials. See in particular Exercise 2.1.16 below.

Exercise 2.1.14. *Show that for every $\lambda \in \mathbb{C}$, the solutions of the differential equation*

$$z u(z) + u'(z) = \lambda u(z), \tag{2.1.18}$$

are entire functions and find them.

Hint: Write $u(z) = e^{-\frac{z^2}{2}} v(z)$ and solve the corresponding equation for the function v.

Remark 2.1.15. The very same change of unknown in the equation

$$-u''(z) + (z^2 + 1)u(z) = \lambda u(z)$$

leads to the differential equation (2.1.22) satisfied by the Hermite polynomials for appropriate values of λ.

Exercise 2.1.16. *Let*

$$H_n(z) = (-1)^n e^{z^2} \left(e^{-z^2}\right)^{(n)}, \quad n = 0, 1, \ldots \tag{2.1.19}$$

(1) Show that the functions (2.1.19) are polynomials.
(2) Show that

$$e^{2uz - u^2} = \sum_{n=0}^{\infty} \frac{H_n(z)}{n!} u^n, \quad z, u \in \mathbb{C}. \tag{2.1.20}$$

Hint: To prove (2.1.20) just consider the Taylor expansion at the point z of an appropriate function.

2.1. Some warm-up exercises

Remark 2.1.17. The polynomials (2.1.19) are called Hermite polynomials. They appear later in the book. See Definition 6.1.12 and Exercise 6.1.11. The expansion (2.1.20) appears (for instance) in Hille's paper [177, (1), p. 431], where the following formulas may also be found (see [177, (6), (7), (8), p. 431]):

$$H_n'(z) - 2nH_{n-1}(z) = 0, \quad n \geq 1, \tag{2.1.21}$$
$$H_n''(z) - 2zH_n'(z) + 2nH_n(z) = 0, \quad n \geq 0, \tag{2.1.22}$$
$$H_n(z) - 2zH_{n-1}(z) + 2(n-1)H_{n-2}(z) = 0, \quad n \geq 2. \tag{2.1.23}$$

The next question is almost a tautology but is very useful in the study of functions in a Hardy space. In the statement, $B(0, R)$ denotes the open disk centered at the origin and with radius R.

Question 2.1.18.

(1) Let f be analytic in the open unit disk $\mathbb{D}$ and vanishing at the point $a \in \mathbb{D}$. Show that f can be written in the form $f(z) = \frac{z-a}{1-z\bar{a}} h(z)$, where h is analytic in $\mathbb{D}$.

(2) Let f be analytic in the open unit disk and non-identically vanishing there, and let $R \in (0, 1)$. Show that f has at most a finite number of zeros $z_1, \ldots, z_N$ in $B(0, R)$. Show that the function

$$\frac{f(z)}{\prod_{n=1}^{N} \left(\dfrac{\dfrac{z}{R} - \dfrac{z_n}{R}}{1 - \dfrac{z \overline{z_n}}{R^2}} \right)}$$

has removable singularities at the points $z_1, \ldots, z_N$. What is its module on $|z| = R$?

In the matrix-valued case the first item in the previous question takes the following form (see for instance [253]). In the statement, a matrix-valued function is said to be analytic if all its entries are analytic. The reader will check readily the equivalence with other definitions which are independent of the coordinates (such as existence of a power series expansion with matrix-valued coefficients).

Exercise 2.1.19. Let F be a $\mathbb{C}^{n \times m}$-valued function analytic in $\mathbb{D}$, and assume that $a^* F(w_0) = 0$ for some $a \in \mathbb{C}^n$ and $w_0 \in \mathbb{D}$. Let $B_{a,w_0}(z)$ be defined by (1.7.17). Show that the function

$$B_{a,w_0}(z)^{-1} F(z)$$

has a removable singularity at the point w_0.

For a proof of the following question we refer to [CABP, Exercise 5.6.9, p. 213].

Question 2.1.20. *Let $R > 0$ and let f be analytic in $|z| < R$. Show that the function*

$$M_2(r) = \frac{1}{2\pi} \int_0^{2\pi} |f(re^{it})|^2 dt, \quad r \in (0, R), \tag{2.1.24}$$

is continuous and increasing.

Hint: Let $f(z) = \sum_{n=0}^{\infty} a_n z^n$ be the power series expansion of f at the origin. One quick way to prove that (2.1.24) is an increasing function of r is to apply Parseval's equality to the Fourier series

$$g(e^{it}) = f(re^{it}) = \sum_{n=0}^{\infty} a_n r^n e^{int}.$$

This last question is important in particular in the proof of Exercise 8.2.2 on equivalent definitions of the Hardy space $\mathbf{H}_2(\mathbb{D})$.

The following result, also important in the theory of Hardy spaces, is a special case of a result on sub-harmonic functions (see [264, Théorème 17.2, p. 318]). Following Julia's book [196, p. 104] we here propose another proof, using Exercise 2.1.20. The result is due to Hardy, and is the key to the theory of Hardy spaces of the unit circle.

Exercise 2.1.21. *Let $p > 0$ and f be analytic in the open unit disk. Show that the function*

$$M_p(f, r) = \left(\int_0^{2\pi} |f(re^{it})|^p dt \right)^{1/p}, \quad r \in (0, 1),$$

is continuous and increasing.

Hint: Assume first that f does not vanish and reduce to the case $p = 2$ using the analytic logarithm of f.

Infinite products play an important role in the theory of analytic functions. They help in building functions with preassigned zeros. The following result can be found in, e.g., [269, (2.2), p. 296 and (2.10), p. 298].

Exercise 2.1.22. *Given a sequence of complex numbers going to infinity in modulus, show that there exists an entire function vanishing at these points, and only at these points.*

Hint: Let $z_1, z_2, \ldots$ be the given sequence. Use the Weierstrass factor

$$E_p(z) = (1 - z)e^{\left(z + \frac{z^2}{2} + \cdots + \frac{z^p}{p}\right)}, \tag{2.1.25}$$

with appropriate p (depending on n), and recall that it holds that

$$|1 - E_p(z)| \le |z|^{p+1}, \quad |z| \le 1, \tag{2.1.26}$$

(see [CABP, p. 110]).

In particular, we have the following important remark:

Remark 2.1.23. Given a sequence $a_1, a_2, \ldots$ of strictly positive numbers such that $\sum_{n=1}^{\infty} a_n < \infty$, the function

$$f(z) = \prod_{n=1}^{\infty} (1 + a_n z^2) \qquad (2.1.27)$$

is entire, with zeros $\frac{\pm i}{\sqrt{a_n}}$, $n = 1, 2, \ldots$. The function

$$\frac{1}{f(t)}$$

is positive definite on the real line, meaning that the kernel $k(t, s) = \frac{1}{f(t-s)}$ is positive definite there. See Definition 7.1.4 (with $1/f$ in place of f) and Remark 7.2.16 for the latter.

Item (2) of the following exercise is also taken from the book of Saks and Zygmund, see [269, Lemma 1.1, p. 171]. It is a nice illustration and easy consequence of integration on a path. Recall that a curve is defined by a continuous and piece-wise differentiable map γ (a parametrization of the curve) from a compact interval $[a, b]$ of the real line, and with values in $\mathbb{C}$. By piecewise differentiable we mean that γ is ruled (i.e., regulated): It may fail to have a derivative at a finite number of points, but then the two components of γ are assumed to have finite left and right derivatives. These points do not intervene in the computations of the various integrals. Ruled functions are also called regulated functions. We refer to, e.g., [111, VIII.6, VIII.7] for more information on these functions and related integration results.

The image $C = \gamma([a, b])$, which should not be confused with the parametrization, is in particular compact (and closed); see Remark 3.4.3 below if need be. Let $g : [a, b] \longrightarrow \mathbb{C}$ be continuous. The integral

$$\int_C g(s)ds = \int_a^b g(\gamma(t))\gamma'(t)dt$$

is independent of the chosen parametrization. For more on line integrals, see for instance [CAPB, p. 193].

Exercise 2.1.24.

(1) *Let C be a curve and let f be a continuous function with domain containing the image of C. Show that the function*

$$F(z) = \int_C \frac{f(s)ds}{z - s} \qquad (2.1.28)$$

is analytic in $\mathbb{C} \setminus C$. Is it analytic at infinity?

(2) *Let K be a compact set and not intersecting C. Show that for every $\epsilon > 0$ there is a rational function R with poles only on C and such that*

$$\sup_{z \in K} \left| \int_C \frac{f(s)ds}{z - s} - R(z) \right| \le \epsilon.$$

Hints: The first question is classic: Use for instance Morera's theorem or prove that there is a power series expansion in a neighborhood of every point of $\mathbb{C} \setminus C$. For the second point, use the fact (as in [269, p. 172]) that the function

$$(t, z) \mapsto \frac{f(\gamma(t))}{z - \gamma(t)}$$

is uniformly continuous on $[a, b] \times K$, and approximate the integral by an appropriate Riemann sum.

Remark 2.1.25. Functions of the form

$$F(z) = \int_I \frac{f(t)dt}{z - t}$$

where I is an interval of the real line (see for instance the function (2.1.28) and Question 2.7.2) are clearly analytic in $\mathbb{C} \setminus \overline{I}$, but in general nothing can be said on the behaviour on $\overline{I}$ (but if, for instance, f vanishes in some open subset of I). Similar remarks hold if one considers measures and functions of the form

$$F(z) = \int_I \frac{d\mu(t)}{z - t}$$

where $d\mu$ is a measure of bounded variation on I. The situation is quite different when one considers double integrals and functions of the form

$$F(z) = \iint_\Omega \frac{f(x, y)}{z - (x + iy)} dx dy, \tag{2.1.29}$$

where Ω is (for instance) a compact subset of the plane (with non-empty interior). The function (2.1.29) is of course analytic in $\mathbb{C} \setminus \Omega$ but it may be defined in $\overset{\circ}{\Omega}$. Then it will not be analytic there. See Question 2.2.8 and formula (2.2.5) for an example.

Exercise 2.1.26. *Given the value $\int_{\mathbb{R}} e^{-\frac{u^2}{2}} du = \sqrt{2\pi}$, prove that*

$$I(t) \overset{\text{def.}}{=} \int_{\mathbb{R}} e^{-\frac{u^2}{2}} e^{-itu} du = \sqrt{2\pi} e^{-\frac{t^2}{2}}. \tag{2.1.30}$$

Hint: For $t = 0$, there are at least two ways to compute this integral. The first one is to write I^2 as a double integral and use Fubini's theorem; see [CAPB, p. 199]. The second one consists in using Cauchy's theorem, with the function $f(z) = e^{-\frac{z^2}{2}}$ on an appropriate contour (see [140, p. 193]). The computation of $I(t)$ is done in [CAPB, (15.9.5), p. 506] using Cauchy's theorem and the value $I(0)$. Here we suggest the following approach. Compute $I'(t)$ in terms of $I(t)$, and solve the corresponding differential equation to find $I(t)$. Here too one needs to know the value of $I(0)$.

In the following exercise, the map (2.1.33) is called the Bargmann (or Bargmann–Segal, or Segal–Bargmann) transform. Formula (2.1.32) appears in [60, (2.2), p. 198] in the setting of $\mathbb{C}^N$. We refer to A. Zayed's paper [325] for a survey on this transform and applications to signal processing. We also refer to [246, Chapter 4]. The proof of the one-to-oneness in the last item uses the Fourier transform and is deferred to Remark 6.1.34.

Exercise 2.1.27.

(1) *Show that formula (2.1.30) can be analytically extended for all $z \in \mathbb{C}$.*

(2) *Let, for $z \in \mathbb{C}$,*

$$h_z(u) = \frac{1}{\pi^{\frac{1}{4}}} e^{\left\{-\frac{1}{2}(\bar{z}^2+u^2)+\sqrt{2}\bar{z}u\right\}}, \quad z \in \mathbb{C}. \tag{2.1.31}$$

Show that

$$e^{z\bar{w}} = \int_{\mathbb{R}} h_w(u)\overline{h_z(u)}du, \quad z, w \in \mathbb{C}. \tag{2.1.32}$$

(3) *Let $f \in \mathbf{L}_2(\mathbb{R}, \mathcal{B}, dx)$ (the Lebesgue space of square summable measurable functions on $\mathbb{R}$). Show that the function*

$$F(z) = \int_{\mathbb{R}} f(u)\overline{h_z(u)}du \tag{2.1.33}$$

is entire, and that

$$F \equiv 0 \iff f = 0.$$

Remarks 2.1.28.

(1) Formula (2.1.32) expresses the function $e^{z\bar{w}}$ as an inner product, and hence as a positive definite function on $\mathbb{C}$. See Exercise 7.1.7. The same conclusion is obtained in a quicker way using the power expansion of the exponential:

$$e^{z\bar{w}} = \sum_{n=0}^{\infty} \frac{z^n \bar{w}^n}{n!}, \quad z, w \in \mathbb{C}.$$

(2) The Bargmann transform plays an important role in the theory of the Fock space. See Chapter 11 for the latter.

(3) In Section 6.1, see Exercise 6.1.11, we solve the equation

$$z^n = \int_{\mathbb{R}} \xi_n(u)\overline{h_z(u)}du \tag{2.1.34}$$

by using the inverse Fourier transform.

Exercise 2.1.29. Let $f \in \mathbf{L}_2(\mathbb{R}_+, \mathcal{B}, dx)$ (the Lebesgue space of square summable measurable functions on $[0, \infty)$).

(1) Show that the formula

$$F(z) = \int_0^\infty e^{iuz} f(u)du \tag{2.1.35}$$

defines a function analytic in the open upper half-plane $\mathbb{C}_+$.

(2) Can F be the restriction to $\mathbb{C}_+$ of a rational function?

(3) Show by an example that F need not have an analytic extension over the real line.

The set of functions of the form (2.1.35) with an appropriate norm is the Hardy space $\mathbf{H}_2(\mathbb{C}_+)$ of the open upper half-plane. See Section 8.6 and in particular Exercise 8.6.5 there.

Laguerre functions and polynomials appear in the study of the space $\mathbf{H}_2(\mathbb{C}_+)$. The following integral is used in their study (see Question 8.6.4).

Question 2.1.30 ([217, (90), p. 474]). Show that

$$\int_{[0,\infty)} u^n e^{-u} e^{iut} = \frac{i^{n+1}n!}{(t+i)^{n+1}}, \quad n = 0, 1, \ldots \tag{2.1.36}$$

Hint: Use integration by parts and induction.

Exercise 2.1.31. Let $\nu > -1$. Prove that the formula

$$F(z) = \int_0^\infty t^\nu e^{-zt} dt \tag{2.1.37}$$

defines a function analytic in the right open half-plane and compute its derivative.

Hint: Differentiating formally under the integral sign one obtains

$$F'(z) = -\int_0^\infty t^{\nu+1} e^{-zt} dt. \tag{2.1.38}$$

Use the dominated convergence theorem to justify the interchange of order of integration and derivation.

The following question is one of the steps in the verification of the functional equation satisfied by Riemann's ζ function, with the choice

$$f(t) = \vartheta(it) - \frac{1}{2},$$

where ϑ is the theta function. See for instance [249, pp. xi–xii].

2.1. Some warm-up exercises

Question 2.1.32. *Let f be a measurable complex-valued function defined on $[1, \infty)$ and such that $|f(t)| = O(e^{-ct})$ for some $c > 0$.*

(1) *Show that the function*

$$F(z) = \int_1^\infty \left(t^{z-1} + t^{-\frac{1}{2}-z}\right) f(t) dt \qquad (2.1.39)$$

is entire.

(2) *Show that*

$$F(z) = F(1/2 - z), \quad z \in \mathbb{C}.$$

In preparation for the next exercise, recall that for $\alpha \in \mathbb{C}$,

$$z^\alpha = e^{\alpha \ln z},$$

where $\ln z$ is the principal determination of the logarithm:

$$\ln z = \ln \rho + i\theta,$$

where $z = \rho e^{i\theta} \in \mathbb{C} \setminus (-\infty, 0]$ (that is, $\rho > 0$ and $\theta \in (-\pi, \pi)$). With this definition of $\ln$ one sets for $\alpha \in \mathbb{C}$,

$$(1 + z)^\alpha = e^{\alpha \ln(1+z)}, \qquad (2.1.40)$$

when $z \in \mathbb{C}_r$. (See also the hint after the exercise.)

Exercise 2.1.33 (see [230, Proposition 3.5, p. 279]). *Let $\nu > -1$ and let*

$$f_n^{(\nu)}(z) = \sqrt{\frac{\Gamma(1+n+\nu)}{n!}} \left(\frac{1-z}{1+z}\right)^n \left(\frac{\sqrt{2}}{1+z}\right)^{1+\nu}, \quad n = 0, 1, \ldots \qquad (2.1.41)$$

where Γ denotes the Gamma function (see (2.1.44) below for the latter). Show that the series

$$\sum_{n=0}^\infty f_n^{(\nu)}(z)\overline{f_n^{(\nu)}(w)} \qquad (2.1.42)$$

converges for all $z, w \in \mathbb{C}_r$ and compute its sum.

Hint: In the proof it is well to recall the following: With

$$\ln(1 - z) = -\sum_{n=1}^\infty \frac{z^n}{n}, \quad |z| < 1,$$

and $\alpha \in \mathbb{C}$, the power series

$$(1 - z)^{-\alpha} = \sum_{n=0}^\infty \frac{\alpha(\alpha+1)\cdots(\alpha+n-1)}{n!} z^n, \quad |z| < 1, \qquad (2.1.43)$$

converges in the open unit disk, where the left-hand side is defined using (2.1.40), that is

$$(1 - z)^\alpha = e^{\alpha \ln(1-z)}.$$

It is also well to recall that the Gamma function

$$\Gamma(z) = \int_0^\infty t^{z-1} e^{-t} dt, \quad \operatorname{Re} z > 0, \tag{2.1.44}$$

satisfies

$$\Gamma(z + 1) = z\Gamma(z).$$

In particular it holds that

$$\Gamma(1 + n + \nu) = (n + \nu) \cdots (1 + \nu)\Gamma(1 + \nu), \tag{2.1.45}$$

where $\nu > -1$ and $n = 1, 2, \ldots$

Remark 2.1.34. The sum (2.1.42) (that is, the function (7.1.11)) is the reproducing kernel of the fractional Hardy space of the open right half-plane.

Remark 2.1.35. The choice $\alpha = -\frac{1}{2}$ in (2.1.43) leads to yet another example of sequence with the Kaluza sign property since

$$\frac{1}{(1 - z)^{-\frac{1}{2}}} = 1 - \underbrace{\left(1 - (1 - z)^{\frac{1}{2}}\right)}_{\text{has positive Taylor coefficients}}.$$

See [211, p. 91].

The following exercise is related to the kernel (1.6.14) and Exercise 1.6.16.

Exercise 2.1.36. *Let f be analytic in a neighborhood of the origin, with power series expansion*

$$f(z) = f_0 + 2f_1 z + 2f_2 z^2 + \cdots, \quad |z| < 1.$$

Assume that $f_0 \in \mathbb{R}$ and define $f_{-1}, f_{-2}, \ldots$ by

$$f_{-n} = \overline{f_n}, \quad n \in \mathbb{Z}.$$

Assume moreover that

$$|f_n| \le K, \quad \forall n \in \mathbb{Z}, \tag{2.1.46}$$

for some $K > 0$. Show that

$$\frac{f(z) + \overline{f(w)}}{2(1 - z\overline{w})} = \sum_{n,m=0}^\infty z^n \overline{w}^m f_{n-m}, \quad z, w \in \mathbb{D}. \tag{2.1.47}$$

Remark 2.1.37. Assume that all the Toeplitz matrices T_N defined in Exercise 1.6.14 have at most κ strictly negative eigenvalues, and exactly κ strictly negative eigenvalues for some N. Then, the kernel (2.1.47) has κ negative squares (is positive definite if $\kappa = 0$; see Definitions 7.1.3 and 7.1.1) and the function f is a generalized Carathéodory function (a Carathéodory function if $\kappa = 0$). When $\kappa > 0$ the condition (2.1.46) need not hold for a kernel of the form (2.1.47) to have a finite number of negative squares. Then the function f may have poles inside $\mathbb{D}$ and (2.1.47) will converge in a neighborhood of the origin; see the paper [189] of Iohvidov and Kreĭn for more information.

2.2 Some important theorems to recall

We begin this section by recalling three important theorems. A proof of the first one is given later in Exercise 3.6.6 using a topological approach.

Theorem 2.2.1. (*the open mapping theorem*): *The image of an open set under a non-constant analytic function is open.*

Theorem 2.2.2. *Assume that f is analytic and one-to-one from the open set Ω into $\mathbb{C}$. Then f^{-1} is analytic from $f(\Omega)$ (which is open by the previous theorem) onto Ω.*

We now recall the following result, called the argument principle, and which is a direct application of the residue theorem.

Theorem 2.2.3. *Let C be a closed smooth simple curve, and let f be a function analytic in a neighborhood of the interior of C, at the possible exception of a finite number of isolated singularities, which are poles. Assume moreover that f does not vanish on C. Then,*

$$\frac{1}{2\pi i} \int_C \frac{f'(z)}{f(z)} dz = Z - P, \qquad (2.2.1)$$

where Z denotes the number of zeros of f inside C and P the number of poles of f inside C (both, included multiplicities).

An immediate consequence of the argument principle is Rouché's theorem.

Question 2.2.4. *Let C be a closed smooth simple curve with interior R, and let f and g denote functions analytic in a neighborhood of the closure $\overline{R}$ of R. Assume that*

$$|g(z)| < |f(z)|, \quad \forall z \in C.$$

Then, f and $f + g$ have the same number of zeros in R.

Hint: Compute

$$\int_C \frac{f'(z) + g'(z)}{f(z) + g(z)} dz - \int_C \frac{f'(z)}{f(z)} dz.$$

Exercise 2.2.5. *Let $\omega_1, \ldots, \omega_N$ be N (not necessarily) distinct points in the open upper half-plane $\mathbb{C}_+$, and let*

$$b(z) = \prod_{n=1}^{N} \frac{z - \omega_n}{z - \overline{\omega_n}}. \tag{2.2.2}$$

Let $\zeta \in \mathbb{D}$. Show that the equation

$$b(z) = \zeta \tag{2.2.3}$$

has N solutions (counting multiplicities) inside $\mathbb{C}_+$.

For a version of the preceding exercise with the open upper half-plane replaced by the open unit disk, see the first item in Exercise 2.3.7.

Remark 2.2.6. Trivially for $\zeta = 0$, and using the fundamental theorem of algebra for $\zeta \neq 0$, we see that (2.2.3) has always N solutions. The point of the exercise is to show that they are all in $\mathbb{C}_+$ when $\zeta \in \mathbb{D}$.

Another important consequence of Theorem 2.2.3 is the following result, due to Hurwitz. See for instance [56, p. 126], [174, Theorem 4.4, p. 180] for various versions of the theorem. The result itself is used in one of the steps in the proof of Riemann's mapping theorem; see [95].

Exercise 2.2.7. *Let $(f_n)_{n \in \mathbb{N}}$ be a sequence of functions analytic in the open unit disk and not vanishing there, and converging uniformly on compact sets to the function f. Then, either $f \equiv 0$ or f does not vanish in the open unit disk.*

Formula (2.2.4) below is (2.1.9) with g instead of $\overline{g}$. See also for instance [70, (37b), p. 57] and see [242, Proposition 3, p. 22] for the second item.

Question 2.2.8.

(1) *Let C denote a Jordan curve, with interior R, and let f and g be continuous and with continuous partial derivatives, in a neighborhood of the closure $B = \overline{R}$ of R. Show that*

$$\frac{1}{2i} \int_C f(z)g(z)dz = \iint_B f(z)\frac{\partial g}{\partial \overline{z}}(z)dxdy + \iint_B \frac{\partial f}{\partial \overline{z}}g(z)dxdy. \tag{2.2.4}$$

(2) *Let G be C^∞ in a neighborhood of $\overline{R}$. Show that there exists a function F also of class C^∞ in a neighborhood of $\overline{R}$ which satisfy*

$$\frac{\partial F}{\partial \overline{z}}(z) = G(z), \quad z \in R.$$

Hint: Define

$$F(z) = \frac{1}{2\pi i} \iint_R \frac{G(z+w)}{w} da db, \quad (\text{where} \quad w = a + ib) \tag{2.2.5}$$

and apply (2.2.4).

The result in item (2) of the previous exercise is very useful in the theory of Riemann surfaces. See for instance [242, p. 22] where it is used to prove that the first cohomology group $H^1(\Omega, \mathcal{O})$ (where $\mathcal{O}$ denotes the sheaf of functions analytic in Ω) is trivial (Mittag-Leffler's theorem).

Question 2.2.9. *Let Ω be an open subset of $\mathbb{C}$ and let $(O_i)_{i \in I}$ denote an open covering of Ω. Assume that we are given analytic functions c_{ij} on $O_i \cap O_j$ such that*

$$c_{ij}(z) + c_{jk}(z) = c_{ik}(z), \quad z \in O_i \cap O_j \cap O_k,$$

when the intersection is non-empty. Finally suppose that there exist functions $g_i \in C^\infty(O_i)$ such that

$$c_{ij}(z) = g_i(z) - g_j(z), \quad z \in O_i \cap O_j,$$

when the intersection is non-empty. Show that the functions g_i may be chosen analytic.

2.3 Cauchy's formula and Laurent expansion

This section contains various results where the Laurent expansion and the calculus of residues play an important role.

Exercise 2.3.1. *Let z and z_0 be complex numbers such that $|z_0| < |z|$. Compute*

$$\int_{C_r} \frac{d\zeta}{(\zeta - z)(\zeta - z_0)^m}, \quad m \in \mathbb{Z}, \tag{2.3.1}$$

first for r such that $|z_0| < r < |z|$, and then for r such that $|z| < r$.

We recall that Cauchy's formula for an annulus, or equivalently the Laurent expansion for an annulus, states the following (see for instance [269, p. 197]):

Proposition 2.3.2. *Let f be analytic in the annulus $r_1 < |z - z_0| < r_2$, and let R_1 and R_2 be such that $r_1 < R_1 < R_2 < r_2$. Then, for every z such that $R_1 < |z - z_0| < R_2$,*

$$f(z) = \underbrace{\frac{1}{2\pi i} \int_{|\zeta - z_0| = R_2} \frac{f(\zeta)}{\zeta - z} d\zeta}_{\text{positive powers in the Laurent expansion}} - \underbrace{\frac{1}{2\pi i} \int_{|\zeta - z_0| = R_1} \frac{f(\zeta)}{\zeta - z} d\zeta}_{\text{principal part in the Laurent expansion}}.$$

$$\tag{2.3.2}$$

Exercise 2.3.3. *Let $R > 0$ and let g be analytic in $|z| > R$ and admitting a limit at infinity. Show that the Laurent expansion of g centered at the origin is of the form*

$$g(z) = L + \sum_{n=1}^{\infty} \frac{b_n}{z^n}, \quad |z| > R,$$

where $L = \lim_{z \to \infty} g(z)$.

The space of functions analytic in the open unit disk and endowed with convergence on compact subsets is a Fréchet space; see Section 5.6. The following result characterizes in fact its topological dual. See Exercise 5.8.1 for more details on the latter.

Exercise 2.3.4. *Let f be analytic in $\mathbb{D}$ and let g be analytic in $|z| > R$ for some $R < 1$, and admitting a limit at infinity. Let*

$$f(z) = \sum_{n=0}^{\infty} a_n z^n \quad and \quad g(z) = \sum_{n=0}^{\infty} \frac{b_n}{z^n}$$

be the corresponding Taylor and Laurent expansions centered at the origin.
(1) Show that the series

$$\sum_{n=0}^{\infty} a_n b_n \tag{2.3.3}$$

converges absolutely.
(2) Express (2.3.3) in terms of an integral involving f and g.

Let F be a rational function, with poles at the points $\nu_1, \ldots, \nu_N \in \mathbb{C}$. Considering the Laurent expansion at each of these poles implies that there exist polynomials $p_1, \ldots, p_N$ such that

$$F(z) - \sum_{n=1}^{N} p_n \left(\frac{1}{z - \nu_n} \right) \tag{2.3.4}$$

is entire, and hence equal to a polynomial, say $p(z)$. The representation of F as

$$F(z) = p(z) + \sum_{n=1}^{N} p_n \left(\frac{1}{z - \nu_n} \right)$$

is called the partial fraction decomposition of F. See for instance [CAPB, p. 329] for more details. As an example, if

$$F(z) = \prod_{n=1}^{N} \frac{z - \omega_n}{z - \overline{\omega_n}}, \tag{2.3.5}$$

where the points ω_n are in $\mathbb{C} \setminus \mathbb{R}$, and assumed pairwise different, we have

$$F(z) = 1 + \sum_{n=1}^{N} \frac{\alpha_n}{z - \overline{\omega_n}},$$

where

$$\alpha_n = \lim_{\substack{z \to \overline{\omega_n} \\ z \neq \overline{\omega_n}}} (z - \overline{\omega_n}) F(z) = (\overline{\omega_n} - \omega_n) \prod_{\substack{j=1 \\ j \neq n}}^{N} \frac{\overline{\omega_n} - \omega_j}{\overline{\omega_n} - \overline{\omega_j}}.$$

Such F (possibly multiplied by a unitary constant) are called finite Blaschke products of the open upper half-plane $\mathbb{C}_+$ when all the ω_j belong to $\mathbb{C}_+$.

More generally, if F not necessarily rational, but still has a finite number of isolated singularities in $\mathbb{C}$, a theorem of Weierstrass states that F can be written as

$$F(z) = H(z) + \sum_{\omega \text{ singularity of } F} H_\omega \left(\frac{1}{z - \omega}\right), \tag{2.3.6}$$

where H and the H_ω are entire functions. See [CABP, Exercise 7.1.9, p. 320], the case of polynomial functions H and H_ω corresponding to the rational case. Such an expansion need not hold when there is an infinite number of singularities. By a result of Mittag-Leffler, (2.3.6) has to be adapted by subtracting to each of the terms a polynomial; see (2.3.7) below. In the next exercise, we outline a result of Cauchy which proves, in certain cases, Mittag-Leffler's result. We base our discussion on the book [269, pp. 306–309] of Saks and Zygmund. We send the reader in particular to that book for a proof of item (4) in the exercise. First we recall a notation: C_R will denote the closed path

$$\gamma(t) = R e^{2\pi i t}, \quad t \in [0, 1],$$

that is, with some abuse of notation, the circle centered at the origin and of radius R. Recall that sometimes we will write, still with some abuse of notation, $\{|\zeta| = R\}$ rather than C_R, and similarly $\{|\zeta - v| = R\}$ to denote the closed path

$$\gamma(t) = v + R e^{2\pi i t}, \quad t \in [0, 1].$$

Exercise 2.3.5.

(1) *Let g be an entire function vanishing at the origin, and let u and v denote complex numbers, $u \neq v$. Show that*

$$\mathrm{Res}\left(\frac{g\left(\frac{1}{z-v}\right)}{z - u}, v\right) = -g\left(\frac{1}{u - v}\right).$$

(2) *Let F be analytic in C_{R_1}, at the possible exception of a (finite) number of isolated singularities. Express*

$$F(z) - \frac{1}{2\pi i} \int_{C_R} \frac{F(s)}{s - z} ds,$$

where $R < R_1$ and where F is analytic in a neighborhood of $R \leq |z| < R_1$, in terms of the principal parts of F at these singularities.

(3) *Let $k \in \mathbb{N}$. Show that*

$$\frac{1}{2\pi i} \int_{C_R} \frac{F(s)}{s - z} ds - \frac{1}{2\pi i} \int_{C_R} \frac{z^k F(s)}{s^k (s - z)} ds$$

is a polynomial.

(4) *Find a condition on F which insures an expression of the form*

$$F(z) = H(z) + \sum_{w \text{ singularity of } F} \left(H_w \left(\frac{1}{z-w} \right) - P_w(z) \right), \qquad (2.3.7)$$

where H and the H_w are entire functions and the P_w are polynomials.

We now give as an exercise an important formula relative to the index. Recall that if C is a closed piece-wise smooth curve and if z_0 does not belong to the image of C, the index (or winding number) of C around the point z_0 is defined by:

$$\text{Ind}_C(z_0) = \frac{1}{2\pi i} \int_C \frac{dz}{z - z_0}. \qquad (2.3.8)$$

It is an integer. For a discussion on the index, see [CAPB, pp. 195–196]. If moreover f does not vanish on C we note that the number

$$\frac{1}{2\pi i} \int_C \frac{f'(z)}{f(z)} dz \qquad (2.3.9)$$

can be seen as the index of the curve $f(C)$ around $z_0 = 0$.

Recall that for $R > 1$ we have denoted by A_R the annulus (2.1.14):

$$A_R = \{ z \in \mathbb{C} \, ; 1 < |z| < R \}.$$

The following exercise is taken from [91, Theorem 10.12 (i), p. 348]. It is used in arguments on the nonconformal equivalence of the annuli A_{R_1} and A_{R_2} when $R_1 \neq R_2$. See Exercises 3.8.25 and 3.8.27 for the latter.

Exercise 2.3.6. *Let $R > 1$, let C_1 and C_2 denote two closed smooth curves in A_R, and let f be analytic in A_R. Then:*

(1) *It holds that*

$$\text{Ind}_{C_1}(0) \int_{C_2} f(z) dz = \text{Ind}_{C_2}(0) \int_{C_1} f(z) dz. \qquad (2.3.10)$$

(2) *If f does not vanish in A_R it holds that:*

$$\text{Ind}_{C_1}(0)\text{Ind}_{f(C_2)}(0) = \text{Ind}_{C_2}(0)\text{Ind}_{f(C_1)}(0). \qquad (2.3.11)$$

Hint: Use the Laurent expansion of f. The second formula is a special case of the first with f'/f instead of f.

The function b defined by (2.3.12) in the following exercise is called a finite Blaschke product of the open unit disk.

Exercise 2.3.7. Let $z_1, \ldots, z_N$ be N *(not necessarily) distinct points in the open unit disk* $\mathbb{D}$, *and let*

$$b(z) = \prod_{n=1}^{N} \frac{z - z_n}{1 - z\overline{z_n}}. \qquad (2.3.12)$$

Let $\zeta \in \mathbb{D}$.

(1) *Show that the equation*

$$b(z) = \zeta \qquad (2.3.13)$$

has N *solutions (counting multiplicities) inside* $\mathbb{D}$.

(2) *Compute the residue of the function*

$$\frac{f(z)b'(z)}{b(z) - \zeta}$$

at a point $s \in \mathbb{D}$ *which is a simple zero of the equation* (2.3.13).

(3) *Let* f *denote a function analytic in the open unit disk. Show that the function*

$$(T_b f)(z) = \frac{1}{N} \left(\sum_{\substack{s \in \mathbb{D} \\ b(s) = z}} f(s) \right) \qquad (2.3.14)$$

is analytic in $\mathbb{D}$.

(4) *Compute* T_b *when* $b(z) = z^N$.

In the case $b(z) = z^N$, the map $f \mapsto T_b f$ is the decimation operator (also called down-sampling operator) from signal processing, which associates to the power series $f(z) = \sum_{n=0}^{\infty} f_n z^n$ the power series

$$D_b f(z) = (T_b(f))(z^N) = \sum_{n=0}^{\infty} f_{nN} z^{nN}.$$

See also [CABP, Exercise 6.2.9, p. 280].

Remark 2.3.8 (see Section 8.4 and Question 8.4.3). We note that the map $f \mapsto D_b f$ is the adjoint of the operator C_b of composition by b in the Hardy space $\mathbf{H}_2(\mathbb{D})$. Indeed, it holds that

$$\langle C_b(f), g \rangle_{\mathbf{H}_2(\mathbb{D})} = \langle f, D_b(g) \rangle_{\mathbf{H}_2(\mathbb{D})}, \qquad (2.3.15)$$

which is equal to

$$\langle C_b(f), g \rangle_{\mathbf{H}_2(\mathbb{D})} = \langle f, D_b(g) \rangle_{\mathbf{H}_2(\mathbb{D})} = \sum_{n=0}^{\infty} f_n \overline{g_{nN}},$$

for $b(z) = z^N$, and with $g(z) = \sum_{n=0}^{\infty} g_n z^n$.

It is also useful to note that (still for a finite Blaschke product b)

$$T_b \circ C_b = I. \qquad (2.3.16)$$

2.4 Star-shaped and simply connected sets

We now briefly discuss the notion of star-shaped and simply connected sets. We here consider the case of subsets of the complex plane. In the proof of Riemann's theorem, open sets of $\mathbb{C}$ (or more precisely from the Riemann sphere) with a key property come into play: sets for which every function analytic there admits an analytic squareroot (see for instance [176, p. 40]). This property is one among the many characterization of an important class of open sets, namely (open) simply connected sets. This notion is of fundamental importance in complex analysis, but often difficult to handle in a first course. Open star-shaped sets (see Definition 2.4.1 just below) form a subclass of open simply connected sets, much easier to introduce and handle. But being star-shaped is not a notion invariant under conformal mapping. On the other hand, simply-connected (open) sets form a class (and in fact, the class, as one sees from Riemann's theorem) invariant under conformal mapping, and in which the main properties of analytic functions (such as the existence of a primitive) hold. We now review some aspects of these two notions.

Definition 2.4.1. A set $E \subset \mathbb{C}$ is called star-shaped if there is a point $z_0 \in E$ such that for every $z \in E$ the closed interval $[z_0, z]$ is included in E:

$$[z_0, z] = \{z_0 + t(z - z_0)\,,\ t \in [0, 1]\} \subset E. \tag{2.4.17}$$

Convex sets are an example of star-shaped sets. It suffices to take for z_0 any point of the given set.

Exercise 2.4.2. *Give an example of a star-shaped set which is not convex.*

Open star-shaped sets are an important class of open sets in which complex analysis works well. For instance one has:

Exercise 2.4.3. *Show that every function analytic in an open star-shaped set admits a primitive there.*

Hint: This is a classical and elementary result. If f is the function at hand, it suffices to take (up to an additive constant)

$$F(z) = \int_{[z_0, z]} f(s)ds.$$

As mentioned above, open star-shaped sets lack an important property: They are not invariant under conformal invariance.

It follows from Exercise 2.4.3 that a power series with an infinite radius of convergence admits a primitive. If now Ω is an open connected subset of $\mathbb{C}$, different from $\mathbb{C}$, and if Ω has the property that any function analytic in Ω admits a primitive there, then Ω is conformally equivalent to the open unit disk. This is Riemann's mapping theorem.

Definition 2.4.4. An open connected set in which every non-vanishing analytic function admits an analytic logarithm is called simply connected.

Exercise 2.4.5. *Give (without using Riemann's theorem!) an example of two open sets, one star-shaped and the other not, which are conformally equivalent.*

The existence of a primitive for every function analytic in a given open set is one of the many facets of simple-connectedness. Other facets of simple-connectedness include existence of an analytic logarithm for every analytic function non-vanishing in the given set, and of a harmonic conjugate for every function harmonic in the given set. Thus (open) simply connected sets are characterized by a number of equivalent properties, each stressing out one facet of complex variable theory, namely:

1. A geometric definition, as an open subset of the Riemann sphere whose complement is connected.
2. Function theory definitions (existence of analytic logarithm, squareroot, and primitive).
3. Harmonic analysis definition: Every function harmonic in the given set has a harmonic conjugate.
4. Homotopy and homology characterizations.
5. Last, but not least, via Riemann's theorem.

See [141, p. 250]. Some of these aspects are considered in Exercise 2.4.9.

Exercise 2.4.6. *Let Ω be a domain with the property that every function analytic and not vanishing in Ω admits an analytic logarithm there. Show that if Ω_1 is conformally equivalent to Ω it satisfies the same property.*

Question 2.4.7. *Assume that the function f is analytic in the open set Ω, and that it has a continuous logarithm g there. Show that g is analytic in Ω.*

In view of the next exercise, recall that two open subsets of $\mathbb{C}$ are homeomorphic if there is a continuous bijection between them, whose inverse is also continuous.

Exercise 2.4.8. *Assume that the open set Ω is such that any non-vanishing continuous function has a continuous logarithm. Prove that the same holds for any set homeomorphic to Ω.*

Exercise 2.4.9. *Let $\Omega \subset \mathbb{C}$ be open and connected, and introduce the statements:*

(1) *Every function analytic in Ω admits a primitive there.*
(2) *Every function analytic and non-vanishing in Ω admits an analytic logarithm there.*
(3) *Every real-valued function harmonic in Ω admits a harmonic conjugate there.*

Show that:

$$(1) \implies (2)$$
$$(1) \implies (3)$$
$$(3) \implies (2).$$

In fact, the statements in the preceding exercise are equivalent, and are three among the characterizations of open simply connected sets discussed above. To prove that (2) implies (1) one needs to go a longer way, and introduce supplementary equivalent conditions. One can for instance use Riemann's mapping theorem to show that (2) implies (and in fact is equivalent) to the fact that Ω is conformally equivalent to $\mathbb{D}$ (when $\Omega \neq \mathbb{C}$).

2.5   Various

Picard's small and big theorems (see for instance [175, Chapter XVI, pp. 329–339], [4, §3, p. 306], [286, Theorem 1.6, p. 66]) allow us to give precise statements on the behavior of an analytic function near an essential singularity. We recall:

Theorem 2.5.1 (Picard's big theorem). *Let f be analytic in $B(z_0, R) \setminus \{z_0\}$. Let z_0 be an essential singularity of the function f. Then, for every $r < R$, the image of $B(z_0, r) \setminus \{z_0\}$ under f is equal to $\mathbb{C}$, from which at most one value has been removed.*

We will not consider this theorem here, but rather prove a much simpler result, namely the Casorati–Weierstrass theorem (see, e.g., [91, Theorem 11.6, p. 361], [95, p. 89], [269, (6.1), p. 144]), which is enough for our purposes.

Exercise 2.5.2 (Casorati–Weierstrass theorem). *Let f be analytic in $B(z_0, R) \setminus \{z_0\}$. Then, z_0 is an essential singularity of the function f if and only if for every $r < R$, the image of $B(z_0, r) \setminus \{z_0\}$ under f is dense in $\mathbb{C}$.*

We note (see for instance [286, Theorem 2.11, p. 490]) that this theorem is used to prove Weierstrass' characterization of meromorphic functions which admit an addition law.

The following result is taken from [4, p. 252]. It states that a conformal map cannot have (isolated) essential singularities. The result is a direct consequence of Picard's big theorem. The elementary proof presented in the present book follows [95, p. 182] and uses the simpler Casorati–Weierstrass theorem.

Exercise 2.5.3. *Let f be analytic and one-to-one from $\mathbb{D} \setminus \{0\}$ into $\mathbb{C}$. Then, 0 is either a pole or a removable singularity.*

In a similar vein (and here we use [136, Theorem 3, p. 2]):

Exercise 2.5.4. *Let $z_1, \ldots, z_N$ be given in $\mathbb{C}$ and let f be analytic and one-to-one from $\mathbb{C} \setminus \{z_1, \ldots, z_N\}$ into $\mathbb{C}$. Show that f is a linear fractional transformation.*

Hint: Use the previous exercise (or the Casorati–Weierstrass theorem) to show that the points $z_1, \ldots, z_N$ are poles or removable singularities.

We recall that a neighborhood of infinity is a set of the form

$$\Omega = \{z \in \mathbb{C} \; ; \; |z| > R\}$$

for some $R > 0$. We also recall that a function f analytic in a neighborhood of infinity has an essential singularity at infinity if, by definition, the function $f(1/z)$ has an essential singularity at the origin.

Exercise 2.5.5 (see [95, p. 182]). *Let f be a one-to-one map from the complex plane into itself (but, at this stage, not necessarily onto $\mathbb{C}$. This last fact will be in fact a consequence of the one-to-oneness).*

(1) *Show that ∞ is not an essential singularity of f.*
(2) *Show that f is a polynomial.*
(3) *Show that f is a degree-one polynomial, and hence is onto.*

2.6 Harmonic and subharmonic functions

We begin with some classical definitions, which can be found in most calculus books. See for instance [49, pp. 385–387]. Let $z(t) = (x(t), y(t))$ be a smooth curve in $\mathbb{R}^2$, where t runs through some open interval I and where x, y are real-valued and have continuous derivatives in I. The tangent vector is defined by

$$T(t) = \frac{1}{\sqrt{(x'(t))^2 + (y'(t))^2}} (x'(t), y'(t)), \tag{2.6.1}$$

and the outer normal is defined by

$$N(t) = \frac{1}{\sqrt{(x'(t))^2 + (y'(t))^2}} (y'(t), -x'(t)). \tag{2.6.2}$$

Furthermore one defines the normal derivative (also called directional derivative in the direction N; see [244, p. 8], [49, p. 386]) of the real-valued smooth function $u(x, y)$ to be

$$\frac{\partial u}{\partial n}(t) \stackrel{\text{def.}}{=} \frac{\partial u}{\partial x}(z(t))y'(t) - \frac{\partial u}{\partial y}(z(t))x'(t), \tag{2.6.3}$$

and, denoting by $s(t) = |z(t)|$ the arc-length,

$$\frac{\partial u}{\partial s}(t) \stackrel{\text{def.}}{=} \frac{\partial u}{\partial x}(z(t))x'(t) + \frac{\partial u}{\partial y}(z(t))y'(t). \tag{2.6.4}$$

The following result is a direct consequence of the Cauchy–Riemann equations:

Question 2.6.1 (see [47, p. 358]). *Assume that the function $F(z) = u(x, y) + iv(x, y)$ is analytic in the open set Ω. Show that*

$$\frac{\partial u}{\partial n} = \frac{\partial v}{\partial s},$$
$$\frac{\partial v}{\partial n} = -\frac{\partial u}{\partial s}.$$

Equation (2.6.3) plays an important role in a reproducing kernel type formula for harmonic functions. See for instance [70, (6), p. 45], [244, p. 25].

Exercise 2.6.2. *Let R be a multiply connected domain, with smooth boundary ∂R, and let u and v be smooth functions defined in a neighborhood of $\overline{R}$, and assume u harmonic there. Show that*

$$\iint_R \left(\frac{\partial u}{\partial x} \frac{\partial v}{\partial x} + \frac{\partial u}{\partial y} \frac{\partial v}{\partial y} \right) (x, y) dx dy = \int_{\partial R} v \frac{\partial u}{\partial n} dz, \qquad (2.6.5)$$

where the last integral is a path integral.

Hint: Use Green's formula (see (2.1.13)) with

$$P(x, y) = v(x, y) \frac{\partial u}{\partial y}(x, y) \quad \text{and} \quad Q(x, y) = -v(x, y) \frac{\partial u}{\partial x}(x, y).$$

For the following exercise, see [266, Ex. 16, p. 292].

Exercise 2.6.3. *The function* $\operatorname{Im} \left(\frac{1+z}{1-z} \right)^2$ *is harmonic in $\mathbb{D}$, has vanishing radial limits on $\mathbb{T} \setminus \{1\}$, but is not 0.*

Exercise 2.6.4. *Let Ω be an open connected subset of $\mathbb{C}$, and let $z_0 \in \Omega$. Let u be a harmonic function in Ω with the following properties: u has a harmonic conjugate (denoted by v) in Ω, and the function $u(x, y) + \ln |z - z_0|$ has a continuous extension to z_0. Show that the function*

$$F(z) = \begin{cases} e^{u(x,y)+iv(x,y)}, & z = x + iy \neq z_0, \\ 0, & z = z_0, \end{cases} \qquad (2.6.6)$$

is analytic in Ω. Show that the point $z = z_0$ is a simple pole of F.

Question 2.6.5. *Let Γ be a Jordan curve, with interior Ω, and assume that there exists a function F analytic in a neighborhood of $\overline{\Omega}$, and with the following properties:*

(a) *It holds that $F(z) = 1$ for $z \in \partial \Omega$.*
(b) *The equation $F(z) = 0$ has N solutions, counting multiplicity.*

Then, show that:

(1) *The equation $F(z) = w$ has N solutions for every $w \in \mathbb{D}$.*
(2) *When $N = 1$, F is a conformal map from Ω onto $\mathbb{D}$.*

2.6. Harmonic and subharmonic functions

Hint: Use the argument principle. See Theorem 2.2.3.

The function F in item (2) in the above exercise provides a (very partial) proof of Riemann's mapping theorem. In particular, the assumption on the analyticity across the boundary is superfluous. To obtain such a function F in general can be done in terms of the Green function of the set Ω. See for instance [310, Theorem 1.17], [174, Exercise 4, p. 97]. See also [70, p. 63] for a similar formula in the case of multiply-connected domains.

The following is [145, Exercise 6, p. 416].

Question 2.6.6. *Let $R \in (0,1)$. Compute Green's function for the annulus $\{R < |z| < R^{-1}\}$ with pole at the point $e^{\frac{2i \ln R}{\pi}}$.*

Question 2.6.7. *Show that a function $u(x,y)$ defined in the open unit disk is harmonic and positive there if and only if it can be written as*

$$u(x,y) = \operatorname{Re} \int_{[0,2\pi]} \frac{e^{it} + z}{e^{it} - z} d\mu(t), \quad z = x + iy \in \mathbb{D}, \qquad (2.6.7)$$

where μ is a positive Borel measure on $[0,2\pi]$.

Hint: The (unique up to an additive constant) function f analytic in $\mathbb{D}$ and with real part u is given by the formula

$$f(z) = \int_{[0,2\pi]} \frac{e^{it} + z}{e^{it} - z} d\mu(t) + ia, \qquad (2.6.8)$$

where $a \in \mathbb{R}$ and μ is a positive Borel measure on $[0,2\pi]$. For a proof, see the discussion in [CAPB, p. 207].

Remark 2.6.8. Formula (2.6.8) is called Herglotz representation formula, and the function f is a Carathéodory function. See Remark 2.1.37 for a related discussion.

For u harmonic in the annulus $R_1 < |z - z_0| < R_2$, and for $r \in (R_1, R_2)$ and $p \in [1, \infty)$, set:

$$J_p(r) = \left(\int_0^{2\pi} |u(z_0 + re^{it})|^p dt \right)^{1/p}. \qquad (2.6.9)$$

Furthermore, set

$$J_\infty(r) = \max_{t \in [0,2\pi]} |u(z_0 + re^{it})|. \qquad (2.6.10)$$

Question 2.6.9. *Show that, for $r \in (R_1, R_2)$:*

$$J_2(r) \leq \frac{\ln R_2/r}{\ln R_2/R_1} J_2(R_1) + \frac{\ln r/R_1}{\ln R_2/R_1} J_2(R_2). \qquad (2.6.11)$$

Inequality (2.6.11) holds in fact for any $p \in [1, \infty]$,

$$J_p(r) \le \frac{\ln R_2/r}{\ln R_2/R_1} J_p(R_1) + \frac{\ln r/R_1}{\ln R_2/R_1} J_p(R_2), \qquad (2.6.12)$$

see [286]. The case $p = \infty$ is called Hadamard three circles theorem. For the case $p = 2$ see [68, Exercise 11, p. 347]. For the cases J_∞ and J_1 we send the student to [176, p. 80].

Recall that a *continuous* real-valued function $u(z)$ defined in an open set Ω of the complex plane is called subharmonic if for every $w_0 \in \Omega$ there exists $r_0 > 0$ such that $\overline{B(w_0, r_0)} \subset \Omega$ and

$$u(w_0) \le \frac{1}{2\pi} \int_0^{2\pi} u(w_0 + re^{it}) dt, \quad \forall r \in (0, r_0).$$

Item (1) in the following exercise is part of any course on subharmonic functions. For item (2) and applications, see [310, p. 1].

Question 2.6.10.

(1) *Prove the maximum principle for subharmonic functions.*
(2) *Let f be analytic in the open set Ω, and $p, q > 0$. Show that $z \mapsto |z|^p |f(z)|^q$ is subharmonic.*
(3) *Let $f_1, \ldots, f_N$ be analytic in Ω. What can you say about the function $z \mapsto \sum_{n=1}^N |f_n(z)|^2$?*

As a corollary one has:

Question 2.6.11. *Let $f_1, \ldots, f_N$ be analytic in the open connected set Ω. Show that the function*

$$M(z) = \sqrt{\sum_{n=1}^N |f_n(z)|^2} \qquad (2.6.13)$$

has no local maximum unless all the functions $f_1, \ldots, f_N$ are constant.

Hint: One can take the square and consider $\sum_{n=1}^N |f_n(z)|^2$. This function has no local maximum unless it is constant. Apply then the formula

$$\Delta |g|^2 = 4|g'(z)|^2$$

(where g is analytic in some open set; see [CAPB, (9.3.3), p. 401]) to get

$$\sum_{n=1}^N |f_n'(z)|^2 \equiv 0$$

when (2.6.13) is constant.

The same method works for the functions

$$\sum_{n=1}^{N} |f_n(z)|^p, \quad p \in \mathbb{R} \setminus \{0\}.$$

One then makes use of the formula

$$\Delta |g|^p = p^2 |g(z)|^{p-2} |g'(z)|^2.$$

See for instance [128, Exercice 8.41, p. 103], and also [128, pp. 152–154] for related exercises on the maximum modulus principle.

2.7 Meromorphic functions in $\mathbb{C}$

Let Ω be a connected open subset of the complex plane. The function f is said to be meromorphic in Ω if it has at most poles in Ω. Equivalently, f is meromorphic in Ω if it is a quotient of two functions analytic in Ω. The idea to prove this equivalence is to gather the poles of the function in a converging infinite product. See [91]. For instance, the function $e^{1/z}$ is not meromorphic in $\mathbb{C}$ since it has $z = 0$ as essential singularity.

Exercise 2.7.1. *The function*

$$\frac{1}{\sin \frac{1}{z}}$$

is not meromorphic in $\mathbb{C}$.

Question 2.7.2. *The function*

$$f(z) = \int_0^1 \frac{dt}{t - z}$$

is not meromorphic in $\mathbb{C}$. *It is analytic in* $\mathbb{C} \setminus [0, 1]$.

Exercise 2.7.3.
(1) *Let f be analytic in a pointed neighborhood V of the point z_0, and assume that z_0 is an essential singularity of f. Show that z_0 is an essential singularity of e^f.*
(2) *What happens if z_0 is not isolated?*

Hint: Use the Casorati–Weierstrass theorem.

Remark 2.7.4. When f has a pole, the above exercise is [CABP, Exercise 7.2.14, p. 322]. There, the proof goes by contradiction as follows: Write

$$f(z) = \frac{h(z)}{(z - z_0)^N}, \quad z \in V,$$

where $N \in \mathbb{N}$ and h is a function analytic in a neighborhood of z_0 and not vanishing at z_0. Proceeding by contradiction, write now

$$e^{\frac{h(z)}{(z-z_0)^N}} = g(z)(z - z_0)^M, \tag{2.7.1}$$

where $g(z)$ is a function analytic in a neighborhood of z_0 and not vanishing at z_0, and $M \in \mathbb{Z}$. Differentiating both sides of (2.7.1) one obtains a contradiction. It follows in particular that if f is an elliptic function, e^f is not an elliptic function.

We recall (see Exercise 2.1.11) that a periodic meromorphic function does not have ∞ as an isolated singularity. As we have already mentioned, meromorphic functions on the Riemann sphere are exactly rational functions. When entire, these are polynomials. In the case of genus one, one has:

Question 2.7.5.

(1) *Show that an entire biperiodic function is constant.*
(2) *Let f be an elliptic function with poles only on the lattice. Show that f is a polynomial in $\wp$, where $\wp$ denotes the associated Weierstrass function (see for instance [141, Proposition V.3.1, p. 275]).*

2.8 Solutions of the exercises

Solution of Exercise 2.1.1: We follow Hardy's book [171, pp. 68–69] on divergent series.

(1) Since $f(0) = 1$ the function $1/f$ is analytic in a neighborhood of the origin and admits a Taylor expansion of the form (2.1.2) there. We have

$$g_1 + f_1 = 0,$$
$$g_2 + g_1 f_1 + f_2 = 0,$$
$$\vdots$$
$$g_n + g_{n-1} f_1 + \cdots + g_1 f_{n-1} + f_n = 0,$$
$$\vdots$$

Multiplying the first equation by f_2 and the second one by f_1 we obtain (2.1.3). More generally, for $n \geq 1$,

$$g_1 f_{n-1} + \cdots + g_{n-1} f_1 + g_n = -f_n, \quad \text{and} \quad g_1 f_n + \cdots + g_n f_1 + g_{n+1} = -f_{n+1},$$

(with the understanding that the first equality is $g_1 = -f_1$ for $n = 1$). Multiplying the first of these equalities by f_{n+1} and the second one by f_n we obtain (2.1.4).

2.8. Solutions of the exercises

(2) From (2.1.4) we have (since the numbers f_n are assumed not being equal to 0)

$$g_{n+1} = \frac{f_{n+1}}{f_n}(g_1 f_{n-1} + \cdots + g_{n-1}f_1 + g_n) - (g_1 f_n + \cdots + g_n f_1)$$

$$= g_1\left(\frac{f_{n+1}f_{n-1}}{f_n} - f_n\right) + g_2\left(\frac{f_{n+1}f_{n-2}}{f_n} - f_{n-1}\right) + \cdots +$$

$$+ g_k\left(\frac{f_{n+1}f_{n-k}}{f_n} - f_{n-k+1}\right) + \cdots + g_n\left(\frac{f_{n+1}}{f_n} - f_1\right)$$

$$= \sum_{k=1}^{n} g_k c_{n,k}.$$

(3) When the $f_n > 0$ and (2.1.6) holds we have

$$c_{n,k} = \frac{f_{n+1}f_{n-k}}{f_n} - f_{n-k+1}$$

$$= f_{n-k}\left(\frac{f_{n+1}}{f_n} - \frac{f_{n-k+1}}{f_{n-k}}\right) \geq 0$$

for $n = 1, 2, \ldots$ and $k = 1, \ldots, n-1$. Equation (2.1.5) allows then to prove by induction that $g_n \leq 0$ for $n = 2, 3, \ldots$ when $g_1 \leq 0$.

(4) Let $x \in [0, 1)$. Since the $f_n > 0$ we can write

$$1 + \sum_{n=1}^{\infty} g_n x^n = \frac{1}{1 + \sum_{n=1}^{\infty} f_n x^n}.$$

Still because $f_n > 0$ we have that

$$\lim_{x \uparrow 1} \frac{1}{1 + \sum_{n=1}^{\infty} f_n x^n}$$

exists (and is possibly equal to 0). It follows that

$$\lim_{x \uparrow 1} 1 + \sum_{n=1}^{\infty} g_n x^n \tag{2.8.1}$$

exists and is nonnegative. Thus

$$\lim_{x \uparrow 1} \sum_{n=1}^{\infty} (-g_n) x^n$$

exists. By elementary results on power series, or by using the monotone convergence theorem, we have

$$\lim_{x \uparrow 1} \sum_{n=1}^{\infty} (-g_n) x^n = \sum_{n=1}^{\infty} (-g_n).$$

Since the limit (2.8.1) is nonnegative, it follows that

$$\sum_{n=1}^{\infty}(-g_n) \leq 1.$$

(5) The case $f_n \equiv 1$ leads to $f(z) = \frac{1}{1-z}$, while the case $f_n = \frac{1}{n+1}$ corresponds to the function

$$f(z) = \begin{cases} -\frac{\ln(1-z)}{z}, & 0 < |z| < 1, \\ 1, & z = 0. \end{cases}$$

$\square$

Solution of Exercise 2.1.5:

(1) Using (2.1.8) we see that formula (2.1.9) is equivalent to the formula

$$\iint_{\mathbb{D}} \frac{\partial H}{\partial \bar{z}}(z)dxdy = \frac{1}{2i}\int_{\mathbb{T}} H(z)dz$$

with $H(z) = f(z)\overline{g(z)}$. The latter is just a rewriting of Green's formula for complex-valued functions. Indeed, setting $H(z) = u(x,y) + iv(x,y)$ we have (see for instance [CAPB, (4.2.10), p. 152]):

$$\iint_{\mathbb{D}} \frac{\partial H}{\partial \bar{z}}(z)dxdy = \iint_{\mathbb{D}} \frac{1}{2}\left\{\left(\frac{\partial u}{\partial x} + i\frac{\partial u}{\partial y}\right) + \left(-\frac{\partial v}{\partial y} + i\frac{\partial v}{\partial x}\right)\right\}dxdy$$

$$= \frac{1}{2}\iint_{\mathbb{D}}\left(\frac{\partial u}{\partial x} - \frac{\partial v}{\partial y}\right)dxdy + \frac{i}{2}\iint_{\mathbb{D}}\left(\frac{\partial v}{\partial x} + \frac{\partial u}{\partial y}\right)dxdy$$

$$= \frac{1}{2}\int_{\mathbb{T}} v(x,y)dx + u(x,y)dy + \frac{i}{2}\int_{\mathbb{T}} -u(x,y)dx + v(x,y)dy$$

$$= \frac{1}{2i}\int_{\mathbb{T}}(u(x,y) + iv(x,y))dx + \frac{1}{2}\int_{\mathbb{T}}(u(x,y) + iv(x,y))dy$$

$$= \frac{1}{2i}\int_{\mathbb{T}} H(z)dz.$$

The proof of (2.1.10) is done in the same way.

(2) To prove (2.1.12), two remarks are of importance. First, it holds that

$$\frac{\partial^a}{\partial \bar{z}^a}(|z|^2 - 1)^b = 0 \tag{2.8.2}$$

for $a > b$, as follows from the formula

$$(|z|^2 - 1)^b = \sum_{n=0}^{b}\binom{b}{n}|z|^{2n}(-1)^{b-n} = \sum_{n=0}^{b}\binom{b}{n}\bar{z}^n z^n(-1)^{b-n}.$$

The second remark is that

$$\frac{\partial^{m_1+m_2}}{\partial z^{m_1}\bar{z}^{m_2}}(|z|^2 - 1)^m\Big|_{|z|=1} = 0 \tag{2.8.3}$$

for $m_1 + m_2 < m$.

2.8. Solutions of the exercises

In the proof of (2.1.12) one distinguishes between various cases. We will consider only part of them, and leave the others to the reader.

Case 1: $j_1 = j_2 = 0$. Then, for $k \in \mathbb{N}_0$,

$$e_{k,0}(z) = \sqrt{\frac{k+1}{\pi}} \cdot \frac{1}{k!} \cdot k! z^k = \sqrt{\frac{k+1}{\pi}} z^k,$$

and the result is an easy computation. We also note that these functions form an orthonormal basis of the Bergman space $\mathcal{B}_2(\mathbb{D})$ (see Remark 10.5.1 for the latter).

Before considering other cases, we remark the following. Setting in (2.1.9)

$$\frac{\partial}{\partial \overline{z}} f(z) = e_{k_1, j_1}(z) \quad \text{and} \quad g(z) = e_{k_2, j_2}(z),$$

and taking into account (2.8.3), we obtain:

$$\iint_{\mathbb{D}} e_{k_1, j_1}(z) \overline{e_{k_2, j_2}(z)} \, dx dy$$

$$= -c \iint_{\mathbb{D}} \frac{\partial^{k_1 + j_1 - 1}}{\partial z^{k_1} \partial \overline{z}^{j_1 - 1}} (|z|^2 - 1)^{j_1 + k_1} \overline{\frac{\partial^{1 + j_2 + k_2}}{\partial z^{k_2} \partial \overline{z}^{1 + j_2}} (|z|^2 - 1)^{j_2 + k_2}} \, dx dy,$$

where

$$c = \sqrt{\frac{k_1 + j_1 + 1}{\pi(k_1 + j_1)!}} \sqrt{\frac{k_2 + j_2 + 1}{\pi(k_2 + j_2)!}}.$$

Iterating this argument, and a similar one with (2.1.10) will lead to the required conclusion More precisely:

Case 2: $k_2 < j_1$. Then (2.1.9) iterated j_1 times leads to

$$\iint_{\mathbb{D}} e_{k_1, j_1}(z) \overline{e_{k_2, j_2}(z)} \, dx dy$$

$$= (-1)^{j_1} c \iint_{\mathbb{D}} \frac{\partial^{k_1}}{\partial z^{k_1}} (|z|^2 - 1)^{j_1 + k_1} \underbrace{\overline{\frac{\partial^{j_1 + j_2 + k_2}}{\partial z^{k_2} \partial \overline{z}^{j_1 + j_2}} (|z|^2 - 1)^{j_2 + k_2}}}_{\text{is equal to 0 since } j_1 + j_2 > j_2 + k_2} \, dx dy$$

$$= 0.$$

Case 3: $k_2 > j_1$. Then (2.1.10) iterated k_2 times leads to

$$\iint_{\mathbb{D}} e_{k_1, j_1}(z) \overline{e_{k_2, j_2}(z)} \, dx dy$$

$$= (-1)^{k_2} c \iint_{\mathbb{D}} \underbrace{\frac{\partial^{j_1 + k_1 + k_2}}{\partial z^{k_1 + k_2} \partial \overline{z}^{j_1}} (|z|^2 - 1)^{k_1 + j_1}}_{\text{is equal to 0 since } k_1 + k_2 > k_1 + j_1} \overline{\frac{\partial^{j_2}}{\partial \overline{z}^{j_2}} (|z|^2 - 1)^{k_2 + j_2}} \, dx dy$$

$$= 0. \qquad \square$$

Solution of Exercise 2.1.6: Applying (2.1.9) for A_R with $f(z) = 1$ and $g(z) = z$, we get

$$
\iint_{A_R} dx\,dy = \frac{1}{2i} \int_{|z|=R} \overline{z}\,dz - \frac{1}{2i} \int_{|z|=1} \overline{z}\,dz
$$
$$
= \frac{1}{2} \int_0^{2\pi} Re^{-it} Re^{it}\,dt - \frac{1}{2} \int_0^{2\pi} e^{-it} e^{it}\,dt
$$
$$
= \pi(R^2 - 1).
$$

$\square$

Solution of Exercise 2.1.7: We first assume that f is a polynomial, that is

$$
f(z) = \sum_{n=0}^{N} f_n z^n.
$$

Then,

$$
f(ze^{it}) = \sum_{n=0}^{N} f_n z^n e^{int},
$$

and for $m = 0, \ldots, N$ we have:

$$
\frac{1}{2\pi} \int_0^{2\pi} \frac{f(ze^{it})}{e^{imt}}\,dt = \sum_{n=0}^{N} \frac{1}{2\pi} \int_0^{2\pi} f_n z^n \frac{e^{int}}{e^{imt}}\,dt
$$
$$
= f_m z^m,
$$

(the latter being a direct computation, and not an application of Cauchy's formula), and so (and using the maximum modulus principle)

$$
|f_m z^m| \le \max_{t \in [0,2\pi]} |f(ze^{it})| = \max_{|\zeta| \le r} |f(\zeta)| = \max_{|\zeta|=r} |f(\zeta)| \le K.
$$

Consider now the case of f analytic in the open disk $B(0, R)$. We follow the proof of Lemma 4 in [102]. The Taylor expansion converges uniformly on compact subsets of $B(0, R)$, and so given $r \in (0, R)$ and $\epsilon > 0$ there exists $N \in \mathbb{N}$ such that

$$
\max_{|\zeta| \le r} \left| f(z) - \sum_{n=0}^{M} f_n z^n \right| \le \epsilon \tag{2.8.4}
$$

for $M \ge N$. Thus

$$
\left| \sum_{n=0}^{M} f_n z^n \right| \le \epsilon + \max_{|\zeta|=r} |f(\zeta)|, \quad \text{for } |z| \le r.
$$

Fix now $m \in \mathbb{N}_0$ and consider $M \ge \max\{m, N\}$. We obtain from the first part of the proof and (2.8.4) that

$$
|f_m z^m| \le \epsilon + \max_{|\zeta|=r} |f(\zeta)|.
$$

The result follows since ϵ is arbitrary.

$\square$

2.8. Solutions of the exercises

Solution of Exercise 2.1.8: If $f \in \mathcal{C}$ is identically vanishing, the result is trivial. If $f \not\equiv 0$, then $1/f \in \mathcal{C}$ since

$$\operatorname{Re} \frac{1}{f} = \frac{\operatorname{Re} f}{|f|^2},$$

and so

$$\frac{1}{f} + g \in \mathcal{C}.$$

Thus

$$\frac{1}{\frac{1}{f} + g} \in \mathcal{C}.$$

To conclude note that

$$\frac{f}{1 + fg} = \frac{1}{\frac{1}{f} + g}. \qquad \square$$

Solution of Exercise 2.1.10: The integral can be rewritten as

$$\frac{T}{2\pi i} \int_0^1 \frac{f'(z_0 + Tt)}{f(z_0 + Tt)} dt = \frac{1}{2\pi i} \int_0^1 \frac{\gamma'(t)}{\gamma(t)} dt$$

with

$$\gamma(t) = f(z_0 + Tt), \quad t \in [0, 1].$$

To conclude, note that γ is a parametrization of a closed smooth path since $f(z_0) = f(z_0 + T)$, and the integral is the index of the origin with respect to this path, and so is an integer. $\qquad \square$

Solution of Exercise 2.1.11:

(1) The function f is not a polynomial, and so its Taylor series at the origin, which has radius of convergence equal to ∞ since f is entire, has an infinite number of nonzero coefficients. So the Laurent series of $f(1/z)$ converges in $\mathbb{C} \setminus \{0\}$ and has an infinite number of nonzero coefficients, and ∞ is an essential singularity of f.

(2) Let T and z_0 be respectively a period and an isolated singularity of the function f. The points $z_n = z_0 + nT$, $n = 1, 2, \ldots$ are isolated singularities of f and so the points $1/z_n$ (except for a possible unique value of n for which $z_n = 0$) are isolated singularities of $f(1/z)$. Since $\lim_{n \to \infty} 1/z_n = 0$, we see that $z = 0$ is not an isolated singularity of $f(1/z)$, and so ∞ is not an isolated singularity of f. $\qquad \square$

(3) is a direct consequence of (2).

Solution of Exercise 2.1.13: Let $\zeta = re^{i\theta}$, where $r \in (0, 1)$ and $\theta \in (-\pi, \pi)$, that is, $\zeta \in \mathbb{D} \setminus (-1, 0]$. Then,

$$\frac{\ln \zeta}{2\pi i} = -i \frac{\ln r}{2\pi} + \frac{\theta}{2\pi} \in \Omega = \left\{ z, \ |\operatorname{Re} z| < \frac{1}{2} \right\} \cap \mathbb{C}_+.$$

We can thus define a function g analytic in $\mathbb{D} \setminus (-1,0]$ via the formula (2.1.17):

$$g(\zeta) = f\left(\frac{\ln \zeta}{2\pi i}\right).$$

To extend g to a function analytic in the pointed open unit disk it is enough to show that g is continuous across the boundary (and then uses for instance Morera's theorem). But the continuity follows from the condition

$$f(z+1) = f(z).$$

Finally, g satisfies (2.1.16) in Ω by construction, and by analytic extension to the whole open upper half-plane.

$\square$

Solution of Exercise 2.1.14: With $u(z) = e^{-\frac{z^2}{2}} v(z)$ we have

$$zu(z) - u'(z) - \lambda u(z) = ze^{-\frac{z^2}{2}} v(z) - ze^{-\frac{z^2}{2}} v(z) + e^{-\frac{z^2}{2}} v'(z) - \lambda e^{-\frac{z^2}{2}} v(z)$$

$$= e^{-\frac{z^2}{2}} \left(v'(z) - \lambda v(z)\right).$$

Thus equation (2.1.18) is equivalent to

$$v'(z) = \lambda v(z).$$

We have $v(z) = ce^{\lambda z}$ where $c = v(0) = u(0) \in \mathbb{C}$, and so the solutions are

$$u(z) = u(0)e^{\lambda z - \frac{z^2}{2}},$$

and in particular are entire functions.

$\square$

Solution of Exercise 2.1.16: The first item is an easy argument. For $n = 0$ we have $H_0(z) \equiv 1$ and the claim is clear. Assume now that

$$H_n(z) = (-1)^n e^{z^2} \left(e^{-z^2}\right)^{(n)}$$

is a polynomial for $n \in \mathbb{N}$. Differentiating both sides of this equality we have

$$H_n'(z) = 2z H_n(z) + (-1)^n e^{z^2} \left(e^{-z^2}\right)^{(n+1)}$$

$$= 2z H_n(z) - (-1)^{n+1} e^{z^2} \left(e^{-z^2}\right)^{(n+1)},$$

which can be rewritten as

$$H_{n+1}(z) = 2z H_n(z) - H_n'(z). \tag{2.8.5}$$

This last equality allows to conclude the induction argument.

2.8. Solutions of the exercises

Let $f(z) = e^{-z^2}$. To prove the second item we write

$$f(z - u) = \sum_{n=0}^{\infty} (-u)^n \frac{f^{(n)}(z)}{n!}$$

$$= \sum_{n=0}^{\infty} \frac{u^n}{n!} (-1)^n \left(e^{-z^2} \right)^{(n)}.$$

Since

$$e^{z^2} f(z - u) = e^{z^2} e^{-(z-u)^2} = e^{2zu - u^2}$$

we have

$$e^{2zu - u^2} = e^{z^2} f(z - u)$$

$$= e^{z^2} \left(\sum_{n=0}^{\infty} \frac{u^n}{n!} (-1)^n \left(e^{-z^2} \right)^{(n)} \right)$$

$$= \sum_{n=0}^{\infty} \frac{H_n(z)}{n!} u^n. \qquad \square$$

Solution of Exercise 2.1.19: We have for $z \neq w$

$$B_{a,w}^{-1}(z) F(z) = \left(I_n - \frac{aa^*}{a^*a} + \frac{1 - z\overline{w_0}}{z - w_0} \frac{aa^*}{a^*a} \right) F(z)$$

$$= \underbrace{\left(I_n - \frac{aa^*}{a^*a} \right) F(z)}_{\text{analytic in } \mathbb{D}} + (1 - z\overline{w_0}) a \frac{a^* F(z)}{z - w_0}.$$

The point $z = w_0$ is a removable pole (for each entry of the function $a^* F(z)$), and this conclude the proof. $\qquad \square$

Solution of Exercise 2.1.21: Assume first that f does not vanish in $\mathbb{D}$, and let g be analytic in $\mathbb{D}$ and such that $f(z) = e^{2g(z)}$ there. Let $h = e^{pg}$. Then, for $r \in (0, 1)$:

$$\int_0^{2\pi} |f(re^{it})|^p dt = \int_0^{2\pi} |e^{pg(re^{it})}|^2 dt$$

and so

$$M_p(f, r) = (M_2(h, r))^{\frac{2}{p}},$$

and the result (including the continuity) follows from Exercise 2.1.20.

More generally, if f does not vanish in $|z| < R$ with $R \in (0, 1)$ we still can write $f(z) = e^{g(z)}$ where g is now analytic in $|z| < R$, and the result still holds for $r \in (0, R)$.

Assume now that f has zeros in the open unit disk. Let $R \in (0,1)$. By the uniqueness theorem for analytic functions there is only a finite number of zeros of f, say $z_1, \ldots, z_N$, in the closed disk $\overline{B(0,R)}$. We can assume that none of them is of modulus R (otherwise, replace R by $R + \epsilon$ for ϵ small enough). Let $r_1, r_2 \in (0, R)$ be such that $r_1 < r_2$ and such that all the zeros of f are inside the open ring defined by the circles $|z| = r_1$ and $|z| = r_2$. Let $b(z)$ denote the Blaschke product with the points $\frac{z_1}{r_2}, \frac{z_2}{r_2}, \ldots, \frac{z_n}{r_2}$ as zeros, with the corresponding multiplicities, and

$$ b_{r_2}(z) = b(z/r_2) = \prod_{n=1}^{N} \frac{\frac{z}{r_2} - \frac{z_n}{r_2}}{1 - \frac{z\overline{z_n}}{r_2^2}}. $$

The points $z_1, \ldots, z_N$ are removable singularities of the function

$$ \frac{f(z)}{b_{r_2}(z)}, $$

and we denote by g the corresponding function analytic in $|z| < R$. We have:

$$ \int_0^{2\pi} |f(r_1 e^{it})|^p dt \leq \int_0^{2\pi} |g(r_1 e^{it})|^p dt $$

(since $|b_{r_2}(z)| < 1$ on $|z| = r_1$)

$$ \leq \int_0^{2\pi} |g(r_2 e^{it})|^p dt $$

(since g has no zeros in $|z| < R$ and so we can apply (1))

$$ = \int_0^{2\pi} |f(r_2 e^{it})|^p dt $$

(since b_{r_2} has modulus 1 on $|z| = r_2$).

This ends the proof when none of the zeros are on the circle $|z| = r_1$ or $|z| = r_2$. To conclude, note that $M_p(f, r)$ is a continuous function of r and that R is arbitrary in $(0, 1)$. $\qquad \square$

Solution of Exercise 2.1.22: We recall the proof appearing in [269, p. 298]. Without loss of generality we can assume that all the points are different from 0. Since $\lim_{n \to \infty} z_n = \infty$ we have:

$$ \forall R > 0, \ \exists \, n(R) \in \mathbb{N} : n \geq n(R) \implies |z_n| > 2R. $$

Thus, for $|z| < R$, and taking into account (2.1.26) and $p = n - 1$,

$$ |1 - E_{n-1}(z/z_n)| \leq \frac{1}{2^n}, $$

2.8. Solutions of the exercises

and hence the infinite product

$$\prod_{n=1}^{\infty} E_{n-1}(z/z_n)$$

converges uniformly in $|z| < R$. Since R is arbitrary, the product converges for all complex numbers z, and the convergence is uniform on compact subsets of the complex plane. □

Remark 2.8.1. As a corollary of the previous argument we have the following result, which is the starting point in the arguments of Saks and Zygmund (see [269, (2.8), pp. 297–298]): Given any sequence $(z_n)_{n\in\mathbb{N}}$ of nonzero complex numbers going to infinity in modulus, there exists a sequence $(p_n)_{n\in\mathbb{N}}$ of integers such that the series

$$\sum_{n=1}^{\infty} \left|\frac{z}{z_n}\right|^{m_n}$$

converges in $\mathbb{C}$, the convergence being uniform on compact sets.

Solution of Exercise 2.1.24:

(1) Let F be as in (2.1.28), and let $z \in \mathbb{C} \setminus C$. Since C is compact the distance $d = d(z, C)$ from z to C is strictly positive, and we have $B(z, d) \subset \mathbb{C} \setminus C$. Let $w \in B(z, d)$. We have

$$\begin{aligned}
F(w) &= \int_C \frac{f(s)ds}{w - s} \\
&= \int_C \frac{f(s)ds}{(w - z + z - s)} \\
&= \int_C \frac{f(s)ds}{(z - s)(1 - \frac{z-w}{z-s})} \\
&= \int_C \left(\frac{f(s)}{z-s}\right)\left(\sum_{n=0}^{\infty}\left(\frac{z-w}{z-s}\right)^n\right) ds \\
&= \sum_{n=0}^{\infty}(z-w)^n F_n,
\end{aligned}$$

with

$$F_n = \int_C \frac{f(s)ds}{(z-s)^{n+1}}.$$

The interchange between the integral and the infinite sum has been done using the dominated convergence theorem, taking into account that

$$d(w) \overset{\text{def.}}{=} \max_{s \in C} \frac{|z - w|}{|z - s|} < 1$$

for $w \in B(z, d)$ and

$$\left| \frac{f(s)}{z-s} \left(\frac{z-w}{z-s} \right)^n \right| \le \frac{M}{d} d(w)^n,$$

where $M = \max_{s \in C} |f(s)|$.

To prove that F is analytic at infinity, one can prove that $\lim_{z \to \infty} F(z)$ exists, or equivalently, that $F(1/z)$ has a removable singularity at the origin. Both approaches are left to the reader.

(2) We follow [269, p. 172]. Set

$$G(z, t) = \frac{f(\gamma(t))}{z - \gamma(t)}, \quad t \in [a, b], \ z \in K.$$

As mentioned in the hint, G is uniformly continuous on $[a, b] \times K$. Therefore given $\epsilon > 0$ there exist $\eta_1 > 0$ and $\eta_2 > 0$ such that

$$|G(t_1, z_1) - G(t_2, z_2)| \le \epsilon$$

as soon as $|t_1 - t_2| < \eta_1$ and $|z_1 - z_2| < \eta_2$. We can in particular take $z_1 = z_2 = z$, and get

$$|t_1 - t_2| < \eta_1 \implies |G(t_1, z) - G(t_2, z)| \le \epsilon.$$

Let now $a = t_0 < t_1 < t_2 < \cdots < t_n = b$ be a subdivision of $[a, b]$ of mesh less than η_1, and let

$$R(z) = \sum_{k=0}^{n-1} \frac{f(\gamma(t_k))}{z - \gamma(t_k)} (\gamma(t_{k+1}) - \gamma(t_k)).$$

The function R is rational and has its poles at the points $\gamma(t_0), \ldots, \gamma(t_{n-1})$, and therefore on C. Set $M = \max_{t \in [a,b]} |\gamma'(t)|$ (recall that γ is ruled). For $z \in K$ we have:

$$|F(z) - R(z)| = \left| \sum_{k=0}^{n-1} \int_{t_k}^{t_{k+1}} (G(t, z) - G(t_k, z)) \gamma'(t) dt \right|$$

$$=\le \epsilon M (b - a).$$

Replacing in the arguments $\epsilon M (b - a)$ by ϵ we obtain the stated result. $\square$

Solution of Exercise 2.1.26: The dominated convergence theorem allows us to prove that

$$\lim_{n \to \infty} \frac{I(t + h_n) - I(t)}{h_n} = -i \int_{\mathbb{R}} u e^{-\frac{u^2}{2}} e^{-itu} du$$

for any sequence of numbers $(h_n)_{n \in \mathbb{N}}$ tending to 0. Thus (recall that in a metric space, limits can be computed on sequences)

$$I'(t) = -i \int_{\mathbb{R}} u e^{-\frac{u^2}{2}} e^{-itu} du.$$

Integrating by part the expression for $I'(t)$ we get

$$I'(t) = i \left\{ \left[e^{-\frac{u^2}{2}} e^{-itu} \right]_{-\infty}^{\infty} - \int_{\mathbb{R}} e^{-\frac{u^2}{2}} (-it) e^{-itu} du \right\}$$

$$= -tI(t).$$

It follows that $I(t) = I(0) e^{-\frac{t^2}{2}} = \sqrt{2\pi} e^{-\frac{t^2}{2}}$, since $\int_{\mathbb{R}} e^{-\frac{u^2}{2}} du = \sqrt{2\pi}$. $\qquad\square$

Solution of Exercise 2.1.27:

(1) The entire function $\sqrt{2\pi} e^{-\frac{z^2}{2}}$ gives the required entire extension.

(2) We have

$$\int_{\mathbb{R}} h_w(u) \overline{h_z(u)} du = \frac{1}{\sqrt{\pi}} e^{-\frac{z^2 + \overline{w}^2}{2}} \int_{\mathbb{R}} e^{-u^2} e^{\sqrt{2}(z+\overline{w})u} du$$

$$= \frac{1}{\sqrt{2\pi}} e^{-\frac{z^2 + \overline{w}^2}{2}} \int_{\mathbb{R}} e^{-\frac{v^2}{2}} e^{(z+\overline{w})v} dv$$

$$= \frac{1}{\sqrt{2\pi}} e^{-\frac{z^2 + \overline{w}^2}{2}} \sqrt{2\pi} e^{\frac{(z+\overline{w})^2}{2}}$$

$$= e^{z\overline{w}},$$

where we have used the change of variable $u = \frac{v}{\sqrt{2}}$ to go from the first to the second line and formula (2.1.30) with $t = -i(z + \overline{w})$ to go from the second to the third line.

(3) The fact that F is entire is proved using Morera's theorem, or as follows:

$$e^{\frac{z^2}{2}} F(z) = \frac{1}{\pi^{\frac{1}{4}}} \int_{\mathbb{R}} f(u) e^{-\frac{u^2}{2}} \left\{ \sum_{n=0}^{\infty} \frac{(\sqrt{2}zu)^n}{n!} \right\} du = \sum_{n=0}^{\infty} F_n z^n, \quad \forall z \in \mathbb{C},$$

where

$$F_n = \frac{1}{\pi^{\frac{1}{4}}} \int_{\mathbb{R}} f(u) e^{-\frac{u^2}{2}} \frac{(\sqrt{2}u)^n}{n!} du, \quad n = 0, 1, \ldots,$$

and where the interchange of summation and integration is done using the dominated convergence theorem.

For the injectivity, see Remark 6.1.34. $\qquad\square$

Solution of Exercise 2.1.29:

(1) One can use Morera's theorem (one then needs to show first that F is continuous) or rely on the same argument as in the proof of Exercise 2.1.31 below.

(2) Take for instance $f(u) = e^{-u}$ to obtain

$$F(z) = \int_0^\infty e^{iuz}e^{-u}du = -\frac{1}{i(z+i)}, \quad z \in \mathbb{C}_+. \tag{2.8.6}$$

(3) Take for instance $f(u) = 1_{[0,1]}(u)$. The corresponding function

$$F(z) = \begin{cases} \frac{e^{iz}-1}{iz}, & z \neq 0, \\ 1, & z = 0, \end{cases}$$

is not rational (otherwise e^{iz} would be rational, but it has an essential singularity at infinity). $\qquad\square$

Solution of Exercise 2.1.31: Let $h \neq 0 \in \mathbb{C}$. We have

$$\frac{e^{-th}-1}{h} = -\int_0^t e^{-hs}ds,$$

and so

$$\left|\frac{e^{-th}-1}{h}\right| \leq \int_0^t e^{-(\operatorname{Re}h)s}ds \leq te^{|\operatorname{Re}h|t}. \tag{2.8.7}$$

Let $z \in \mathbb{C}_r$ and let $h \neq 0$ be such that

$$|\operatorname{Re}h| < \frac{\operatorname{Re}z}{2}. \tag{2.8.8}$$

Then, in view of (2.8.7),

$$\left|t^\nu\frac{e^{-(z+h)t}-e^{-zt}}{h}\right| = \left|t^\nu e^{-zt}\frac{-e^{-ht}-1}{h}\right|$$

$$\leq t^{\nu+1}e^{-(\operatorname{Re}z-|\operatorname{Re}h|)t}$$

$$\leq t^{\nu+1}e^{-\frac{(\operatorname{Re}z)t}{2}}.$$

Let now $(h_n)_{n\in\mathbb{N}}$ be a sequence of numbers subject to (2.8.8) and tending to 0, and let

$$g_n(z) = t^\nu\frac{e^{-(z+h_n)t}-e^{-zt}}{h_n}, \quad n \in \mathbb{N},$$

$$g(t) = t^{\nu+1}e^{-\frac{(\operatorname{Re}z)t}{2}}.$$

From the above discussion we have $|g_n(t)| \leq g(t)$. Furthermore, g is summable and

$$\lim_{n\to\infty} g_n(t) = -t^{\nu+1}e^{-zt}, \quad t > 0.$$

The dominated convergence theorem allows then to interchange integration and limit in

$$\lim_{n\to\infty}\int_0^\infty g_n(t)dt$$

and obtain the formula (2.1.38) for $F'(z)$. $\quad\square$

Solution of Exercise 2.1.33: We follow the computation in [230, p. 280]. Since

$$1-\left|\frac{1-z}{1+z}\right|^2=\frac{4\operatorname{Re}z}{|1+z|^2}>0,\quad z\in\mathbb{C}_r,$$

the convergence of the series for z,w in the open right half-plane is clear. Furthermore, still for $z,w\in\mathbb{C}_r$, and taking into account (2.1.45), we have:

$$\sum_{n=0}^\infty f_n^{(\nu)}(z)\overline{f_n^{(\nu)}(w)}$$

$$=\left(\frac{2^{1+\nu}}{(1+z)^{1+\nu}(1+\overline{w})^{1+\nu}}\right)\sum_{n=0}^\infty\frac{\Gamma(1+n+\nu)}{n!}\left(\frac{1-z}{1+z}\right)^n\left(\frac{1-\overline{w}}{1+\overline{w}}\right)^n$$

$$=\left(\frac{2^{1+\nu}}{(1+z)^{1+\nu}(1+\overline{w})^{1+\nu}}\right)\Gamma(1+\nu)$$

$$\times\sum_{n=0}^\infty\frac{(1+\nu)\cdots(n+\nu)}{n!}\left(\frac{1-z}{1+z}\right)^n\left(\frac{1-\overline{w}}{1+\overline{w}}\right)^n$$

$$=\Gamma(1+\nu)\left(\frac{2^{1+\nu}}{(1+z)^{1+\nu}(1+\overline{w})^{1+\nu}}\right)\left(1-\left(\frac{1-z}{1+z}\right)\left(\frac{1-\overline{w}}{1+\overline{w}}\right)\right)^{-1-\nu}$$

$$=\Gamma(1+\nu)\left(\frac{2^{1+\nu}}{(1+z)^{1+\nu}(1+\overline{w})^{1+\nu}}\right)\left(\frac{2(z+\overline{w})}{(1+z)(1+\overline{w})}\right)^{-1-\nu},$$

where we have used the power series expansion (2.1.43). We want to show that this expression is equal to

$$\frac{\Gamma(1+\nu)}{(z+\overline{w})^{1+\nu}},\quad z,w\in\mathbb{C}_r.$$

For $z,w\in(0,\infty)$ the result is clear. To extend it to $z,w\in\mathbb{C}_r$ we use analytic continuation as follows: Fix first $u\in(0,\infty)$. Then for $x>0$ it holds that

$$\frac{1}{(x+u)^{1+\nu}}=\left(\frac{1}{(1+x)^{1+\nu}(1+u)^{1+\nu}}\right)\left(1-\left(\frac{1-x}{1+x}\right)\left(\frac{1-u}{1+u}\right)\right)^{-1-\nu}.$$

Both sides are restricted to $z=x$ of functions analytic in z in $\mathbb{C}_r$ and therefore

$$\frac{1}{(z+u)^{1+\nu}}=\left(\frac{1}{(1+z)^{1+\nu}(1+u)^{1+\nu}}\right)\left(1-\left(\frac{1-z}{1+z}\right)\left(\frac{1-u}{1+u}\right)\right)^{-1-\nu},\quad z\in\mathbb{C}_r.$$

Fixing now $z \in \mathbb{C}_r$ we extend analytically this expression from $w = u$ to all w in $\mathbb{C}_r$ to obtain the required equality, with w instead of $\overline{w}$. Since $\mathbb{C}_r$ is invariant under complex conjugation the result follows.

$\qquad\qquad\qquad\qquad\qquad\qquad\qquad\qquad\qquad\qquad\qquad\qquad\qquad$ $\square$

Solution of Exercise 2.1.36: For $z, w \in \mathbb{D}$ we have

$$\sum_{n=0}^{\infty} |z|^n \left(\sum_{m=0}^{\infty} |w|^m |f_{n-m}| \right) \leq \sum_{n=0}^{\infty} |z|^n \left(\sum_{m=0}^{\infty} K |w|^m \right)$$

$$\leq \frac{K}{(1 - |z|)(1 - |w|)} < \infty, \quad z, w \in \mathbb{D}.$$

By a classical result on a summable family (see, e.g., [CAPB, Theorem 12.3.1, p. 454]) we can therefore compute the right-hand side of (2.1.47) with any ordering of the indices. Considering successively the cases $n = m$, $n = m+1$, $n = m+2, \ldots$, and $m = n + 1$, $m = n + 2, \ldots$ we obtain:

$$\sum_{n,m=0}^{\infty} z^n \overline{w}^m f_{n-m} = \frac{f_0}{1 - z\overline{w}} + \frac{f_1 z}{1 - z\overline{w}} + \cdots +$$

$$+ \frac{f_{-1}\overline{w}}{1 - z\overline{w}} + \frac{f_{-2}\overline{w}^2}{1 - z\overline{w}} + \cdots$$

$$= \frac{f(z) + \overline{f(w)}}{2(1 - z\overline{w})}.$$

$\qquad\qquad\qquad\qquad\qquad\qquad\qquad\qquad\qquad\qquad\qquad\qquad\qquad$ $\square$

Solution of Exercise 2.2.5: For $r > 0$ we denote by C_r the closed contour

$$C_r = [-r, r] \cup \{re^{it} \, ; \, t \in [0, \pi]\} \, .$$

Let $R > 0$ be such that

$$|z| \geq R \implies |b(z)| > |\zeta|.$$

Then, applying Rouché's theorem (Question 2.2.4) to

$$f(z) = b(z) \quad \text{and} \quad g(z) = -\zeta$$

we see that the equation (2.2.3) has exactly N solutions inside C_R for every R such that

$$R > \max_{n=1,\ldots,N} |w_n|.$$

$\qquad\qquad\qquad\qquad\qquad\qquad\qquad\qquad\qquad\qquad\qquad\qquad\qquad$ $\square$

Solution of Exercise 2.2.7: The function f is analytic in the open unit disk (since the sequence $(f_n)_{n \in \mathbb{N}}$ converges uniformly on compact subsets of $\mathbb{D}$ to f). So either f vanishes identically, or its zeros (if any) are isolated. Assume that f vanishes at z_0 and that $f \not\equiv 0$. Then, there is $r > 0$ such that $B(z_0, r) \subset \mathbb{D}$ and

$$0 < |z - z_0| < r \implies f(z) \neq 0.$$

Furthermore the sequence $(f'_n)_{n\in\mathbb{N}}$ converges uniformly on compact subsets to f'. Thus $\left(\frac{f'_n}{f_n}\right)_{n\in\mathbb{N}}$ converges uniformly to $\frac{f'}{f}$ on compact subsets of $\mathbb{D}$ where f does not vanish. Taking as compact subset of $\mathbb{D}$ the circle $|z| = \frac{r}{2}$ we have:

$$\lim_{n\to\infty} \frac{1}{2\pi i} \int_{|z|=r/2} \frac{f'_n(z)}{f_n(z)} dz = \frac{1}{2\pi i} \int_{|z|=r/2} \frac{f'(z)}{f(z)} dz.$$

By the argument principle

$$\frac{1}{2\pi i} \int_{|z|=r/2} \frac{f'_n(z)}{f_n(z)} dz = 0, \quad \forall n \in \mathbb{N},$$

while

$$\frac{1}{2\pi i} \int_{|z|=r/2} \frac{f'(z)}{f(z)} dz$$

is equal to the order of the zero z_0 and so is strictly positive. We thus have a contradiction and f does not vanish in $\mathbb{D}$, unless $f \equiv 0$ there. $\qquad\square$

Solution of Exercise 2.3.1: We first assume $|z_0| < r < |z|$. For $m \le 0$, Cauchy's theorem implies that the integral is equal to 0. For $m > 0$, Cauchy's formula applied to the function $f(\zeta) = 1/(\zeta - z)$ gives

$$\int_{C_r} \frac{d\zeta}{(\zeta - z)(\zeta - z_0)^m} = \frac{2\pi i (m-1)!(-1)^{m-1}}{(z_0 - z)^m}$$

$$= -\frac{2\pi i (m-1)!}{(z - z_0)^m}, \quad m = 1, 2, \dots \qquad (2.8.9)$$

Consider now the case $|z| < r$. For $m \le 0$, Cauchy's formula applied now to the function $f(\zeta) = (\zeta - z_0)^{-m}$ gives

$$\int_{C_r} \frac{d\zeta}{(\zeta - z)(\zeta - z_0)^m} = 2\pi i (z - z_0)^{-m}, \quad m = 0, -1, -2, \dots$$

Let now

$$h(\zeta) = \frac{1}{(\zeta - z)(\zeta - z_0)^m}.$$

When $m > 0$, the exactity relation (see [CAPB, Exercise 7.3.6, p. 326]) states that

$$\mathrm{Res}\,(h, z) + \mathrm{Res}\,(h, z_0) = 0.$$

By the residue theorem, the integral (2.3.1) is then equal to 0. $\qquad\square$

Solution of Exercise 2.3.3: Let R_1 and R_2 be such that $R < R_1 < R_2$. Then, for any given z such that $R_1 < |z| < R_2$ we have from formula (2.3.2) with $z_0 = 0$:

$$g(z) = \frac{1}{2\pi i} \int_{|\zeta|=R_2} \frac{g(\zeta)}{\zeta - z} d\zeta - \frac{1}{2\pi i} \int_{|\zeta|=R_1} \frac{g(\zeta)}{\zeta - z} d\zeta.$$

Furthermore, with $L = \lim_{z \to \infty} g(z)$, we have:

$$\lim_{R_2 \to \infty} \int_{|\zeta|=R_2} \frac{g(\zeta) - L}{\zeta - z} d\zeta = 0.$$

Hence,

$$\lim_{R_2 \to \infty} \int_{|\zeta|=R_2} \frac{g(\zeta)}{\zeta - z} d\zeta = \lim_{R_2 \to \infty} \int_{|\zeta|=R_2} \frac{g(\zeta) - L}{\zeta - z} d\zeta + \lim_{R_2 \to \infty} \int_{|\zeta|=R_2} \frac{L}{\zeta - z} d\zeta$$

$$= 2\pi i L.$$

Thus formula (2.3.2) (with $z_0 = 0$) reduces to

$$g(z) = L - \underbrace{\frac{1}{2\pi i} \int_{|\zeta|=R_1} \frac{g(\zeta)}{\zeta - z} d\zeta}_{\text{principal part in the Laurent expansion}},$$

from which the result follows.

$\square$

Solution of Exercise 2.3.4: We first note that the Laurent expansion is of the asserted type with $b_0 = \lim_{z \to \infty} g(z)$. See the previous exercise.

(1) Let $R_1 \in (R, 1)$. The formula for the coefficients of the singular part of the Laurent expansion gives

$$b_n = \frac{1}{2\pi i} \int_{|\zeta|=R_1} \zeta^{n-1} g(\zeta) d\zeta, \quad n = 1, 2, \ldots,$$

and in particular, with $K_{R_1} = \max_{|\zeta|=R_1} |g(\zeta)|$,

$$|b_n| \leq K_{R_1} R_1^n, \quad n = 1, 2, \ldots$$

Let $R_2 \in (R_1, 1)$. The formula for the coefficients of a Taylor series gives

$$a_n = \frac{1}{2\pi i} \int_{|\zeta|=R_2} \frac{f(\zeta)}{\zeta^{n+1}} d\zeta, \quad n = 0, 2, \ldots,$$

and in particular, with $K_{R_2} = \max_{|\zeta|=R_2} |f(\zeta)|$,

$$|a_n| \leq \frac{K_{R_2}}{R_2^n}, \quad n = 0, 2, \ldots$$

Since $\frac{R_1}{R_2} < 1$ we have that

$$\left| \sum_{n=0}^{\infty} a_n b_n \right| \leq K_{R_1} K_{R_2} \sum_{n=0}^{\infty} \left(\frac{R_1}{R_2} \right)^n < \infty.$$

2.8. *Solutions of the exercises*

(2) Set

$$g(z) = \sum_{m=0}^{\infty} \frac{b_m}{z^m}. \qquad (2.8.10)$$

The power series defining g converges in $|z| > R$, and the dominated convergence theorem allows us to prove that, for any $R_1 \in (R, 1)$:

$$\frac{1}{2\pi i} \int_{|z|=R_1} \frac{f(z)g(z)}{z} dz = \sum_{n=0}^{\infty} a_n \left(\frac{1}{2\pi i} \int_{|z|=R_1} z^{n-1} g(z) dz \right)$$

$$= \sum_{n=0}^{\infty} a_n b_n. \qquad \square$$

Solution of Exercise 2.3.5:

(1) Let $g(z) = \sum_{n=1}^{\infty} g_n z^n$ denote the power series expansion of g centered at the origin. Then

$$g \left(\frac{1}{z-v} \right) = \sum_{n=1}^{\infty} \frac{g_n}{(z-v)^n}, \quad z \neq v.$$

Then, with r small enough (i.e., such that $u \notin \overline{B(v,r)}$),

$$\mathrm{Res} \left(\frac{g\left(\frac{1}{z-v}\right)}{z-u}, v \right) = \frac{1}{2\pi i} \int_{|\zeta-v|=r} \frac{g\left(\frac{1}{\zeta-v}\right)}{\zeta-u} d\zeta$$

$$= \sum_{n=1}^{\infty} g_n \frac{1}{2\pi i} \int_{|\zeta-v|=r} \frac{d\zeta}{(\zeta-v)^n(\zeta-u)}$$

$$= \sum_{n=1}^{\infty} g_n \mathrm{Res} \left(\frac{1}{(\zeta-v)^n(\zeta-u)}, v \right)$$

$$= -\sum_{n=1}^{\infty} g_n \underbrace{\mathrm{Res} \left(\frac{1}{(\zeta-v)^n(\zeta-u)}, u \right)}_{\text{by the exactity relation (see [CAPB, Exercise 7.3.6, p. 326]),}}$$

$$= -\sum_{n=1}^{\infty} g_n \frac{1}{(u-v)^n}$$

$$= -g \left(\frac{1}{u-v} \right).$$

(2) From the residue theorem we obtain the formula (see [269, (4.1), p. 307])

$$\frac{1}{2\pi i} \int_{C_R} \frac{F(s)}{s-z} ds = F(z) + \sum_{\omega \text{ singularity of } F} H_\omega \left(\frac{1}{z-\omega} \right),$$

where the H_ω are entire functions vanishing at the origin.

(3) We have

$$\frac{1}{2\pi i}\int_{C_R}\frac{F(s)}{s-z}ds - \frac{1}{2\pi i}\int_{C_R}\frac{z^k F(s)}{s^k(s-z)}ds = \frac{1}{2\pi i}\int_{C_R}\left(\sum_{a=1}^{k}\frac{z^{a-1}}{s^a}\right)F(s)ds$$

$$= \sum_{a=1}^{k}z^{a-1}\left(\frac{1}{2\pi i}\int_{C_R}\frac{F(s)}{s^a}ds\right)$$

since

$$\frac{1-\frac{z^k}{s^k}}{s-z} = \frac{1}{s^k}\left(s^{k-1}+s^{k-2}z+\cdots+z^{k-1}\right) = \frac{\sum_{a=1}^{k}s^{k-a}z^{a-1}}{s^k}.$$

As mentioned before the statement of the exercise, the proof of (4) is omitted. □

Solution of Exercise 2.3.6:

(1) We consider the Laurent expansion of f in A_R (see [CAPB, p. 317]):

$$f(z) = \sum_{n=0}^{\infty}a_n z^n + \sum_{n=1}^{\infty}\frac{b_n}{z^n}.$$

For $z, w \in A_R$ we have:

$$K(z,w) \stackrel{\text{def.}}{=} \frac{f(w)}{z} - \frac{f(z)}{w} = \sum_{n=0}^{\infty}a_n\left(\frac{w^n}{z}-\frac{z^n}{w}\right) + \sum_{n=2}^{\infty}b_n\left(\frac{1}{w^n z}-\frac{1}{z^n w}\right).$$

Let now $(\gamma_1(t), t \in [0,1])$ and $(\gamma_2(t), t \in [0,1])$ be parametrizations of C_1 and C_2. We have (recall that γ_1 and γ_2 are ruled, and that there may be a finite number of points at which the components of γ_1 and γ_2 have different but finite derivatives from the left and from the right; these points do not affect the integrals computed below):

$$\iint_{[0,1]^2}K(\gamma_1(t),\gamma_2(s))\gamma_1'(t)\gamma_2'(s)dtds$$

$$= \sum_{n=0}^{\infty}a_n\iint_{[0,1]^2}\left(\frac{\gamma_2(s)^n}{\gamma_1(t)}-\frac{\gamma_1(t)^n}{\gamma_2(s)}\right)\gamma_1'(t)\gamma_2'(s)dtds$$

$$+ \sum_{n=2}^{\infty}b_n\iint_{[0,1]^2}\left(\frac{1}{\gamma_2(s)^n\gamma_1(t)}-\frac{1}{\gamma_1(t)^n\gamma_2(s)}\right)\gamma_1'(t)\gamma_2'(s)dtds$$

$$= \sum_{n=0}^{\infty}a_n\left\{\left(\int_{C_2}w^n dw\right)\left(\int_{C_1}\frac{dz}{z}\right)-\left(\int_{C_1}z^n dz\right)\left(\int_{C_2}\frac{dw}{w}\right)\right\}$$

$$+ \sum_{n=2}^{\infty}b_n\left\{\left(\int_{C_2}\frac{dw}{w^n}\right)\left(\int_{C_1}\frac{dz}{z}\right)-\left(\int_{C_1}\frac{dz}{z^n}\right)\left(\int_{C_2}\frac{dw}{w}\right)\right\}$$

$$\stackrel{.}{=} 0,$$

since the second sum begins at $n = 2$, and since, thanks to Cauchy's theorem, the various integrals

$$\int_{C_2} w^m dw \quad \text{and} \quad \int_{C_1} z^m dz$$

vanish for $m \in \mathbb{Z}$ different from -1.

On the other hand, from the definition of K we have:

$$\iint_{[0,1]^2} K(\gamma_1(t), \gamma_2(s))\gamma_1'(t)\gamma_2'(s)dtds$$

$$= \left(\int_{[0,1]} f(\gamma_2(s))\gamma_2'(s)ds \right) \left(\int_{[0,1]} \frac{\gamma_1'(t)dt}{\gamma_1(t)} \right)$$

$$- \left(\int_{[0,1]} f(\gamma_1(t))\gamma_1'(t)dt \right) \left(\int_{[0,1]} \frac{\gamma_2'(s)ds}{\gamma_2(s)} \right)$$

$$= \left(\int_{C_2} f(w)dw \right) (\mathrm{Ind}_{C_1}(0)) - \left(\int_{C_1} f(z)dz \right) (\mathrm{Ind}_{C_2}(0)),$$

and hence the result. Since the Laurent expansion converges absolutely and uniformly on compact subsets of A_R the various interchange of integrals and infinite sums in the above equations are legitimate, thanks to Weierstrass' theorem (see [CAPB, p. 456] for the latter).

(2) As mentioned in the hint, the second formula is a special case of the first one with f'/f instead of f. □

Solution of Exercise 2.3.7:

(1) This is a direct application of Rouché's theorem (Theorem 2.2.4), as in Exercise 2.2.5, with

$$f(z) = b(z) \quad \text{and} \quad g(z) = -\zeta$$

and with contour the circle centered at the origin and of radius $R \in (0,1)$ such that

$$R > \max_{n=1,\dots,N} |w_n|. \tag{2.8.11}$$

For more on this question, see Remark 2.8.2 after the solution of the exercise.

(2) When $s \in \mathbb{D}$ is a simple zero of the equation $b(z) = \zeta$, applying the well-known formula

$$\mathrm{Res}\left(\frac{n}{m}, z_0 \right) = \frac{n(z_0)}{m'(z_0)} \tag{2.8.12}$$

where n and m are analytic in a neighborhood of the point z_0, the latter being a simple zero of m (see, e.g., [CAPB, (7.3.2), p. 325]), gives

$$\mathrm{Res}\left(\frac{f(z)b'(z)}{b(z) - \zeta}, s \right) = \frac{f(z)b'(z)}{b'(z)}\Big|_{z=s} = f(s).$$

(3) We first note that there are only finitely many points, independent of $z \in \mathbb{D}$, for which equation (2.3.13) possibly has nonsimple roots (namely these points are among the roots of the equation $b'(u) = 0$). We denote by E the set of these roots.

Let now R as in (2.8.11). Let $z \in \mathbb{D}$ such that the equation $b(u) = z$ has N different roots, say $u_1(z), \ldots, u_N(z)$. By the preceding item we have

$$\frac{1}{2\pi i} \int_{|\zeta|=R} \frac{f(u)b'(u)}{b(u) - z} du = \sum_{j=1}^{N} \operatorname{Res}\left(\frac{f(u)b'(u)}{b(u) - z}, u_n(z)\right)$$

$$= \sum_{n=1}^{N} f(u_n(z)), \quad z \in B(0, R) \setminus E.$$

The function

$$\frac{1}{2\pi i} \int_{|\zeta|=R} \frac{f(u)b'(u)}{b(u) - z} du$$

does not depend on $r > R$ and defines a function analytic in $\mathbb{D}$. It follows that the left side of (2.3.14) has an analytic extension to $\mathbb{D}$.

(4) Let $f(z) = \sum_{n=0}^{\infty} f_n z^n$ be the Taylor expansion of the function f at the origin. We have:

$$\frac{1}{N} \sum_{\substack{u \in \mathbb{D} \\ u^N = z}} f(u) = \sum_{n=0}^{\infty} f_n \left(\frac{1}{N} \sum_{\substack{u \in \mathbb{D} \\ u^N = z}} u^n\right)$$

$$= \sum_{k=0}^{\infty} f_{kN} z^k,$$

since

$$\frac{1}{N} \sum_{\substack{u \in \mathbb{D} \\ u^N = z}} u^n = \begin{cases} z^k, & \text{if } n = kN, \\ 0, & \text{otherwise.} \end{cases}$$

See [CAPB, Exercise 6.2.9, p. 280].

$\square$

Remark 2.8.2. The first item can also be solved in the following ways, which are longer, but of independent interest. First one can use the fundamental theorem of algebra. Fix $z \in \mathbb{D}$ and let

$$p(\zeta) = \prod_{n=1}^{N}(\zeta - z_n) - z \prod_{n=1}^{N}(1 - \zeta \overline{z_n}). \tag{2.8.13}$$

Since $1 - z \prod_{n=1}^{N}(-\overline{z_n}) \neq 0$, we see that p is a polynomial of degree N in the variable ζ. By the fundamental theorem of algebra. the polynomial equation

$$\prod_{n=1}^{N}(\zeta - z_n) - z \prod_{n=1}^{N}(1 - \zeta \overline{z_n}) = 0 \tag{2.8.14}$$

has N solutions, counting multiplicities. When $z_n \neq 0$, the number $\overline{z_n}^{-1}$ is not a solution of $p(\zeta) = 0$, and therefore equation (2.8.14) is equivalent to $b(\zeta) = 0$, and this last equation has N solutions, counting multiplicities. All the roots lie in $\mathbb{D}$ since $|b(\zeta)| < 1$ if and only if $|\zeta| < 1$. Failing to notice that, one could use (as in the presented proof) Rouché's theorem to get to the same conclusion, or, equivalently, the argument principle (see Theorem 2.2.3 above). Indeed, using Weierstrass theorem to justify the interchange of sum and integration, we can write

$$\frac{1}{2\pi i} \int_{|\zeta|=1} \frac{(b(\zeta) - z)'}{b(\zeta) - z} d\zeta = \frac{1}{2\pi i} \int_{|\zeta|=1} \frac{b'(\zeta)}{b(\zeta) - z} d\zeta$$

$$= \frac{1}{2\pi i} \int_{|\zeta|=1} \frac{b'(\zeta)}{b(\zeta)(1 - \frac{z}{b(\zeta)})} d\zeta$$

$$= \frac{1}{2\pi i} \int_{|\zeta|=1} \frac{b'(\zeta)}{b(\zeta)} \left(\sum_{j=0}^{\infty} \left(\frac{z}{b(\zeta)} \right)^j \right) d\zeta$$

$$= \sum_{j=0}^{\infty} \frac{1}{2\pi i} z^j \int_{|\zeta|=1} \frac{b'(\zeta)}{b(\zeta)^{j+1}} d\zeta.$$

For $j = 1, 2, \ldots$ each of the integrals $\int_{|\zeta|=1} \frac{b'(\zeta)}{b(\zeta)^{j+1}} d\zeta$ is equal to 0 since

$$\frac{b'(\zeta)}{b(\zeta)^{j+1}} = \frac{1}{-j} \left(\frac{1}{b(\zeta)^j} \right)',$$

while for $j = 0$, the argument principle gives

$$\frac{1}{2\pi i} \int_{|\zeta|=1} \frac{b'(\zeta)}{b(\zeta)^{j+1}} d\zeta = \frac{1}{2\pi i} \int_{|\zeta|=1} \frac{b'(\zeta)}{b(\zeta)} d\zeta = N.$$

Solution of Exercise 2.4.2: It suffices to take $\Omega = \mathbb{C} \setminus (-\infty, 0]$. The points $z_{\pm} = -1 \pm i \in \Omega$ but the interval $[-1 - i, -1 + i]$ meets the negative axis at the point -1, and thus is not included in Ω. Hence Ω is not convex. It is star-shaped with respect to any point $x_0 \neq 0$ on the positive real axis $\qquad \square$

Solution of Exercise 2.4.3: This is a well-known fact from a first course in complex variables, and we recall briefly the argument. Let $\Omega \subset \mathbb{C}$ be open and star-shaped, and let $z_0 \in \Omega$ be such that (2.4.17) holds. Define

$$F(z) = \int_{[z_0, z]} f(s) ds = (z - z_0) \int_0^1 f(z_0 + t(z - z_0)) dt, \quad z \in \Omega.$$

Let $\epsilon > 0$ be such that $B(z, \epsilon) \subset \Omega$. We have for $w \in B(z, \epsilon) \setminus \{z\}$:

$$\frac{F(w) - F(z)}{w - z} - f(z) = \frac{\int_{[z_0, w]} f(s)ds - \int_{[z_0, z]} f(s)ds}{w - z} - f(z)$$

$$= \frac{\int_{[z, w]} f(s)ds}{z - w} - f(z),$$

where we have used the Cauchy–Goursat theorem (recall that $[z, w] \subset \Omega$),

$$= \int_0^1 (f(z + t(w - z)) - f(z))dt,$$

and this last expression tends to 0 as w tends to z, as follows from the continuity of f.

$\square$

Solution of Exercise 2.4.5: Recall that we denote by $\mathbb{C}_+$ the open upper half-plane. The set

$$\mathbb{D}_+ = \mathbb{C}_+ \cap \mathbb{D}$$

is convex, and in particular star-shaped. Its image under the map $-1/z$ is equal to $\mathbb{C}_+ \setminus \overline{\mathbb{D}}$, which is not star-shaped.

$\square$

In relation with the previous exercise, we note that the application

$$\varphi(z) = \frac{z - i(z^2 + 1)}{z + i(z^2 + 1)} = \frac{1 - i(z + \frac{1}{z})}{1 + i(z + \frac{1}{z})}$$

maps conformally $\mathbb{D}_+ = \mathbb{D} \cap \mathbb{C}_+$ onto $\mathbb{D}$. See [CABP, Exercise 10.3.6, p. 425]. Since the map $z \mapsto -\frac{1}{z}$ sends conformally $\mathbb{C}_+ \setminus \overline{\mathbb{D}}$ onto $\mathbb{D}_+ = \mathbb{D} \cap \mathbb{C}_+$, and since $z \mapsto \frac{1 + iz}{1 - iz}$ is conformal from $\mathbb{D}$ onto $\mathbb{C}_+$, we see that the map $f(z) = z + \frac{1}{z}$ maps conformally $\mathbb{C}_+ \setminus \overline{\mathbb{D}}$ onto $\mathbb{C}_+$, and $\mathbb{C}_+ \setminus \overline{\mathbb{D}}$ is not star-shaped.

Solution of Exercise 2.4.6: Let φ denote a conformal map from Ω onto Ω_1, and let F be analytic in Ω_1. The function $F \circ \varphi$ is analytic and does not vanish in Ω. So there exists a function g analytic in Ω and such that

$$F(\varphi(z)) = e^{g(z)}, \quad z \in \Omega.$$

Thus

$$F(z) = e^{g(\varphi^{-1}(z))}, \quad z \in \Omega_1,$$

and this finishes the proof since $g \circ \varphi^{-1}$ is analytic in Ω_1.

$\square$

Solution of Exercise 2.4.8: Let Ω_1 be homeomorphic to Ω, and let φ be a homeomorphism from Ω onto Ω_1. Let g be a non-vanishing continuous function on Ω_1.

Then, $f = g \circ \varphi$ is a non-vanishing function on Ω. By hypothesis, there exists a function h continuous on Ω and such that

$$e^{h(z)} = g(\varphi(z)), \quad \forall z \in \Omega.$$

Replacing z by $\varphi^{-1}(w)$ (with $w \in \Omega_1$) we get

$$e^{h(\varphi^{-1}(w))} = g(w), \quad \forall w \in \Omega_1,$$

which ends the proof since $h \circ \varphi^{-1}$ is continuous on Ω_1. $\square$

Solution of Exercise 2.4.9: Assume that (1) is in force, and let f be analytic and not vanishing in Ω. Let $z_0 \in \Omega$. Since $f(z_0) \neq 0$, there exists $K_0 \in \mathbb{C}$ such that $f(z_0) = e^{K_0}$. Furthermore, the function $g = \frac{f'}{f}$ is analytic in Ω. Let G be its primitive which takes the value K_0 at the point z_0. Then, the function $z \mapsto f(z)e^{-G(z)}$ has value 1 at z_0 and has identically vanishing derivative in Ω. Thus, it holds that

$$f(z) = e^{G(z)}, \quad z \in \Omega,$$

and so (2) is in force.

Still assuming that (1) holds, let u be harmonic in Ω and define

$$g(z) = \frac{\partial u}{\partial x}(x, y) - i \frac{\partial u}{\partial y}(x, y).$$

The function g is analytic in Ω since u is harmonic (use for instance that the Cauchy–Riemann equations hold and that the real and imaginary parts of g are differentiable, since u is harmonic). Let $G = U + iV$ be a primitive of g in Ω. Then,

$$G'(z) = \frac{\partial U}{\partial x}(x, y) - i \frac{\partial U}{\partial y}(x, y),$$

so that

$$\frac{\partial U}{\partial x}(x, y) = \frac{\partial u}{\partial x}(x, y) \quad \text{and} \quad \frac{\partial U}{\partial y}(x, y) = \frac{\partial u}{\partial y}(x, y).$$

Thus $U(x, y) = u(x, y) + K$ for some constant $K \in \mathbb{R}$, and $V(x, y)$ is a harmonic conjugate to u (it is unique up to a real constant).

Assume now that (3) is in force, and let $f = u + iv$ be analytic and not vanishing in Ω. Then, the function $a = \frac{1}{2}\ln(u^2 + v^2)$ is harmonic (see for instance [CAPB, pp. 396 and 404] for a proof). Let b denote a harmonic conjugate of a in

Ω. Using the Cauchy–Riemann equations one sees that

$$
\begin{aligned}
\frac{\partial a}{\partial x} - i\frac{\partial a}{\partial y} &= \frac{\frac{\partial u}{\partial x}u + \frac{\partial v}{\partial x}v}{u^2 + v^2} - i\frac{\frac{\partial u}{\partial y}u + \frac{\partial v}{\partial y}v}{u^2 + v^2} \\
&= \frac{\frac{\partial u}{\partial x}u - \frac{\partial u}{\partial y}v - i\frac{\partial u}{\partial y}u - i\frac{\partial u}{\partial x}v}{u^2 + v^2} \\
&= \left(\frac{\partial u}{\partial x} - i\frac{\partial u}{\partial y}\right)\frac{u - iv}{u^2 + v^2} \\
&= \frac{\frac{\partial u}{\partial x} - i\frac{\partial u}{\partial y}}{u + iv} \\
&= \frac{f'}{f}.
\end{aligned}
\tag{2.8.15}
$$

Fix $z_0 \in \Omega$. There is a constant $K \in \mathbb{C}$ such that the function

$$
h(z) = f(z)e^{-(a(x,y)+ib(x,y)+K)}
$$

is equal to 1 at z_0. Furthermore,

$$
\begin{aligned}
h'(z) &= \left(f'(z) - f(z)\left(\frac{\partial a}{\partial x}(x,y) - i\frac{\partial a}{\partial y}(x,y)\right)\right)e^{-(a(x,y)+ib(x,y)+K)} \\
&= 0,
\end{aligned}
$$

in view of (2.8.15). So $h(z) \equiv 1$ and $a + ib + K$ is an analytic logarithm of f in Ω.

$\square$

Solution of Exercise 2.5.2: We assume that f is defined in a pointed neighborhood $B(z_0, R) \setminus \{z_0\}$ of z_0, and that for some $r \in (0, R)$ the image of $B(z_0, r) \setminus \{z_0\}$ is not dense in $\mathbb{C}$. There exist then $w_0 \in \mathbb{C}$ and $\rho_0 > 0$ such that

$$
f((B(z_0, r) \setminus \{z_0\}) \cap B(w_0, \rho_0) = \emptyset.
$$

The function $\frac{1}{f(z)-w_0}$ is bounded in $B(z_0, r) \setminus \{z_0\}$. By Riemann's removable singularity theorem

$$
\ell \stackrel{\text{def.}}{=} \lim_{z \to z_0} \frac{1}{f(z) - w_0}
$$

exists, and the function

$$
g(z) = \begin{cases} \frac{1}{f(z)-w_0}, & z \neq z_0, \\ \ell, & z = z_0, \end{cases}
$$

is analytic in $B(z_0, r)$. It follows that

$$
\lim_{z \to z_0} f(z)
$$

exists (and is ∞ when $\ell = 0$). Therefore z_0 is not an essential singularity.

2.8. Solutions of the exercises

Conversely, if z_0 is not an essential singularity, then $\lim_{z \to z_0} f(z)$ exists (if z_0 is removable), or is ∞ (if z_0 is a pole). In either case, the image of $B(z_0, r) \setminus \{z_0\}$ is not dense in $\mathbb{C}$ for r small enough. $\quad\square$

Solution of Exercise 2.5.3: We follow an argument in [95, p. 182] and use the Casorati–Weierstrass theorem. Let $z_0 \in \mathbb{D} \setminus \{0\}$, and let $r > 0$ be such that $0 \notin B(z_0, r)$. The image of $B(z_0, r)$ is open (recall Theorem 2.2.1) and so there exists $\epsilon > 0$ such that

$$B(f(z_0), \epsilon) \subset f(B(z_0, r)) \subset f(\mathbb{D} \setminus \{0\}).$$

On the other hand, let ρ be such that $(B(0, \rho) \setminus \{0\}) \cap B(z_0, r) = \emptyset$. Since f is conformal,

$$f(B(0, \rho) \setminus \{0\}) \cap f(B(z_0, r)) = \emptyset$$

and in particular

$$f(B(0, \rho) \setminus \{0\}) \cap B(f(z_0), \epsilon) = \emptyset.$$

Thus, $f(B(0, \rho) \setminus \{0\})$ is not dense in $\mathbb{C}$. The Casorati–Weierstrass theorem (see Exercise 2.5.2) implies then that z_0 is not an essential singularity, that is z_0 is either a pole or a removable singularity. $\quad\square$

Solution of Exercise 2.5.4: By the previous exercise, the points $z_1, \ldots, z_N$ are removable singularities or poles. Furthermore, the point ∞ is also either a removable singularity or a pole. Otherwise, the function $f(1/z)$ would have an essential singularity at the origin, and so will not be one-to-one, contradicting the fact that f is one-to-one. The function f has only poles as possible singularities, including at the point ∞, and so is a rational function. Let thus

$$f(z) = \frac{p(z)}{q(z)},$$

where p and q are polynomials, not both constant, and with no common zeros. We want to show that p and q have at most degree 1. Assume that this is not the case, and let $c \in \mathbb{C} \setminus \{0\}$ be such that the degree of the polynomial $cq(z) - p(z)$ is strictly bigger than 1. By the fundamental theorem of algebra, the equation $cq(z) = p(z)$ has then more than one solution. These solutions are not poles of f since p and q have no common zeros. Thus the equation

$$f(z) = c$$

has more than one solution and f will not be one-to-one. $\quad\square$

Solution of Exercise 2.5.5:

(1) The point $z = 0$ is an isolated singularity of the function $g(z) = f(1/z)$. From the hypothesis on f the function g is one-to-one. Exercise 2.5.3 asserts that 0 is not an essential singularity of g, and hence ∞ is not an essential singularity of f.

(2) In view of (1) the origin is either a removable singularity or a pole (say of order N) of the function $f(1/z)$. Let

$$f(1/z) = \sum_{n=0}^{\infty} a_n z^n + \frac{b_1}{z} + \cdots + \frac{b_N}{z^N}$$

be the Laurent expansion of $f(1/z)$ at the origin. Then

$$f(z) = b_1 z + \cdots + b_N z^N + \sum_{n=0}^{\infty} \frac{a_n}{z^n}, \quad z \neq 0.$$

Since f is entire we have $a_1 = a_2 = \cdots = 0$, and hence f is polynomial. Since f is one-to-one, it is a polynomial of degree 1 and we have $N = 1$ (and hence $b_1 \neq 0$).
□

Solution of Exercise 2.6.2: With P and Q as in the hint we have

$$\frac{\partial Q}{\partial x} = -\frac{\partial v}{\partial x}\frac{\partial u}{\partial x} - v\frac{\partial^2 u}{\partial x^2}$$

$$\frac{\partial P}{\partial y} = \frac{\partial v}{\partial y}\frac{\partial u}{\partial y} + v\frac{\partial^2 u}{\partial y^2}.$$

Since $\Delta u = 0$ we have:

$$\frac{\partial Q}{\partial x} - \frac{\partial P}{\partial y} = -\frac{\partial v}{\partial x}\frac{\partial u}{\partial x} - \frac{\partial v}{\partial y}\frac{\partial u}{\partial y},$$

and Green's theorem gives

$$\iint_R \left(\frac{\partial u}{\partial x}\frac{\partial v}{\partial x} + \frac{\partial u}{\partial y}\frac{\partial v}{\partial y}\right)(x,y)dxdy = -\int_{\partial R}(P(x,y)dx + Q(x,y)dy)$$

$$= -\int_{\partial R} v(x,y)\left(\frac{\partial u}{\partial y}(x,y)dx - \frac{\partial u}{\partial x}(x,y)dy\right).$$

This ends the proof in view of the definition (2.6.3) of $\frac{\partial u}{\partial n}$.
□

Solution of Exercise 2.6.3: The function is harmonic in $\mathbb{D}$ (and in fact in $\mathbb{C} \setminus \{1\}$) since it is the real part of a function analytic there.

Let now $t \in (0, 2\pi)$ (and thus $e^{it} \neq 1$). Then

$$\frac{1 + e^{it}}{1 - e^{it}} = \frac{e^{-it/2} + e^{it/2}}{e^{-it/2} - e^{it/2}} = \frac{\cos(t/2)}{-i\sin(t/2)},$$

and so

$$\left(\frac{1 + e^{it}}{1 - e^{it}}\right)^2 = -\cot^2(t/2),$$

which belongs to $\mathbb{R}$.
□

2.8. *Solutions of the exercises*

Solution of Exercise 2.6.4: We assume $z_0 = 0$. For $z \neq 0$ we have

$$|F(z)|^{-1} = e^{-u(x,y)}$$
$$= e^{-(u(x,y)+\ln|z|)+\ln|z|}$$
$$= |z|e^{-(u(x,y)+\ln|z|)}$$
$$\to 0, \quad \text{as} \quad z \to 0,$$

since $\lim_{z\to 0} (u(x,y) + \ln|z|)$ exists. Thus, F is continuous at the origin. Since it is analytic in $\Omega \setminus \{0\}$, Riemann's removable singularity theorem insures that F is analytic in all Ω. $\qquad\square$

Solution of Exercise 2.7.1: The function $f(z) = \frac{1}{\sin\frac{1}{z}}$ is analytic in $\mathbb{C}\setminus\{0, \frac{1}{\pi}, \frac{1}{2\pi}, \dots\}$. Therefore 0 is not an isolated singularity of f, and f is not meromorphic in $\mathbb{C}$. $\qquad\square$

Solution of Exercise 2.7.3:

(1) We will use the inequality, valid for every choice of $a, b \in \mathbb{C}$:

$$|e^b - e^a| \leq e^{|b-a|}|b - a|. \qquad (2.8.16)$$

For completeness we recall the proof. Let

$$\gamma(t) = a + t(b - a), \quad t \in [0, 1]$$

denote a parametrization of the closed interval $[a, b]$. We have

$$e^b - e^a = \int_0^1 e^{\gamma(t)}\gamma'(t)dt$$
$$= (b - a)\int_0^1 e^{t(b-a)}dt,$$

and hence the result by taking absolute values.

Let now $\alpha \in \mathbb{C}$, and first assume that $\alpha \neq 0$. There exists $\beta \in \mathbb{C}$ such that $\alpha = e^\beta$. By the Casorati–Weierstrass theorem there exists a sequence of points $\omega_0, \omega_1, \dots \in V$ such that

$$\lim_{n\to\infty} |f(\omega_n) - \beta| = 0.$$

Using (2.8.16) we have:

$$|e^{f(\omega_n)} - \alpha| \leq M \cdot |f(\omega_n) - \beta|,$$

where M is a constant such that

$$e^{|f(\omega_n)-\beta|} \leq M, \quad \forall n \in \mathbb{N}_0.$$

Let now $\alpha = 0$. By the Casorati–Weierstrass theorem, there exists a sequence of points $\omega_0, \omega_1, \ldots \in V$ such that

$$\lim_{n\to\infty} \operatorname{Re} f(\omega_n) = -\infty,$$

and so

$$\lim_{n\to\infty} \left| e^{f(\omega_n)} - 0 \right| = 0.$$

Thus, the image of V under e^f is dense in $\mathbb{C}$, and so z_0 is an essential singularity of e^f.

(2) Essential singularities are isolated by definition. The example $f(z) = \ln z$ and $V = \mathbb{C} \setminus (-\infty, 0]$ shows that one can obtain a removable singularity after applying the exponential function to a function with non-isolated singularities. □

Part II

Topology and Functional Analysis

Chapter 3

Topological Spaces

> *Les premiers faits topologiques rencontrés en théorie des fonctions analytiques ont apparu avec l'introduction des surfaces de Riemann.*
>
> S. Stoïlow, Leçons sur les principes topologiques de la théorie des fonctions analytiques, [299, Préface, p. v].

Topological and metric spaces permeate complex analysis in numerous places. For instance, the space $\mathcal{H}(\Omega)$ of functions analytic in some open domain Ω is a Fréchet space (and moreover, a Montel space). Hilbert and Banach spaces of analytic functions also play an important role. These examples are in fact instances of topological vector spaces. As another example, Riemann surfaces are in particular analytic manifolds, and as such are special cases of topological spaces, and in fact of metric spaces. It is therefore important to have a good command of elementary topological notions to understand forthcoming exercises. In this chapter we review various topological facts needed in the sequel, and discuss various families of metric spaces. Among the topics considered we mention in particular the quotient topology and the notions of manifold and surface. Most of the exercises will be used later in the text, but not all of them have a direct complex analysis flavor. Still, one already meets future key players, such as the Riemann sphere (see Exercise 3.8.8), lattices (see Exercise 3.8.30), and a characterization of the boundary of a simply connected set (see Exercise 8.1.2). We conclude this introduction with the following question: Can one characterize analytic functions purely in topological terms? Surprisingly, the answer is yes, as was proved by S. Stoïlow [299]. We discuss his results in some of the exercises.

3.1 Topological spaces

We assume familiarity with topological spaces, but briefly review some definitions and basic results in this section. For more information and complements we send the reader to [78, 99].

Definition 3.1.1. Let X be a non-empty set and $\mathcal{O}$ be a subset of $\mathcal{P}(X)$. The family $\mathcal{O}$ is called a topology if:

1. For every family $(O_i)_{i \in I}$ of elements in $\mathcal{O}$, the union $\cup_{i \in I} O_i$ belongs to $\mathcal{O}$.
2. For every finite family $O_{i_1}, \ldots, O_{i_N}$ of elements in $\mathcal{O}$, the intersection $\cap_{j=1}^{N} O_{i_j}$ belongs to $\mathcal{O}$.
3. The sets X and $\emptyset$ belong to $\mathcal{O}$.

The elements of a topology $\mathcal{O}$ are called the open sets of X (in the given topology), and the pair $(X, \mathcal{O})$ (or just X, if $\mathcal{O}$ is understood from the context) is called a topological space. A set F is called closed (in the given topology) if its complement in X is an open set. Open and closed sets will be given characterizations in Exercises 3.1.8 and 3.1.13 below. A third characterization in terms of semi-continuous functions is given in Exercise 3.4.9.

Recall that the set of real numbers (the real line) is denoted by $\mathbb{R}$. Furthermore, $[-\infty, +\infty]$ (also denoted by $\overline{\mathbb{R}}$; the overline refers to a closure, see (3.1.3)) denotes the extended real line, that is the real line to which have been added two points denoted by $\pm\infty$ (and which do not belong to $\mathbb{R}$). We now present two topologies on $\mathbb{R}$ and recall the standard topology of $\overline{\mathbb{R}}$.

Exercise 3.1.2. *Let $X = \mathbb{R}$ and denote by $\mathcal{O} \subset \mathcal{P}(\mathbb{R})$ the family which consists of the empty set and of all the non-empty subsets O of $\mathbb{R}$ with the following property:*

$$\forall x \in O, \ \exists r_x > 0 \ such \ that \ (x - r_x, x + r_x) \subset O.$$

(1) *Show that $\mathcal{O}$ is a topology.*
(2) *Show that every non-empty open subset of $\mathbb{R}$ in this topology can be written as a countable union of disjoint open intervals, that is, of sets of the form (a, b), with $a < b$, or $(-\infty, a)$, or (a, ∞).*

The next question and its follow up, Question 3.1.9, are [Exercise 13.1.7, p. 465, CAPB], to which we refer the reader for a proof.

Question 3.1.3 (see [265, p. 7]). *Define a set O to be open in $[-\infty, +\infty]$ if it is the empty set or if it is a (not necessarily disjoint) union of sets of the following forms:*

(i) *O open in $\mathbb{R}$.*
(ii) *$\{-\infty\} \cup (-\infty, a)$ where $a \in \mathbb{R}$.*
(iii) *$\{+\infty\} \cup (b, \infty)$.*

Show that this defines a topology.

3.1. *Topological spaces*

Exercise 3.1.4. *Consider in* $\mathbb{R}$ *the family* $\mathcal{T}$, *which consists of the empty set, to-gether with all sets of the form* $\mathbb{R} \setminus U$, *where* U *has a finite number of elements or* $U = \emptyset$.

(1) *Show that this defines a topology.*
(2) *Show that two non-empty open sets always intersect.*

The topology in the preceding exercise is called the Zariski topology. It plays an important role in algebraic geometry.

Let X be a non-empty set. The sets $\{\emptyset, X\}$ and $\mathcal{P}(X)$ are clearly topologies, and any topology $\mathcal{O}$ satisfies

$$\{\emptyset, X\} \subset \mathcal{O} \subset \mathcal{P}(X).$$

The topology $\{\emptyset, X\}$ is called the coarse topology, and $\mathcal{P}(X)$ is called the discrete topology. Let $\mathcal{O}_1$ and $\mathcal{O}_2$ be two topologies on X. Then $\mathcal{O}_1$ is coarser (or weaker) (resp. finer, stronger) than $\mathcal{O}_2$ if $\mathcal{O}_1 \subset \mathcal{O}_2$ (resp. $\mathcal{O}_2 \subset \mathcal{O}_1$). See for instance Exercise 4.2.3 for an important example.

Exercise 3.1.5. *Given a set* X *and* $\mathscr{B} \subset \mathcal{P}(X)$, *show that there is a unique topology* $\mathcal{O}(\mathscr{B})$ *on* X *with the following properties:* $\mathscr{B} \subset \mathcal{O}(\mathscr{B})$, *and every topology which contains* $\mathscr{B}$ *already contains* $\mathcal{O}(\mathscr{B})$.

If $A \subset X$, the induced topology $\mathcal{O}_A$ consists of the sets of the form $A \cap O$ with O running through $\mathcal{O}$.

Exercise 3.1.6. *Check that* $\mathcal{O}_A$ *is indeed a topology, and give a necessary and sufficient condition for the inclusion* $\mathcal{O}_A \subset \mathcal{O}$ *to hold.*

The induced topology appears in a number of places in this book. See for instance Exercise 4.1.10.

Remark 3.1.7. We refer to Exercise 3.8.9 for an interesting result, whose proof uses the induced topology, the quotient topology and properties of continuous functions.

Recall that $V \subset X$ is called a neighborhood of the point x if there is an open set included in V which contains x.

Exercise 3.1.8. *Let* $(X, \mathcal{O})$ *be a topological space. Show that* $A \subset X$ *is open if and only if it is a neighborhood of each of its points.*

We will denote by $\mathcal{N}(x)$ the family of neighborhoods of x. A family $\mathscr{B} \subset \mathcal{N}(x)$ is a basis of neighborhoods of x if for every $N \in \mathcal{N}(x)$ there exists $B \in \mathscr{B}$ such that $B \subset N$. The space is *Hausdorff* if every two distinct points have disjoint neighborhoods. For instance, $\mathbb{R}$ is Hausdorff when endowed with the topology of Exercise 3.1.2, while it is not Hausdorff when endowed with the topology of Exercise 3.1.4. As mentioned above Question 3.1.3 we refer to [Exercise 13.1.7, p. 465, CAPB] for a proof of the following question.

Question 3.1.9. *Let $[-\infty, \infty]$ be defined as in Question 3.1.3. Is $[-\infty, \infty]$ Hausdorff with the topology defined in that question?*

In the following exercise one meets the first of various characterizations of one of the key players in complex analysis, namely the extended complex plane, or Riemann sphere $\widehat{\mathbb{C}}$. In the statement, and more generally in the book, we denote by $B(z_0, r)$ (where $z_0 \in \mathbb{C}$ and $r > 0$) the open ball

$$B(z_0, r) = \{z \in \mathbb{C} \; ; \; |z - z_0| < r\}. \tag{3.1.1}$$

The symbol ∞ appearing in the exercise is called the point at infinity, and

$$B(\infty, R) = \{z \in \mathbb{C} \; ; \; |z| > R\} \cup \{\infty\}, \quad R > 0, \tag{3.1.2}$$

denotes a basis of neighborhoods of infinity.

Exercise 3.1.10.

(1) *Define in $\mathbb{C} = \mathbb{R}^2$ a set O to be open if it is empty or if for every $z \in O$ there exists $\epsilon_z > 0$ such that $B(z, \epsilon_z) \subset O$. Show that this defines a topology, say $\mathcal{T}$, on the complex plane, which is Hausdorff.*

(2) *Denote by ∞ a point not in $\mathbb{C}$ and on $\widehat{\mathbb{C}} = \mathbb{C} \cup \{\infty\}$ define $\mathcal{B} \subset \mathcal{P}(\widehat{\mathbb{C}})$ to be*

$$\mathcal{B} = \{B(z, \epsilon), B(\infty, R) \; ; \; \epsilon, R > 0\}.$$

Show that the topology generated by $\mathcal{B}$ (see Exercise 3.1.5 for the latter) is Hausdorff.

We will see in Exercise 3.2.7 that $\widehat{\mathbb{C}}$ is moreover compact (see Definition 3.2.1 below for the later).

We mention that there exists a hierarchy of separation axioms, denoted $T_0, \ldots, T_5$, of which Hausdorff is called T_2. See for instance [297]. We will recall the axiom T_4, which is used in particular in Exercise 3.3.3.

Definition 3.1.11. A topological space is said to satisfy the separation axiom T_4 if the following condition is in force: Given any two disjoint closed sets F_1 and F_2 there exist disjoint open sets O_1 and O_2 such that $F_1 \subset O_1$ and $F_2 \subset O_2$.

For future reference (see Theorem 3.9.14 below) we mention (see for instance [284, p. 32]):

Definition 3.1.12. The topological space is called *regular* if every point has a basis of neighborhoods consisting of closed sets.

In view of Exercise 3.1.13, recall that an open covering of a subset Y of X is a family of open sets $(O_i)_{i \in I}$ such that $Y \subset \cup_{i \in I} O_i$.

Exercise 3.1.13 (see [172, Proof of Lemma 3.1, p. 15]). *Let X be a topological space. Then $F \subset X$ is closed if and only if there is an open covering $(O_i)_{i \in I}$ of X such that $O_i \cap F$ is closed in O_i for every $i \in I$.*

We denote by $\overset{\circ}{E}$ and $\overline{E}$ respectively the interior and the closure of the set E, that is the greatest open set inside E (respectively, the smallest closed set containing E):

$$\overset{\circ}{E} = \bigcup_{\substack{U \text{ open} \\ U \subset E}} U, \quad \text{and} \quad \overline{E} = \bigcap_{\substack{F \text{ closed} \\ F \supset E}} F. \tag{3.1.3}$$

These are well defined, since the union (resp. the intersection) of an arbitrary family of open (resp. closed) sets is open (resp. closed). We also recall

$$\overset{\circ}{\widehat{E \setminus A}} = E \setminus \overline{A}. \tag{3.1.4}$$

It is readily seen that

$$\widehat{A \cap B} = \overset{\circ}{A} \cap \overset{\circ}{B},$$
$$\overline{A \cup B} = \overline{A} \cup \overline{B},$$
$$\overset{\circ}{A} \cup \overset{\circ}{B} \subset \widehat{A \cup B},$$
$$\overline{A \cap B} \subset \overline{A} \cap \overline{B}.$$

These are of course still true for finite collections of sets. The first two equalities will not hold in general for infinite collections of sets. See Exercise 3.1.14. In particular, one may have an infinite collection of sets, each one with an empty interior, and whose union has a non-empty interior. In connection with this last remark, we note that Baire's theorem (see Theorems 3.2.14 and 3.9.12) gives important conditions for equality to hold for countable collections of sets in the two last inclusions.

Exercise 3.1.14. *Let $(A_i)_{i \in I}$ be a family of subsets of the topological set $(X, \mathcal{O})$. Show that*

$$\cup_{i \in I} \overline{A_i} \subset \overline{\cup_{i \in I} A_i} \tag{3.1.5}$$

and equality holds in (3.1.5) when I is a finite set. Show by an example that the inclusion in (3.1.5) may be strict.

The closure of a set is in general larger than the limits of sequences of points in that set, as illustrated by the following exercise, taken from [166, p. 4]. In the definition, it is useful to remark that any finite intersection of sets of the form (3.1.6) with fixed v_0 contains a set of the same form.

Exercise 3.1.15. *Consider the space $\mathcal{V}$ of real-valued functions bounded on $[0, 1]$, endowed with the topology generated by the sets*

$$N(v_0, x_1, \ldots, x_N, \epsilon)$$

parametrized by $v_0 \in \mathcal{V}$, $N \in \mathbb{N}$, $x_1, \ldots, x_N \in [0,1]$ and $\epsilon > 0$ and defined by:

$$N(v_0, x_1, \ldots, x_N, \epsilon) = \{v \in \mathcal{V}; \ |v(x_k) - v_0(x_k)| < \epsilon, \ k = 1, \ldots, N\}. \qquad (3.1.6)$$

Consider the set E of functions equal to 1, at the possible exception of a finite number of points where they vanish. Show that the function identically equal to 0 is in the closure of E, but is not the limit of a sequence of elements of E.

Let X be a topological space. The boundary of $E \subset X$, denoted by ∂E is the closed set

$$\partial E = \overline{E} \cap \overline{X \setminus E} = \overline{E} \cap (X \setminus \overset{\circ}{E}). \qquad (3.1.7)$$

Exercise 3.1.16. *Assume that Ω is open. Show that $\partial \Omega \cap \Omega = \emptyset$.*

Exercise 3.1.17. *Let X be a topological space and $E \subset X$. Then:*

$$\overset{\circ}{E} \cap \partial E = \emptyset, \qquad (3.1.8)$$

and

$$\overline{E} = \overset{\circ}{E} \cup \partial E. \qquad (3.1.9)$$

For another exercise on the boundary, see Exercise 8.1.2 below.

We now turn to the notion of completeness. The present paragraph alludes to various notions defined only in the sequel, but we think the present discussion is nevertheless useful for the student. For a typical undergraduate student the notion of completeness is related to metric spaces, and is defined in terms of Cauchy sequences. See Definition 3.9.10 below. There is a need to define completeness in a more general setting, in particular for the definition of a Fréchet space, that is, of a locally convex topological vector space (see Definition 5.1.13) which is metrizable and complete. Metrizable means that the topology can be given in terms of a metric (see Section 3.9). But a topological vector space can be metrizable with respect to two different metrics, and be complete (in the metric space sense) with respect to one of the metrics and not complete with respect to the second. So completeness should be defined in an intrinsic way. This is the reason why we introduce in this book completeness in spaces which *a priori* are not endowed with a metric. First two definitions (see for instance [307, p. 6], [119, p. 3]). For the second one, recall that a directed set is a set, say $\mathcal{J}$, endowed with a partial order such that

$$\forall a, b \in \mathcal{J}, \ \exists c \in \mathcal{J} \ \text{such that} \ \begin{cases} c \geq a & \text{and} \\ c \geq b. \end{cases}$$

Definition 3.1.18 (see, e.g., [307, pp. 6–7, and p. 10]). Let X be a space. A family $\mathcal{A}$ of elements of $\mathcal{P}(X)$ is called a filter if it does not contain the empty set, is closed under finite intersection and if, for any given set in $\mathcal{A}$, all sets which contain it are also in $\mathcal{A}$. It is called a basis of filter if it consists of non-empty sets such that the intersection of two sets in $\mathcal{A}$ contains an element of $\mathcal{A}$.

Assume now that X is endowed with a topology. The filter is said to converge to the point $x \in X$ is every neighborhood of x belongs to the filter.

Exercise 3.1.19. *Let $\mathcal{V}$ be an infinite-dimensional vector space (say, over the complex numbers). Show that the family of vector subspaces of $\mathcal{V}$ of finite codimension is a basis of filter.*

Definition 3.1.20. A net in a set X is a family of elements of X indexed by a directed set.

The following basis of filters on continuous functions on $[0,1]$ is used to make $C[0,1]$ into a topological vector space which is not locally convex. See [183, Example 2, p. 86] and Exercise 5.1.14.

Question 3.1.21. *Let $C[0,1]$ denote the vector space of continuous functions on $[0,1]$. Let $f_0 \in C[0,1]$. Show that the sets*

$$V_{\epsilon,\eta}(f_0) = \{f \in C[0,1] \text{ such that }$$
$$|f(t) - f_0(t)| \leq \epsilon \text{ outside some open set of } [0,1] \text{ of length less or equal to } \eta\}$$

is a basis of filter for the point f_0.

Hint: As remarked in [183],

$$V_{\epsilon_1,\eta_1}(f_0) \cap V_{\epsilon_2,\eta_2}(f_0) \supset V_{\epsilon,\eta}(f_0)$$

with $\epsilon = \epsilon_1 \wedge \epsilon_2$ and $\eta = \eta_1 \wedge \eta_2$.

3.2 Compact sets

We now consider compact sets and their various properties.

Definition 3.2.1. A Hausdorff space X is compact if every cover of X with open subsets $X = \cup_{i \in I} O_i$ admits a finite subcover $X = \cup_{k=1}^{N} O_{i_k}$ (the Heine–Borel–Lebesgue property; see [284, p. 81]).

It is useful to recall the following classical fact (see for instance [100, Proposition 11-6, p. 35]):

Proposition 3.2.2. *Every infinite subset of a compact set has one accumulation point.*

The product of two compact sets with the product topology is compact. More generally, Tychonoff's theorem asserts:

Theorem 3.2.3. *Any product of compact sets endowed with the product topology is compact.*

We recall (see, e.g., [307, Corollary 1, p. 53 and Proposition 6.1, p. 50]):

Theorem 3.2.4. *A compact set is closed and a closed subspace of a compact space is compact.*

Compact spaces appear in a number of important occasions in complex analysis. For instance, in the space $\mathcal{H}(\Omega)$ of functions analytic in a connected set Ω with the topology defined by uniform convergence on compact sets, sets which are bounded (in the sense of boundedness in a topological vector space; see Definition 5.1.17) and closed are compact, and hence sequentially compact since $\mathcal{H}(\Omega)$ is metrizable. See [284, p. 86] and Exercise 3.9.21.

The following question deals with the Cantor set. See [50, p. 78]. The solution can be found in [CAPB, p. 477], and is omitted here.

Question 3.2.5. *Let*

$$U_0 = \left(\frac{1}{3}, \frac{2}{3} \right),$$

and for $n \geq 1$,

$$U_n = \cup_{(\epsilon_1,\ldots,\epsilon_n)\in\{0,2\}^n} \left(3^{-n-1} + \sum_{k=1}^{n} \frac{\epsilon_k}{3^k}, 2 \cdot 3^{-n-1} + \sum_{k=1}^{n} \frac{\epsilon_k}{3^k} \right).$$

Let

$$C = [0,1] \setminus \cup_{n\in\mathbb{N}_0} U_n.$$

Show that C is compact, not countable, but that the total length of the U_n is equal to 1.

Another example of compact metric space is presented in Exercise 3.9.19 below.

The intersection of an arbitrary family of compact sets is compact, when it is non-empty. This follows directly from the fact that a compact set is closed and that a closed set in a compact set is compact. On the other hand, a decreasing family of closed sets can be empty (for instance $F_n = [n, \infty)$ in $\mathbb{R}$ endowed with its usual topology). The following exercise lies at the intersection of these facts, and is also a direct consequence of the definition of compactness.

Exercise 3.2.6. *Show that a decreasing family of compact sets has a non-empty intersection.*

In a metric space (see Definition 3.9.1 below), compactness is equivalent to sequential compactness; this is the Bolzano–Weierstrass property; see [284, p. 86] and Exercise 3.9.21 below. For general topological vector spaces the two notions need not be related. If the range space is Hausdorff, the continuous image of a compact space is also compact. The projective space $\mathbb{P}^n$, the Riemann sphere and the torus (see Exercises 3.2.7, 3.8.8 and 3.8.30 below) are first examples of compact spaces, and in fact are compact Riemann surfaces of genus 0 and 1 respectively. When $n = 1$, the projective space is also denoted by $\mathbb{S}$ (and then its name is the Riemann sphere) or by $\widehat{\mathbb{C}}$ (the extended complex plane).

Exercise 3.2.7. *Let $\widehat{\mathbb{C}}$ be as in Exercise 3.1.10, and let $\mathscr{B}$ denote the family of sets of the form $B(z_0, r)$, when z_0 runs through $\mathbb{C}$ and r runs through $(0, +\infty)$, together with the sets of the form $B(\infty, R)$ (see (3.1.2) for the latter), when R runs through $(0, +\infty)$. Let $\mathcal{O}(\mathscr{B})$ denote the corresponding topology, as in Exercise 3.1.5. Show that $\widehat{\mathbb{C}}$ is compact in this topology.*

The above exercise presented $\widehat{\mathbb{C}}$ as the one point compactification of the complex plane. We note that $\widehat{\mathbb{C}}$ is in fact a manifold, and more precisely, the simplest example of a compact Riemann surface. As a compact manifold, it is a metrizable space, and the corresponding metric is presented in Exercise 3.9.27.

Exercise 3.2.8. *Let X be a compact topological space and let F_1 and F_2 be two disjoint closed subspaces of X. Show that there exist disjoint open sets O_1 and O_2 such that $F_1 \subset O_1$ and $F_2 \subset O_2$.*

The following result is taken from Forster's book on Riemann surfaces [138, §4. 20, p. 28], where it is used to note that a proper map is closed; see Exercise 3.6.5.

Exercise 3.2.9. *Let X be a locally compact topological space. Then, $A \subset X$ is closed if and only if its intersection with every compact subset of X is compact (when non-empty).*

We now discuss Baire spaces and Baire's theorem. We first recall that a subset E of the topological space X is called *nowhere dense* if

$$\overset{\circ}{\overline{E}} = \emptyset, \tag{3.2.1}$$

and *meager* (or of first category) if it is a countable union of nowhere dense sets. It is second category if it is not of first category. The set of functions analytic in the open unit disk and which fail to have a radial limit at all points is a second category set. This is a result due to Kierst and Szpilrajn, and mentioned in [269, Exercise 6, p. 179 and Exercise 4, p. 178]. This is the content of Question 5.6.5. Of course, one needs to know which topology to put on the space $\mathcal{H}(\mathbb{D})$ of functions analytic in $\mathbb{D}$ to make the previous statement precise. The topology of $\mathcal{H}(\mathbb{D})$ is discussed in Section 5.6.

We mention two questions where the notion of sets of first and second category intervene. In Question 5.2.7 one shows that in a countably normed space which is not a normed space (for instance, the space of functions analytic in an open subset of $\mathbb{C}$ in its natural topology), a bounded set is nowhere dense. In Question 3.2.11 (which presents a result of M. Zorn [328], answering a question of Bochner) one uses the fact that a complete metric space is a set of second category to prove a result on convergence of power series in two complex variables. First a definition:

Definition 3.2.10. A function $f(z, w)$ defined in an open subset Ω of $\mathbb{C} \times \mathbb{C}$ is said to be analytic if it admits a power expansion in a neighborhood of every point of Ω: If $(z_0, w_0) \in \Omega$ there exist $h > 0$ and $k > 0$ such that in $B(z_0, h) \times B(w_0, k)$ we have

$$f(z, w) = \sum_{n,m=0}^{\infty} f_{n,m}(z - z_0)^n (w - w_0)^m.$$

Question 3.2.11. *Let for* $n = 0, 1, 2, \ldots$

$$P_n(u, v) = \sum_{s+t=n} a_{s,t} u^s v^t$$

be a homogeneous polynomial of degree n, and assume that, for every choice of (u, v), the power series in z

$$\sum_{n=0}^{\infty} P_n(u, v) z^n$$

has always a strictly positive radius of convergence (which a priori depends on (u, v)). Show that the power series

$$\sum_{s,t=0}^{\infty} a_{s,t} u^s v^t \tag{3.2.2}$$

converges absolutely in a neighborhood of the origin.

Hint: Show that the set of pairs (u, v) such that the power series (3.2.2) converges absolutely is of second category.

Definition 3.2.12. A Baire space is a topological Hausdorff space in which every countable union of nowhere dense closed sets has an empty interior.

Exercise 3.2.13. *Show that an equivalent statement to the definition of a Baire space is: if $(O_n)_{n \in \mathbb{N}}$ is a family of open sets such that $\overline{O_n} = E$ for $n = 1, 2, \ldots$. Then $\bigcap_{n \in \mathbb{N}} O_n = E$.*

We now turn to Baire's theorem (see, e.g., [284, p. 324]). In the statement, recall that a Hausdorff space is called *locally compact* if every point has a basis of compact neighborhoods.

Theorem 3.2.14. *A locally compact space is a Baire space.*

Remark 3.2.15. Another family of Baire spaces is presented in Section 3.9 devoted to metric spaces. We mention that, in the setting of locally convex topological vector spaces, Baire's theorem is used to prove the open mapping theorem, and hence also (as its corollaries) the inverse mapping theorem and the closed graph theorems. We also mention that the uniform boundedness theorem (that is, the Banach–Steinhaus theorem), can be set for barreled spaces or for Baire spaces. In the case of linear continuous applications defined on a topological vector space which is a Baire space, the range space is a general topological vector space. In the case of applications defined on a barreled space the range space is assumed locally convex; see for instance the discussion in [285, p. 335].

To conclude this section, we mention the one point compactification theorem (see Exercise 3.2.7 for the case of the plane). See for instance [1, Exercice 17, p. 51] for this classical and important result.

Question 3.2.16. *Let X be a locally compact space, and let ∞ denote an element not in X. On $X \cup \{\infty\}$ we define an open set to be a set O of one of the following forms:*

(1) $O = X \cup \{\infty\}$.
(2) $O \subset X$ *is an open set in the topology of X.*
(3) $O = (X \setminus K) \cup \{\infty\}$, *where K is compact.*

Show that these sets (together with X itself and the empty set) define indeed a topology and that $X \cup \{\infty\}$ is compact in this topology.

Question 3.2.17. *Show that the space $\mathbb{R}^{\mathbb{N}}$ is not locally compact.*

3.3 Connected sets

Path-connected sets use the notion of continuous functions and are considered at a later stage. See Exercise 3.4.12. A topological space X is connected if it does not admit a non-trivial partition into open subsets: If $X = O_1 \cup O_2$ where O_1 and O_2 are open and do not intersect, then, $O_1 = X$ or $O_2 = X$. Recall that the continuous image of a connected set is connected. Recall also that the union of a family $(C_j)_{j \in J}$ of connected sets such that the intersection $\cap_{j \in J} C_j \neq \emptyset$ is still connected. It is also useful to rewrite the connectedness condition with an inclusion rather than an equality:

Exercise 3.3.1. *The non-empty set $E \subset X$ is connected if and only if for any inclusion:*

$$E \subset O_1 \cup O_2, \tag{3.3.1}$$

where O_1 and O_2 are open sets of X such that

$$E \cap O_1 \cap O_2 = \emptyset, \tag{3.3.2}$$

then E is included in either O_1 or O_2.

Example 3.3.2. *Let X be a topological space and let A, B be subsets of X. Assume that $A \subset B \subset \overline{A}$ and A connected. Show that B is connected.*

Exercise 3.3.3. *Let X be a topological space, which is Hausdorff and meets the separation axiom T_4 (see Definition 3.1.11), let $(C_n)_{n \in \mathbb{N}}$ be a decreasing sequence of (non-empty) connected sets, and let $C = \cap_{n \in \mathbb{N}} C_n$.*

(1) *Show by an example that C may be empty. Assume that $C \neq \emptyset$. Show by an example that C need not be connected.*

(2) *Assume that the space X is compact and the C_n closed (and hence compact). Show that $C \neq \emptyset$ and is connected.*

3.4 Continuous maps

Topology is important in particular to define continuous maps. Given two topological spaces X and Y the function $f : X \longrightarrow Y$ is continuous at the point $x \in X$ if the inverse image of every neighborhood of $f(x)$ is a neighborhood of x. One can, and usually does, consider a basis of neighborhoods of $f(x)$ to check continuity. A map between two topological spaces is called continuous if the inverse image of every open set is an open set. The reader should prove easily that a map between two topological spaces is continuous if and only if it is continuous at every point. Another characterization of continuity, using the notion of closure of a set, is presented in Exercise 3.4.4 below.

Exercise 3.4.1. *Let X be a space and let $(Y, \mathcal{T})$ be a topological space. Let f be a map from X into Y.*

(1) *Show that the family $\mathcal{T}_f$ of sets of the form $f^{-1}(O)$ where O runs through $\mathcal{T}$ defines a topology on X.*

(2) *Show that the topology $\mathcal{T}_f$ is the smallest topology on X for which f is continuous.*

Exercise 3.4.2. *Let X and Y be two topological spaces, and let $X = \cup_{i \in I} U_i$ be an open covering of X. Then, a function f from X into Y is continuous if and only if the restriction of f to each U_i is continuous (in the induced topology; see Exercise 3.1.6 for the latter).*

Remark 3.4.3. We recall that the continuous image of a compact space is compact. More precisely, let X be a compact space, Y be a Hausdorff space and let f be continuous from X into Y. Then, $f(X)$ is compact in Y. This is used in part (2) of the following exercise. For part (1) of this exercise, see, e.g., [100, Théorème 7.5, p. 19]. For part (2) (given for metric spaces), see [288, Exercise 18, p. 198], and see [1, Exercice 17, p. 33], [77, Proposition 9, p. TG I.35] for (3).

Example 3.4.4.

(1) *Let $(X, \mathcal{T}_X)$ and $(Y, \mathcal{T}_Y)$ be two topological spaces and let f be a function from X into Y. Prove that f is continuous if and only if*

$$\forall A \subset X, \quad f(\overline{A}) \subset \overline{f(A)}. \tag{3.4.3}$$

(2) *Assume that X is compact and Y Hausdorff, and let f be a continuous map from X into Y. Show that:*

$$f(\overline{A}) = \overline{f(A)}, \quad \forall A \subset X. \tag{3.4.4}$$

(3) *Let f be a continuous from X into Y. Show that*

$$f(\overline{A}) = \overline{f(A)}, \quad \forall A \subset X \tag{3.4.5}$$

holds if and only f sends closed sets into closed sets.

Remarks 3.4.5.

(1) It trivially follows from (3) that the inverse of a closed bijection (see Section 3.6 for the definition of a closed map) is continuous. See for instance [1, Exercice 17 (b), p. 33], which is repeated as Exercise 3.6.2 below.

(2) The first condition in the exercise is in terms of arbitrary sets of the original space. In terms of arbitrary sets of the range space, one has the equivalent condition (see [120, Lemma 16, p. 13])

$$\forall B \subset Y, \quad f^{-1}(\overline{B}) \subset \overline{f^{-1}(B)}.$$

Exercise 3.4.6. *In the setting of Exercise* 3.1.10, *show that the function*

$$i_P(z) = \begin{cases} \infty, & z = 0, \\ \frac{1}{z}, & z \neq \infty, \\ 0, & z = \infty \end{cases}$$

is continuous.

We now consider a sequence $(f_n)_{n \in \mathbb{N}}$ of real-valued continuous functions defined on a topological space X. Even if the function $\sup_{n \in \mathbb{N}} f_n(x)$ is finite for every $x \in X$, it need not be continuous, but it is lower semi-continuous. We now recall the definition of this notion.

Definition 3.4.7. Let f be a function from the topological space X with values in $(-\infty, \infty]$.

(1) f is called lower semi-continuous at the point $x \in X$ if for every $a \in (-\infty, \infty)$ such that $a < f(x)$ there exists a neighborhood V of x such that $f(V) \subset (a, \infty]$.

(2) It is called lower semi-continuous on X if for every $a \in \mathbb{R}$ the set $f^{-1}(a, \infty]$ is open.

(3) f is called upper semi-continuous at the point $x \in X$ if for every $a \in (-\infty, \infty)$ such that $a > f(x)$ there exists a neighborhood V of x such that $f(V) \subset [-\infty, a)$.

(4) It is called upper semi-continuous on X if for every $a \in \mathbb{R}$ the set $f^{-1}[-\infty, a)$ is open.

Recall that we defined the characteristic function of a set in Definition 1.1.1.

Question 3.4.8 (see [100, p. 133]). *The function $1_{\mathbb{R}\setminus\mathbb{Q}}$ is lower semi-continuous at every rational point and upper semi-continuous at every irrational point.*

We will present only the proof of (1) in the next exercise.

Exercise 3.4.9.

(1) *The function is lower semi-continuous on X if and only if it is lower semi-continuous at every point of X.*
(2) *The subset $A \subset X$ is open if and only if the characteristic function 1_A is lower semi-continuous.*
(3) *The subset $A \subset X$ is closed if and only if the characteristic function 1_A is upper semi-continuous.*
(4) *The supremum of any family of continuous functions is lower semi-continuous.*

Another exercise on lower semi-continuity is given in Section 4.2; see Exercise 4.2.18 there.

The notion of lower semi-continuity will be used in particular in Exercise 5.3.6, where one proves that a lower semi-continuous semi-norm in a Fréchet space is continuous. This last exercise in turn plays an important role in proving that an everywhere defined operator in a Hilbert space has an adjoint if and only if it is continuous. See Exercise 4.2.19. In fact, locally convex topological vector spaces for which lower semi-continuous semi-norms are continuous form an important class of spaces, called *barreled spaces*. See [307]. See for instance Question 5.2.13.

The following exercise is inspired from a technical step in the proof of the implicit function theorem for analytic functions of two variables; see [242, p. 3] for the latter.

Exercise 3.4.10. *Let X be a topological space, and let f be a complex-valued continuous function which does not vanish on a given compact set K. Show that f does not vanish in a neighborhood of K.*

Connected sets behave well under continuous functions. More precisely:

Theorem 3.4.11. *The image of a connected set under a continuous map is connected.*

To define path-connected spaces, recall first that a (continuous) path in a topological space X is a continuous map from $[0,1]$ (or more generally, from a compact interval of $\mathbb{R}$) into X.

Definition 3.4.12. The space is called path-connected if every two points in X can be linked by a path.

Exercise 3.4.13. *Show that a path-connected set is connected. What about the converse?*

3.4. Continuous maps

Recall that a bijection between two topological spaces which is continuous and whose inverse is continuous is called a homeomorphism. The quest for all topological spaces homeomorphic to a given one seems to have no reasonable solution, but in special cases, such as surfaces. The following criteria is used, for instance, to prove that the map

$$z \mapsto \begin{cases} (\wp(z), \wp'(z), 1), & \text{if } z \notin \Lambda \\ (0, 1, 0), & \text{if } z \in \Lambda, \end{cases}$$

where $\wp$ denotes the Weierstrass function associated with the lattice Λ, allows to define a homeomorphism between a torus and the associated cubic. See [204, p. 124].

Question 3.4.14. *A continuous bijection from a compact set onto a Hausdorff space is a homeomorphism.*

An important notion is now introduced (see for instance [91, p. 24]), already used in previous exercises (see for instance Exercise 1.5.5).

Definition 3.4.15. A Jordan curve is a subset of $\mathbb{R}^2$ homeomorphic to the unit circle $\mathbb{T}$.

The following exercise gives a characterization of the boundary in certain important cases.

Exercise 3.4.16. *Let Ω be an open subset of the complex plane, and let f be a one-to-one and onto continuous map from the open unit disk $\mathbb{D}$ onto Ω. Assume that f^{-1} is also continuous.*

(1) *Show that*

$$\partial\Omega = \cap_{r \in (0,1)} \overline{f(r < |z| < 1)}. \tag{3.4.6}$$

(2) *Assume that Ω is bounded. Show that $\partial\Omega \neq \emptyset$ and that it is connected.*

The material of the following exercise is taken from [310, Proof of Theorem 1.2]. In the proof, use is made of the fact that an open connected subset of $\mathbb{C}$ is path-connected. This is the reason why the exercise is put in this section and not in the previous one.

Exercise 3.4.17. *Let Ω be an open connected subset of $\mathbb{C}$, and let E be a non-empty closed set, and assume that*

$$\emptyset \subsetneq E \cap \Omega \subsetneq \Omega. \tag{3.4.7}$$

Show that $\partial E \cap \Omega \neq \emptyset$.

Remark 3.4.18. When (3.4.7) does not hold, then the intersection $\partial E \cap \Omega$ can be empty, as is illustrated by the examples $\Omega = \mathbb{D}$ and $E = \overline{\mathbb{D}}$ and $E = \mathbb{C} \setminus \mathbb{D}$ respectively.

For other examples of path-connected sets, see Exercises 3.8.8 and 3.8.33 below.

We conclude this section with an example appearing in topological analysis. Recall that integration theory plays a key role in the classical approach to
the theory of functions of a complex variable, in particular in the proof of the
Cauchy–Goursat theorem. In such an approach, this theorem is followed by the
power series expansion theorem, and one can derive then numerous corollaries such
as the maximum modulus principle. A different approach exists, called *topological analysis*, which does not resort to integration, but uses topological tools. See
Whyburn's book [320] and the papers [252, 102] of Connell and Porcelli. In that
approach, one proves first that a function f differentiable in an open set $\Omega \subset \mathbb{C}$ is
open. Then one proves the maximum modulus principle for $|f|$, and then the fact
that f has derivatives of all order there, and then one obtains the power series
expansion.

Exercise 3.4.19 (see [252, Lemma 1, p. 177]). *Let $\Omega \subset \mathbb{C}$ be a bounded open set, and
let f be open in Ω and continuous on the closure of Ω. Let C denote a connected
component of $\mathbb{C} \setminus f(\overline{\Omega} \setminus \Omega)$ and assume that $f(\Omega) \cap C \neq \emptyset$. Show that $C \subset f(\Omega)$.*

3.5   Products

The next exercise deals with the cartesian product and its dual (in the sense of
categories; we will not discuss categories here), namely the disjoint union. The
results are standard, and we used as a source [228, pp. 17–18].

Exercise 3.5.1. *Let $(X_i)_{i \in I}$ be a family of topological spaces.*

(1) *The cartesian product $\prod_{i \in I} X_i$ is the set of sequences $x = (x_i)_{i \in I}$ where $x_i \in$
    X_i. Describe the smallest topology on $\prod_{i \in I} X_i$ for which the maps $p_i(x) = x_i$
    are all continuous.*

(2) *Show that a map f from a topological space Y into the cartesian product
    endowed with the above topology is continuous if and only if all the maps
    $p_i \circ f$ are continuous.*

(3) *Whether the X_i intersect or not, the spaces $\{i\} \times X_i$ do not. The disjoint
    union of the X_i is*

$$\coprod_{i \in I} X_i = \bigcup_{i \in I} \{i\} \times X_i.$$

Let $i_0 \in I$. We set for $x \in X_{i_0}$

$$q_{i_0}(x) = (i_0, x) \in \coprod_{i \in I} X_i.$$

*Describe the biggest (finest) topology on the disjoint union for which all the
maps q_i are continuous.*

(4) *Let f be a map from the disjoint union (endowed with the above topology)
    into a topological space Y, and let $f_i = f \circ q_i$. Show that f is continuous if
    and only if all the maps f_i are continuous.*

The topology in item (1) of Exercise 3.5.1 is called the product topology. It is generated by sets of the form

$$U = \prod_{i \in I} U_i \qquad (3.5.1)$$

where for all but a finite number of indices $U_i = X_i$ and U_i is an open subset of X_i for the other indices. Note that the intersection of two such sets is of the same form. We refer for instance to [1, Proposition 35, p. 22] for the following result. The proof makes use of (1.1.1). The result itself is used, for instance, in the proof of Exercise 4.1.10.

Question 3.5.2. *Let f be a map from the topological space X into the product of topological spaces $\prod_{i \in I} X_i$, the latter endowed with the product topology. Show that f is continuous if and only if the inverse image of every set of the form (3.5.1) (where for all but a finite number of indices $U_i = X_i$ and U_i is an open subset of X_i for the other indices) is open in X.*

We conclude with the classical result:

Exercise 3.5.3. *Let X be a Hausdorff space. Show that the diagonal*

$$\Delta = \{(x,x)\,,\, x \in X\}$$

is closed in $X \times X$.

3.6 Closed, open and proper maps

Definition 3.6.1. A continuous map between topological spaces X and Y is called *closed* if the image of a closed set is closed. It is called *open* if the image of an open set is open.

Following the discussion after Exercise 3.4.4 we ask:

Exercise 3.6.2. *Show that the inverse of a continuous one-to-one onto closed map is continuous.*

Hint: Use the third item in Exercise 3.4.4.

Examples of closed maps are presented in particular in Exercises 3.7.6 and 3.6.5. Examples of open maps are presented in Exercises 3.6.6, 4.2.7 and Question 3.7.5. In functional analysis, the *open mapping theorem* gives an important class of open linear maps between Banach spaces. See Theorem 4.2.15.

As already recalled, the continuous image of a compact space is compact (when the range is inside a Hausdorff space). On the other hand, let X be a Hausdorff space and Y be a compact space, and let f be continuous from X into

Y. Then, $f^{-1}(Y)$ need not be compact. A continuous map between Hausdorff spaces X and Y is called *proper* if the inverse image of every compact in Y is compact in X. The notion is of interest when X is not compact. The case of compact X is considered in the next exercise.

Exercise 3.6.3. *Let X and Y be two Hausdorff spaces, and let f be a continuous map from X into Y. Assume that X is compact. Show that f is proper.*

The key in the following classical exercise is that in $\mathbb{R}$, a set is compact if and only if it is closed and bounded. The exercise is taken from [139, p. 210].

Exercise 3.6.4. *Let f be a continuous map from $\mathbb{R}$ into itself. Show that (a) and (b) are equivalent:*

(a) $\lim_{x \to \pm\infty} |f(x)| = +\infty$.
(b) *For every compact $K \subset \mathbb{R}$, the set $f^{-1}(K)$ is compact.*

The notion of proper map is important in the construction of the Riemann surface associated with an algebraic curve; see [242, Chapters 1 and 4]. In particular use is made of the following result (see [138, §4.20, p. 28]):

Exercise 3.6.5. *Let X and Y be two locally compact spaces, and let f be a proper map from X into Y. Show that f is closed.*

Hint: Use Exercise 3.2.9.

Before presenting Exercise 3.6.6 we make a number of remarks. By definition of a continuous function, the inverse image of an open set is open. The direct image of an open set under a continuous function need not be open. Consider a function f from an open subset Ω of $\mathbb{R}^n$ into $\mathbb{R}^n$, and assume that it is one-to-one, of class C_1, and that its Jacobian matrix is invertible at all points of Ω. Then, $f(\Omega)$ is open. This is the global inverse function theorem. If the Jacobian matrix is invertible at a given point of Ω, then f is a local homeomorphism, and in particular maps open sets into open sets. Without the above invertibility property, it is not true that the image of an open set is an open set. These remarks put into perspective the open mapping theorem for analytic functions, which states that the continuous image of an open set of the complex plane under an analytic function is open. Two standard proofs are available, which we propose in the next exercise and in Exercise 2.2.1. See for instance the discussion in [140, p. 55].

Exercise 3.6.6.

(1) *Give examples of functions continuous on an open set of $\mathbb{R}^2$, with range in $\mathbb{R}^2$ and $\mathbb{R}$ respectively, and whose images are not open.*
(2) *Prove the open mapping theorem using the implicit function theorem of calculus (see also Exercise 2.2.1 for another proof): A function analytic in an open subset of $\mathbb{C}$ maps open sets into open sets.*

Exercise 3.6.7. *Let* Ω, Ω_1 *and* Ω_2 *denote open sets of* $\mathbb{C}$*. Let* f *be analytic in* Ω*. Furthermore, let* f_1 *denote a homeomorphism from* Ω_1 *onto* Ω*, and let* f_2 *denote a homeomorphism from* $f(\Omega)$ *onto* Ω_2*. Show that the map*

$$f_2 \circ f \circ f_1 \tag{3.6.1}$$

is open.

The function (3.6.1) is topologically equivalent (that is, homeomorphic) to an analytic function, and one can ask whether this is a characterization of open maps. This is not the case, as already shown in Stoïlow's book [299, pp. 104–106], where the following counterexample can be found:

Exercise 3.6.8 ([298, p. 381], [299, p. 105]). *For* $z = x + iy \in \mathbb{C}_r$ *(the open right half-plane) define a map* $f(z)$ *by:*

$$f(z) = \begin{cases} \left(e^{-\frac{1}{y^2}} + xy^2\right) e^{\frac{2\pi i}{y}}, & if \quad y \neq 0, \\ 0, & if \quad y = 0. \end{cases}$$

Then, f *is open but* f *is not topologically equivalent to a function analytic in* $\mathbb{C}_r$*.*

Hint: (following Stoïlow)

(1) What is the image of $(0, \infty)$ under f?
(2) Show that the image of an open subset lying completely in the right upper quarter plane or in the right lower quarter plane is open.
(3) Let $x_0 > 0$, and let I_n denote the interval with endpoints $(x_0 - \epsilon, y_n)$ and $(x_0 + \epsilon, y_n)$, where the y_n are appropriately chosen and satisfy:

$$\frac{1}{y_{n+1}} - \frac{1}{y_n} = 1, \quad n = 0, 1, \dots . \tag{3.6.2}$$

Compute for ϵ small enough $\cup_{n \in \mathbb{N}} f(I_n)$, and conclude that there is a neighborhood of the origin which is in the image of $\mathbb{C}_r$.

We note that Stoïlow's theorem establishes when a function is topologically equivalent to an analytic function. The statement involves manifolds and Riemann surfaces (see [320, p. 103]).

In Exercise 3.2.7 the inclusion map is not a closed map from $\mathbb{R}^2$ into $\widehat{\mathbb{C}}$. For instance the set of points $(x, y) \in \mathbb{R}^2$ such that $x^2 + y^2 \geq R$ is closed in $\mathbb{R}^2$ but not in $\widehat{\mathbb{C}}$. The following exercise, taken from [175, p. 2], clarifies the issue.

Exercise 3.6.9. *A subset of* $\mathbb{R}^2$ *is closed in the topology of* $\widehat{\mathbb{C}}$ *if and only if it is closed and bounded in* $\mathbb{R}^2$*.*

3.7   Quotient topology

There are numerous ways to build new topological spaces from given ones. In this section we consider the notion of quotient topology. It is used in particular to build surfaces from surfaces.

An important example can be given from measure theory. Let $(\Omega, \mathcal{A}, \mu)$ be a measured space. Two measurable functions are called equivalent if the set where they differ is of zero measure. The following key example appears in the book of E. Nelson (see [245, p. 84]).

Definition 3.7.1. Let $\mathcal{C}$ denote the set of positive Borel measures μ on the real line such that

$$\int_{\mathbb{R}} \frac{d\mu(t)}{t^2 + 1} < \infty. \tag{3.7.1}$$

Let f, g be measurable functions and let $\mu, \nu \in \mathcal{C}$. The pairs (f, μ) and (g, ν) are called equivalent if there exists a measure λ such that $\mu \ll \lambda$ and $\nu \ll \lambda$ and

$$f\sqrt{\frac{d\mu}{d\lambda}} = g\sqrt{\frac{d\nu}{d\lambda}}, \quad \lambda \text{ a.e.}$$

The equivalent classes in the above example are denoted $f\sqrt{d\mu}$.

We now turn to the continuity property of the canonical projection associated with an equivalence relation. From the basic definition of continuity one obtains:

Theorem 3.7.2. *There is a topology on the quotient space, called the quotient topology, uniquely characterized by one of the following equivalent properties:*

(a) *U is an open set in $X/\sim$ if and only if $q^{-1}(U)$ is an open set in the original topology.*

(b) *F is a closed set in $X/\sim$ if and only if $q^{-1}(F)$ is a closed set in the original topology.*

(c) *The quotient topology is the finest (that is, largest) topology on $X/\sim$ for which q is continuous.*

The topology defined in the previous theorem has the following universal property:

Theorem 3.7.3. *Let X be a topological space. There exists on $X/\sim$ a unique topology with the following two properties:*

1. *The canonical projection q is continuous.*
2. *A function h is continuous from $X/\sim$ into a topological space Y if and only if $h \circ q$ is continuous from X into Y.*

Exercise 3.7.4. *In the notation of Theorem 3.7.3, let f from X into Y be compatible with $\sim$ (i.e., $x \sim y \Longrightarrow f(x) = f(y)$). Then the associated map $\tilde{f} : X/\sim \to Y$ is continuous.*

The equivalence relation is called open (resp. closed) if the canonical projection q is open (resp. closed), that is, if its image of an open (resp. closed) set is open (resp. closed).

We note that even if the original space is Hausdorff, the quotient space need not be a Hausdorff space when endowed with the quotient topology. We now present conditions which ensure the quotient space to be separated.

Question 3.7.5. *Let X be a Hausdorff topological space, and let R be an equivalence relation on X. Assume that:*

(1) *The projection q is an open map.*
(2) *The graph of the relation is closed in $X \times X$.*

Then, the quotient space is Hausdorff.

See Exercise 4.2.7 for an example of a quotient topology, with open associated projection map.

In the case of compact spaces one has the following result (see [152, Théorème 4.8, p. 25]):

Exercise 3.7.6. *Let X be a compact topological space, and let R be an equivalence relation on X. Then, the following are equivalent:*

(1) *The projection q is a closed map.*
(2) *The quotient space is Hausdorff.*
(3) *The graph of the relation is closed in $X \times X$.*

The proof follows the arguments in [152, p. 25]. We now give a number of hints.

Hints to the proof of Exercise 3.7.6:

To prove (1) $\Longrightarrow$ (2): Let u and v in the quotient space, and take U and V disjoint neighborhoods of $q^{-1}(u)$ and $q^{-1}(v)$, and set:

$$F = q^{-1}(q(X \setminus U)) \quad \text{and} \quad G = q^{-1}(q(X \setminus V)). \qquad (3.7.2)$$

Then $q(X \setminus F)$ and $q(X \setminus G)$ are disjoint neighborhoods of u and v respectively.

To prove (2) $\Longrightarrow$ (3): Use the fact that the diagonal of $(X/\sim) \times (X/\sim)$ is closed (see Exercise 3.5.3 for the latter).

To prove (3) $\Longrightarrow$ (1): Let Γ denote the graph of the relation, and let F be an open subset of X. Use the fact that $(F \times X) \cap \Gamma$ is compact to prove that $q^{-1}(q(F))$ is closed.

As mentioned earlier (see Remark 3.1.7), an interesting result which uses the induced topology and the quotient topology is presented in the sequel in Exercise 3.8.9.

Manifolds are the topic of the next section. The quotient topology plays an important role in part of the exposition.

3.8 Manifolds and surfaces

Definition 3.8.1. A topological space X is called a real topological manifold if it is Hausdorff, and if every point of X has an open neighborhood homeomorphic to an open subset of $\mathbb{R}^{n(x)}$ (for some $n(x) \in \mathbb{N}$).

The number $n(x)$ is uniquely defined, thanks to Brouwer's invariance of the domain theorem (see for instance [294, Theorem 16, p. 199]). If $n(x) = 2$ the manifold is called a topological surface.

Theorem 3.8.2. (*Brouwer's invariance of the domain theorem*). *If $U \subset \mathbb{R}^n$ is open and f is a continuous one-to-one mapping from U into $\mathbb{R}^n$, then $f(U)$ is open and f^{-1} is continuous.*

Question 3.8.3. *Let X be a topological manifold. Using the invariance of the domain theorem, show that the number $n(x)$ is independent of x in a given connected component of X.*

When X is connected, the number $n = n(x)$ is called the dimension of the topological manifold.

It may happen that every point of a topological space X has an open neighborhood homeomorphic to an open subset of $\mathbb{R}^{n(x)}$, while X is not a Hausdorff space. See for instance Spivak's book [295, Appendix A, p. 623] where the following example is given.

Question 3.8.4. *Consider the space $X = \mathbb{R} \cup \{\infty\}$, and let $\mathcal{O}$ denote the topology generated by the sets U satisfying the following two conditions:*

(1) *$U \cap \mathbb{R}$ is an open subset of $\mathbb{R}$ (the latter with the usual topology).*
(2) *If $\infty \in U$, the set $(U \cap \mathbb{R}) \cup \{0\}$ is a neighborhood of 0.*

Show that X is not Hausdorff, but locally homeomorphic to $\mathbb{R}$.

Other examples include the following, taken from Daniel Leborgne's book [216].

Question 3.8.5 (see [216, p. 16]). *Let u and v denote two objects not belonging to $\mathbb{R}$, and let $X = \mathbb{R} \cup \{u, v\}$ with the topology generated by the open subsets of the real line, together with the sets of the form*

$$\{\mathbb{R} \setminus \{x\}\} \cup \{u\} \quad and \quad \{\mathbb{R} \setminus \{x\}\} \cup \{v\}, \quad x \in \mathbb{R}.$$

Show that X is locally homeomorphic to an open subset of $\mathbb{R}$, but is not Hausdorff.

The topic of Exercise 3.8.7 is the construction of topological manifolds from a given one using the quotient topology. First some definitions. Consider a topological space X, and denote by Aut (X) the group of homeomorphisms from X onto itself. Recall that group actions were introduced in Definition 1.2.6.

Definition 3.8.6. A group action defines a relation in $X \times X$ by:

$$x \sim y \quad \Longleftrightarrow \quad \exists \, g \in G \quad \text{such that} \quad x = a(g)y. \tag{3.8.1}$$

The group G is said to act totally discontinuously on X if the following condition holds: Every $x \in X$ has a neighborhood V such that $a(g)V \cap a(g')V = \emptyset$ when $g \neq g'$. It is said to be separating on X if for every pair of points x and x' in X with $x \nsim x'$ there exist neighborhoods U and U' of x and x' such that $a(g)U \cap a(g')U' = \emptyset$ for every choice of $g, g' \in G$.

Exercise 3.8.7 (see [218, Proposition 0.2.1, p. 18]). *Let G be a group, let X be a topological manifold of dimension n, and let $a : G \mapsto \mathrm{Aut}\,(X)$ denote an automorphism of group.*

(1) *Show that* (3.8.1) *defines an equivalence relation.*
(2) *Assume that a is separating and totally discontinuous. Show that $X/\sim$ is a topological manifold of dimension n.*

Exercise 3.8.8. *The projective space $\mathbb{P}^n$: We consider $\mathbb{C}^{n+1} \setminus \{(0,0,\ldots,0)\}$, with the equivalence relation:*

$$x \sim y \quad \Longleftrightarrow \quad \exists \lambda \in \mathbb{C} \setminus \{0\} \quad \text{such that} \quad x = \lambda y.$$

The quotient space $\mathbb{P}^n = \mathbb{C}^{n+1} \setminus \{(0,0,\ldots,0)\} / \sim$ is called the complex projective space. Show that it is Hausdorff and compact when endowed with the quotient topology. Show that it is path-connected.

In the following exercise a homogeneous polynomial of three complex variables is a non-identically vanishing polynomial $p(z_1, z_2, z_3)$ with the following property: There exists $m \in \mathbb{N}$ such that:

$$p(\lambda z_1, \lambda z_2, \lambda z_3) = \lambda^m p(z_1, z_2, z_3), \quad \forall (z_1, z_2, z_3) \in \mathbb{C}^3 \quad \text{and} \quad \forall \lambda \in \mathbb{C}. \tag{3.8.2}$$

Note that the condition (3.8.2) allows us to set the condition $p(z) = 0$ for $z \in \mathbb{P}^2$.

Exercise 3.8.9. *Let p be a homogeneous polynomial. Show that the set*

$$C_p = \{z \in \mathbb{P}^2 \, : \, p(z) = 0\}$$

is compact.

Remark 3.8.10. The set C_p is called a projective (algebraic) curve. We follow the approach of Kirwan [204, p. 42] in the proof.

Definition 3.8.11. Let X be a topological manifold. A chart is a pair (U, φ), where U is an open set of X and φ is a homeomorphism from U onto an open subset of $\mathbb{R}^n$. Let $x \in X$. The chart (U, φ) is called a chart at the point x if $x \in U$.

A family of charts $(U_i, \varphi_i)_{i \in I}$ such that $\cup_{I \in I} U_i = X$ is called an atlas. Given two charts (U_1, φ_1) and (U_2, φ_2) such that $U_1 \cap U_2 \neq \emptyset$, the map

$$\varphi_2 \circ \varphi_1^{-1} : \; \varphi_1(U_1 \cap U_2) \longrightarrow \varphi_2(U_1 \cap U_2)$$

is called the transition map. It is continuous. Since both $\varphi_1(U_1 \cap U_2)$ and $\varphi_2(U_1 \cap U_2)$ are open subsets of $\mathbb{R}^n$, it makes sense to ask the transition maps to be differentiable, or analytic when n is even. This remark is used below in the definition of differentiable functions between manifolds.

Question 3.8.12. *Let X and Y be two connected topological manifolds, of dimension n and m respectively, and with given atlases $\mathcal{A}$ and $\mathcal{B}$. Let f be a function from X into Y.*

(1) *Show that f is continuous at the point x if and only if for every charts $(U, \varphi) \in \mathcal{A}$ and $(V, \psi) \in \mathcal{B}$ at the points x and $f(x)$ respectively, the map*

$$\psi \circ f \circ \varphi^{-1} \tag{3.8.3}$$

is continuous.

(2) *Show that the expression of the continuity in terms of local charts does not depend on the given atlases.*

Since the maps (3.8.3) are defined between open subsets of $\mathbb{R}^n$ and $\mathbb{R}^m$, it is natural to require that they be smooth, or when n and m are even, analytic. For the definition to make sense one has to require the transition maps to be differentiable, or analytic. The atlas is called differentiable if all the transition maps are smooth. In opposition with the continuous case, the definition of differentiable functions will now depend on the given atlases, up to the following equivalence relation (see [250, §1.1, p. 109]).

Definition 3.8.13. Two differentiable atlases are called equivalent if their union is still a differential atlas.

The above indeed defines an equivalence relation among differential atlases. An equivalence class of the above equivalence relation (see Definition 3.8.13) is called a differential structure. A given manifold may lack any differential structure, or have several non-equivalent ones. See the work of Milnor for the latter. The notion of differential manifold was introduced by Riemann in 1851; see for instance [210, Introduction, p. 7].

Definition 3.8.14. A manifold endowed with a differential structure will be called a differential manifold. A map between two differential manifolds will be called a diffeomorphism if it is one-to-one onto, and if it is, as well as its inverse, smooth. Two differential manifolds will be called diffeomorphic if there is a diffeomorphism between them.

Exercise 3.8.15. *Show that the unit sphere of $\mathbb{R}^{n+1}$ is a differential manifold.*

Exercise 3.8.16. *Show that the open unit* $\mathbb{D}$ *and the ellipse*

$$\mathcal{E} = \left\{ (x, y) \in \mathbb{R}\,;\, \frac{x^2}{a^2} + \frac{y^2}{b^2} \right\}$$

are diffeomorphic.

We note that these two sets are in fact conformally equivalent (that is, there exists a one-to-one onto analytic map between the two given open sets), as follows from Riemann's mapping theorem. The problem of finding a conformal map between the two is much more difficult than finding a diffeomorphism. The reader will check that the diffeomorphism proposed in the solution of Exercise 3.8.16 is not analytic.

Exercise 3.8.17. *Let* $R \in (1, \infty)$ *and let* A_R *denote the annulus defined by* (2.1.14):

$$A_R = \{ z \in \mathbb{C}\,;\, 1 < |z| < R \}.$$

Show that for any choice of R_1 *and* R_2 *in* $(1, \infty)$ *the annuli* A_{R_1} *and* A_{R_2} *are diffeomorphic.*

The previous exercise is far from being innocent. It says that from the differential geometry point of view, all annuli are the same. On the other hand, and as we will see below (Theorem 3.8.24 and Exercise 3.8.27), A_{R_1} and A_{R_2} will be conformally equivalent (that is, there exists a one-to-one onto analytic map from A_{R_1} onto A_{R_2}) if and only if $R_1 = R_2$. Then the map is a rotation.

Remark 3.8.18. More generally, any A_R is also diffeomorphic to $\mathbb{C} \setminus \{0\}$, or to $\mathbb{D} \setminus \{0\}$. But see Theorem 3.8.23 below for a bémol.

When n is even, one can require, in the equivalence relation given in Definition 3.8.13, the transition maps to be analytic rather than differentiable. An equivalence class of the new equivalence relation will be called an analytic structure, and the number $n/2$ is called the complex dimension of the manifold.

Exercise 3.8.19. *Show that the complex projective space* (*defined in Exercise* 3.8.8) *admits a complex analytic structure.*

Definition 3.8.20. A Riemann surface is an analytic manifold of complex dimension 1.

The Riemann sphere (or extended complex plane) is the simplest example of a compact Riemann surface.

Exercise 3.8.21. *Show that the extended complex plane is a compact Riemann surface.*

Definition 3.8.22. Two analytic manifolds are conformally equivalent if there is a homeomorphism between them which is moreover analytic.

For $R > 1$ recall that we have denoted by A_R the annulus

$$A_R = \{z \in \mathbb{C} \, ; 1 < |z| < R\}$$

(see (2.1.14)). In relation to Remark 3.8.18 we mention now:

Theorem 3.8.23 (see, for instance, [306, Théorème 2.7.3, p. 219]). *Let S be a Riemann surface homeomorphic to an annulus A_R. Then S is conformally equivalent to an annulus, to $\mathbb{C} \setminus \{0\}$, or to $\mathbb{D} \setminus \{0\}$.*

Bergman, see [70, p. 61], quotes Julia, see [195], for the following result. A proof based on elementary facts is proposed in Exercise 3.8.27.

Theorem 3.8.24. *Let R_1 and R_2 belong to $(1, \infty)$. The annuli A_{R_1} and A_{R_2} are not conformally equivalent for $R_1 \neq R_2$.*

The following exercise is taken from [91, Theorem 10.12 (i), p. 348]. It is used in the arguments on the nonconformal equivalence of the annuli A_{R_1} and A_{R_2} when $R_1 \neq R_2$.

Exercise 3.8.25. *Let $R > 1$ and let f be analytic and not vanishing in A_R. Show (using Exercise 2.3.6, and in particular without using notions of homotopy or homology of curves) that there exist $n \in \mathbb{Z}$ and g analytic in A_R such that*

$$f(z) = z^n e^{g(z)}, \quad z \in A_R. \tag{3.8.4}$$

Exercise 3.8.26.

(1) *Show that the function z has no analytic logarithm in A_R.*
(2) *Let f be a conformal map from A_{R_1} onto A_{R_2}. Show that f has no analytic logarithm.*

Hint: In fact $f(z) = z$ has no continuous logarithm in A_R. A clever elementary argument in [140] shows that z has no continuous squareroot in $\mathbb{C} \setminus \{0\}$; see also [CAPB, Exercise 4.1.9, p. 147]. This fact implies in particular that z has no continuous logarithm there. The arguments are the same when $\mathbb{C} \setminus \{0\}$ is replaced by A_R. For the second item see [91, Corollary 10.14, p. 349].

Exercise 3.8.27. *Prove Theorem 3.8.24: A_{R_1} is conformally equivalent to A_{R_2} if and only if $R_1 = R_2$.*

Hint: Assume that $R_1 > R_2$, and recall that a conformal map from A_{R_1} onto A_{R_2} has no analytic logarithm (see Exercise 3.8.26). Let f be a conformal mapping from A_{R_1} onto A_{R_2}, with representation (3.8.4). Then $n \neq 0$. Show that there exists a constant A such that:

$$0 \leq n \ln r + A \leq \ln R_2, \quad \forall r \in (1, R_1). \tag{3.8.5}$$

Conclude that $|n| = 1$ and $R_1 = R_2$ (in fact, $n = 1$ and $f(z) = cz$ where $c \in \mathbb{T}$, as is explained in the sequel).

The hint given above follows the proof in [91, Corollary 10.14, p. 349], and uses in particular Exercises 2.3.6 and 3.8.26. The third item in the hint is taken from [258]; see [91, Remark 6. 21, p. 202]. A much shorter argument can be found in [244, Chapter VII, p. 133]. It uses the boundary values of the function $\ln |f|$, a fact which we wanted to avoid, in order to keep the arguments elementary. Another shorter argument can be found in [286, p. 52], under the assumption that the function f is continuous on the boundary. The proof there uses Schwarz' reflection principle and the Casorati–Weierstrass theorem. See also [266, p. 337] for an argument where f is assumed continuous on the boundary. An even shorter argument uses the conformal invariance of Abelian integrals. This will not be discussed here.

The hint in fact gives the proof of the following result, which also follows from [91, Theorem 6.20, p. 201].

Theorem 3.8.28. *Let $R_1, R_2 > 1$ and assume that there exists an analytic map f from A_{R_1} into A_{R_2}.*

(1) *If $R_1 > R_2$, f has a logarithm, or equivalently,*
(2) *If f has no logarithm, $R_1 = R_2$.*

Much more is true, and the function f in Exercise 3.8.26 is equal to cz, where $|c| = 1$. See [91, Theorem 6.19, p. 200 and Theorem 6.20, p. 201]. The proof in [91] uses properties of analytic functions sending the strip $\mathbb{R} \times (-\frac{\pi}{2}, \frac{\pi}{2})$ into itself, and Schwarz' lemma. The fact that the form of f is fixed by the underlying geometry is an example of a rigidity theorem. See the discussion in [213] for instance.

Finally we recall (see for instance [244]):

Theorem 3.8.29. *Any doubly connected domain whose boundary consists of two smooth Jordan curves is conformally equivalent to some A_R.*

See also the discussion [272, pp. 142 and 169]

Exercise 3.8.30. (*The torus*) *Let $\tau \in \mathbb{C}$ be such that $\mathrm{Im}\ \tau > 0$ and let Γ denote the lattice*

$$\Gamma = \{z = n + \tau m, \ where \ n, m \in \mathbb{Z}\}.$$

Define a relation in $\mathbb{C}$ by:

$$z \sim w \quad \Longleftrightarrow \quad z - w \in \Gamma.$$

(1) *Show that $\mathbb{C}_\tau \overset{\mathrm{def.}}{=} \mathbb{C}/\sim$ is compact.*
(2) *Show that $\mathbb{C}_\tau$ is a complex manifold.*

Exercise 3.8.31. *Show that two torii are always diffeomorphic.*

Note that one can also build a torus as the "double" of an annulus. In a way similar to Exercise 3.8.27 one can ask when are two torii conformally equivalent. We will not present this as an exercise but recall the result (see for instance [140,

p. 315]: Two torii C_τ and $C_{\tau'}$ are conformally equivalent if and only if there exists a matrix

$$M = \begin{pmatrix} a & b \\ c & d \end{pmatrix} \in \mathbb{Z}^{2 \times 2}$$

such that $ad - bc = 1$ (that is $M \in SL(2, \mathbb{Z})$) and such that

$$\tau' = \frac{a\tau + b}{c\tau + d}.$$

Exercise 3.8.32. *Recall that the extended plane $\widehat{\mathbb{C}}$ has been defined in Exercise 3.2.7, and let $\mathbb{C}_\tau$ be as in the previous exercise.*

(1) *Show that there is a one-to-one correspondence between continuous functions from $\mathbb{C}_\tau$ into $\widehat{\mathbb{C}}$ and continuous biperiodic functions from $\mathbb{C}$ into $\widehat{\mathbb{C}}$, that is continuous functions such that*

$$f(z + n + \tau m) = f(z), \quad \forall z \in \mathbb{C} \quad and \quad n, m \in \mathbb{Z}. \tag{3.8.6}$$

(2) *Show that there is a one-to-one correspondence between analytic functions from $\mathbb{C}_\tau$ into $\widehat{\mathbb{C}}$ and meromorphic biperiodic functions from $\mathbb{C}$ into $\widehat{\mathbb{C}}$ (that is, elliptic functions).*

More generally than the torus one has:

Exercise 3.8.33. *Let $A \in \mathbb{C}^{g \times g}$ be such that $\operatorname{Im} A > 0$, and let Γ denote the lattice $\mathbb{Z}^g + A\mathbb{Z}^g$.*

(1) *Show that the quotient space $\mathbb{C}^g/\Gamma$ is Hausdorff, compact and path-connected.*
(2) *Show that $\mathbb{C}^g/\Gamma$ is a complex manifold.*

Not every manifold can have a Riemann surface structure. It must be triangulable and orientable. But:

Exercise 3.8.34. *Let X denote a topological manifold of dimension 2, and let Y denote a Riemann surface. Let p denote a local homeomorphism from X into Y. There is a unique complex structure on Y such that p is analytic.*

3.9 Metric spaces

We recall that a metric (we will also say *distance*) on a space E is a map d from the Cartesian product $E \times E$ into $[0, \infty)$ with the following properties: For all $x, y, z \in E$ it holds that:

(1) $d(x, y) = 0$ if and only if $x = y$,   (non-degeneracy)
(2) $d(x, y) = d(y, x)$,   (symmetry)
(3) $d(x, y) \leq d(x, z) + d(z, y)$ (triangle inequality).

3.9. Metric spaces

When d satisfies (2) and (3), but with (1) replaced by the weaker requirement:

(1)' $x = y \implies d(x, y) = 0$,

we will call d a semi-metric.

The metric (distance) d is called ultra-metric (or an ultra-metric distance) if satisfies the condition

(3)' $d(x, z) \leq \max \{d(x, y), d(y, z)\}$,

which is stronger than (3).

Definition 3.9.1. A metric space is a pair (E, d), where d is a metric on E. An ultra-metric space is a pair (E, d), where d is a metric on E which is moreover ultra-metric.

Every non-empty set can be given a (not very interesting) metric structure, as is explained in the following exercise.

Exercise 3.9.2. *Let E be a non-empty set. Show that*

$$d(x, y) = \begin{cases} 0 & \text{if } x = y, \\ 1 & \text{if } x \neq y, \end{cases}$$

defines a metric on E.

Exercise 3.9.3. *Let d be a semi-metric (resp. a metric) on the space E. Then, $\frac{d}{1+d}$ is also a semi-metric (resp. a metric) on E. In particular, if $(d_n)_{n \in \mathbb{N}}$ is a family of semi-metrics on E, so is*

$$d(x, y) = \sum_{n=1}^{\infty} \frac{1}{2^n} \frac{d_n(x, y)}{1 + d_n(x, y)}. \tag{3.9.1}$$

Give a condition on the d_n's for d defined by (3.9.1) to be a metric.

The functions $\varphi(d) = \frac{d}{1+d}$ and $\varphi(d) = \min\{1, d\}$ satisfy $\varphi(0) = 0$ and

$$c \leq a + b \implies \varphi(c) \leq \varphi(a) + \varphi(b), \quad \forall a, b, c \in [0, \infty), \tag{3.9.2}$$

and they define metrics when d is a metric (see the previous exercise for the case $d/(1 + d)$). More generally:

Exercise 3.9.4. *Let (E, d) be metric space and let φ be a function from $\mathbb{R}_+$ into itself vanishing at the origin and satisfying (3.9.2). Show that $(E, \varphi(d))$ is a semi-metric space. Give a condition on φ for $\varphi(d)$ to define a metric.*

We note that the functions

$$|x|^h, \quad (h \in (0, 1)), \quad \ln(1 + x)$$

and

$$\varphi(x) = \begin{cases} \sin(x), & x \in [0, \pi/2), \\ 1, & x \geq \pi/2 \end{cases}$$

satisfy (3.9.2).

We use the notation

$$B(x, r) = \{y \in E \, ; \, d(x, y) < r\}.$$

$B(x, r)$ is called the open ball with center x and radius r. The metric d defines a topology on the metric space (E, d), namely the topology generated by the balls $B(x, r)$. It is the smallest topology for which d is continuous.

Question 3.9.5 (see, e.g., [1, p. 60]). *Define the closed ball with center x and radius r to be*

$$B_c(x, r) = \{y \in E \, ; \, d(x, y) \leq r\}.$$

Show that $B_c(x, r)$ is closed in the topology defined by d, but that it need not be equal to the closure of $B(x, r)$, and that its interior need not be $B(x, r)$.

Hint: Use Exercise 3.9.2 to build counterexamples.

Question 3.9.6 (see [1, p. 77]). *Let (E, d) be a metric space and assume that d is ultra-metric. Show the following:*

(1) *An open ball is both open and closed in the topology defined by d.*
(2) *Show that a series $\sum_{n=0}^{\infty} x_n$ converges if and only if $\lim_{n \to \infty} x_n = 0$.*
(3) *Describe the triangles of an ultra-metric space.*

Question 3.9.7. *Consider the field $\mathbb{C}(X)$ of rational functions and for $r \neq 0 \in \mathbb{C}(X)$ let $\nu(r)$ be defined by*

$$r(X) = X^{\nu(r)} \frac{p(X)}{q(X)},$$

where p and q are polynomials not divisible by X. Define for r_1 and r_2 in $\mathbb{C}(X)$:

$$d(r_1, r_2) = \begin{cases} 0, & \text{if } r_1 = r_2, \\ 2^{-\nu(r_1 - r_2)}, & \text{if } r_1 \neq r_2. \end{cases}$$

(1) *Show that d defines an ultra-metric distance.*
(2) *Does the series $\sum_{n=0}^{\infty} X^n$ converge with respect to this metric?*
(3) *Does the series $\sum_{n=0}^{\infty} X^{-n}$ converge with respect to this metric?*

In quite a number of cases of interest, a metric does not appear alone, but one is given a countable collection of metrics. The following exercise pertains to this case.

3.9. Metric spaces

Question 3.9.8. *Let* $d_1, d_2, \ldots$ *be a sequence of semi-metrics. Show that the smallest topology with respect to which these semi-metrics are continuous is defined by the semi-metric* (3.9.1):

$$d(x, y) = \sum_{n=1}^{\infty} \frac{1}{2^n} \frac{d_n(x, y)}{1 + d_n(x, y)}.$$

The space ℓ_∞ of bounded complex sequences indexed by $\mathbb{N}_0$, with the supremum norm is a Banach space, but its closed unit ball $\overline{\mathbb{D}}^{\mathbb{N}_0}$ is not compact (recall Riesz' theorem which asserts that the closed unit ball of a normed vector space is compact if and only if the space is finite dimensional). On the other hand the set $\overline{\mathbb{D}}^{\mathbb{N}}$ endowed with entry-wise convergence plays an important role in the study of convergence of sequences of functions in the open unit disk.

Exercise 3.9.9. *The space* $\overline{\mathbb{D}}^{\mathbb{N}}$ *endowed with the metric*

$$d(z, w) = \sum_{n=1}^{\infty} \frac{1}{2^n} \frac{|z_n - w_n|}{1 + |z_n - w_n|}, \tag{3.9.3}$$

where $z = (z_n)_{n \in \mathbb{N}}$ *and* $w = (w_n)_{n \in \mathbb{N}}$ *are in* $\overline{\mathbb{D}}^{\mathbb{N}}$, *is a compact metric space. Furthermore, the sequence* $(z^{(p)})_{p \in \mathbb{N}}$ *of sequences in* $\overline{\mathbb{D}}^{\mathbb{N}}$ *converges to the sequence* $z \in \overline{\mathbb{D}}^{\mathbb{N}}$ *if and only if*

$$\lim_{p \to \infty} z_n^{(p)} = z_n, \quad \forall n \in \mathbb{N}.$$

In a similar vein, the space $\mathbb{R}^{\mathbb{N}}$ endowed with the metric (3.9.3) plays an important role in the study of probability distributions on linear space. See Vakhania's book [312, Chapter 1]. Question 5.4.5 is dedicated to this example.

Definition 3.9.10. Let (E, d) be a metric space. The sequence $(a_n)_{n \in \mathbb{N}}$ of elements of E is called a Cauchy sequence if:

$$\forall \epsilon > 0, \ \exists N \in \mathbb{N}, \ n, m > N \implies d(a_n, a_m) < \epsilon.$$

The metric space is called complete if every Cauchy sequence has a limit (which is then unique).

Note that completeness as introduced in the preceding definition is a metric notion. One can also define completeness in the setting of topological vector spaces using Cauchy filters rather than Cauchy sequences.

Two different, and non-equivalent metrics, may define the same topology, the space being complete with respect to one of the metrics and not complete with respect to the other one. This is illustrated by the following classical example (see for instance [173, p. 162]).

Exercise 3.9.11. *Let $E = (0, 1]$ and define*

$$d_1(x, y) = |x - y| \quad and \quad d_2(x, y) = \left| \frac{1}{x} - \frac{1}{y} \right|.$$

Show that d_1 and d_2 define the same topology on E and that (E, d_1) is not complete while (E, d_2) is complete.

In contrast with Theorem 3.2.14, the following two theorems connect the fact of being second category (which is a topological notion) to a non-topological notion, namely completeness of a metric space. Among the numerous applications of Theorem 3.9.12, we mention the fact that a Fréchet space is barreled. See Question 5.2.13.

Theorem 3.9.12. *A complete metric space is a Baire space.*

More generally (see for instance [114, (12.6.1), p. 82]).

Theorem 3.9.13. *Let E be a topological metric space in which every point has a neighborhood homeomorphic to a complete metric space. Then, E is a Baire space (i.e., is second category).*

In other words, a complete metric space, or a space as in Theorem 3.9.13, cannot be written as a countable union of nowhere dense sets.

A topological space is called *metrizable* if its topology can be defined by a metric. Not all the spaces considered here will be metrizable spaces. See Section 5.8 for instance. The Nagata–Smirnov theorem characterizes metrizable spaces. See [78, TG IX.109, Exercice 32]. We will not recall its precise formulation here, but only mention the following corollary for future reference. Recall that regular space has been defined in Definition 3.1.12.

Theorem 3.9.14. *Every regular space with a countable basis of neighborhoods is metrizable.*

It follows from Theorem 3.9.14 that manifolds, and in particular Riemann surfaces, are metrizable spaces. We give the following complement to Theorem 3.9.14 as a question (see [78, Proposition 16, TG IX.21]). This question allows us also to make a transition to our next topic, namely compact metric spaces.

Question 3.9.15. *A compact space is metrizable if and only if its topology admits a countable basis.*

The metric space E is called *totally bounded* if the following condition holds:

$$\forall R > 0, \ \exists N \in \mathbb{N} \text{ and } x_1, \ldots, x_N \in E \text{ such that } E = \cup_{n=1}^{N} B(x_n, R).$$

Question 3.9.16. *Is a compact metric space totally bounded?*

Theorem 3.9.17.
  (1) *A compact metric space is complete.*
  (2) *A closed subset of a complete space is complete.*
  (3) *A complete subset of a metric space is closed.*

The following exercise appears in [269, p. 29] in the setting of the Riemann sphere $\widehat{\mathbb{C}}$. It is a special case of Exercise 3.2.8.

Exercise 3.9.18. *Let F_1 and F_2 denote disjoint closed sets in the compact metric space (E, d). Show that there are open sets O_1 and O_2 such that*

$$F_1 \subset O_1 \quad and \quad F_2 \subset O_2,$$

and
$$\overline{O_1} \cap \overline{O_2} = \emptyset. \tag{3.9.4}$$

Assume that $E = \widehat{\mathbb{C}}$. Does the result remain true if one replaces $\widehat{\mathbb{C}}$ by $\mathbb{C}$?

For the next exercise, recall that ℓ_2 denotes the space of sequences of complex numbers $a = (a_n)_{n \in \mathbb{N}}$ such that

$$\sum_{n=1}^{\infty} |a_n|^2 < \infty. \tag{3.9.5}$$

The expression $\|a\| = \sqrt{\sum_{n=1}^{\infty} |a_n|^2}$ is then a norm (see Definition 4.1.1), which defines the topology of ℓ_2 via the metric

$$d(a, b) = \sqrt{\sum_{n=1}^{\infty} |a_n - b_n|^2}, \quad a, b \in \ell_2.$$

Exercise 3.9.19. *Show that the set*

$$C = \left\{ a \in \ell_2 \; ; \sum_{n=1}^{\infty} (n+1)^2 |a_n|^2 \leq 1 \right\}$$

is compact in ℓ_2.

Hint: Use the diagonal process.

More generally, we have:

Exercise 3.9.20. *Let $\alpha_n > 0$ for $n = 1, 2, \ldots$. The set*

$$C(\alpha_1, \alpha_2, \ldots) = \left\{ a \in \ell_2 \; ; \sum_{n=1}^{\infty} \alpha_n |a_n|^2 \leq 1 \right\}$$

is compact if and only if

$$\sum_{n=1}^{\infty} \frac{1}{\alpha_n} < \infty.$$

Exercise 3.9.21. *Show that in a metric space compactness and sequential compact-ness are equivalent.*

Hint: Let X be the given metric space. Show first that given an open covering $X = \cup_{i \in I} O_i$ there exists $\epsilon > 0$ such that

$$\forall x \in X, \; \exists i \in I, \; B(x, \epsilon) \subset O_i. \tag{3.9.6}$$

Based on the preceding claim, show that X admits finite coverings made of open balls of arbitrary small radius.

We note that (half of) the previous exercise is used in the proof that a subset of $\mathcal{H}(\Omega)$ is compact if and only if it is closed and bounded (and in particular is a normal family). See [95, pp. 166–170]. The boundedness has to be understood in the topology of $\mathscr{H}(\Omega)$; see Definition 5.1.17. We will go back to normal families in Section 5.7, but already send the interested reader to [4, pp. 219–227].

If f is a function between two topological spaces E and F, then continuity of f at a point $t \in E$ implies sequential continuity, but the converse is not true in general. When the domain is a metric space, the situation is nicer, as shown in the following two exercises; see, e.g., [146, Théorème 2, p. 92] and (in the setting of vector spaces) [307, Proposition 8.5, p. 76].

Exercise 3.9.22. *Let (E, d) be a metric space, and let $(F, \mathcal{T})$ be a topological space. Show that the function f from E into F is continuous at the point $t \in E$ if and only if for any sequence of points $(t_n)_{n \in \mathbb{N}}$ of E such that $\lim_{n \to \infty} d(t_n, t) = 0$ it holds that*

$$\lim_{n \to \infty} f(t_n) = f(t). \tag{3.9.7}$$

Continuity between metric spaces has a special property, when the domain space is compact.

Theorem 3.9.23. *Let (E_1, d_1) and (E_2, d_2) be two metric spaces, and assume that E_1 is compact. Let f be continuous from E_1 into E_2. Then, f is uniformly con-tinuous, that is:*

$$\forall \epsilon > 0, \; \exists \eta > 0, \; \forall x, y \in E_1, \; d_1(x, y) < \eta \implies d_2(f(x), f(y)) < \epsilon. \tag{3.9.8}$$

Exercise 3.9.24. *Let (E, d) be a compact metric space, and let f be a map from E into itself which preserves distance:*

$$d(f(x), f(y)) = d(x, y), \quad \forall x, y \in E. \tag{3.9.9}$$

Show that f is onto.

We note that in [78, Exercice 10, p. TG IX.92], the following stronger result is proposed as an exercise: Condition (3.9.9) is replaced by the weaker requirement

$$d(f(x), f(y)) \geq d(x, y). \tag{3.9.10}$$

Compactness of the space implies that f preserves distance and is onto, that is, is an isometry of metric spaces.

Let (E, d) be a metric space, and let CS be the set of Cauchy sequences in E. The elements $(x_n)_{n \in \mathbb{N}}$ and $(y_n)_{n \in \mathbb{N}}$ in CS are said to be equivalent if

$$\lim_{n \to \infty} d(x_n, y_n) = 0.$$

This defines indeed an equivalence class $\sim$, and the completion E, which is unique up to a unitary mapping which preserves E, can be defined as $CS/\sim$, with the distance

$$D(a, b) = \lim_{n \to \infty} d(x_n, y_n), \quad a, b \in CS/\sim,$$

where $(x_n) \in a$ and $(y_n) \in b$.

See Exercise 5.1.3 for an illustration on completion of pre-Hilbert spaces.

The following discussion and example are taken, with minor changes, from [166, §2.2, p. 13–14]. The question is as follows: Consider a vector space E, with two norms $\|x\|_1$ and $\|x\|_2$ such that

$$\|x\|_1 \leq \|x\|_2, \quad \forall x \in E. \tag{3.9.11}$$

Let E_1 and E_2 be the completions of E with respect to the norms $\|x\|_1$ and $\|x\|_2$ respectively. For $x = (x_n)_{n \in \mathbb{N}}$ a Cauchy sequence in E, let $e_1(x) \in E_1$ and $e_2(x) \in E_2$ be the associated equivalence classes in E_1 and E_2 respectively. Any Cauchy sequence with respect to $\| \cdot \|_2$ is a Cauchy sequence with respect to $\| \cdot \|_1$, and it would be tempting to identify $e_2(x)$ with $e_1(x)$, and say that $E_2 \subset E_1$. But this is not quite so. The map i which to $e_2(x)$ associates $e_1(x)$

$$i(e_2(x)) = e_1(x)$$

is indeed well defined, but the problem is that it need not be one-to-one. Two norms for which the map i is one-to-one are called compatible. An important case of such norms is when the spaces are reproducing kernel Hilbert spaces.

Exercise 3.9.25. *Show that the map i is well defined.*

Exercise 3.9.26 (see also Exercise 5.2.3). *Let E denote the space of continuous functions on $[0, 1]$ which admit a right derivative at the origin, endowed with the following two norms:*

$$\|x\|_1 = \max_{t \in [0,1]} |x(t)| \quad \text{and} \quad \|x\|_2 = \max_{t \in [0,1]} |x(t)| + |x'(0)|,$$

and let E_1 and E_2 denote the completions of E with respect to $\| \cdot \|_1$ and $\| \cdot \|_2$ respectively. Show that the map which to $e_2(x)$ associates $e_1(x)$ is not one-to-one.

As we already mentioned, the Riemann sphere can be seen as the one point compactification $\mathbb{C} \cup \{\infty\}$ of the complex plane. This is stressed out in the following question, where we denote by

$$\mathbb{S}_2 = \{(x_1, x_2, x_3) \in \mathbb{R}^3 \; ; \; x_1^2 + x_2^2 + x_3^2 = 1\}$$

the unit sphere of $\mathbb{R}^3$.

Question 3.9.27. *For* $(x_1, x_2, x_3) \in \mathbb{S}_2 \setminus \{(0, 0, 1)\}$, *define* $\varphi(x_1, x_2, x_3)$ *to be the intersection of the line defined by the points* $(0, 0, 1)$ *and* (x_1, x_2, x_3) *with the complex plane. Show that*

$$\varphi(x_1, x_2, x_3) = \frac{x_1 + ix_2}{1 - x_3}, \tag{3.9.12}$$

and that φ *is a bijection between* $\mathbb{S}_2 \setminus \{(0, 0, 1)\}$ *and* $\mathbb{C}$, *with inverse given by*

$$\varphi^{-1}(u + iv) = \left(\frac{2u}{u^2 + v^2 + 1}, \frac{2v}{u^2 + v^2 + 1}, \frac{u^2 + v^2 - 1}{u^2 + v^2 + 1} \right). \tag{3.9.13}$$

Setting $z = u + iv$, (3.9.13) may be rewritten as:

$$\varphi^{-1}(z) = \left(\frac{z + \bar{z}}{|z|^2 + 1}, \frac{z - \bar{z}}{i(|z|^2 + 1)}, \frac{|z|^2 - 1}{|z|^2 + 1} \right).$$

The map (3.9.12) is called the *stereographic projection*.

The geometrical interpretation of the point at infinity is as follows: The map φ is extended to the point $(0, 0, 1)$ by

$$\varphi(0, 0, 1) = \infty, \tag{3.9.14}$$

and going to ∞ on the complex plane means going to $(0, 0, 1)$ on the Riemann sphere. More precisely, recall that, by definition, a sequence of complex numbers $(z_n)_{n \in \mathbb{N}}$ tends to infinity if

$$\lim_{n \to \infty} |z_n| = +\infty, \tag{3.9.15}$$

that is, if and only if

$$\lim_{n \to \infty} \varphi^{-1}(z_n) = (0, 0, 1), \tag{3.9.16}$$

where this last limit can be understood in two equivalent ways: The first, and simplest, is just to say that the limit is coordinate-wise in $\mathbb{R}^3$. The second is to view $\mathbb{S}_2$ as a topological manifold, and see the limit in the corresponding topology. See also Question 3.9.27 above, where φ allows to define a metric on the Riemann sphere, called the stereographic metric.

Theorem 3.9.28. *The function*

$$d(z,w) = \|\varphi^{-1}(z) - \varphi^{-1}(w)\|_{\mathbb{R}^3} \tag{3.9.17}$$

$$= \begin{cases} \dfrac{2|z-w|}{\sqrt{1+|z|^2}\sqrt{1+|w|^2}}, & z, w \in \mathbb{C}, \\[2mm] \dfrac{2|w|}{\sqrt{1+|w|^2}}, & z = \infty \quad \text{and} \quad w \in \mathbb{C}, \\[2mm] 0, & z = w = \infty, \end{cases} \tag{3.9.18}$$

where φ is defined by (3.9.12) and (3.9.14) is a metric on the Riemann sphere.

In the thirties of the previous century, K. Menger considered the following problem (see [233], [279, p. 788]).

Problem 3.9.29. *When can a metric space be isometrically imbedded in ℓ_2.*

This problem has interesting connections with positive definite kernels (or, more precisely, with the related notion of conditionally negative kernel). See Theorem 7.3.7.

3.10 Solutions of the exercises

Solution of Exercise 3.1.2:

(1) The empty set belongs to $\mathcal{O}$ by definition, and $\mathbb{R}$ belongs also to $\mathcal{O}$ since, for instance

$$(x-1, x+1) \subset \mathbb{R}, \quad \forall x \in \mathbb{R}.$$

Let now $(O_i)_{i \in I}$ denote a family of sets in $\mathcal{O}$, and let $O = \cup_{i \in I} O_i$. Let $x \in O$, and let $i \in I$ be such that $x \in O_i$. There exists $r_{x,i} > 0$ such that

$$(x - r_{x,i}, x + r_{x,i}) \subset O_i. \tag{3.10.1}$$

In particular, $(x - r_{x,i}, x + r_{x,i}) \subset O$ and this shows that $O \in \mathcal{O}$.

Assume now I finite, $I = \{1, \ldots, N\}$, let $x \in \cap_{i=1}^N O_i$, and let $r_{x,1}, \ldots, r_{x,N}$ be strictly positive numbers such that (3.10.1) holds. Let

$$r = \min\{r_{x,1}, \ldots, r_{x,N}\}.$$

Then, $(x - r, x + r) \subset \cap_{i=1}^N O_i$, and so $\cap_{i=1}^N O_i \in \mathcal{O}$.

(2) Let O be a non-empty open subset of $\mathbb{R}$, and let $x \in O$. There exists an open interval $(x - \mu, x + \nu)$ containing x and included in O. Let μ fixed, and define

$$I_x = \{\eta > 0 \,;\, (x - \mu, x + \eta) \subset O\}.$$

If I_x is not bounded from above, then

$$I_x = (x - \mu, \infty).$$

Otherwise, since every bounded set of real numbers has an upper bound, we can define r_x to be the supremum of the set I_x. Using a similar argument for the numbers μ we see that one, and only one of the following non-overlapping cases can happen:

(1) $I_x = \mathbb{R}$.
(2) $I_x = (x - l_x, \infty)$.
(3) $I_x = (-\infty, x + r_x)$.
(4) $I_x = (x - l_x, x + r_x)$.

The first case is clear: Then, $O = \mathbb{R}$. We now turn to the case where only intervals of the form (4) appear. We claim that if x and y are in O, either $I_x = I_y$ or $I_x \cap I_y = \{\emptyset\}$. Indeed, suppose that $I_x \cap I_y \neq \{\emptyset\}$. Then

$$I_x \cup I_y = (\min(x - l_x, y - l_y), \max(x + r_x, y + r_y)).$$

Assume that $x + r_x < y + r_y$. Then we obtain a contradiction with the fact that r_x is maximum. The other cases are treated similarly, and hence $I_x = I_y$.

We can therefore write O as a union of disjoint intervals. To check that this union is countable, it suffices to pick up a rational point in each interval. One obtains this way a one-to-one map from the intervals composing O into the rational numbers.

$\square$

Solution of Exercise 3.1.4: Let $O_i = \mathbb{R} \setminus U_i, i \in I$ be a family of sets of $\mathcal{T}$, where I is some index set. Then,

$$\cup_{i \in I} O_i = \mathbb{R} \setminus \cap_{i \in I} U_i$$
$$\cap_{i \in I} O_i = \mathbb{R} \setminus \cup_{i \in I} U_i.$$

For an *arbitrary* set of index I, the intersection $\cap_{i \in I} U_i$ of finite sets is still finite or empty. Thus, an arbitrary union of sets in $\mathcal{T}$ is still in $\mathcal{T}$. On the other hand, we need the set index to be finite to insure that the union $\cup_{i \in I} U_i$ of finite sets is always a finite set. Thus a finite intersection of sets in $\mathcal{T}$ is still in $\mathcal{T}$, and we have a topology since $\mathbb{R}$ and $\emptyset$ are also in $\mathcal{T}$.

$\square$

As already mentioned, the topology in the preceding exercise is called the Zariski topology. It is not separated because any two non-empty open sets always intersect.

Solution of Exercise 3.1.5: We first show that an arbitrary intersection of topologies on X is still a topology. Let $(\mathcal{O}_j)_{j \in J}$ be an arbitrary family of topologies on X. Thus

$$\forall j \in J, \quad \mathcal{O}_j \subset \mathcal{P}(X).$$

Since X and $\emptyset$ belong to every $\mathcal{O}_j$ they belong to the intersection $\cap_{j \in J} \mathcal{O}_j$. Let now a family $(O_i)_{i \in I}$ of sets of X such that

$$\forall i \in I, \forall j \in J, \ O_i \in \mathcal{O}_j.$$

3.10. Solutions of the exercises

In particular, $\cup_{i \in I} O_i \in \mathcal{O}_j$ for every $j \in J$ and so

$$\cup_{i \in I} O_i \in \cap_{j \in J} \mathcal{O}_j.$$

Let now $N \in \mathbb{N}$ and $O_1, \ldots, O_N$ be subsets of X which belong to every topology $\mathcal{O}_j$ $(j \in J)$. Thus $\cap_{i=1}^N O_i \in \mathcal{O}_j$ for every $j \in J$, which implies that $\cap_{i=1}^N O_i \in \cap_{j \in J} \mathcal{O}_j$. $\square$

Solution of Exercise 3.1.6: Let $(A_i)_{i \in I}$ be an arbitrary family of sets in $\mathcal{O}_A$. By definition, there exists a family $(O_i)_{i \in I}$ of open sets of X such that

$$A_i = A \cap O_i, \quad i \in I.$$

Thus

$$\cup_{i \in I} A_i = A \cap (\cup_{i \in I} O_i),$$

which belongs to $\mathcal{O}_A$ since $\cup_{i \in I} O_i \in \mathcal{O}$. Similarly,

$$\cap_{i \in I} A_i = A \cap (\cap_{i \in I} O_i),$$

belongs to $\mathcal{O}_A$ as soon as I is finite since then $\cap_{i \in I} O_i \in \mathcal{O}$. Finally, $A = A \cap X$ and $\emptyset = A \cap \emptyset$ belong to $\mathcal{O}_A$ since both X and $\emptyset$ belong to $\mathcal{O}$.

Taking $O = X$ we have $A \cap O = A$, and thus a necessary condition for $\mathcal{O}_A \subset \mathcal{O}$ to hold is that A is open. This condition is also sufficient since the intersection of two open sets is open. $\square$

Solution of Exercise 3.1.8: Assume A open, and let $a \in A$. Then, A is a neighborhood of a by the very definition of a neighborhood. Conversely, assume that A is a neighborhood of each of its points. Then, for every $a \in A$ there exists an open set $O_a \subset A$ such that $a \in O_a$. Thus

$$A \subset \cup_{a \in A} O_a \subset A,$$

and so $A = \cup_{a \in A} O_a$ is open as a union of open sets. $\square$

Solution of Exercise 3.1.10:

(1) We note that a subset of $\mathbb{R}^2$ is in $\mathcal{T}$ if and only if it can be written as a union of set of the form $B(z, r)$. Thus, any union of elements in $\mathcal{T}$ is still in $\mathcal{T}$. Let now $O_1, \ldots, O_N$ be N elements of $\mathcal{T}$ with a non-empty intersection. Let $z \in \cap_{j=1}^N O_j$. There exist strictly positive numbers $r_1, \ldots, r_N$ such that

$$B(z, r_j) \subset O_j, \quad j = 1, \ldots, N,$$

and so, with $r = \min\{r_1, \ldots, r_N\}$, we have $B(z, r) \subset \cap_{j=1}^N O_j$. We also note that $\emptyset$ and $\mathbb{R}^2$ belong to $\mathcal{T}$. The topological space $(\mathbb{R}^2, \mathcal{T})$ is Hausdorff since for z_1 and z_2 two different points of $\mathbb{R}^2$ we have $B(z_1, \frac{|z_1 - z_2|}{2}) \cap B(z_2, \frac{|z_1 - z_2|}{2}) = \emptyset$.

(2) It suffices to see that any point in $\mathbb{R}^2$ and ∞ have disjoint neighborhoods, but this is clear from

$$B(z_0, 1) \cap B(\infty, |z_0| + 1) = \emptyset.$$

Indeed, for $z \in B(z_0, 1)$ we have

$$|z| = |z_0 + (z - z_0)| \le |z_0| + |z - z_0| < |z_0| + 1,$$

while $|z| > |z_0| + 1$ for $z \in B(\infty, |z_0| + 1)$.

$\square$

Solution of Exercise 3.1.13: The direct implication is clear. It suffices to take as open covering of X the set X itself. To prove the converse statement, assume that $(O_i)_{i \in I}$ is a covering of F with the required property: $F \cap O_i$ is closed in O_i (in the induced topology) for every $i \in I$. Let $x \notin F$, and let $i \in I$ be such that $x \in O_i$. We have

$$x \in O_i \setminus (O_i \cap F).$$

By hypothesis, the set $O_i \cap F$ is closed in O_i, and so $O_i \setminus (O_i \cap F)$ is open in O_i, and so open in X. Hence, $O_i \setminus (O_i \cap F)$ is a neighborhood of x. Since $O_i \setminus (O_i \cap F) \subset X \setminus F$, we have $X \setminus F$ is open, and F is closed.

$\square$

The above result is used in [172, Proof of Lemma 3.1, p. 15] to prove that a regular function on a quasi-affine variety is continuous in the Zarisky topology.

Solution of Exercise 3.1.14: Since

$$A_i \subset \cup_{i \in I} A_i \subset \overline{\cup_{i \in I} A_i},$$

we have by definition of the closure

$$\overline{A_i} \subset \overline{\cup_{i \in I} A_i},$$

and so

$$\cup_{i \in I} \overline{A_i} \subset \overline{\cup_{i \in I} A_i}.$$

Assume now that I is finite, say $I = \{1, 2, \ldots, N\}$. Since

$$A_n \subset \overline{A_n}$$

we have

$$\cup_{n=1}^{N} A_n \subset \cup_{n=1}^{N} \overline{A_n}.$$

Since we have a *finite* number of sets, $\cup_{n=1}^{N} \overline{A_n}$ is closed, and it follows from the definition of the closure that

$$\overline{\cup_{n=1}^{N} A_n} \subset \cup_{n=1}^{N} \overline{A_n},$$

and hence equality holds in (3.1.5).

That equality in (3.1.5) need not hold for an infinite number of sets can be seen by taking $A_n = \{r_n\}$, where $(r_n)_{n \in \mathbb{N}}$ is an enumeration of the rational numbers. Then, $\cup_{n \in \mathbb{N}} A_n = \mathbb{Q}$, and in the usual topology of $\mathbb{R}$,

$$\overline{\cup_{n \in \mathbb{N}} A_n} = \mathbb{R},$$

while

$$\cup_{n \in \mathbb{N}} \overline{A_n} = \cup_{n \in \mathbb{N}} A_n = \mathbb{Q}. \qquad \square$$

Solution of Exercise 3.1.15: (see [166, p. 4]) Let $N(0, x_1, \ldots, x_N, \epsilon)$ be a neighborhood of the function $f(x) \equiv 0$ (which we denote by 0). It contains functions vanishing at the points $x_1, \ldots, x_N$, for instance the function

$$f(x) = \begin{cases} 0, & \text{if } x \in \{x_1, \ldots, x_N\}, \\ 1, & \text{otherwise.} \end{cases}$$

So $0 \in \overline{E}$.

On the other hand, assume that there is a sequence $f_1, f_2, \ldots$ of elements of E such that

$$0 = \lim_{n \to \infty} f_n. \qquad (3.10.2)$$

The set of points Ω where at least one of the functions f_n vanishes is countable. Let $y \in [0, 1] \setminus \Omega$. Then $f_n(y) = 1$ for $n = 1, 2, \ldots$ and $f_n \notin N(0, y, 1/2)$, which contradicts (3.10.2). $\qquad \square$

Solution of Exercise 3.1.16: Let E be the underlying topological space. We have

$$\partial\Omega = \overline{\Omega} \cap \left(\overline{E \setminus \Omega}\right) = \overline{\Omega} \cap (E \setminus \Omega) \quad \text{since } \Omega \text{ is open.}$$

Thus

$$\partial\Omega \cap \Omega = \overline{\Omega} \cap (E \setminus \Omega) \cap \Omega = \emptyset. \qquad \square$$

Solution of Exercise 3.1.17: The first claim follows from

$$\overset{\circ}{E} \cap \partial E = \overset{\circ}{E} \cap (\overline{E} \cap (X \setminus \overset{\circ}{E})),$$

while the second claim follows from

$$\overset{\circ}{E} \cup \partial E = \overset{\circ}{E} \cup (\overline{E} \cap (X \setminus \overset{\circ}{E})) = \overline{E} \cap X = \overline{E}. \qquad \square$$

Solution of Exercise 3.1.19: We first recall (see Section 1.3) that an algebraic supplement always exists. Also recall the notation (1.3.1). Let F_1 and F_2 be two linear subspaces of $\mathcal{V}$, of finite codimension, with finite-dimensional direct supplements E_1 and E_2:

$$\mathcal{V} = F_1 \dotplus E_1 = F_2 \dotplus E_2.$$

We will show that $F_1 \cap F_2$, which is a linear subspace of $\mathcal{V}$, has also finite codimension. Let G_1 and G_2 be linear subspaces of $\mathcal{V}$ such that

$$F_1 = (F_1 \cap F_2) \dotplus G_1$$
$$F_2 = (F_1 \cap F_2) \dotplus G_2.$$

We have $G_1 \cap G_2 = \{0\}$, and so the sum

$$(F_1 \cap F_2) + G_1 + G_2$$

is direct. Let G_3 be such that

$$\underbrace{(F_1 \cap F_2) \dotplus G_1}_{\text{equal to } F_1} \dotplus G_2 \dotplus G_3 = \underbrace{(F_1 \cap F_2) \dotplus G_2}_{\text{equal to } F_2} \dotplus G_1 \dotplus G_3 = \mathcal{V}. \tag{3.10.3}$$

If one of the spaces G_1, G_2 or G_3 is infinite dimensional, (3.10.3) implies that F_1 or F_2 has infinite codimension. So the three spaces are finite dimensional and $F_1 \cap F_2$ has finite codimension.

$\square$

Solution of Exercise 3.2.6: Let E denote a topological Hausdorff space, let J be an ordered set (with order relation denoted by $\leq$) and let $(K_j)_{j \in J}$ be a decreasing sequence of compact subsets of E indexed by J:

$$i \leq j \quad \Longleftrightarrow \quad K_j \subseteq K_i,$$

and assume that $\cap_{j \in J} K_j = \emptyset$. Then,

$$E = \bigcup_{j \in J} (E \setminus K_j).$$

Fix $j_0 \in J$. We have

$$K_{j_0} = \bigcup_{j \in J} \left((E \setminus K_j) \cap K_{j_0} \right).$$

Since K_{j_0} is compact, there are $j_1, \ldots, j_N \in J$, which we can assume ordered,

$$j_1 \leq j_2 \leq \cdots \leq j_N,$$

such that

$$K_{j_0} = \bigcup_{k=1}^{N} \left((E \setminus K_{j_k}) \cap K_{j_0} \right).$$

Thus

$$K_{j_0} = (E \setminus K_{j_1}) \cap K_{j_0},$$

so that

$$K_{j_0} \subseteq E \setminus K_{j_1}. \tag{3.10.4}$$

If $j_0 = j_1$, there is clearly a contradiction. More generally, the inclusion (3.10.4) is impossible since the sequence is decreasing. Indeed, if $j_0 \le j_1$, then $K_{j_1} \subseteq K_{j_0}$ and so (3.10.4) leads to

$$K_{j_1} \subseteq K_{j_0} \subseteq E \setminus K_{j_1},$$

which is not possible. When $j_1 \le j_0$ we have $E \setminus K_{j_1} \subseteq E \setminus K_{j_0}$ and (3.10.4) leads to

$$K_{j_0} \subseteq E \setminus K_{j_1} \subseteq E \setminus K_{j_0},$$

which is also impossible. $\qquad\square$

Solution of Exercise 3.2.7: By definition of the topology, sets $B(z,r)$ are open neighborhood of points in $\mathbb{C}$ and sets $B(\infty, R)$ (that is, of the form (3.1.2)) are open neighborhoods of ∞. Thus every open set is a union of such sets. Let now $(O_i)_{i \in I}$ be an open covering of $\widehat{\mathbb{C}}$, and write each O_i as a union $O_i = \cup_{j \in J} O_{i,j}$, where the $O_{i,j}$ are of the form $B(z,r)$ or $B(\infty, R)$. At least one of the $O_{i,j}$ is of the form $B(\infty, R)$ for some $R > 0$. Let R_0 be the infimum of such R (possibly $R_0 = 0$). Assume first $R_0 > 0$. Then,

$$\overline{B(0, R_0)} \subset \cup \{O_{i,j} \; ; \; O_{i,j} \text{ of the form } B(z,r)\}.$$

Since $\overline{B(0, R_0)}$ is compact, it is a finite union of such $O_{i,j}$. Replacing each such $O_{i,j}$ by O_i we obtain a finite subcovering of $\widehat{\mathbb{C}}$. When $R_0 = 0$ one replaces $\overline{B(0, R_0)}$ by $\{0\}$ to get to the same conclusion. $\qquad\square$

Solution of Exercise 3.2.8: First remark that the sets F_1 and F_2 are compact (see Theorem 3.2.4). Fix $x \in F_1$. For every $y \in F_2$ there exist non-intersecting open sets $U_{x,y}$ and $V_{x,y}$ such that $x \in U_{x,y}$ and $y \in V_{x,y}$. Since F_2 is compact there is an open cover of F_2 of the form

$$Z_x = \cup_{j=1}^{n(x)} V_{x,y_j}, \quad \text{where the points } y_1, \ldots, y_{n(x)} \text{ depend on } x.$$

Set $W_x = \cap_{j=1}^{n(x)} U_{x,y_j}$. We have

$$Z_x \cap W_x = \emptyset, \quad \forall\, x \in F_1. \tag{3.10.5}$$

The sets W_x are open and form a covering of F_1 as x runs through F_1. By compactness, we obtain an open covering of F_1 of the form $W_{x_1} \cup \cdots \cup W_{x_M}$ for some points $x_1, \ldots, x_M \in F_1$. The sets $U_1 = W_{x_1} \cup \cdots \cup W_{x_M}$ and $U_2 = \cap_{t=1}^{M} Z_{x_t}$ are open, contain respectively F_1 and F_2. They do not intersect, as is seen from (3.10.5). $\qquad\square$

Solution of Exercise 3.2.9: (see also [283]). Let $A \subset X$ with the assumed property, and let $a \in \overline{A}$. Then, for every neighborhood V of a we have $V \cap A \ne \emptyset$. Let now

W denote another neighborhood of a. Then, $V \cap W$ is also a neighborhood of a and therefore $(V \cap W) \cap A \neq \emptyset$. Writing

$$(V \cap W) \cap A = W \cap (V \cap A) \neq \emptyset$$

we see that $a \in \overline{V \cap A}$. Since X is locally compact, we can take V compact. Then, by hypothesis $\overline{V \cap A} = V \cap A$ and in particular $a \in A$. Thus $\overline{A} \subset A$, and A is closed.

Conversely, let A be closed and K be compact. Then K is in particular closed and so $A \cap K$ is closed, and hence compact, since a closed subset of a compact set in compact.

$\square$

Solution of Exercise 3.2.13: This is a direct consequence of (3.1.4). If the nowhere dense set C is closed then its (open) complement O is dense:

$$\emptyset = \overset{\circ}{C} = E \setminus \overline{O}.$$

$\square$

Solution of Exercise 3.3.1: Assume first E connected, and let O_1 and O_2 be two open sets of X such that (3.3.1) and (3.3.2) hold. Then,

$$E = (E \cap O_1) \cup (E \cap O_2)$$

is the decomposition of E into two disjoint subsets of E, which are open in the induced topology. Therefore, one of them is empty and the conclusion follows.

Conversely, suppose that the condition of the exercise is in force, and let $E = U_1 \cup U_2$ be the decomposition of E into a disjoint union of two open sets of E. Thus $U_j = E \cap O_j$, $j = 1, 2$, where O_1 and O_2 are open sets of X, and we have

$$E = (E \cap O_1) \cup (E \cap O_2).$$

So

$$E \subset O_1 \cup O_2.$$

Since $E \cap O_1 \cap O_2 = U_1 \cap U_2 = \emptyset$, the hypothesis implies that $E \subset O_1$ or $E \subset O_2$, that is $E = U_1$ or $E = U_2$.

$\square$

Solution of Exercise 3.3.2: Let O_1 and O_2 be two open sets such that

$$B = (B \cap O_1) \cup (B \cap O_2).$$

Then,

$$A = A \cap B = (A \cap O_1) \cup (A \cap O_2).$$

Since A is connected we have either $A \cap O_1 = \emptyset$ or $A \cap O_2 = \emptyset$. Without loss of generality we may assume that $A \cap O_2 = \emptyset$. Then

$$A \subset X \setminus O_2.$$

Since $X \setminus O_2$ is a closed set, we also have

$$\overline{A} \subset X \setminus O_2.$$

Thus

$$\overline{A} \cap O_2 = \emptyset,$$

and hence

$$B \cap O_2 \subset \overline{A} \cap O_2 = \emptyset.$$

Therefore $B \cap O_2 = \emptyset$, and B is connected. $\qquad\square$

Solution of Exercise 3.3.3:

(1) It suffices to take $C_n = [n, \infty)$, $n = 0, 1, \ldots$ to answer the first item of the question. The example

$$C_n = \mathbb{D} \setminus (-1 + 1/n, 1 - 1/n)$$

will answer the second item.

(2) Assume the intersection empty. Then X is covered by the complements $X \setminus C_i$, and hence, by compactness, has a finite subcover. From $X = \cup_{j=1}^{N} (X \setminus C_{i_j})$ we get $\emptyset = \cap_{j=1}^{N} C_{i_j}$, which is not possible since the sequence $(C_n)_{n \in \mathbb{N}}$ is decreasing and made of non-empty sets.

Assume now that C is not connected, and write $C = F_1 \cup F_2$, where F_1 and F_2 are closed and do not intersect. Since the space satisfies the separation axiom T_4, we can find two disjoint open subsets O_1 and O_2, containing F_1 and F_2 respectively. Since the C_n are connected we have $C_n \subset O_1$ or $C_n \subset O_2$ for every n. Since the sequence $(C_n)_{n \in \mathbb{N}}$ is decreasing, they all belong to the same set, say O_1, for n large enough. It follows that $C = F_1$ and so C is connected. $\qquad\square$

Solution of Exercise 3.4.1:

(1) We have that X and $\emptyset$ belong to $\mathcal{T}_f$ since

$$f^{-1}(Y) = \{x \in X \, , \; f(x) \in Y\} = X,$$

and

$$f^{-1}(\emptyset) = \{x \in X \, , \; f(x) \in \emptyset\} = \emptyset.$$

Let $(O_i)_{i \in I}$ be a family of open subsets of $\mathcal{T}$. The formula

$$f^{-1}(\cup_{i \in I} O_i) = \cup_{i \in I} f^{-1}(O_i)$$

shows that $\cup_{i \in I} f^{-1}(O_i) \in \mathcal{T}_f$.

Let now $O_1, \ldots, O_N$ be a finite number of open subsets of $\mathcal{T}$. Then, the equality

$$f^{-1}(\cap_{j=1}^{N} O_j) = \cap_{j=1}^{N} f^{-1}(O_j)$$

shows that $\cap_{j=1}^{N} f^{-1}(O_j) \in \mathcal{T}_f$. Thus we have a topology. $\qquad\square$

Solution of Exercise 3.4.2: Assume first f continuous from X to Y, and let U denote an open set of Y. Let $i \in I$ and set $f_i = f|_{U_i}$. We have:

$$f_i^{-1}(U) = \{x \in U_i \text{ such that } f(x) \in U\}$$
$$= f^{-1}(U) \cap U_i,$$

which is open in U_i by definition of the induced topology, since $f^{-1}(U)$ is open. *Note that this direction does not use the fact that the U_i are open.*

Conversely, assume that all the restrictions f_i are continuous. Let $U \subset Y$ be an open set. Then,

$$f^{-1}(U) = \{x \in X \text{ such that } f(x) \in U\}$$
$$= \cup_{i \in I} \{x \in U_i \text{ such that } f(x) \in U\}$$
$$= \cup_{i \in I} f_i^{-1}(U).$$

The set $f_i^{-1}(U)$ is open in the induced topology since f_i is assumed continuous. Thus, *and here we use the fact that we have an open covering*, $f_i^{-1}(U)$ is open in X (see Exercise 3.1.6). Hence $f^{-1}(U) = \cup_{i \in I} f_i^{-1}(U)$ is open. $\qquad \square$

Solution of Exercise 3.4.4: To begin with, we note that

$$\underbrace{A \subset f^{-1}(f(A))}_{\text{use (1.1.6)}} \subset f^{-1}\left(\overline{f(A)}\right), \quad \forall A \subset X. \tag{3.10.6}$$

(1) We first assume f continuous. Since $\overline{f(A)}$ is closed, this implies that the set $f^{-1}\left(\overline{f(A)}\right)$ is closed. So the inequality

$$A \subset f^{-1}\left(\overline{f(A)}\right)$$

implies that

$$\overline{A} \subset f^{-1}\left(\overline{f(A)}\right), \tag{3.10.7}$$

and so (3.4.3) holds, as in seen by applying f on both sides of (3.10.7) and using (1.1.5).

Conversely, assume that (3.4.3) is in force. Let B be a closed subset of Y. We want to show that $f^{-1}(B)$ is closed. Let $A = f^{-1}(B)$. Then, by hypothesis,

$$f(\overline{A}) \subset \overline{f(A)} = \underbrace{\overline{f(f^{-1}(B))} \subset \overline{B}}_{\text{use (1.1.5)}} = B.$$

Thus,

$$\overline{A} \subset f^{-1}(B) = A,$$

and so $\overline{A} = A$ and A is closed.

3.10. Solutions of the exercises

(2) Let $A \subset X$. In view of (1), it is enough to show that

$$\overline{f(A)} \subset f\left(\overline{A}\right).$$

Trivially one has $f(A) \subset f(\overline{A})$. So it is enough to show that $f(\overline{A})$ is closed in Y. Since X is compact the closure $\overline{A}$ is compact. Thus, $f(\overline{A})$ is compact, and hence closed.

(3) Assume first that (3.4.5) is in force. Then for $A = \overline{A}$ we have

$$f(A) = f(\overline{A}) = \overline{f(\overline{A})},$$

which is closed. So closed sets are sent to closed sets.

Conversely, assume that f is closed. Since f is continuous we have (3.4.3):

$$f(\overline{A}) \subset \overline{f(A)}, \quad A \subset X.$$

Since f is closed we have $f(\overline{A}) = \overline{f(\overline{A})}$, and (3.4.3) can be rewritten as

$$\overline{f(\overline{A})} = f(\overline{A}) \subset \overline{f(A)},$$

and so (3.4.5) holds since $\overline{f(A)} \subset \overline{f(\overline{A})}$. □

Solution of Exercise 3.4.6: We only need to check continuity at the origin and infinity. To check continuity at the point ∞, for $\epsilon > 0$ take $R > 0$ such that $1/R < \epsilon$. Then

$$i_P(B(\infty, R)) \subset B(0, \epsilon).$$

To check continuity at the origin, for $R > 0$ take $\epsilon > 0$ such that $1/\epsilon > R$. Then,

$$i_P(B(0, \epsilon)) \subset B(\infty, R).$$ □

Solution of Exercise 3.4.9: Assume first that f is lower semi-continuous at every point of X, and let $x \in X$ and $a < f(x)$. Then, $V = f^{-1}(a, \infty]$ is by assumption a neighborhood of x and so X is a neighborhood of each of its points, and hence open.

Conversely, let f be lower semi-continuous on X and fix $x \in X$ and $a < f(x)$. Then $V = f^{-1}(a, \infty]$ contains x and is open, and so is a neighborhood of x. Hence (1) is in force. □

Solution of Exercise 3.4.10: Since f is continuous we can find for every $x \in K$ an open neighborhood U_x of x in which f does not vanish. The open set $\cup_{x \in K} U_x$ is an open covering of K. By compactness, there exist $N \in \mathbb{N}$ and $x_1, \ldots, x_N \in K$ such that

$$K \subset \cup_{j=1}^N U_{x_j}.$$

By (1.1.3)

$$f(K) \subset \cup_{j=1}^{N} f(U_{x_j})$$

and so f does not vanish in the open neighborhood $\cup_{j=1}^{N} U_{x_j}$ of K. $\qquad \square$

Solution of Exercise 3.4.13: Let us denote by A the set, and let us fix a point $a_0 \in A$. For every $a \in A$ there is a path γ_a linking a_0 to a, and thus:

$$A = \cup_{a \in A} \gamma_a([0, 1])$$

is connected, since $a_0 \in \cap_{a \in A} \gamma_a([0, 1])$.

Conversely, it is not true that a connected set is path-connected as the well-known example of the graph of the function $\sin(1/x)$ $(x > 0)$ together with the closed interval joining the two points $(0, -1)$ and $(0, 1)$ illustrates. $\qquad \square$

Solution of Exercise 3.4.16:

(1) We proceed in a number of steps.

STEP 1: $w \in \partial\Omega$ *if and only if there exist a point* $v \in \mathbb{T}$ *and a sequence of points* $(z_n)_{n \in \mathbb{N}}$ *in* $\mathbb{D}$ *such that*

$$\lim_{n \to \infty} z_n = v, \qquad (3.10.8)$$

$$\lim_{n \to \infty} f(z_n) = w. \qquad (3.10.9)$$

Indeed, let $w \in \partial\Omega$. In particular, $w \in \overline{\Omega}$, and there exists a sequence $(w_n)_{n \in \mathbb{N}}$ of points in Ω such that

$$\lim_{n \to \infty} |w - w_n| = 0. \qquad (3.10.10)$$

Since f is assumed onto, there exist points $z_1, z_2, \ldots \in \mathbb{D}$ such that

$$w_n = f(z_n), \quad n = 1, 2, \ldots$$

By maybe taking a subsequence, we may assume that $\lim_{n \to} z_n$ exists. Let us denote by v this limit. We claim that $v \in \mathbb{T}$. Assume by contradiction that $v \in \mathbb{D}$. Since f is continuous, we have $w = f(v)$, and thus $w \in \partial\Omega \cap \Omega$. But this cannot be since Ω is open. See Exercise 3.1.16. Thus (3.10.8)–(3.10.9) are in force.

Conversely, assume that (3.10.8)–(3.10.9) hold. Then (3.10.10) holds with $w_n = f(z_n)$, and so $w \in \overline{\Omega}$. Assume that $w \in \Omega$. Then there is $\alpha \in \mathbb{D}$ such that $f(\alpha) = w$. In view of (3.10.8), there exist $\epsilon > 0$ and $n_0 \in \mathbb{N}$ such that

$$B(\alpha, \epsilon) \cap \{z_{n_0}, z_{n_0+1}, \ldots\} = \emptyset.$$

Since f is assumed one-to-one,

$$f(B(\alpha, \epsilon)) \cap \{f(z_{n_0}), f(z_{n_0+1}), \ldots\} = \emptyset.$$

Since f^{-1} is continuous the set $f(B(\alpha, \epsilon))$ is open and so contains an open ball $B(w, \epsilon)$ for some $\epsilon > 0$, contradicting (3.10.9).

STEP 2: *We prove the direct inclusion.*

Let $w \in \partial\Omega$ and let $r \in (0,1)$. By (3.10.8) there exist points z_n such that $r < |z_n| < 1$ and such that $f(z_n) \longrightarrow w$. Thus, $w \in \overline{f(r < |z| < 1)}$. Hence, $w \in \cap_{r \in (0,1)} \overline{f(r < |z| < 1)}$.

STEP 3: *We prove the converse inclusion.*

Let $w \in \cap_{r \in (0,1)} \overline{f(r < |z| < 1)}$. For every $r \in (0,1)$ and every $\epsilon > 0$ there exists a point $z_{r,\epsilon} \in \mathbb{D}$ such that

$$|f(z_{r,\epsilon}) - w| < \epsilon \quad \text{and} \quad r < |z_{r,\epsilon}| < 1.$$

Thus there exists a sequence $(z_n)_{n \in \mathbb{N}}$ of numbers of $\mathbb{D}$ such that $\lim_{n \to \infty} z_n$ exists and belongs to $\mathbb{T}$ and $\lim_{n \to \infty} f(z_n) = w$. In view of Step 1, $w \in \partial\Omega$.

(2) The set $f(r < |z| < 1)$ is connected as a continuous image of a connected set, and so is its closure $C_r = \overline{f(r < |z| < 1)}$. The sets C_r are compact and

$$C_r \subset C_s \quad \text{for} \quad s \geq r, \ r,s \in (0,1).$$

Thus $\cap_{r \in (0,1)} C_r \neq \emptyset$. See Exercise 3.2.6. But a decreasing family of connected sets whose intersection is non-empty is connected. See Exercise 3.3.3. Thus $\partial\Omega$ is connected. $\qquad\square$

Solution of Exercise 3.4.17: If E has an empty interior, then it is equal to its boundary (since it is closed; see (3.1.9) in Exercise 3.1.17), and the result holds in view of (3.4.7). Assume now that $\overset{\circ}{E} \neq \emptyset$, and let

$$\Omega_1 = \Omega \cap \overset{\circ}{E} \quad \text{and} \quad \Omega_2 = \Omega \setminus E.$$

Both Ω_1 and Ω_2 are open. From (3.1.9) we have the equality

$$\emptyset \neq \Omega \cap E = (\Omega \cap \overset{\circ}{E}) \cup (\Omega \cap \partial E).$$

Thus if $\Omega_1 = \emptyset$ we have $\Omega \cap \partial E \neq \emptyset$, and the result is proved. We now suppose $\Omega_1 \neq \emptyset$. We also remark that the set Ω_2 is non-empty. Otherwise we would have $E \cap \Omega = \Omega$, but $E \cap \Omega \subsetneqq \Omega$ by hypothesis.

Let now $z_1 \in \Omega_1$ and $z_2 \in \Omega_2$, and let γ denote a continuous path in Ω linking z_1 to z_2, with parametrization $\gamma(t), t \in [0,1]$. Then, the sets

$$\gamma^{-1}(\Omega_1) \quad \text{and} \quad \gamma^{-1}(\Omega_2)$$

are open subsets of $[0,1]$. Since $[0,1]$ is connected, there exists $t_0 \in [0,1]$ which does not belong to $\gamma^{-1}(\Omega_1) \cup \gamma^{-1}(\Omega_2)$. For such a t_0, we have (use Question 1.1.2 if need be)

$$\gamma(t_0) \in (\Omega \setminus \Omega_1) \cap (\Omega \setminus \Omega_2) = (\Omega \cap (\mathbb{C} \setminus \overset{\circ}{E})) \cap (\Omega \cap E)$$
$$= \Omega \cap \partial E,$$

since, E being closed, we have $\partial E = (\mathbb{C} \setminus \overset{\circ}{E}) \cap E$. $\qquad\square$

Solution of Exercise 3.4.19: We follow the argument in [252, p. 177]. By hypothesis f is open and so $f(\Omega)$ is an open subset of $\mathbb{C}$, and $f(\Omega) \cap C$ is open in C in the induced topology. On the other hand,

$$f(\overline{\Omega}) = f((\overline{\Omega} \setminus \Omega) \cup \Omega) = f(\overline{\Omega} \setminus \Omega) \cup f(\Omega).$$

By hypothesis, $f(\overline{\Omega} \setminus \Omega) \cap C = \emptyset$, and so

$$f(\overline{\Omega}) \cap C = f(\Omega) \cap C.$$

We note that $\overline{\Omega}$ is closed and bounded, and so is compact in $\mathbb{C}$. It follows that $f(\overline{\Omega})$ is compact, and hence closed, and $f(\Omega) \cap C$ is closed in C in the induced topology. Thus $f(\Omega) \cap C$ is closed and open in C. Since C is connected we have $C \subset f(\Omega)$.

$\qquad\qquad\qquad\qquad\qquad\qquad\qquad\qquad\qquad\qquad\qquad\qquad\qquad\qquad$ □

The above lemma allows us to prove a removable singularity theorem: If f is open in $\Omega \setminus \{z_0\}$ and continuous in Ω, then f is open in Ω. This is Lemma 2 in the above-mentioned paper of Porcelli and Connell.

Solution of Exercise 3.5.1:

(1) Let us denote by $\mathcal{T}_\Pi$ the smallest (coarsest) topology for which all the applications p_i are continuous. Let $i_0 \in I$, and let U be an open subset of X_{i_0}. The set

$$p_{i_0}^{-1}(U) = \{(x_i)_{i \in I} \text{ such that } x_{i_0} \in U\} \in \mathcal{T}_\Pi. \qquad (3.10.11)$$

Let now $N \in \mathbb{N}$, $i_1, \ldots, i_N \in I$ and $U_1 \in X_{i_1}, \ldots, U_N \in X_{i_N}$ be open sets. The set

$$\cap_{j=1}^{N} p_{i_j}^{-1}(U_j) = \{(x_i)_{i \in I} \text{ such that } x_{i_j} \in U_j, \; j = 1, \ldots, N\}$$

is still in $\mathcal{T}_\Pi$. Together with the empty set, arbitrary unions of such sets form exactly $\mathcal{T}_\Pi$.

(2) Let now f denote a map from the topological space Y into $\prod_{i \in I} X_i$, the latter being endowed with the topology $\mathcal{T}_\Pi$. If f is assumed continuous then clearly the maps $p_i \circ f$ are also continuous (and this, for any topology for which the p_i are continuous). Conversely, assume that all the maps $p_i \circ f$ are continuous. Take first $V \in \mathcal{T}_\Pi$ of the form (3.10.11). Then,

$$\begin{aligned} f^{-1}(V) &= \{y \in Y \text{ such that } f(y) \in V\} \\ &= \{y \in Y \text{ such that } p_{i_0}(f(y)) \in U\} \\ &= (p_{i_0} \circ f)^{-1}(U), \end{aligned}$$

which is open since $p_{i_0} \circ f$ is assumed continuous. It follows that $f^{-1}(U)$ is open for every open subset of $\prod_{i \in I} X_i$ in view of the characterization of these open sets and using Exercise 1.1.4.

(3) Let $\mathcal{T}_{\amalg}$ denote the finest (largest) topology for which the maps q_i are continuous, and let $O \in \mathcal{T}_{\amalg}$. Then, for $i_0 \in I$ the set

$$
\begin{aligned}
q_{i_0}^{-1}(O) &= \{x \in X_{i_0} \text{ such that } q_{i_0}(x) \in O\} \\
&= \{x \in X_{i_0} \text{ such that } (i_0, x) \in O \cap (\{i_0\} \times X_{i_0})\}
\end{aligned}
$$

is open in X_{i_0}. In other words, the set O contains the sets (i_0, x) where x runs through an open set of X_{i_0}. The topology $\mathcal{T}_{\amalg}$ is generated by these sets.
(4) Let now f denote a map from the disjoint union $\amalg_{i \in I} X_i$, the latter being endowed with the topology $\mathcal{T}_{\amalg}$, into the topological space Y. If f is assumed continuous then clearly the maps $f \circ q_i$ are also continuous (and this, for any topology for which the q_i are continuous). Conversely, assume that all the maps $f \circ q_i$ are continuous. Let U be an open set of Y. Then,

$$
\begin{aligned}
f^{-1}(U) &= \left\{ x \in \coprod_{i \in I} X_i \text{ such that } f(x) \in U \right\} \\
&= \bigcup_{i \in I} \{x \in X_i \text{ such that } f(q_i(x)) \in U\} \\
&= \bigcup_{i \in I} (f \circ q_i)^{-1}(U),
\end{aligned}
$$

which is open since the maps $f \circ q_i$ are assumed continuous. $\qquad\square$

Solution of Exercise 3.5.3: Let $(x, y) \in X \times X \setminus \Delta$. Thus $x \neq y$. Since the space is Hausdorff, there exist open sets U and V such that

$$
x \in U, \quad y \in V \text{ and } U \cap V = \emptyset.
$$

The set $U \times V$ is a neighborhood of (x, y) in the product topology. It does not intersect Δ since any element $(w, w) \in (U \times V) \cap \Delta$ would be such that $w \in U \cap V$. $\qquad\square$

Solution of Exercise 3.6.2: We use the notation of Exercise 3.4.4. The function f^{-1} is continuous if and only if $(f^{-1})^{-1}(F)$ is closed for every closed set F, i.e., if and only if

$$
f(F) = \overline{f(F)}
$$

for every closed set F. Since f is continuous, this is just (3.4.5) with $A = F$ since $F = \overline{F}$. $\qquad\square$

Solution of Exercise 3.6.3: We use the fact that a compact set is closed and that a closed subset of a compact set is compact. Let $K \subset Y$ be compact. Then, it is closed in Y and so $f^{-1}(K)$ is closed in X. Since X is compact, $f^{-1}(K)$ is compact in X. $\qquad\square$

Solution of Exercise 3.6.4: Assume that (a) is in force, and let K be a compact subset of $\mathbb{R}$. Then K is closed (this is true in *any* topological separated space, and in particular in $\mathbb{R}$). The continuity of f implies that $f^{-1}(K)$ is closed. Assume that $f^{-1}(K)$ is not bounded. Then there is a sequence $(x_n)_{n \in \mathbb{N}}$ in $f^{-1}(K)$ with $|x_n| \to \infty$. But then, by (a), we have that $|f(x_n)| \to \infty$ as $n \to \infty$, and this is not possible since $f(x_n) \in K$ and K is bounded.

Conversely, assume that (b) is in force. For every $N > 0$ the set $f^{-1}([-N, N])$ is compact, and hence bounded. Thus there exists $M_N > 0$ such that

$$f^{-1}([-N, N]) \subset [-M_N, M_N].$$

Therefore,

$$|x| > M_N \Longrightarrow |f(x)| > N,$$

and thus (a) holds.

$\square$

Solution of Exercise 3.6.5: (see also [283]) We use Exercise 3.2.9. Let $F \subset X$ be a closed set, and let V be a compact subset of Y, such that $f(F) \cap V \neq \emptyset$. Note that

$$f(F) \cap V = f\left(F \cap f^{-1}(V)\right). \tag{3.10.12}$$

Indeed, let $y \in f(F) \cap V$. Then $y = f(x)$ for some $x \in F$. Since $f(x) \in V$, we have $x \in f^{-1}(V)$ and so $x \in F \cap f^{-1}(V)$ and $y \in f\left(F \cap f^{-1}(V)\right)$. Conversely, let $y \in f\left(F \cap f^{-1}(V)\right)$. There is $x \in F \cap f^{-1}(V)$ such that $y = f(x)$. So $y \in f(F) \cap V$, and (3.10.12) holds.

Since f is proper, $f^{-1}(V)$ is compact, and so $F \cap f^{-1}(V)$ is compact. Since the continuous image of a compact set is compact we have from (3.10.12) that $f(F) \cap V$ is compact. By Exercise 3.2.9 we have that $f(F)$ is closed. $\square$

Solution of Exercise 3.6.6:

(1) Any real-valued function will have its image not open in $\mathbb{R}^2$. If the image is considered as a subset of $\mathbb{R}$, consider the following function:

$$f(x, y) = \begin{cases} e^{-\frac{1}{1-(x^2+y^2)}}, & \text{if } x^2 + y^2 < 1, \\ 0, & \text{if } x^2 + y^2 \geq 1. \end{cases}$$

Then, the range of f is the closed interval $[0, 1]$.

Another example is given by the function $f(x) = x^2$ from $\mathbb{R}$ into $\mathbb{R}$. We have $f(-1, 1) = [0, 1)$, which is not open in $\mathbb{R}$.

(2) It is enough to show that the image of an open disk $B(a, r)$ is open, or even, to show that the image of $\mathbb{D}$ is open. If f is one-to-one in $\mathbb{D}$ this is a direct consequence of the corresponding result from calculus. If $f(0) = 0$ we can write

$$f(z) = z^N g(z),$$

where $N \in \mathbb{N}$ is the order of the zero $z = 0$ and where g is analytic and non-vanishing in $\mathbb{D}$. As such, g admits an analytic Nth root, say h, and we have

$$f(z) = (zh(z))^N, \quad z \in \mathbb{D}.$$

The function $j(z) = zh(z)$ is analytic and $j'(0) = h(0) \neq 0$. So it is one-to-one in a neighborhood, say $B(0, r)$ of the origin, and the image of $B(0, r)$ is open, a property conserved by the map $w \mapsto w^N$. $\qquad \square$

Solution of Exercise 3.6.7: The result is a direct consequence of the open mapping theorem for an analytic function (see the previous exercise for the latter). $\qquad \square$

Solution of Exercise 3.6.8: We basically repeat the arguments of Stoilow given in [298, p. 381], [299, p. 105], with some extra explanations. First note that, by construction, $f(0, \infty) = \{0\}$. Therefore, the function f cannot be topologically equivalent to a function analytic in the half-plane $\mathbb{C}_r = \{z \in \mathbb{C} \mid \operatorname{Re} z > 0\}$. Indeed, assume $f = \varphi_2 \circ g \circ \varphi_1$ where the function φ_1 is a homeomorphism from $\mathbb{C}_r$ onto $\varphi_1(\mathbb{C}_r)$, the function g is analytic from $\varphi_1(\mathbb{C}_r)$ onto its image $g(\varphi_1(\mathbb{C}_r))$, and φ_2 is a homeomorphism from $g(\varphi_1(\mathbb{C}_r))$ onto its image. We have

$$g(\varphi_1(0, \infty)) = \varphi_2^{-1}(0).$$

The image of $(0, \infty)$ under the homeomorphism φ_1 is not countable. Thus the analytic function $g(z) - \varphi_2^{-1}(0)$ has a noncountable family of zeros, contradicting the fact that g is not a constant.

We now show that f is open. Write $f(z) = u(x, y) + iv(x, y)$ with

$$u(x, y) = \begin{cases} \left(e^{-\frac{1}{y^2}} + xy^2\right)\cos\left(\frac{2\pi}{y}\right), & \text{if } y \neq 0, \\ 0, & \text{if } y = 0, \end{cases}$$

and

$$v(x, y) = \begin{cases} \left(e^{-\frac{1}{y^2}} + xy^2\right)\sin\left(\frac{2\pi}{y}\right), & \text{if } y \neq 0, \\ 0, & \text{if } y = 0. \end{cases}$$

Note that, for $y \neq 0$ and $x > 0$,

$$\lim_{y \to 0} \frac{u(x, y) - u(x, 0)}{y} = \lim_{y \to 0} \left(\frac{e^{-\frac{1}{y^2}}}{y} + xy\right)\cos\left(\frac{2\pi}{y}\right) = 0,$$

and so $\frac{\partial u}{\partial x}(x, 0) = 0$, and similarly (still for $x > 0$) we have

$$\frac{\partial u}{\partial y}(x, 0) = \frac{\partial v}{\partial x}(x, 0) = \frac{\partial v}{\partial y}(x, 0) = 0.$$

For $y \neq 0$ we have:

$$\frac{\partial u}{\partial x}(x,y) = y^2 \cos\left(\frac{2\pi}{y}\right)$$

$$\frac{\partial u}{\partial y}(x,y) = \left(\frac{2}{y^3}e^{-\frac{1}{y^2}} + 2xy\right)\cos\left(\frac{2\pi}{y}\right) + \frac{2\pi}{y^2}\left(e^{-\frac{1}{y^2}} + xy^2\right)\sin\left(\frac{2\pi}{y}\right),$$

$$\frac{\partial v}{\partial x}(x,y) = y^2 \sin\left(\frac{2\pi}{y}\right)$$

and

$$\frac{\partial v}{\partial y}(x,y) = \left(\frac{2}{y^3}e^{-\frac{1}{y^2}} + 2xy\right)\sin\left(\frac{2\pi}{y}\right) - \frac{2\pi}{y^2}\left(e^{-\frac{1}{y^2}} + xy^2\right)\cos\left(\frac{2\pi}{y}\right).$$

Therefore the Jacobian

$$J(x,y) = \det \begin{pmatrix} \frac{\partial u}{\partial x} & \frac{\partial u}{\partial y} \\ \frac{\partial v}{\partial x} & \frac{\partial v}{\partial y} \end{pmatrix}(x,y)$$

at a point (x,y) is equal to:

$$J(x,y) = \begin{cases} 0, & \text{if } y = 0, \\ -2\pi\left(e^{-\frac{1}{y^2}} + xy^2\right), & \text{if } y \neq 0. \end{cases}$$

The function $e^{-\frac{1}{y^2}} + xy^2$ does not vanish in $\mathbb{C}_r \setminus (0, \infty)$. By the inverse function theorem, the function f, or more precisely the map

$$(x,y) \quad \mapsto \quad (u(x,y), v(x,y))$$

is a local homeomorphism in a neighborhood of every point (x,y) off the positive real axis, and so the image of any open set included in the upper or lower open right quarter-plane will be open. We now consider the case of an open set, say Ω, which intersects the positive real axis, and define:

$$\Omega_+ = \{z = x + iy \in \Omega \; ; \; y > 0\},$$
$$\Omega_- = \{z = x + iy \in \Omega \; ; \; y < 0\},$$
$$\Omega_0 = \{z = x + iy \in \Omega \; ; \; y = 0\} = \bar{\Omega} \cap \mathbb{R}_+.$$

Both Ω_+ and Ω_- are open, and hence, by the above discussion, their respective images under f are open. If $\Omega_0 = \emptyset$ this ends the proof. We now assume that $\Omega_0 \neq \emptyset$.

The image of Ω_0 is just the origin. To show that $f(\Omega)$ is open it is enough to show that there is a neighborhood of the origin which is in the image of Ω.

3.10. Solutions of the exercises

Fix some $x_0 \in \Omega_0$, and choose $\epsilon_0 > 0$ such that the open square C_{ϵ_0} with center x_0 and vertices $(x_0 - \epsilon_0, \pm\epsilon_0), (x_0 + \epsilon_0, \pm\epsilon_0)$ lies in Ω. *The idea is to show that the image C_{ϵ_0} contains an open ball $B(0, r_0)$.* The result will then follow since

$$\Omega = \Omega_+ \cup \Omega_- \cup \Omega_0 = \Omega_+ \cup \Omega_- \cup C_{\epsilon_0},$$

and so, by (1.1.3),

$$\begin{aligned} f(\Omega) &= f(\Omega_+) \cup f(\Omega_-) \cup f(\Omega_0) \cup f(C_{\epsilon_0}) \\ &= f(\Omega_+) \cup f(\Omega_-) \cup \underbrace{f(\Omega_0) \cup B(0, r_0)}_{\text{equals } B(0, r_0)}. \end{aligned}$$

As mentioned earlier we closely follow Stoilow's arguments, and divide the sequel of the proof in a number of steps. We write

$$B(0, r_0) = \bigcup_{\theta \in [0, 2\pi)} e^{i\theta}[0, r_0),$$

and will show that for every $\theta \in [0, 2\pi)$ (and for r_0 to be chosen as mentioned above), the ray $e^{i\theta}[0, r_0)$ belongs to the image of $f(\Omega)$.

We first define for $\theta \in [0, \pi]$ and $n \in \mathbb{N}$,

$$y_n = \frac{2\pi}{2\pi n + \theta}, \quad n = 1, 2, \ldots$$

Remark that (3.6.2) holds:

$$\frac{1}{y_{n+1}} - \frac{1}{y_n} = 1, \quad n = 1, 2 \ldots$$

and in particular

$$e^{\frac{2\pi i}{y_n}} = e^{\frac{2\pi i}{y_{n+1}}} = e^{i\theta}.$$

STEP 1: *We prove:*

$$\exists n_0 \in \mathbb{N}, \ \forall \theta \in [0, \pi], \ n \geq n_0 \implies y_n \in (0, \epsilon_0).$$

Indeed, we have

$$0 \leq y_n \leq \frac{1}{n}$$

which will be less than ϵ_0 for $n \geq [\frac{1}{\epsilon_0}] + 1$.

STEP 2: *Let $\theta \in [0, \pi]$, let y_n and n_0 be as in the previous step, and let $I_n(\theta)$ denote the interval with endpoints $(x_0 - \epsilon_0, y_n), (x_0 + \epsilon_0, y_n)$. Then, $f(I_n(\theta))$ is the interval with endpoints*

$$a_n = e^{i\theta}(e^{-\frac{1}{y_n^2}} + (x_0 - \epsilon_0)y_n^2) \quad \text{and} \quad b_n = e^{i\theta}(e^{-\frac{1}{y_n^2}} + (x_0 + \epsilon_0)y_n^2).$$

This follows directly from the definition of f.

STEP 3: *There exists an integer $N_0 \geq n_0$ (which depends on x_0 and ϵ_0) such that, uniformly in $\theta \in [0, \pi]$:*

$$n \geq N_0 \implies |a_n| < |b_{n+1}|. \tag{3.10.13}$$

Indeed, the above-required inequality

$$e^{-\frac{1}{y_n^2}} + (x_0 - \epsilon_0)y_n^2 < e^{-\frac{1}{y_{n+1}^2}} + (x_0 + \epsilon_0)y_{n+1}^2$$

is equivalent to

$$x_0 < \epsilon_0 \frac{y_n^2 + y_{n+1}^2}{y_n^2 - y_{n+1}^2} + \frac{e^{-\frac{1}{y_{n+1}^2}} - e^{-\frac{1}{y_n^2}}}{y_n^2 - y_{n+1}^2},$$

that is,

$$x_0 < \epsilon_0 \frac{(2\pi n + \theta)^2 + (2\pi(n+1) + \theta)^2}{2\pi(2\pi(2n+1) + 2\theta)}$$
$$+ \frac{(2\pi n + \theta)^2(2\pi(n+1) + \theta)^2}{(2\pi)^3(2\pi(2n+1) + 2\theta)} \left(e^{-\frac{(2\pi(n+1)+\theta)^2}{(2\pi)^2}} - e^{-\frac{(2\pi n+\theta)^2}{(2\pi)^2}} \right).$$

We have

$$\frac{(2\pi n + \theta)^2 + (2\pi(n+1) + \theta)^2}{2\pi(2\pi(2n+1) + 2\theta)} > \frac{n^2 + (n+1)^2}{2n+2}, \quad \forall \theta \in [0, \pi].$$

Thus there exists N_0 such that, for $n \geq N_0$

$$\frac{(2\pi n + \theta)^2 + (2\pi(n+1) + \theta)^2}{2\pi(2\pi(2n+1) + 2\theta)} > \frac{2x_0}{\epsilon_0}.$$

Similarly

$$e^{-\frac{(2\pi n+\theta)^2}{(2\pi)^2}} - e^{-\frac{(2\pi(n+1)+\theta)^2}{(2\pi)^2}} \leq e^{-n^2}, \quad n \in \mathbb{N},$$

and

$$\frac{((2\pi n + \theta)^2)((2\pi(n+1) + \theta)^2)}{(2\pi)^3(2\pi(2n+1) + 2\theta)} \leq \frac{((2\pi n + \pi)^2)((2\pi(n+1) + \pi)^2)}{(2\pi)^4(2n+1)}, \quad \forall \theta \in [0, \pi],$$

and one can assume that for the same N_0 we have for $n \geq N_0$

$$\frac{((2\pi n + \theta)^2)((2\pi(n+1) + \theta)^2)}{(2\pi)^3(2\pi(2n+1) + 2\theta)} \left(e^{-\frac{(2\pi n+\theta)^2}{(2\pi)^2}} - e^{-\frac{(2\pi(n+1)+\theta)^2}{(2\pi)^2}} \right) < \frac{x_0}{\epsilon_0}.$$

The result follows.

STEP 4: *We have (with N_0 as in STEP 3)*

$$f(C_{\epsilon_0}) \supset B(0, b_{N_0}).$$

Indeed, for $n \geq N_0$,
$$(a_{n+1}, b_{n+1}) \cap (a_n, b_n) \neq \emptyset,$$

and therefore (since $\lim_{n\to\infty} a_n = 0$)
$$\bigcup_{n \geq N_0} (a_n, b_n) = (0, b_{N_0}).$$

Thus,
$$f(C_\epsilon) \supset \bigcup_{\theta \in (0,\pi)} \bigcup_{n > N_0} f(I_n(\theta))$$
$$\supset \bigcup_{\theta \in (0,\pi)} e^{i\theta} \bigcup_{n > N_0} (a_n, b_n)$$
$$\supset \bigcup_{\theta \in (0,\pi)} e^{i\theta} (0, b_{N_0}).$$

We obtain the result by considering by symmetry also the part of C_ϵ in the open lower half-plane. $\qquad\square$

Remark 3.10.1. As already mentioned we followed [299, pp. 105–106]. In particular, (3.10.13) is equation (6), p. 172, in this reference.

Solution of Exercise 3.6.9: Assume $F \subset \mathbb{R}^2$ and closed in the topology of $\widehat{\mathbb{C}}$. Then, its complement, which contains the point ∞, is open, and thus contains a set of the form $B(\infty, R)$. Hence $F \subset \overline{B(0, R)}$ and F is bounded. It is closed since the topology of $\mathbb{R}^2$ coincides with that of $\widehat{\mathbb{C}}$ in $\mathbb{R}^2$. Conversely, assume that F is bounded and closed in $\mathbb{R}^2$. Note that $\infty \in \widehat{\mathbb{C}} \setminus F$. Since F is bounded, there is an R such that $F \subset \overline{B(0, R)}$, and so $\infty \in B(\infty, R)$. As for points in $\mathbb{R}^2 \setminus F$, they have neighborhoods of the form $B(z, r)$ inside $\mathbb{R}^2 \setminus F$. Thus $\widehat{\mathbb{C}} \setminus F$ is open. $\qquad\square$

Solution of Exercise 3.7.4: Let Y be a topological space and let f be a map from X into Y, compatible with the equivalence relation. Let U be an open subset of Y. Then,
$$\tilde{f}^{-1}(U) = \left\{ \tilde{x} \in X/\sim \; ; \; \tilde{f}(\tilde{x}) \in U \right\}$$
$$= \left\{ \tilde{x} \in X/\sim \; ; \; \tilde{f}(q(x)) \in U \right\},$$

and so $q^{-1}\left(\tilde{f}^{-1}(U)\right) = f^{-1}(U)$, which is open since f is assumed continuous. It follows that $\tilde{f}^{-1}(U)$ is open (see the first item in Theorem 3.7.2) and hence $\tilde{f}$ is continuous. $\qquad\square$

Solution of Exercise 3.7.6: As we mentioned in the hints to the exercise we follow here [152, pp. 25–26].

$(1) \implies (2)$: The space X is compact and in particular Hausdorff. Thus singletons are closed in X. Let $x \in X$ be such that $u = q(x)$. Since q is closed it follows that $\{u\}$ is closed in the quotient space and so $q^{-1}(u) = q^{-1}(q(x))$ is closed in X since q is continuous (and similarly for $q^{-1}(v)$).

The closed sets $q^{-1}(u)$ and $q^{-1}(v)$ do not intersect and so by Exercise 3.2.8 there exist disjoint open sets U and V such that $q^{-1}(u) \subset U$ and $q^{-1}(v) \subset V$. Since q is closed and continuous, the sets defined by (3.7.2) are closed. They are saturated and so their respective complement sets are also saturated: Thus

$$q^{-1}(q(X \setminus F)) = X \setminus F \quad \text{and} \quad q^{-1}(q(X \setminus G)) = X \setminus G.$$

Hence the sets $q(X \setminus F)$ and $q(X \setminus G)$ are open (recall that a set in the quotient space is open if and only if its inverse image under q is open in X). To show that the quotient space is Hausdorff we prove that these two sets are disjoint, and respectively contain u and v.

Assume that

$$q(X \setminus F) \cap q(X \setminus G) \neq \emptyset.$$

There exist then $a \in X \setminus F$ and $b \in X \setminus G$ such that $q(a) = q(b)$. By definition of F and G we have (see (1.1.6) if need be)

$$F \supset X \setminus U \quad \text{and} \quad G \supset X \setminus V.$$

Hence $a \in U$ and $b \in V$. On the other hand, the set

$$X \setminus F = \{x \in X \: : \: q(x) \notin q(X \setminus U)\}$$

contains all elements equivalent to a together with a. So $b \in X \setminus F$, contradicting $U \cap V = \emptyset$.

We now check that $u \in q(X \setminus F)$. Assume not. Then there exists $a \in F$ such that $q(a) = u$. Since all points equivalent to a are in F, this contradicts that $q^{-1}(u) \subset U$.

$(2) \implies (3)$: Since

$$(a, b) \in \Gamma \iff q(a) = q(b),$$

the graph of the relation is the inverse image of the diagonal under the map

$$(a, b) \mapsto (q(a), q(b)). \tag{3.10.14}$$

The diagonal of $(X/\sim) \times (X/\sim)$ is closed in the product topology since the quotient space is Hausdorff (see Exercise 3.5.3). So the graph is closed.

$(3) \implies (1)$: Let F be a closed subset of X. We want to show that $q(F)$ is closed in the quotient space, or equivalently that $q^{-1}(q(F))$ is closed in X. But

$$
\begin{aligned}
q^{-1}(q(F)) &= \{y \in X \: q(y) \in q(F)\} \\
&= \{y \in X \: \exists x \in F \text{ such that } q(y) = q(x)\} \\
&= \{y \in X \: \exists x \in F \text{ such that } (x, y) \in \Gamma \cap (F \times E)\}.
\end{aligned}
$$

3.10. Solutions of the exercises

The set $\Gamma \cap (F \times E)$ is compact and so is $q^{-1}(q(F))$ since the projections associated with the product topology are continuous. This ends the proof since a compact set is closed. $\square$

Remark 3.10.2. We note that the implication (2) $\implies$ (3) does not use the compactness of the underlying space.

Solution of Exercise 3.8.7:

(1) Since $a(e) = I$ we have $x = a(e)x$ and so $x \sim x$. The relation is reflexive. In view of (1.2.10) we have

$$x = a(g)y \iff y = a(g^{-1})x$$

and so $\sim$ is symmetric. Finally if $x, y, z \in X$ are such that $x = a(g)y$ and $y = a(h)z$ for some $g, h \in G$, we have

$$x = a(g)a(h)z = a(gh)z$$

and hence $\sim$ is transitive.

(2) We follow the arguments from [218, Proposition 0.2.1, p. 18]. Let $x \in X$ and let U be an open neighborhood of x such that all the images $a(g)U$ are pairwise distinct when g runs through G. The canonical projection p associated with $\sim$ is then one-to-one on U. Its restriction $p|_U$ is a homeomorphism from U onto $p(U)$ (and from any $a(g)U$ onto $p(U)$ for that matter). By hypothesis the point x has a neighborhood, say V, homeomorphic to $\mathbb{R}^n$. Let φ be the homeomorphism from V onto $\mathbb{R}^n$. We can choose r such that $\varphi^{-1}(B(\varphi(x), r)) \subset (V \cap U)$, where we denote by $B(\varphi(x), r)$ the open ball in $\mathbb{R}^n$ with center $\varphi(x)$ and radius r. Thus, $p(\varphi^{-1}B(\varphi(x), r))$ is homeomorphic to $B(\varphi(x), r)$ (and hence to $\mathbb{R}^n$).

We just used the fact that a acts totally discontinuously on X. We now use the fact that it is separating, to show that the quotient space is Hausdorff. Let thus $p(x)$ and $p(y)$ be two different points in the quotient space, where x and y are in X. Of course, x and y are not uniquely determined, and one could take any element in the respective equivalence classes. Take two open neighborhoods U_x and U_y of x and y with the property that

$$a(g)U_x \cap a(h)U_y = \emptyset, \quad \forall x, y \in X.$$

Then $p(U_x)$ and $p(U_y)$ are open sets in the quotient space and contain respectively $p(x)$ and $p(y)$. $\square$

Solution of Exercise 3.8.8: Let q denote the canonical projection from the space $\mathbb{C}^{n+1} \setminus \{0\}$ onto $\mathbb{P}^n$, and let $z \in \mathbb{C}^{n+1} \setminus \{0\}$. Denote by φ the map

$$\varphi(z) = \frac{z}{\|z\|}, \quad \mathbb{C}^{n+1} \setminus \{0\} \to \mathbb{S}_n \subset \mathbb{C}^{n+1} \setminus \{0\}.$$

We note that

$$q(z) = q\left(\frac{z}{\|z\|}\right) = q \circ \varphi(z).$$

So

$$q(\mathbb{C}^{n+1} \setminus \{0\}) = q(\mathbb{S}_n).$$

Since $\mathbb{S}_n$ is compact and q is continuous it follows that $\mathbb{P}^n$ is compact. □

A direct proof that $\mathbb{P}^n$ is Hausdorff can be found for instance in [204, p. 39]. More generally, if V is any vector space on any field K, the associated projective space $P(V)$ is the set of one-dimensional subspaces of V.

Solution of Exercise 3.8.9: We follow [204, p. 42]. Denote by q the canonical projection from $\mathbb{C}^3 \setminus \{(0,0,0)\}$ onto $\mathbb{P}^2$. The set

$$Z_p = \left\{(z_1, z_2, z_3) \in \mathbb{C}^3 \setminus \{(0,0,0)\} \ : \ p(z_1, z_2, z_3) = 0\right\}$$

is closed (in the induced topology). Since $Z_p = q^{-1}(C_p)$, C_p is closed by definition of the quotient topology. Since $\mathbb{P}^2$ is compact (see Exercise 3.8.8) C_p is compact by Theorem 3.2.4. □

Solution of Exercise 3.8.15: We define

$$N = (1, \underbrace{0, 0, \ldots, 0}_{n \text{ times } 0}) \quad \text{and} \quad S = (-1, \underbrace{0, 0, \ldots, 0}_{n \text{ times } 0}),$$

and

$$U_N = \mathbb{S}_n \setminus \{N\} \quad \text{and} \quad U_S = \mathbb{S}_n \setminus \{S\}$$

respectively, and the maps:

$$i_N(x) = \frac{1}{1 - x_0}(x_1, \ldots, x_n), \quad U_N \longrightarrow \mathbb{R}^n,$$

and

$$i_S(x) = \frac{1}{1 + x_0}(x_1, \ldots, x_n), \quad U_S \longrightarrow \mathbb{R}^n.$$

The maps i_N and i_S are homeomorphisms, and their respective inverses are given by:

$$i_N^{-1}(y) = \frac{1}{\|y\|^2 + 1}(\|y\|^2 - 1, 2y_1, \ldots, 2y_n),$$

and

$$i_S^{-1}(y) = \frac{1}{\|y\|^2 + 1}(-\|y\|^2 + 1, 2y_1, \ldots, 2y_n).$$

To check these equations, remark that

$$\frac{1}{1 - x_0}(x_1, \ldots, x_n) = (y_1, \ldots, y_n)$$

implies that

$$\frac{\sum_{u=1}^{n} x_u^2}{(1-x_0)^2} = \|y\|^2$$

and hence

$$\|y\|^2 + 1 = \frac{\sum_{u=1}^{n} x_u^2 + x_0^2 + 1 - 2x_0}{(1-x_0)^2} = \frac{2}{1-x_0} \quad \text{and} \quad \|y\|^2 - 1 = \frac{2x_0}{1-x_0}.$$

(U_N, i_N) and (U_S, i_S) form an atlas of the unit sphere, and it holds that

$$i_S \circ i_N^{-1}(y) = \frac{y}{\|y\|^2} \qquad \mathbb{R}^n \setminus \{(0,0,\ldots,0)\} \longrightarrow \mathbb{R}^n \setminus \{(0,0,\ldots,0)\}.$$

Thus the unit sphere of $\mathbb{R}^{n+1}$ is a smooth manifold of dimension n. $\qquad\square$

Note that the previous exercise does not say anything about an analytic structure when n is even.

Solution of Exercise 3.8.16: It suffices to consider the map

$$\varphi(x,y) = (ax, by) = (x_1, y_1), \tag{3.10.15}$$

since

$$x^2 + y^2 < 1 \quad \Longleftrightarrow \quad \frac{x_1^2}{a^2} + \frac{y_1^2}{b^2} < 1. \qquad\square$$

As remarked after the statement of the exercise, the map (3.10.15) is not analytic (when $a \neq b$) as is seen for instance by writing down the Cauchy–Riemann equations.

Solution of Exercise 3.8.17: Intuitively, a natural way is to spread out (when $R_1 \geq R_2$) or to contract (when $R_1 \leq R_2$) along radial directions, A_{R_1} onto A_{R_2}. This suggests that we consider the map:

$$f(z) = (1 + a(|z| - 1)) \frac{z}{|z|},$$

with

$$a = \frac{R_2 - 1}{R_1 - 1}.$$

Since $z \neq 0$, the function f is clearly differentiable since $\operatorname{Re} f$ and $\operatorname{Im} f$ are smooth functions of x and y (with $z = x + iy$). Furthermore, its inverse is given by:

$$h(z) = \left(1 + \frac{|z| - 1}{a}\right) \frac{z}{|z|}.$$

To see this, set $f(z) = w$ and note that (since $|z| > 1$)

$$|w| = 1 + a(|z| - 1),$$

so that

$$h \circ f(z) = \left(1 + \frac{|w| - 1}{a}\right) \frac{w}{|w|}$$

$$= |z| \frac{(1 + a(|z| - 1)) \frac{z}{|z|}}{1 + a(|z| - 1)}$$

$$= z.$$

By symmetry, we also have $f \circ h(z) = z$ for $z \in A_{R_2}$, and this concludes the proof.

$\square$

Solution of Exercise 3.8.19: We denote by $\overset{\circ}{z}$ the equivalence classes, that is the elements of $\mathbb{P}^n$. Let $i \in \{0, \ldots, n\}$ and let $U_i \subset \mathbb{P}^n$ denote the set of equivalence classes of vectors $(z_0, z_1, \ldots, z_n)$ such that $z_i \neq 0$. This last condition does not depend on the given representative in the equivalence class. Define φ_i from U_i into $\mathbb{C}^n$ by

$$\varphi_i(\overset{\circ}{z}) = (\frac{z_0}{z_i}, \ldots, \frac{z_{i-1}}{z_i}, \frac{z_{i+1}}{z_i}, \ldots, \frac{z_N}{z_i}), \tag{3.10.16}$$

with obvious changes when $i = 0$ or $i = n$. The functions φ_i are continuous, and so the $\{(U_i, \varphi_i), 0 = 1, \ldots, n\}$ define an atlas of $\mathbb{P}^n$. We now compute the transition maps. Let $i < j$ (the case $j < i$ is treated in the same way), and first remark the following. An equivalence class $\overset{\circ}{z}$ belongs to the intersection $U_i \cap U_j$ if and only if both $z_i \neq 0$ and $z_j \neq 0$ in some (and hence in every) representative $z \in \overset{\circ}{z}$. In (3.10.16), the component $z_j \in z$ is at the $(j - 1)$ slot. In (3.10.16) with j instead of i, the component z_i is not shifted since $i < j$. Thus we have:

$$\varphi_i(U_i \cap U_j) = \{z \in \mathbb{C}^n \; ; \; z_{j-1} \neq 0\},$$
$$\varphi_j(U_i \cap U_j) = \{z \in \mathbb{C}^n \; ; \; z_i \neq 0\},$$

and, for $z \in \varphi_i(U_i \cap U_j)$,

$$\varphi_j \circ \varphi_i^{-1}(z) = \frac{z_i}{z_{j-1}} z.$$

The transition maps are analytic, and thus $\mathbb{P}^n$ is an analytic manifold. $\square$

Solution of Exercise 3.8.21: We take an atlas made of the two following charts:

$$U_N = \mathbb{C} \quad \text{and} \quad i_N(z) = z,$$

and

$$U_P = \widehat{\mathbb{C}} \setminus \{0\} \quad \text{and} \quad i_P(z) = \begin{cases} \frac{1}{z}, & z \neq \infty \\ 0, & z = \infty. \end{cases}$$

3.10. Solutions of the exercises

Then, i_N is trivially continuous and so is i_P by Exercise 3.2.7. Furthermore, the transition function

$$i_N \circ i_S^{-1}(z) = \frac{1}{z}, \quad \mathbb{C} \setminus \{0\} \longrightarrow \mathbb{C} \setminus \{0\},$$

is analytic, and this ends the proof. □

Solution of Exercise 3.8.25: We take $r \in (1, R)$ and denote by

$$n = \frac{1}{2\pi i} \int_{|z|=r} \frac{f'(z)}{f(z)} dz.$$

Since $\frac{f'}{f}$ is analytic in A_R, Green's theorem (for instance) will show that n does not depend on r but only on f. We claim that the function $h(z) = z^{-n} f(z)$ has an analytic logarithm in A_R. It is equivalent to prove that

$$\int_C \frac{h'(z)}{h(z)} dz = 0 \tag{3.10.17}$$

for every closed smooth curve, say C, in A_R. But (3.10.17) is equivalent to

$$-n \int_C \frac{dz}{z} + \int_C \frac{f'(z)}{f(z)} dz = 0. \tag{3.10.18}$$

Since $\mathrm{Ind}_{|z|=r}(0) = 1$, equation (3.10.18) in turn is exactly (2.3.11), that is (2.3.10) with $\frac{f'}{f}$ instead of f and the curves $C_1 = \{|z| = r\}$ and $C_2 = C$. Thus

$$z^{-n} f(z) = e^{g(z)}$$

where g is analytic in A_R. □

Solution of Exercise 3.8.26:

(1) We proceed by contradiction. Let g denote an analytic logarithm of the function $f(z) = z$ in A_R. Then,

$$e^{g(z)} = z, \quad \forall z \in A_R,$$

and so

$$g'(z) e^{g(z)} = g'(z) z = 1, \quad z \in A_R.$$

Thus for any $r \in (1, R)$,

$$0 = \int_{|z|=r} g'(z) dz = \int_{|z|=r} \frac{dz}{z} = 2\pi,$$

which cannot be.

(2) Let f denote a conformal map from A_{R_1} onto A_{R_2}, with inverse g, and assume by contradiction that there exists a function w analytic in A_{R_1} and such that $f(z) = e^{w(z)}$ for $z \in A_{R_1}$. Then,

$$e^{w(g(z))} = z, \quad z \in A_{R_2},$$

and thus the function z would have an analytic logarithm in A_{R_2}. This cannot be in view of (1).

$\square$

Solution of Exercise 3.8.27: We follow the hint. Assume that A_{R_1} is conformally equivalent to A_{R_2}, with conformal mapping f. We can always assume $R_2 \leq R_1$ (if $R_1 \leq R_2$, consider f^{-1}). By Exercise 3.8.25, f admits a representation of the form (3.8.4), that is:

$$f(z) = z^n e^{g(z)}, \quad z \in A_{R_1},$$

where $n \in \mathbb{Z}$ and g is analytic in A_{R_1}. By Exercise 3.8.26, f has no analytic logarithm, and hence $n \neq 0$. From (3.8.4) we have:

$$\ln |f(z)| = n \ln |z| + u(x, y),$$

where $u(x, y) = \operatorname{Re} g(z)$. Let $g(z) = \sum_{m=0}^{\infty} a_m z^m + \sum_{m=1}^{\infty} \frac{b_m}{z^m}$ denote the Laurent expansion of g in A_{R_1}, and let $R \in (1, R_1)$. We have:

$$\int_0^{2\pi} u(R \cos t, R \sin t) dt = \operatorname{Re} \int_0^{2\pi} g(Re^{it}) dt$$

$$= \operatorname{Re} \left(\sum_{m=0}^{\infty} a_m \int_0^{2\pi} e^{imt} dt + \sum_{m=1}^{\infty} b_m \int_0^{2\pi} e^{-imt} dt \right)$$

$$= 2\pi \operatorname{Re} (a_0)$$

$$\overset{\text{def.}}{=} 2\pi A.$$

Thus

$$\frac{1}{2\pi} \int_0^{2\pi} \ln |f(Re^{it})| dt = n \ln R + \frac{1}{2\pi} \int_0^{2\pi} u(R \cos t, R \sin t) dt = n \ln R + A.$$

Since $|f| \in [1, R_2]$ we have:

$$0 \leq n \ln R + A \leq \ln R_2, \quad \forall R \in (1, R_1).$$

From $R \to 1$ we get

$$A \in (0, \ln R_2), \tag{3.10.19}$$

while $R \to R_1$ leads to

$$0 \leq n \ln R_1 + A \leq \ln R_2. \tag{3.10.20}$$

When $n \geq 0$ and since $A \geq 0$, this last inequality leads to

$$n \leq \frac{\ln R_2}{\ln R_1} \leq 1,$$

and hence $n = 1$ since $n \neq 0$. It follows that $R_1 = R_2$.

Assume now $n \leq 0$. The first inequality in (3.10.20) implies

$$-n \ln R_1 \leq A,$$

and so, in view of (3.10.19)

$$-n \ln R_1 \leq \ln R_2,$$

and so

$$-n \leq \frac{\ln R_2}{\ln R_1}.$$

Since $-n > 0$ and $R_2 \leq R_1$, we here too get $R_1 = R_2$. $\qquad\square$

Solution of Exercise 3.8.30: This exercise is a special case of Exercise 3.8.33 with $g = 1$ and $A = \tau$. We send the reader to this last exercise for more details. $\qquad\square$

Solution of Exercise 3.8.31: Let τ_1 and τ_2 denote two points in the open upper half-plane, and let C_{τ_1} and C_{τ_2} be the associated torii, with canonical projections q_1 and q_2 respectively. We claim that the map

$$X(q_1(z)) = q_2(z)$$

defines a homeomorphism from C_{τ_1} onto C_{τ_2}. Let $z \in \mathbb{C}$ and let ϵ be such that both q_1 and q_2 are one-to-one on $B(z, \epsilon)$. The set $q_1(B(z, \epsilon))$ is open by definition of the quotient topology and on $q_1(B(z, \epsilon))$ we have

$$X = q_2 \circ (q_1|_{B(z,\epsilon)})^{-1}.$$

Let $U \subset q_2(B(z, \epsilon))$ be open (in the induced topology or in the quotient topology; these are the same here since $q_2(B(z, \epsilon))$ is open). The set $V = q_2^{-1}(U)$ is an open subset of the complex plane, and so $q_1(V)$ is open in C_{τ_1}, and

$$X^{-1}(U) = q_1(V)$$

is open in C_{τ_1}. Thus X is continuous on $q_1(B(z, \epsilon))$. The open sets $q_1(B(z, \epsilon))$ cover all of C_{τ_1} and hence X is continuous; see Exercise 3.4.2 if need be. Interchanging the role of τ_1 and τ_2 we see that X^{-1} is also continuous. $\qquad\square$

Solution of Exercise 3.8.32: Consider first a continuous function F from $\mathbb{C}_\tau$ into $\widehat{\mathbb{C}}$, and denote by q the canonical projection from $\mathbb{C}$ onto $\mathbb{C}_\tau$. The function

$$f(z) = F(q(z))$$

is then continuous, and satisfies (3.8.6):

$$f(z + n + \tau m) = f(z), \quad \forall z \in \mathbb{C} \quad \text{and} \quad n, m \in \mathbb{Z},$$

since

$$q(z + n + \tau m) = q(z), \quad \forall z \in \mathbb{C} \quad \text{and} \quad n, m \in \mathbb{Z}.$$

Conversely, start from f continuous from $\mathbb{C}$ into $\widehat{\mathbb{C}}$ and satisfying (3.8.6). The function $\breve{f}(\pi(z)) = f(z)$ is continuous in view of Theorem 3.7.3. $\qquad\square$

Solution of Exercise 3.8.33: We use Exercise 3.8.7 with $X = \mathbb{C}^g$ and $G = \mathbb{Z}^g + A\mathbb{Z}^g$ which acts on $\mathbb{C}$ via translation: For $g = m + An \in G$ (with $m, n \in \mathbb{Z}^g$)

$$a(g)(z) = m + An + z, \quad z \in \mathbb{C}^g. \tag{3.10.21}$$

We proceed in four steps.

STEP 1: *It holds that*

$$\min_{g \in G, g \neq 0} \|g\|_{\mathbb{C}^g} > 0. \tag{3.10.22}$$

Indeed, let $A = A_R + iA_I$, where $A_R = \frac{A+A^*}{2}$ and $A_I + \frac{A-A^*}{2i}$. By hypothesis, there is $\epsilon > 0$ such that

$$\langle n, A_I n \rangle_{\mathbb{C}^g} \geq \epsilon \langle n, n \rangle_{\mathbb{C}^g}. \tag{3.10.23}$$

Assume that the minimum in (3.10.22) is equal to 0. Then, there exists a sequence $(m_k, n_k)_{k \in \mathbb{N}} \in (\mathbb{Z}^g)^2$ such that $(m_k, n_k) \neq (0, 0)$ for all k and

$$\lim_{k \to \infty} \|m_k + A_R n_k + iA_I n_k\| = 0, \tag{3.10.24}$$

where $\|\cdot\|$ denotes the Euclidean norm. We remark that $n_k \neq 0$ for an infinite number of indices (otherwise we would have $\lim_{k \to \infty} m_k = 0$ and so $m_k = 0$ for all but a finite number of indices). Without loss of generality we will assume that $n_k \neq 0$ for all $k \in \mathbb{N}$. By the Cauchy–Schwarz inequality we then have

$$\lim_{k \to \infty} \frac{\langle m_k + A_R n_k + iA_I n_k, n_k \rangle_{\mathbb{C}^g}}{\langle n_k, n_k \rangle_{\mathbb{C}^g}} = 0.$$

But

$$\frac{\langle m_k + A_R n_k + iA_I n_k, n_k \rangle_{\mathbb{C}^g}}{\langle n_k, n_k \rangle_{\mathbb{C}^g}} = \frac{\langle A_R n_k, n_k \rangle_{\mathbb{C}^g}}{\langle n_k, n_k \rangle_{\mathbb{C}^g}} + \frac{\langle m_k, n_k \rangle_{\mathbb{C}^g}}{\langle n_k, n_k \rangle_{\mathbb{C}^g}}$$
$$+ i\frac{\langle A_I n_k, n_k \rangle_{\mathbb{C}^g}}{\langle n_k, n_k \rangle_{\mathbb{C}^g}}.$$

This expression cannot go to 0 since

$$\frac{\langle A_R n_k, n_k \rangle_{\mathbb{C}^g}}{\langle n_k, n_k \rangle_{\mathbb{C}^g}} + \frac{\langle m_k, n_k \rangle_{\mathbb{C}^g}}{\langle n_k, n_k \rangle_{\mathbb{C}^g}} \in \mathbb{R}, \quad \text{and} \quad \frac{\langle A_I n_k, n_k \rangle_{\mathbb{C}^g}}{\langle n_k, n_k \rangle_{\mathbb{C}^g}} \geq \epsilon.$$

3.10. Solutions of the exercises

STEP 2: *G acts totally discontinuously on* $\mathbb{C}^g$: Let $z_0 \in \mathbb{C}^g$ and let U denote the open ball of center z_0 and radius $\frac{\eta}{2}$, where η is the minimum in (3.10.22). Then, (with $a(g)$ defined by (3.10.21))

$$a(g_1)U \cap a(g_2)U = \emptyset, \quad \forall g_1, g_2 \in G \quad \text{with} \quad g_1 \neq g_2.$$

Indeed, if there is z_1 in the intersection, we have:

$$\|g_1 - g_2\| = \|(z_0 + g_1 - z_1) - (z_0 + g_2 - z_1)\| \leq \|z_0 + g_1 - z_1\| + \|z_0 + g_2 - z_1\| < \eta,$$

contradicting (3.10.22).

STEP 3: *G is separating on* $\mathbb{C}^g$: Let z and w be two points in $\mathbb{C}^g$ such that $z - w \notin G$, and let $e = \min_{g \in G} \|z - w - g\|$. Let U denote the open ball with center z and radius $r = \frac{e}{2}$, and W be the open ball with center w and same radius r. We claim that

$$a(g_1)U \cap a(g_2)W = \emptyset, \quad \forall g_1, g_2 \in G.$$

Indeed if there is $z_1 \in a(g_1)U \cap a(g_2)W$, we have:

$$\|z - w + g_1 - g_2\| = \|(z + g_1 - z_1) - (w + g_2 - z_1)\| \leq \|z + g_1 - z_1\| + \|w + g_2 - z_1\| < e,$$

which contradicts the definition of e.

STEP 4: *The quotient space is compact:* We note first that the columns of A together with the standard basis form a basis of $\mathbb{C}^g$ over the real numbers. Indeed, let $x, y \in \mathbb{R}^g$ be such that $x + Ay = 0$. Then,

$$\langle x + A_R y, y \rangle_{\mathbb{C}^g} + i \langle A_I y, y \rangle_{\mathbb{C}^g} = 0.$$

Since $\langle x + A_R y, y \rangle_{\mathbb{C}^g} \in \mathbb{R}$ and $\langle A_I y, y \rangle_{\mathbb{C}^g} \geq \epsilon \langle y, y \rangle_{\mathbb{C}^g}$ we conclude that $x = y = 0$. Let now $z \in \mathbb{C}$. It can be written as

$$z = m + x + A(n + y)$$

where $m, n \in \mathbb{Z}^g$ and where the entries of x and y are in $[0, 1)$. Hence, every $z \in \mathbb{C}^g$ is equivalent to a point in the ball $\|z\| \leq (1 + \|A\|)\sqrt{g}$. Since the canonical projection onto $\mathbb{C}^g/G$ is continuous, and since $\|z\| \leq (1 + \|A\|)\sqrt{g}$ is compact, we conclude that $\mathbb{C}^g/G$ is compact. $\square$

Solution of Exercise 3.8.34: We first assume that such a complex structure exists, and show what it should be. This will show existence and uniqueness of a complex structure on X with the required property.

Since Y is a Riemann surface, every point $y \in Y$ has an open neighborhood, say U_y, diffeomorphic to an open subspace of $\mathbb{C}$, via a map

$$\varphi_y : U_y \longrightarrow \varphi_y(U_y),$$

and the chart (U_y, φ_y) belongs to the complex structure of Y. This neighborhood can be assumed small enough so that p is a homeomorphism from $p^{-1}(U_y)$ onto U_y. We can assume that U_y is chosen also small enough so that there exists a homeomorphism ψ_y from $p^{-1}(U_y)$ onto $\psi_y(p^{-1}(U_y)) \subset \mathbb{C}$, and such that the chart $(p^{-1}(U_y), \psi_y)$ belongs to the complex structure of X, if such a structure exists. The map p will be analytic if and only if the map

$$T_y = \varphi_y \circ p \circ \psi_y^{-1}, \quad \psi_y(p^{-1}(U_y)) \longrightarrow \varphi_y(U_y)$$

is analytic. We claim that under this constraint, the charts $(p^{-1}(U_y), \psi_y)$ define a complex structure on X. Indeed, assume that two different sets U_y and U_z intersect. Then the transition map

$$\psi_y \circ \psi_z^{-1} : \quad \psi_z(p^{-1}(U_y \cap U_z)) \longrightarrow \psi_y(p^{-1}(U_y \cap U_z))$$

is analytic since

$$\psi_y \circ \psi_z^{-1} = T_y^{-1} \circ \varphi_y \circ p \circ (p|_{U_y \cap U_z})^{-1} \circ \varphi_z^{-1} \circ T_z = T_y^{-1} \circ \varphi_y \circ \varphi_z^{-1} \circ T_z$$

is analytic. So the complex structure on X is necessarily defined by an atlas of the form

$$T_y^{-1} \circ \varphi_y \circ (p|_{U_y})^{-1}, \tag{3.10.25}$$

where T_y is analytic and one-to-one on $\varphi_y(U_y)$. $\square$

The previous result is of special importance when Y is taken to be the Riemann sphere.

Solution of Exercise 3.9.2: The function d is symmetric and positive. By definition, $d(x, y) = 0$ if and only if $x = y$. The triangle inequality is directly checked by inspection. If all three elements x, y, z are pairwise different, it reduces to $1 \leq 2$, and similar remarks hold for the other cases. $\square$

Solution of Exercise 3.9.3: Let $D = \frac{d}{1+d}$. Since $d(x, y) = d(y, x)$ and $d(x, x) \geq 0$ for all $x, y \in E$ the function D is clearly symmetric and positive. To prove the triangle inequality one uses the following standard argument. Let $u, v, w \in [0, \infty)$ be such that $u \leq v + w$. Since $v + w \leq v + w + vw$ and since the function $a(x) = \frac{x}{1+x}$ is increasing, we have:

$$u \leq v + w \Longrightarrow \frac{u}{1+u} \leq \frac{v+w+vw}{1+v+w+vw}. \tag{3.10.26}$$

Hence,

$$\frac{u}{1+u} \leq \frac{v+w+vw}{1+v+w+vw} \leq \frac{v+w+2vw}{1+v+w+vw} = \frac{v}{1+v} + \frac{w}{1+w}.$$

Thus,

$$u \leq v + w \implies \frac{u}{1+u} \leq \frac{v}{1+v} + \frac{w}{1+w}. \tag{3.10.27}$$

Applying (3.10.27) to

$$u = d(x, y), \quad v = d(x, z) \quad \text{and} \quad w = d(z, y) \quad (x, y, z \in E)$$

we see that D satisfies the triangle inequality, and hence is a semi-metric. Finally, $D(x, x) = 0$ if and only if $d(x, x) = 0$, and so D will be a metric if and only if d is a metric.

Since a (possibly infinite, with non-negative weights) converging sum of semi-metrics is still a semi-metric, we have that (3.9.1) is a semi-metric. It will be a metric if and only if the following condition is in force:

$$\forall x, y \in E \text{ be such } x \neq y, \quad \exists n \in \mathbb{N} \text{ such that } d_n(x, y) \neq 0. \qquad \square$$

Solution of Exercise 3.9.4: Conditions $(1)'$ and (2) in the definition of a semi-metric are inherited from the corresponding conditions (1) and (2) on d (one could even assume d to be a semi-metric). The triangle inequality follows directly from (3.9.2) with

$$c = d(x, y) \leq a + b, \quad \text{with} \quad a = d(x, z) \quad \text{and} \quad b = d(z, y).$$

Finally $\varphi(d)$ will be a metric if and only if φ vanishes only at the origin. $\qquad \square$

Solution of Exercise 3.9.9: The fact that d is a metric follows from Exercise 3.9.3, or from the previous exercise. Let now $(z^{(p)})_{p \in \mathbb{N}}$ be a Cauchy sequence in (E, d):

$$\forall \epsilon > 0, \exists P \in \mathbb{N}, p, q \geq P \implies \sum_{n=1}^{\infty} \frac{1}{2^n} \frac{|z_n^{(p)} - z_n^{(q)}|}{1 + |z_n^{(p)} - z_n^{(q)}|} \leq \epsilon.$$

In particular, we have:

$$\forall \epsilon > 0, \exists P \in \mathbb{N}, p, q \geq P \implies \forall n \in \mathbb{N}, \frac{1}{2^n} \frac{|z_n^{(p)} - z_n^{(q)}|}{1 + |z_n^{(p)} - z_n^{(q)}|} \leq \epsilon.$$

For ϵ such that $2^n \epsilon < 1$ this means that

$$|z_n^{(p)} - z_n^{(q)}| \leq \frac{\epsilon 2^n}{1 - \epsilon 2^n},$$

and in particular for every $n \in \mathbb{N}$, the sequence $(z_n^{(p)})_{n \in \mathbb{N}}$ is a Cauchy sequence of numbers in $\overline{\mathbb{D}}$. Let

$$z_n = \lim_{p \to \infty} z_n^{(p)}, \quad \text{and} \quad z = (z_n)_{n \in \mathbb{N}}.$$

We claim that $\lim_{p\to\infty} d(z, z^{(p)}) = 0$, and hence that (E, d) is complete. Indeed, let $\epsilon > 0$ and take N such that $\sum_{n=N+1}^{\infty} \frac{1}{2^n} < \epsilon$. For every $n \in \{1, \ldots, N\}$ there exists a number P_n (which depends on n) such that

$$p \geq P_n \implies \frac{|z_n^{(p)} - z_n|}{1 + |z_n^{(p)} - z_n|} \leq \epsilon/N.$$

Therefore, for $p \geq \max_{n=1,2,\ldots,N} P_n$ we have

$$d(z, z^{(p)}) \leq 2\epsilon.$$

$\square$

Solution of Exercise 3.9.11: Consider first $x \in (0, 1)$ and $\rho > 0$. Let $y \in E$. Then, $y \in B_2(x, \rho)$ if and only if $|x - y| < |xy|\rho$, that is, if and only if

$$(x - y)^2 < \rho^2 x^2 y^2,$$

i.e., if and only if

$$y^2(1 - x^2\rho^2) - 2yx + x^2 < 0. \tag{3.10.28}$$

We now assume that

$$1 - x^2\rho^2 > 0. \tag{3.10.29}$$

Then, (3.10.28) is equivalent to

$$\left(y - \frac{x}{1 - x^2\rho^2}\right)^2 < \frac{x^2}{(1 - x^2\rho^2)^2} - \frac{x^2}{(1 - x^2\rho^2)} = \frac{x^4\rho^2}{(1 - x^2\rho^2)^2}.$$

Thus, when (3.10.29) is in force we have:

$$B_2(x, \rho) = B_1(x_1, \rho_1), \tag{3.10.30}$$

where

$$x_1 = \frac{x}{1 - x^2\rho^2} \quad \text{and} \quad \rho_1 = \frac{x^2\rho}{(1 - x^2\rho^2)}, \tag{3.10.31}$$

when ρ is assumed such that $B_1(x_1, \rho_1) \subset E$. When $x = 1$ we look for ρ_1 such that

$$B_2(1, \rho) = B_1(1, \rho_1). \tag{3.10.32}$$

But $y \in B_2(1, \rho)$ if and only if it belongs to E and satisfies

$$\left(y - \frac{1}{1 - \rho^2}\right)^2 < \frac{\rho^2}{(1 - \rho^2)^2}, \tag{3.10.33}$$

that is, if and only if $y \in (\frac{1}{1+\rho}, 1]$. Thus we need to have

$$1 - \rho_1 = \frac{1}{1 + \rho}, \tag{3.10.34}$$

and $\rho_1 = \frac{\rho}{1+\rho}$. But any open set in $\mathcal{T}_2$ can be written as a union of open balls of the forms (3.10.30) and (3.10.32), and thus $\mathcal{T}_2 \subset \mathcal{T}_1$.

To prove the converse inclusion, we invert the system of equations (3.10.31). For $x_1 \neq 1 \in E$ and $\rho_1 > 0$ given we have

$$\frac{\rho_1}{x_1} = x\rho,$$

and thus

$$x = x_1(1 - x^2\rho^2) = x_1\left(1 - \left(\frac{\rho_1}{x_1}\right)^2\right).$$

We choose ρ_1 such that

$$x = x_1\left(1 - \left(\frac{\rho_1}{x_1}\right)^2\right) \in E \quad \text{and} \quad B_1(x, \rho_1) \subset E. \tag{3.10.35}$$

Then,

$$\rho = \frac{\rho_1}{x_1 x} = \frac{\rho_1}{x_1^2\left(1 - \left(\frac{\rho_1}{x_1}\right)^2\right)}.$$

Thus, for any $x_1 \in E$ and ρ_1 such that (3.10.35) is in force we have

$$B_1(x_1, \rho_1) = B_2(x, \rho),$$

where

$$x = x_1\left(1 - \left(\frac{\rho_1}{x_1}\right)^2\right) \quad \text{and} \quad \rho = \frac{\rho_1}{x_1^2\left(1 - \left(\frac{\rho_1}{x_1}\right)^2\right)}.$$

Such open balls are therefore open in $\mathcal{T}_2$. The case $x_1 = 1$ is considered as follows. Given a small enough $\rho_1 > 0$, we look for ρ such that

$$B_1(1, \rho_1) = B_2(1, \rho). \tag{3.10.36}$$

But $x \in (0, 1]$ is such that

$$|x - 1| < \rho|x| \quad \text{(that is, belongs to $B_2(1, \rho)$)}$$

if and only if (3.10.33) holds. It follows that $B_1(1, \rho_1) = B_1(\frac{1}{1-\rho^2}, \frac{\rho}{1-\rho^2}) \cap E$, and as above we need to have (3.10.34), where now the unknown is ρ. Equation (3.10.36) follows with $\rho = \frac{\rho_1}{1-\rho_1}$. This implies that $\mathcal{T}_1 \subset \mathcal{T}_2$ since every open subset of $\mathcal{T}_1$ is a union of sets of the form $B_1(x, \rho_x)$ for ρ_x small enough.

The sequence $x_n = 1/n$, $n = 1, 2, \ldots$ is a Cauchy sequence for the metric d_1, and has no limit in E. Thus (E, d_1) is not complete. Note that $(x_n)_{n \in \mathbb{N}}$ is *not* a Cauchy sequence for the metric d_2 since

$$d_2(x_n, x_{n+1}) \equiv 1.$$

We now check that (E, d_2) is complete. To that purpose, consider the space $[1, \infty)$ with the metric d_1. It is complete. On the other hand, the map

$$x \mapsto \frac{1}{x}$$

is an isometry of metric spaces between (E, d_2) and $([1, \infty), d_1)$, and so (E, d_2) is complete.

$\square$

Solution of Exercise 3.9.18: We follow the (quite standard) argument in [269, p. 29]. The two sets F_1 and F_2 are closed and so their distance is strictly positive:

$$\rho = d(F_1, F_2) = \inf_{\substack{x_1 \in F_1, \\ x_2 \in F_2}} d(x_1, x_2) > 0,$$

and in fact the above infimum is a minimum. Let

$$U_i = \cup_{x \in F_i} B(x, \rho/3), \quad i = 1, 2.$$

U_1 and U_2 are open and do not intersect. Still it is not clear that their closure will not intersect. We here use the fact that E is compact, and hence F_1 and F_2 are also compact, to define open subsets of U_1 and U_2 which will satisfy (3.9.4). By compactness,

$$F_1 \subset O_1 = \cup_{n=1}^{N} B(x_n, \rho/3) \quad \text{and} \quad F_1 \subset O_2 = \cup_{n=1}^{M} B(y_n, \rho/3)$$

for some $x_1, \ldots, x_N \in F_1$ and $y_1, \ldots, y_M \in F_2$. We check that $\overline{O_1} \cap \overline{O_2} = \emptyset$. In view of Exercise 3.1.14, it is enough to show that

$$\overline{B(x, \rho/3)} \cap \overline{B(y, \rho/3)} = \emptyset \quad \text{for} \quad x \in F_1 \text{ and } y \in F_2.$$

Let z belong to the intersection, assumed non-empty, and let $(v_p)_{p \in \mathbb{N}}$ and $(w_p)_{p \in \mathbb{N}}$ be sequences of elements in F_1 and F_2 respectively, and such that

$$\lim_{p \to \infty} d(v_p, z) = \lim_{p \to \infty} d(w_p, z) = 0.$$

The inequality

$$d(x, y) \leq d(x, v_p) + d(v_p, z) + d(z, w_p) + d(w_p, y)$$

leads to

$$d(x, y) \leq \frac{2\rho}{3}$$

which is not possible since the distance between F_1 and F_2 is ρ.

$\square$

Solution of Exercise 3.9.19: Take a sequence $(a^{(p)})_{p \in \mathbb{N}}$ of elements in C. Then,

$$|a_n^{(p)}| \leq \frac{1}{n}, \quad n = 1, 2, \ldots, \quad p = 1, 2, \ldots$$

3.10. *Solutions of the exercises*

Let $n = 1$. There exists a subsequence $(a^{(j(1,k))})_{k \in \mathbb{N}}$ such that

$$\lim_{k \to \infty} a_1^{(j(1,k))}$$

exists. Let a_1 denote this limit. We have $|a_1| \leq 1$. Consider now the sequence $a_2^{(j(1,k))}$. It is bounded by $1/2$ in modulus, and therefore there is a subsequence $(j(2,k))_{k \in \mathbb{N}}$ of the sequence $(j(1,k))_{k \in \mathbb{N}}$ such that

$$\lim_{k \to \infty} a_2^{(j(2,k))}$$

exists. Let a_2 denote this limit. We have $|a_2| \leq 1/2$. Note that we also have

$$\lim_{k \to \infty} a_1^{(j(2,k))} = a_1.$$

Iteratively, we build this way a sequence of sequences $((j(n,k))_{k \in \mathbb{N}})_{n \in \mathbb{N}}$ and a sequence of numbers $(a_n)_{n \in \mathbb{N}}$ such that

$$|a_n| \leq 1/n, \quad n = 1, 2, \ldots$$

and

$$\forall m \in \{1, \ldots, n\}, \quad \lim_{k \to \infty} a_m^{(j(n,k))} = a_m.$$

We now use the diagonal process, and form a new sequence $p_1, p_2, \ldots$ by defining p_t to be the tth element of the sequence $(j(t,k))_{k \in \mathbb{N}}$.

The sequence $(p_t)_{t \in \mathbb{N}}$ is *not* a subsequence of all the sequences $(j(n,k))_{k \in \mathbb{N}}$, $j = 1, 2, \ldots$, but the following is true: By construction, for every n, the sequence $(p_t)_{t \geq n}$ is a subsequence of the sequence $(j(n,k))_{k \geq k_n}$ for some $k_n \in \mathbb{N}$. Therefore, all the limits

$$\lim_{t \to \infty} a_n^{(p_t)} = a_n, \quad n = 1, 2, \ldots.$$

hold. We now prove that:

(1) $(a_n)_{n \in \mathbb{N}} \in C$.
(2) The limit

$$\lim_{t \to \infty} d(a^{(p_t)}, a) = 0 \qquad (3.10.37)$$

holds. This last claim will prove the compactness of C.

To prove (1) we note that, for every $M \in \mathbb{N}$,

$$\sum_{n=1}^{M} n^2 |a_n^{(p_t)}|^2 \leq 1.$$

Letting $t \to \infty$ we have

$$\sum_{n=1}^{M} n^2 |a_n|^2 \leq 1.$$

It suffices now to let $M \to \infty$ to obtain that $(a_n)_{n \in \mathbb{N}}$ belongs to C. Up to this stage, one could replace C by any set of the form

$$C(\alpha_1, \ldots) = \left\{ a \in \ell_2, \sum_{n=1}^{\infty} \alpha_n |a_n|^2 \leq 1 \right\}$$

where the α_n are pre-assigned numbers greater than or equal to 1, this last condition insuring that

$$C(\alpha_1, \ldots) \subset \ell_2.$$

Note that weaker conditions are possible to ensure this inclusion.

We know that the closed unit ball of ℓ_2 (which corresponds to $\alpha_n \equiv 1$) is not compact. To prove (3.10.37) we will make use of the fact that $\alpha_n = n^2$. Let $\epsilon > 0$ and let $M \in \mathbb{N}$ be such that

$$\sum_{n=M}^{\infty} \frac{1}{n^2} \leq \epsilon.$$

We note that

$$|a_n^{(p_j)} - a_n|^2 \leq 2(|a_n^{(p_j)}|^2 + |a_n|^2) \leq \frac{4}{n^2}, \quad n = 1, 2, \ldots$$

Thus, we have

$$d(a^{(p_j)}, a)^2 = \left(\sum_{n=1}^{M-1} |a_n^{(p_j)} - a_n|^2 \right) + \left(\sum_{n=M}^{\infty} |a_n^{(p_j)} - a_n|^2 \right)$$

$$\leq \left(\sum_{n=1}^{M-1} |a_n^{(p_j)} - a_n|^2 \right) + 4 \sum_{n=M}^{\infty} \frac{1}{n^2}$$

$$\leq \left(\sum_{n=1}^{M-1} |a_n^{(p_j)} - a_n|^2 \right) + 4\epsilon.$$

To conclude, it suffices to take j such that

$$\left(\sum_{n=1}^{M-1} |a_n^{(p_j)} - a_n|^2 \right) \leq \epsilon.$$

This is possible since we have a finite sum.

$\square$

Remark 3.10.3. To illustrate the diagonal argument above, consider the case where the sequences are as follows:

$$
\begin{array}{llllllll}
j(1,k) & : & 1 & 2 & 3 & 4 & 5 & \cdots \\
j(2,k) & : &   & 2 & 3 & 4 & 5 & \cdots \\
j(3,k) & : &   &   & 3 & 4 & 5 & \cdots
\end{array}
$$

Then the sequence $(p_t)_{t \in \mathbb{N}}$ coincides with the first sequence.

3.10. Solutions of the exercises

Solution of Exercise 3.9.20: The proof of the previous exercise goes through till the argument to prove (3.10.37), which has to be changed as follows: let $\epsilon > 0$. There exists $M \in \mathbb{N}$ such that

$$\sum_{n=M}^{\infty} \frac{1}{\alpha_n} \le \epsilon.$$

Thus

$$
\begin{aligned}
d(a^{(p_j)}, a)^2 &= \left(\sum_{n=1}^{M-1} |a_n^{(p_j)} - a_n|^2 \right) + \left(\sum_{n=M}^{\infty} |a_n^{(p_j)} - a_n|^2 \right) \\
&= \left(\sum_{n=1}^{M-1} |a_n^{(p_j)} - a_n|^2 \right) + \left(\sum_{n=M}^{\infty} \frac{1}{\alpha_n} \alpha_n |a_n^{(p_j)} - a_n|^2 \right) \\
&\le \left(\sum_{n=1}^{M-1} |a_n^{(p_j)} - a_n|^2 \right) + 2 \left(\sum_{n=M}^{\infty} \frac{1}{\alpha_n} \left(\alpha_n |a_n^{(p_j)}|^2 + \alpha_n |a_n|^2 \right) \right) \\
&\le \left(\sum_{n=1}^{M-1} |a_n^{(p_j)} - a_n|^2 \right) + 4\epsilon,
\end{aligned}
$$

since

$$\alpha_n |a_n^{(p_j)}|^2 \le 1 \quad \text{and} \quad \alpha_n |a_n|^2 \le 1. \qquad \square$$

Solution of Exercise 3.9.21: We first assume that the metric space X is sequentially compact, and follow the arguments from [95, pp. 166–167] and [284, pp. 86–87] to prove that X is compact. Let thus $X = \cup_{i \in I} O_i$ be an open covering of X. We prove (3.9.6), that is

$$\exists \epsilon \; \forall x \in X, \; \exists i \in I, \; B(x, \epsilon) \subset O_i,$$

by contradiction. Assume no such $\epsilon > 0$ exists such that (3.9.6) holds. Then, there exists a sequence $x_1, x_2, \dots$ of points of X such that

$$\forall i \in I, \quad B\left(x_n, \frac{1}{n}\right) \not\subset O_i. \tag{3.10.38}$$

By hypothesis, there is a subsequence of points of X, which we still denote by $x_1, x_2, \dots$, which converges in the metric of X to some point $x \in X$. There exists $\iota \in I$ and $\rho > 0$ such that $B(x, \rho) \subset O_\iota$, and this contradicts (3.10.38) since $B(x_n, \frac{1}{n}) \subset B(x, \rho)$ for n large enough.

We also prove the second claim by contradiction. Suppose that there is $\epsilon > 0$ such that

$$X \ne \cup_{n=1}^{N} B(x_n, \epsilon), \quad N = 1, 2, \dots$$

for any choice of $x_1, x_2, \ldots$. One then builds by induction a sequence of points of X such that $d(x_n, x_m) \geq \epsilon$ for $n \neq m$, and this contradicts the sequential compactness hypothesis.

$\square$

Solution of Exercise 3.9.22: We first show that condition (3.9.7) implies the continuity of f at the point t. We proceed by contradiction. Assume that f is not continuous at t. Then, there is a neighborhood V of $f(t)$ such that for every neighborhoods W of t,

$$f(W) \not\subset V.$$

We now use the hypothesis that E is a metric space, and choose

$$W = B\left(t, \frac{1}{n}\right), \quad n = 1, 2, \ldots$$

For every $n \in \mathbb{N}$ there exists $t_n \in B(t, \frac{1}{n})$ such that $f(t_n) \notin V$. In particular, $f(t_n) \nrightarrow f(t)$. But $d(t, t_n) < \frac{1}{n}$, and this contradicts the hypothesis.

Conversely, assume that f is continuous at the point t. By definition, for every neighborhood V of $f(t)$ there exists a neighborhood W of t such that $f(W) \subset V$. Since E is a metric space, there exists $\epsilon > 0$ such that $B(t, \epsilon) \subset W$. Let now $(t_n)_{n \in \mathbb{N}}$ be a sequence such that $\lim_{n \to \infty} d(t, t_n) = 0$. Let N be such that

$$n \geq N \implies d(t, t_n) < \epsilon.$$

Therefore

$$n \geq N \implies t_n \in W,$$

and it follows that (3.9.7) holds.

$\square$

Remark 3.10.4. We note that the converse assertion is true in fact for a general topological space. We also note that the direct assertion in the preceding result remains true when E is not necessarily a metric space, but the given point of E is then required to have a countable basis of neighborhoods.

Solution of Exercise 3.9.24: The function f is in particular continuous, and so $f(E)$ is compact. Since E is compact, it is in particular a Hausdorff space, and $f(E)$ is closed in E, since any compact subspace of a Hausdorff space is closed. Assume now by contradiction that

$$f(E) \subsetneq E,$$

and let $e \in E \setminus f(E)$. Set $\epsilon = d(e, f(E))$. We have that $\epsilon > 0$ since $f(E)$ is closed (recall that the set of elements with distance 0 to a given set is its closure). Define the diameter of $J \subset E$ by

$$\operatorname{diam} J = \sup_{x, y \in J} d(x, y).$$

3.10. Solutions of the exercises

Let N be the minimal number of open sets of diameter less than or equal to $\frac{2\epsilon}{3}$ which cover E.

$$E = \cup_{i=1}^{N} J_i.$$

Such a number exists since E is compact and can be covered by a finite number of open balls of radius $\frac{\epsilon}{3}$, and the diameter of such balls is less than or equal[1] to $\frac{2\epsilon}{3}$. There is a i_0 such that $e \in J_{i_0}$. We claim that $J_{i_0} \cap f(E) = \emptyset$. Indeed, if $f(t) \in J_{i_0} \cap f(E)$, we have

$$d(e, f(t)) \le \frac{2\epsilon}{3},$$

but $d(e, f(t)) \ge \epsilon$ by definition of the distance. Without loss of generality, we assume that $i_0 = N$. Therefore

$$f(E) \subset \cup_{i=1}^{N-1} J_i,$$

and (see (1.1.6) for the first inequality and (1.1.1) for the second one)

$$E \subset f^{-1}(f(E)) \subset \cup_{i=1}^{N-1} f^{-1}(J_i).$$

The sets $f^{-1}(J_i)$ are open since f is continuous. We now check that

$$\operatorname{diam} f^{-1}(J_i) \le \operatorname{diam} J_i, \quad i = 1, \ldots, N-1.$$

Indeed, let $u, v \in f^{-1}(J_i)$. Then $f(u)$ and $f(v)$ both belong to J_i and $d(f(u), f(v)) \le \frac{2\epsilon}{3}$. Since f preserves distance we have

$$d(u, v) = d(f(u), f(v)),$$

and so

$$\operatorname{diam} f^{-1}(J_i) = \sup_{u,v \in f^{-1}(J_i)} d(u, v) = \sup_{u,v \in f^{-1}(J_i)} d(f(u), f(v)) \le \operatorname{diam} J_i.$$

$$(3.10.39)$$

$\square$

This contradicts the minimality of N.

Remark 3.10.5. We note that the same proof works for the condition (3.9.10). Now (3.10.39) becomes

$$\operatorname{diam} f^{-1}(J_i) = \sup_{u,v \in f^{-1}(J_i)} d(u, v) \le \sup_{u,v \in f^{-1}(J_i)} d(f(u), f(v)) \le \operatorname{diam} J_i.$$

$$(3.10.40)$$

Solution of Exercise 3.9.25: We have to show that i sends the (equivalence class) of the zero element in E_2 to the (equivalence class) of the zero element in E_1. Let $0 \in E_2$ and let $(r_n)_{n \in \mathbb{N}}$ be a Cauchy sequence in the $\| \cdot \|_2$ norm and tending to 0 in E_2. Using (3.9.11) we see that $(r_n)_{n \in \mathbb{N}}$ be a Cauchy sequence in the $\| \cdot \|_1$ norm and tends to 0 in E_1. So $i(0) = 0$ and i is well defined. $\square$

---

[1] It could be strictly less than $\frac{2\epsilon}{3}$, for instance in an ultra-metric space.

Solution of Exercise 3.9.26: Consider the sequence of functions $r = (r_n)_{n \in \mathbb{N}}$ defined by

$$r_n(t) = \begin{cases} t & \text{if} \quad t \in [0, \frac{1}{n+1}], \\ \frac{1-t}{n} & \text{if} \quad t \in [\frac{1}{n+1}, 1]. \end{cases}$$

$(r_n)_{n \in \mathbb{N}}$ is a Cauchy sequence with respect to the norms $\| \cdot \|_1$ and $\| \cdot \|_2$. It tends to 0 for $\| \cdot \|_1$ but not for $\| \cdot \|_2$ since

$$\|r_n\|_2 = \frac{1}{n+1} + 1.$$

Let now x be a Cauchy sequence for $\| \cdot \|_2$. Then

$$e_1(x + r) = e_1(x)$$

since r tends to 0 in the norm $\| \cdot \|_1$, while

$$e_2(x + r) \neq e_2(x)$$

since r does not tend to 0 in the norm $\| \cdot \|_2$. $\qquad\qquad \Box$

Chapter 4

Normed Spaces

In the present chapter we discuss important classes of normed spaces, and in particular, Hilbert spaces and Banach spaces. We also discuss indefinite inner product spaces. A distinguished class of Hilbert spaces, namely reproducing kernel Hilbert spaces, is studied at a later stage in Chapter 7. These various notions play an important role in the third part of the book, dedicated to spaces of analytic functions and their operators.

4.1 Normed Banach and Hilbert spaces

In this section we present various exercises related to normed spaces, and in particular Banach spaces and Hilbert spaces. We mention the Bergman space $\mathcal{B}_2(\Omega)$ (see Exercise 4.1.17), which is the first instance of reproducing kernel Hilbert space appearing in this book (and historically, certainly one of the first explicit instances, if not the first, of reproducing kernel Hilbert space; see [70]). We also point out the exercises on operator ranges and complementation; see Exercises 4.2.40, 4.2.41. We first recall a number of definitions and results.

Definition 4.1.1. A semi-norm in a vector space $\mathcal{V}$ (over the real or the complex numbers) is a map p from $\mathcal{V}$ into $[0, \infty)$ with the following two properties:

(1) For all $u, v \in \mathcal{V}$:

$$p(u + v) \le p(u) + p(v). \tag{4.1.1}$$

(2) For all $\lambda \in \mathbb{C}$ (or $\mathbb{R}$, if $\mathcal{V}$ is a real vector space) and all $v \in \mathcal{V}$:

$$p(\lambda v) = |\lambda| \cdot p(v). \tag{4.1.2}$$

If moreover it holds that:

(3) $p(v) = 0$ if and only if $v = 0$,

the semi-norm is called a norm.

Given a semi-norm p, the set

$$\{v \in \mathcal{V}\,;\, p(v) \leq 1\} \tag{4.1.3}$$

has special properties. To explain this, we need first some definitions.

Definition 4.1.2. A subset $A \subset \mathcal{V}$ is called absorbing if for every $v \in \mathcal{V}$ there exists $t_v > 0$ such that (with real or complex z depending on the setting)

$$|z| \leq t_v \implies zv \in A.$$

It is called balanced if for every $v \in A$ and $z \in \overline{\mathbb{D}}$ (or $z \in [-1,1]$ in the real case) the vector $zv \in A$.

Remark 4.1.3. In the definition of an absorbing set, $v \in \mathcal{V}$ while $v \in A$ in the definition of a balanced set.

Question 4.1.4. *Let p be a semi-norm on a vector space $\mathcal{V}$ (see Definition 4.1.1). Then the set (4.1.3) is absorbing, convex and balanced.*

Exercise 4.1.5. *Given an example of a balanced but not absorbing set, and of an absorbing, but not balanced, set.*

Question 4.1.6. *Let A be an absorbing balanced convex set. Then*

$$p_A(v) = \inf\{t > 0\,:\, v \in tA\}$$

is a semi-norm (called the Minkowski functional).

Hint: The fact that A is absorbing guarantees that p_A is finite. The convexity implies the sub-additivity property (4.1.1), while property (4.1.2) holds since A is assumed balanced.

Remark 4.1.7. When there is an underlying suitable topology, a closed absorbing balanced convex set is called a barrel. See Definition 5.1.9 below.

Exercise 4.1.8. *Let $\Omega \subset \mathbb{C}$ be open and connected, and let $K \subset \Omega$ be a compact set with non-empty interior (for instance,*

$$K = \{z \in \mathbb{C}\,;\, |z - z_0| \leq r_0\}$$

for some $z_0 \in \Omega$ and $r_0 > 0$). Show that

$$p_K(f) = \max_{z \in K} |f(z)| \tag{4.1.4}$$

is a semi-norm on the space $\mathcal{C}(\Omega)$ of functions continuous on Ω, and is a norm on $\mathcal{H}(\Omega)$.

Although p_K is a norm on $\mathcal{H}(\Omega)$, it does not quite take into account the structure of the latter. A natural path is to cover Ω by an increasing sequence of compact sets $K_0, K_1, \ldots$ such that

$$K_0 \subset \overset{\circ}{K_1} \subset K_1 \subset \overset{\circ}{K_2} \subset \cdots,$$

and in particular, with non-empty interiors,

$$\Omega = \bigcup_{n=0}^{\infty} K_n, \tag{4.1.5}$$

see Exercise 5.6.2, and consider *all* the associated norms. These norms define a topology, defined by a metric. See Exercise 5.6.3.

A similar remark holds for the semi-norms p_K. Note in particular that if a continuous function on Ω is such that $p_K(f) = 0$ for all compact subsets of Ω, then $f \equiv 0$.

A norm defines a metric via the formula $d(u, v) = \|u - v\|$, and hence induces an associated topology, called the strong topology. We denote by $\mathcal{V}'$ the topological dual of $\mathcal{V}$, that is the family of linear functionals continuous with respect to this topology. Recall that $\mathcal{V}'$ endowed with the norm

$$\|\varphi\|' = \sup_{\|v\| \leq 1} |\varphi(v)| \tag{4.1.6}$$

is a Banach space.

Definition 4.1.9. Let $\mathcal{V}$ be a normed space. The $\sigma(\mathcal{V}', \mathcal{V})$ topology on $\mathcal{V}'$ (also called the weak-$*$ topology) is the coarsest (that is, smallest) topology for which all the linear functionals

$$\varphi \mapsto \varphi(v), \quad v \in \mathcal{V}, \tag{4.1.7}$$

are continuous.

Exercise 4.1.10. *Show that the map* (4.1.7) *defines a homeomorphism* X:

$$(X(\varphi))(v) = \varphi(v), \quad v \in \mathcal{V}, \tag{4.1.8}$$

from $\mathcal{V}'$ *endowed with the* $\sigma(\mathcal{V}', \mathcal{V})$ *topology onto its image* $X(\mathcal{V}') \subset \mathbb{C}^{\mathcal{V}}$ *endowed with the topology induced by the product topology.*

As is well known from the Hilbert space case, the norm (4.1.6) need not be continuous in the $\sigma(\mathcal{V}', \mathcal{V})$ topology. Indeed, consider an infinite-dimensional (say, separable) Hilbert space $\mathcal{H}$ with inner product $\langle \cdot, \cdot \rangle$ and associated norm $\| \cdot \|$. If $(e_n)_{n \in \mathbb{N}}$ is an orthonormal basis of $\mathcal{H}$, Parseval's inequality implies

$$\lim_{n \to \infty} \langle h, e_n \rangle = 0$$

while $\|e_n\| = 1$ for all $n \in \mathbb{N}$. Thus the sequence of functionals

$$h \mapsto \langle h, e_n \rangle$$

goes to 0 weakly while the sequence of their norms does not tend to 0. The following related exercise is taken from [114, (12.15.8) and (12.15.9), p. 76]. See Definition 3.4.7 for the notion of a lower semi-continuous function.

Exercise 4.1.11. *Let $\mathcal{V}$ be a normed space. Then:*

(1) *The norm $\|\cdot\|'$ is lower semi-continuous with respect to the $\sigma(\mathcal{V}', \mathcal{V})$ topology.*
(2) *Let $\mathcal{V}$ denote a separable normed space. Then every closed ball in $\mathcal{V}'$ is metrizable and compact in the $\sigma(\mathcal{V}', \mathcal{V})$ topology.*

Hint: (see for instance [114, p. 75]) For (2), consider a countable dense subset $v_1, v_2, \ldots$ of the closed unit ball of $\mathcal{V}$ and show that, on the closed ball

$$\{\varphi \in \mathcal{V}; \|\varphi - \varphi_0\|' \le r\}, \tag{4.1.9}$$

the $\sigma(\mathcal{V}', \mathcal{V})$ topology coincides with the topology generated by the semi-norms

$$d_n(\varphi) = |\varphi(v_n)|, \quad n = 1, 2, \ldots$$

This latter topology is metrizable.

A slight adaptation of the arguments in the proof of this exercise allows us to solve the following exercise (see for instance [153, p. 36]):

Exercise 4.1.12. *Let $\mathcal{V}$ be a Banach space. Show that in $\mathcal{V}'$ endowed with the $\sigma(\mathcal{V}', \mathcal{V})$ topology, a closed set which is bounded in the norm $\|\cdot\|_{\mathcal{V}'}$ (defined by (4.1.6)) is compact.*

Hint: Use Exercise 4.1.10 to work in $\mathbb{C}^{\mathcal{V}}$ and use Tychonoff's theorem (see Theorem 3.2.3).

Remark 4.1.13. In the previous exercises, compactness and closedness are in the weak sense, while boundedness is with respect to the topology defined by the norm. These two topologies are different, unless the space is finite dimensional. It is interesting to ask whether there are examples of infinite-dimensional vector spaces, endowed with a topology such that compactness is equivalent to being closed and bounded (the latter in an appropriate sense). The answer is yes, these are Montel spaces (more precisely, a Montel space is a locally convex Hausdorff barreled space in which compactness is equivalent to being closed and bounded) and one example of such space is the space of functions analytic in an open set. See Section 5.6.

The following exercise is true in fact in the more general setting of Fréchet spaces; see Question 5.3.6. It appears in the setting of Hilbert spaces in [6, Lemme, p. 64], where the result is attributed to Gelfand [147]. In the present book the exercise is used in particular in the solution of Exercise 4.2.19.

Exercise 4.1.14. *A lower semi-continuous semi-norm in a normed space is contin-uous.*

Hint: (see [114, (12.16.3), p. 84]) Let p be a lower semi-continuous semi-norm in the normed space $\mathcal{V}$ and let

$$C_n = \{v \in \mathcal{V} \,:\, p(v) \leq n\}, \quad n \in \mathbb{N}.$$

Remark that $\mathcal{V} = \cup_{n \in \mathbb{N}} C_n$ and apply Baire's theorem (see Theorem 3.9.13).

The following question, taken from [90, Proposition I.5, p. 4], is one of the first steps in the statement and proof of the various geometrical versions of the Hahn–Banach theorem.

Question 4.1.15. *Let $\mathcal{V}$ be a normed space, and let $\lambda \not\equiv 0$ be a linear functional on $\mathcal{V}$. Let $z \in \mathbb{C}$. Then the set*

$$\{v \in \mathcal{V} \,:\, \lambda(v) = z\} \tag{4.1.10}$$

is closed if and only if λ is continuous.

Remark 4.1.16. The set (4.1.10) is called an affine hyperplane. For $z = 0$, a hyperplane of a given vector space $\mathcal{V}$ can be defined equivalently as a maximal proper linear subspace of $\mathcal{V}$. See for instance [183, p. 41] for a discussion.

A complete normed space is called a Banach space. For instance the space $\mathbf{H}_\infty(\mathbb{D})$ of functions analytic and bounded in the open unit disk endowed with the norm

$$\|f\|_\infty = \sup_{z \in \mathbb{D}} |f(z)|,$$

and more generally, the Hardy spaces $\mathbf{H}_p(\mathbb{D})$ for $p \in [1, \infty)$, are Banach spaces.

Still with $p \in [1, \infty)$, and given a measured space $(\Omega, \mathcal{B}, \mu)$, recall that the Lebesgue space $\mathbf{L}_p(\Omega, \mathcal{B}, \mu)$ consisting of (equivalence classes) of measurable functions such that

$$\int_\Omega |f(w)|^p d\mu(w) < \infty$$

is a Banach space when endowed with the norm

$$\|f\|_p = \left(\int_\Omega |f(w)|^p d\mu(w) \right)^{1/p}.$$

When $p = \infty$, one replaces the above norm by

$$\|f\|_\infty = \text{ess.sup}_{w \in \Omega} |f(w)|.$$

The following two inequalities (Hölder's and Minkowski's inequalities) play an important role. In particular, Minkowski's inequality shows that $\|f\|_p$ is a norm. We begin with Hölder's inequality. Let $p, q \in (1, \infty)$ be such that

$$\frac{1}{p} + \frac{1}{q} = 1,$$

and let $f \in \mathbf{L}_p(\Omega, \mathcal{B}, \mu)$ and $g \in \mathbf{L}_q(\Omega, \mathcal{B}, \mu)$. Then, $fg \in \mathbf{L}_1(\Omega, \mathcal{B}, \mu)$ and it holds that

$$\|fg\|_1 \leq \|f\|_p \|g\|_q. \tag{4.1.11}$$

Let now $f, g \in \mathbf{L}_p(\Omega, \mathcal{B}, \mu)$. Then, $f + g \in \mathbf{L}_p(\Omega, \mathcal{B}, d\mu)$ and it holds that

$$\|f + g\|_p \leq \|f\|_p + \|g\|_p. \tag{4.1.12}$$

Exercise 4.1.17. *Given $p \in [1, \infty)$ and given an open set $\Omega \subset \mathbb{C}$, show that the space $\mathcal{B}_p(\Omega)$ of functions analytic in Ω and such that*

$$\|f\|_p = \left(\int_\Omega |f(z)|^p dxdy \right)^{1/p} < \infty$$

is a Banach space. Show that the point evaluations

$$f \mapsto f(w), \quad w \in \Omega, \tag{4.1.13}$$

are continuous.

The case $p = 2$ in the above exercise is of special importance. It corresponds to the Bergman space. The fact that the point evaluations are continuous means that the corresponding Hilbert space is a reproducing kernel Hilbert space. For a follow-up of this exercise when $p = 2$, see Exercise 10.1.2.

The following question is taken from [235, §8, p. 194].

Question 4.1.18. *Let $\mathcal{B}$ denote the space of functions f analytic in the open upper half-plane $\mathbb{C}_+$ and such that*

$$\|f\| = |f(i)| + \iint_{\mathbb{C}_+} |f'(z)| dxdy < \infty : \tag{4.1.14}$$

(1) *Show that (4.1.14) defines a norm.*
(2) *Show that $\mathcal{B}$ is a Banach space and compute its dual.*

Let us now assume that $\mathcal{V}$ is a Hilbert space (which we will denote by $\mathcal{H}$) with inner product $\langle \cdot, \cdot \rangle_{\mathcal{H}}$ and associated norm $\| \cdot \|_{\mathcal{H}}$. As we saw above, the strong topology is the topology defined by this norm. Recall now the Riesz representation theorem which characterizes linear functionals in a Hilbert space which are continuous in the strong topology.

Theorem 4.1.19. *Let $\mathcal{H}$ be a Hilbert space. A map $h \mapsto \varphi(h)$ is a linear continuous functional on $\mathcal{H}$ if and only if it can be written as*

$$\varphi(h) = \langle h, h_\varphi \rangle_\mathcal{H}$$

for some uniquely determined $h_\varphi \in \mathcal{H}$. Furthermore

$$\|\varphi\| = \|h_\varphi\|_\mathcal{H}. \tag{4.1.15}$$

This theorem allows us to identify $\mathcal{H}$ with its dual and the weak topology is the weakest topology for which the functionals

$$h \mapsto \langle h, u \rangle_\mathcal{H}, \quad u \in \mathcal{H}$$

are all continuous. By the Cauchy–Schwarz inequality these functionals are continuous with respect to the strong topology, and therefore the weak topology is weaker (that is, smaller) than the strong topology (see Remark 4.1.22 for more on this). The weak topology is thus the $\sigma(\mathcal{H}', \mathcal{H})$ topology after identifying $\mathcal{H}$ and $\mathcal{H}'$ via the antilinear map which to $\varphi \in \mathcal{H}$ associates the functional $h \mapsto \langle h, h_\varphi \rangle_\mathcal{H}$ (but see the warning [283, p. 121] on this identification).

Remark 4.1.20. Denote by $\mathcal{H}^*$ the space of continuous antilinear maps from $\mathcal{H}$ into $\mathbb{C}$. Then, $\psi \in \mathcal{H}^*$ if and only if it is of the form

$$\psi(h) = \langle h_\psi, h \rangle_\mathcal{H}$$

for some uniquely defined $h_\psi \in \mathcal{H}$. Thus

$$\langle \psi, h \rangle_{\mathcal{H}^*, \mathcal{H}} = \langle h_\psi, h \rangle_\mathcal{H},$$

where the first set of brackets denotes the duality between $\mathcal{H}$ and $\mathcal{H}^*$.

A basis of neighborhoods of the origin of the strong topology consists of sets of the form

$$\{h \in \mathcal{H}; \|h\|_\mathcal{H} < \epsilon\}, \tag{4.1.16}$$

where $\epsilon > 0$, while the sets

$$\{h \in \mathcal{H}; |\langle h, h_j \rangle_\mathcal{H}| < \epsilon_j, \ j = 1, \ldots, N\}, \tag{4.1.17}$$

with $N \in \mathbb{N}$, $h_1, \ldots, h_N \in \mathcal{H}$ and $\epsilon_1, \ldots, \epsilon_N \in \mathbb{R}_+$, form a basis of neighborhoods of the origin of the weak topology.

Exercise 4.1.21. *Let $\mathcal{H}$ be an infinite-dimensional Hilbert space.*

(1) *Show that sets of the form (4.1.17) contain elements of arbitrary large norm.*
(2) *Give an example of a strongly closed set which is not weakly closed.*
(3) *Show that a closed and convex set is weakly closed.*

Hint: In (3), use both the basic separation theorem (see Theorem 1.3.3) and the following result, used here in the setting of Hilbert spaces, but stated in the setting of topological vector spaces; see Section 5.1 for the latter.

Remark 4.1.22. It follows from (4.1.16) and (4.1.17) that every weak neighborhood contains a strong neighborhood. On the other hand, (1) in the previous exercise shows that no strong neighborhood of the form (4.1.16) contains a weak neighborhood. Thus (as is clear from the very definitions), weak continuity will imply strong continuity, while strong continuity will not imply weak continuity.

Proposition 4.1.23 (see [120, Lemme 7, p. 417]). *A linear functional on a topological vector space which separates two sets, one of them having an interior point, is continuous.*

We recall that every Hilbert space $\mathcal{H}$ has an orthonormal Hilbert basis, that is, an orthonormal family $(e_a)_{a \in A}$ such that every element $h \in \mathcal{H}$ can be written, in a unique way, as

$$h = \sum_{a \in A} \langle h, a \rangle e_a$$

where the convergence is in the norm of $\mathcal{H}$. We recall Parseval's equality

$$\|h\|^2 = \sum_{a \in A} |\langle h, e_a \rangle|^2. \tag{4.1.18}$$

We barely touched the geometry of Hilbert spaces (let alone Banach spaces) in this section. We conclude this section with a discussion on the sum of closed subspaces in a Hilbert space.

Theorem 4.1.24 (see for instance [79, Proposition 9, p. EVT V-14] for a more complete statement). *Let $\mathcal{H}$ be a Hilbert space and let $\mathcal{H}_1$ and $\mathcal{H}_2$ be two Hilbert spaces isometrically included in $\mathcal{H}$. Let $p_{\mathcal{H}_1}$ and $p_{\mathcal{H}_2}$ denote the orthogonal projections on $\mathcal{H}_1$ and $\mathcal{H}_2$ respectively. Then, the following are equivalent:*

(a) *It holds that $p_{\mathcal{H}_1} p_{\mathcal{H}_2} = p_{\mathcal{H}_2} p_{\mathcal{H}_1}$.*
(b) *We have*

$$\mathcal{H}_1 = (\mathcal{H}_1 \cap \mathcal{H}_2) + (\mathcal{H}_1 \cap \mathcal{H}_2^{\perp}).$$

When these conditions are in force, the space $\mathcal{H}_1 + \mathcal{H}_2$ is closed and the projection on it is equal to

$$p_{\mathcal{H}_1 + \mathcal{H}_2} = p_{\mathcal{H}_1} + p_{\mathcal{H}_2} - p_{\mathcal{H}_1} p_{\mathcal{H}_2}.$$

More generally, one can find in Aronszajn's paper [52, §12, p. 375] a formula (in form of an infinite series) for the orthogonal projection on the sum of two closed subspaces, when the latter is closed. Namely it holds that (see [52, formula (7), p. 375]):

$$P = P_0 + \sum_{k=1}^{\infty} \left(P_1 (P_2 P_1)^{k-1} + P_2 (P_1 P_2)^{k-1} - (P_2 P_1)^k - (P_1 P_2)^k \right),$$

where P_0 denotes the orthogonal projection onto the intersection of the two spaces. A discussion of the projection on the intersection of two subspaces can be found in [170, Exercise 122, p. 64].

When the two spaces have a zero intersection, a necessary and sufficient condition for the sum to be closed can also be found there and is as follows:

Theorem 4.1.25. *Assume that $\mathcal{H}_1 \cap \mathcal{H}_2 = \{0\}$. A necessary and sufficient condition for the sum $\mathcal{H}_1 + \mathcal{H}_2$ to be closed is that there exists $c > 0$ such that*

$$\|h_1 + h_2\| \geq c\|h_1\|, \quad \forall h_1 \in \mathcal{H}_1 \text{ and } \forall h_2 \in \mathcal{H}_2. \tag{4.1.19}$$

See Aronszajn's paper [52, p. 378], who attributes the result to H. Kober ; see [205, Theorem 1, p. 135 and Theorem 1a, p. 136 (the latter for more than two spaces)].

For an application of this result see Exercise 10.4.3.

4.2   Operators on normed spaces

We denote by $\mathbf{L}(\mathcal{V}_1, \mathcal{V}_2)$ the space of linear continuous operators from the normed space $\mathcal{V}_1$ into the normed space $\mathcal{V}_2$, and set $\mathbf{L}(\mathcal{V}, \mathcal{V}) = \mathbf{L}(\mathcal{V})$ when $\mathcal{V}_1 = \mathcal{V}_2 (= \mathcal{V})$. The strong operator topology on $\mathbf{L}(\mathcal{V}_1, \mathcal{V}_2)$ is the topology generated by the semi-norms

$$T \mapsto \|Tu\|, \quad u \in \mathcal{V}_1.$$

The weak operator topology is the topology generated by the semi-norms

$$T \mapsto |\varphi(Tu)|, \quad u \in \mathcal{V}_1, \ \varphi \in \mathcal{V}_2'.$$

In the case of a Hilbert space, the weak operator topology is the topology generated by the semi-norms

$$T \mapsto \langle Tu, v \rangle, \quad u, v \in \mathcal{V}.$$

Note that the weak topology is coarser than the strong topology, that is, is included in the strong topology. To see this (in the setting of some Hilbert space $\mathcal{H}$; see also Exercise 4.2.23), let O be an open set in the weak topology. Then, every $T_0 \in O$ is contained in an open set of the form

$$V = \{T \in \mathcal{H} ; \ |\langle (T - T_0)f, g \rangle| < \epsilon\}$$

for some $f, g \in \mathcal{H}$ and $\epsilon > 0$. But $T_0 \in W$, where $W \subset V$ is the set

$$\left\{ T \in \mathcal{H} ; \ \|T - T_0\| < \frac{\epsilon}{\|f\| \cdot \|g\|} \right\},$$

so O is open in the strong topology.

Definition 4.2.1. Let $\mathcal{V}_1$ and $\mathcal{V}_2$ be two normed spaces. A linear operator from $\mathcal{V}_1$ into $\mathcal{V}_2$ is called bounded if it maps bounded sets into bounded sets.

Item (2) in the following question is false in the setting of general topological vector spaces. Being bounded and being continuous are not equivalent in general in that general setting. More precisely (see, e.g., [307, Proposition 14.2, p. 138]), a continuous map between topological spaces is bounded, but the converse need not hold.

Question 4.2.2.

(1) *In the notation of Definition 4.2.1 a linear operator T is bounded if and only if there is $K > 0$ such that*

$$\|Tv_1\|_2 \le K\|v_1\|_1, \quad \forall v_1 \in \mathcal{V}_1.$$

(2) *Show that a linear operator between two normed spaces is continuous if and only if it is bounded.*

Exercise 4.2.3. *Let $(X_1, \|\cdot\|_1)$ and $(X_2, \|\cdot\|_2)$ be two normed spaces, and assume that $X_1 \subset X_2$ and that the inclusion map from X_1 into X_2 is continuous. Show that the topology induced in X_1 by $\|\cdot\|_2$ is coarser than the topology of X_1.*

Remark 4.2.4. The preceding exercise applies in particular when

$$\|x\|_2 \le \|x\|_1, \quad \forall x \in X_1.$$

One can ask the following question: Given an increasing family of normed spaces $(X_n, \|\cdot\|_n)$, $n = 1, 2, \ldots$, with decreasing norms. Is there a natural topology on $\cup_{n \in \mathbb{N}} X_n$, which takes into account this structure? The inductive topology answers the question.

We now turn to operators in Banach spaces. As already mentioned (see Remark 3.2.15), Baire's theorem is used in the proof of two important results, the uniform boundedness theorem (that is, the Banach–Steinhaus theorem) and the open mapping theorem; see Theorem 4.2.15 for the latter and Theorem 4.2.6 for one of its important corollaries. The uniform bounded theorem takes the following form in Banach spaces:

Theorem 4.2.5. *Let $(T_a)_{a \in A}$ be a family of linear bounded operators from a Banach space $\mathcal{B}_1$ into a Banach space $\mathcal{B}_2$, and assume that*

$$\sup_{a \in A} \|T_a b_1\| < \infty, \quad \forall b_1 \in \mathcal{B}_1.$$

Then,

$$\sup_{a \in A} \|T_a\| < \infty.$$

4.2. Operators on normed spaces

Everywhere defined linear operators in a Banach (or a Hilbert space; see Exercise 4.2.19) need not be continuous. The closed graph theorem gives an important characterization of everywhere defined operators which are continuous, which is useful in the setting of reproducing kernel Hilbert spaces (see for instance Question 7.6.18 for the latter).

Theorem 4.2.6 (The closed graph theorem). *Let $\mathcal{B}_1$ and $\mathcal{B}_2$ be two Banach spaces, and let A denote an everywhere defined linear operator from $\mathcal{B}_1$ into $\mathcal{B}_2$. Then A is continuous if and only if the graph of A*

$$\Gamma(A) = \{(b, Ab) \, ; \, b \in \mathcal{B}_1\}$$

is closed in $\mathcal{B}_1 \times \mathcal{B}_2$.

We refer to [257] for a proof. The result also holds in more general settings. See [307]. The equivalence relation appearing in the next exercise is defined in Question 1.1.11, and the notion of open map is given in Definition 3.6.1. We will give only the answer to the second item in the exercise. We give two hints for solving the first item.

Exercise 4.2.7 (see, e.g., [119, p. 21]).

(1) *Let $(\mathcal{B}, \|\cdot\|)$ be a Banach space and let $\mathcal{M}$ be a closed subspace of $\mathcal{B}$. Show that*

$$\|\widetilde{b}\|_{\mathcal{B}/\mathcal{M}} = \inf_{m \in \mathcal{M}} \|b + m\|$$

defines a norm on $\mathcal{B}/\mathcal{M}$ and that $(\mathcal{B}/\mathcal{M}, \|\cdot\|_{\mathcal{B}/\mathcal{M}})$ is a Banach space.
(2) *Show that the canonical projection q from $\mathcal{B}$ onto $\mathcal{B}/\mathcal{M}$ is open.*

Hints: Use the completeness of $\mathcal{M}$ to show that $\|\widetilde{f}\|_{\mathcal{B}/\mathcal{M}} = 0$ implies that $\widetilde{f} = 0$. To prove completeness of the quotient space, given a Cauchy sequence $(\widetilde{f}_n)_{n \in \mathbb{N}}$, build a sequence $(g_n)_{n \in \mathbb{N}}$ of elements of $\mathcal{B}$ such that

$$g_n \in \widetilde{f}_n \quad \text{and} \quad \|g_{n+1} - g_n\| < \frac{1}{2^n}.$$

Show that the Cauchy sequence converges to $\widetilde{g}$ where $g = \sum_{n=1}^{\infty}(g_{n+1} - g_n)$. To prove the second item, consider first an open ball and then make use of (1.1.3).

Recall that the resolvent set and the spectrum of a linear bounded operator T in a Banach space $\mathcal{B}$ are defined as follows:

Definition 4.2.8. The resolvent set is the set of complex numbers such that the operator $\lambda I_{\mathcal{B}} - T$ is boundedly invertible, and is denoted by $\rho(T)$. The complement of the resolvent set is called the spectrum, and is denoted by $\sigma(T)$.

When $\lambda \in \sigma(T)$ the operator $\lambda I_{\mathcal{B}} - B$ is in particular one-to-one onto its range $\mathrm{ran}\,(\lambda I_{\mathcal{B}} - T)$. A point λ can be in the spectrum for one of the three following reasons:

(1) The operator $\lambda I_{\mathcal{B}} - T$ is not one-to-one. Such λ's form the point spectrum and are called eigenvalues.

In the two other cases, the operator $\lambda I_{\mathcal{B}} - T$ is one-to-one onto its range, but two cases occur:

(2) The operator $\lambda I_{\mathcal{B}} - T$ has dense range. Such λ's form the essential spectrum.
(3) The operator $\lambda I_{\mathcal{B}} - T$ has not dense range. Such λ's form the residual spectrum.

Remark 4.2.9. In the second case we cannot have $\mathrm{ran}\,(\lambda I_{\mathcal{B}} - T) = \mathcal{B}$. This is a consequence of the open mapping theorem. See Theorem 4.2.15.

Question 4.2.10. *In the notation of Definition 4.2.8 let $\lambda, \mu \in \rho(T)$ and set $R(\lambda) = (T - \lambda I_{\mathcal{B}})^{-1}$. Show that the operators $R(\lambda)$ satisfy the resolvent identity* (1.7.4):

$$R(\lambda) - R(\mu) = (\lambda - \mu)R(\lambda)R(\mu).$$

The operator $R(\lambda)$ is called the resolvent function of T, and (1.7.4) is called the resolvent equation.

Theorem 4.2.11. *Let T be a bounded operator in a Banach space $\mathcal{B}$. Then,*

$$r(T) = \limsup_{n \to \infty} \|T^n\|^{1/n}$$

exists and the operator $T - \lambda I_{\mathcal{B}}$ is boundedly invertible for all λ such that $|\lambda| > r(T)$.

Definition 4.2.12. The number $r(T)$ is called the spectral radius of the operator.

The spectra of the forward and backward shift operators in the Hardy space $\mathbf{H}_2(\mathbb{D})$ are computed Exercise 8.3.3.

In the next exercise one computes the integral of the operator-valued function $(\lambda I_{\mathcal{B}} - T)^{-1}$ along a curve which lies in the resolvent set of T. The reader should check (using for instance the resolvent equation) that this function is continuous in the operator topology. The integral can then be defined as converging Riemann sums in that topology.

Exercise 4.2.13. *Let T be a bounded operator from the Banach space $\mathcal{B}$ into itself, and let C denote a closed simple contour which does not intersect the spectrum of T. Show that*

$$P = \int_C (zI_{\mathcal{B}} - T)^{-1}dz \tag{4.2.1}$$

is a projection operator.

Hint: See the discussion after Exercise 1.5.5 (which follows the argument in [155, pp. 9–10]).

Definition 4.2.14. The operator P in (4.2.1) is called the Riesz projection corresponding to the spectrum of T inside C.

Recall that a bounded operator between Banach spaces is compact if the closure of the image of the closed unit ball is compact. For example, if Ω_1 and Ω_2 are two open subsets of $\mathbb{C}$ such that $\overline{\Omega_1} \subset \Omega_2$, and with $p \in [1, \infty)$, the restriction map from the Bergman space $\mathcal{B}_p(\Omega_2)$ into the Bergman $\mathcal{B}_p(\Omega_1)$ is compact (recall that the Bergman space has been defined in Exercise 4.1.17). See for instance [168, p. 61, p. 65], where this fact is used in arguments toward the computation of cohomology groups associated with a compact Riemann surface. See also Exercise 5.7.2 for a similar result.

In view of Exercise 4.2.17 it is well to recall the following two results:

Theorem 4.2.15 (The open mapping theorem). *Let $\mathcal{B}_1$ and $\mathcal{B}_2$ be two Banach spaces and let T be a continuous map from $\mathcal{B}_1$ onto $\mathcal{B}_2$. Then T is open. If T is moreover one-to-one, then T^{-1} is continuous.*

The proof of this result is instructive (see for instance [90], and, in the setting of topological vector spaces, [307, Theorem 17.1, p. 170]). It is divided into two steps, which use the hypothesis of the theorem in a dichotomic way. The first step makes use of Baire's theorem and of the fact that the range space is complete. The other step uses the completeness of the domain space. The theorem holds in the setting of metrizable (not necessarily locally convex) vector spaces. See [79, Corollaire 1, p. EVT I.19] and the discussion after Remark 5.2.9.

Theorem 4.2.16. *The closed unit ball of a Banach space is compact if and only if the space is finite dimensional.*

Exercise 4.2.17. *Let T be a compact operator from the Banach space $\mathcal{B}_1$ into the Banach space $\mathcal{B}_2$. Assume that T is onto. Show that $\mathcal{B}_2$ is finite dimensional.*

The following result is taken from [200, p. 9].

Exercise 4.2.18. *Let $\mathcal{B}$ be a Banach space and let $t \mapsto X(t)$ be a strongly continuous function from $[0, 1]$ into $\mathbf{L}(\mathcal{B})$.*
(1) *Show that the function $t \mapsto \|X(t)\|$ is bounded.*
(2) *Show that it is lower semi-continuous.*

Hint: Let $k \in \mathbb{R}$. We have $\|X(t)\| > k$ if and only if there is a $b \in \mathcal{B}$ such that $\|X(t)b\| > k\|b\|$. Use then the assumed strong continuity of the function X.

In preparation of the following two exercises we recall that in Hilbert spaces there are everywhere defined operators which are unbounded. Constructing such an operator requires the axiom of choice. See Exercise 7.5.14. We also recall that the adjoint T^* of an operator T with domain $\mathrm{Dom}\,(T) \subset \mathcal{H}_1$ and range in $\mathcal{H}_2$, where

$\mathcal{H}_1$ and $\mathcal{H}_2$ are two Hilbert spaces, is defined as follows. The domain $\mathrm{Dom}\,(T^*)$ of T^* is the set of elements $h_2 \in \mathcal{H}_2$ such that the map $h_1 \mapsto \langle Th_1, h_2 \rangle_{\mathcal{H}_2}$ is continuous (or more precisely, extends to a continuous map from $\mathcal{H}_1$ into $\mathbb{C}$). One defines then

$$\langle Th_1, h_2 \rangle_{\mathcal{H}_2} = \langle h_1, T^*h_2 \rangle_{\mathcal{H}_1}.$$

The vector T^*h_2 is well defined in view of Riesz representation theorem (see Theorem 4.1.19).

The first of the two exercises alluded to above is a classical result, which can be found for instance in [114, Theorem 12.16.7, p. 87].

Exercise 4.2.19. *Let $\mathcal{H}$ be a Hilbert space and let A be an everywhere defined operator in $\mathcal{H}$.*

(1) *Assume A weakly continuous. Show that A has a continuous adjoint and is continuous.*

(2) *Assume that A is self-adjoint. Show that A is continuous.*

Hint: In (1), to show that the adjoint is continuous, remark that the map

$$g \mapsto \sup_{\substack{h \in \mathcal{H} \\ \|h\| \leq 1}} |\langle Ah, g \rangle| \tag{4.2.2}$$

is a lower semi-continuous semi-norm, and use Exercise 4.1.14.

A slight rewriting of the preceding exercise gives:

Exercise 4.2.20. *An everywhere defined operator A in a Hilbert space $\mathcal{H}$ is bounded if and only if it has an everywhere defined adjoint A^*. Then,*

$$c^2 I_\mathcal{H} - A^*A \geq 0$$

for some $c > 0$. The smallest such c is equal to the norm of A.

Question 4.2.21. *Let T be a bounded operator from the Hilbert space $\mathcal{H}$ into itself and let*

$$A = \frac{T + T^*}{2} \quad and \quad B = \frac{T - T^*}{2i}$$

denote respectively its real and imaginary part. Show that

$$\|A\| \leq \|T\| \quad and \quad \|B\| \leq \|T\|.$$

Hint: Use the fact that $\|T\| = \|T^*\|$.

The operators appearing in the following question are the Hilbert–Schmidt operators. We refer for instance to [257] for the prime definition of these operators.

4.2. Operators on normed spaces

Question 4.2.22. *Let $\mathcal{H}$ be a separable Hilbert space, and let $(e_n)_{n\in\mathbb{N}}$ and $(f_n)_{n\in\mathbb{N}}$ be two orthonormal basis of $\mathcal{H}$. Let $(\lambda_n)_{n\in\mathbb{N}}$ be a sequence of numbers in $\ell_2(\mathbb{N})$. Show that the formula*

$$Kf = \sum_{n=1}^{\infty} \lambda_n \langle f, e_n \rangle f_n \qquad (4.2.3)$$

defines a bounded operator from $\mathcal{H}$ into itself.

We recall that a sequence of bounded operators in a Hilbert space $\mathcal{H}$ converges weakly to the bounded operator A if

$$\lim_{n\to\infty} \langle A_n f, g \rangle = \langle Af, g \rangle, \quad \forall f, g \in \mathcal{H},$$

and it converges strongly to A if

$$\lim_{n\to\infty} \|A_n f - Af\| = 0, \quad \forall f \in \mathcal{H}.$$

Exercise 4.2.23. *Show that the weak topology is coarser than the strong topology, which is itself coarser than the operator topology.*

Question 4.2.24. *Let $\mathcal{H}_1, \mathcal{H}_2$ and $\mathcal{H}_3$ be three Hilbert spaces. Show that a bilinear form from $\mathcal{H}_1 \times \mathcal{H}_2$ into $\mathcal{H}_3$ which is separately continuous is jointly continuous.*

The preceding result is in fact true in much more general situations, in the general setting of topological vector spaces. See for instance [307, Corollary, p. 354 and Theorem 41.1, p. 421].

Multiplication of operators in Hilbert space is trivially jointly continuous in the operator norm topology, but (as is required to be proved in Exercise 4.2.25) fails to be jointly continuous in the weak and strong operator topologies. The reason is that $\mathbf{L}(\mathcal{H})$ is not metrizable in these topologies (but see Exercise 4.2.26 for more results on metrizability). To find a proof that multiplication is not jointly continuous in the strong topology is not that easy, notwithstanding von Neumann's stating in [191, p. 382] that ... *auch hier sind Gegenbeispiele so einfach, dass wir sie übergehen*. To quote Halmos in [170, p. 61] about the counterexamples, *the quickest ones require unfair trickery*. One approach uses the strong density of operators with vanishing square ($T \in \mathbf{L}(\mathcal{H})$ such that $T^2 = 0$); see [170, Problem 111, p. 61, p. 152 and p. 248]. We here follow another approach, taken from [116, Exercise 2, pp. 45–46] and [67, pp. 116–117]. Hints are given after the exercise.

Exercise 4.2.25. *Let $\mathcal{H}$ be an infinite-dimensional Hilbert space. Show that the multiplication operator fails to be jointly continuous, when the domain is endowed with the strong operator topology, and the range is endowed with either the weak or the strong operator topology.*

Hints: To prove that multiplication is not continuous one can proceed as follows (see [116, Exercise 2, pp. 45–46] and [67, pp. 116–117]). Take $h, h_1, \ldots, h_N, k \in \mathcal{H}$ and set

$$\mathcal{U} = \{T \in \mathbf{L}(\mathcal{H}) \quad \text{such that} \quad |\langle Th, k \rangle| < 1\},$$

and
$$W = \{T \in \mathbf{L}(\mathcal{H}) \quad \text{such that} \quad \|Th_j\| < 1, \quad j = 1, \dots, N\}.$$
Then:

(a) 'Show that there is $m \in \mathcal{H}$ of norm 1 and such that
$$\sup_{T \in W} \|Tm\| = \infty. \tag{4.2.4}$$

(b) Build $T_1 \in W$ such that $T_1 h = \lambda m$ for some $\lambda > 0$.
(c) Build a contraction $T_2 \in W$ such that $\|T_2 T_1 h\| > 1$.
(d) Build a contraction T_3 such that $\langle T_3 T_2 T_1 h, k \rangle \geq 1$.

This ends the hints to Exercise 4.2.25 for proving the lack of joint continuity of the product in the strong operator topology. We also note that the map $(T, S) \mapsto TS$ in $\mathbf{L}(\mathcal{H})$ is separately weakly or strongly continuous.

As a consequence of Exercise 4.2.26 below, we have that multiplication is sequentially jointly continuous in $\mathbf{L}(\mathcal{H})$ (see [67, Exercise 6, p. 116]). Note that Exercise 4.2.26 below is a direct consequence of Exercise 5.2.1 if one recalls that $\mathbf{L}(\mathcal{H})$ is the dual of the space of trace class operators on $\mathcal{H}$. See for instance [67, Theorem 2.1, p. 101] for this well-known fact. Here a direct proof without Exercise 5.2.1 is sought for. For references to Exercise 4.2.26 we send the reader for instance to [116, p. 30], [79, Chap. III] and [170, Chap. 14, p. 63].

Exercise 4.2.26.

(1) *Let $\mathcal{H}$ be a separable Hilbert space. Show that the closed unit ball of $\mathbf{L}(\mathcal{H})$ is not compact in the strong operator topology, while it is compact in the weak operator topology.*
(2) *Show that it is metrizable in the weak operator topology.*
(3) *Show that multiplication is sequentially jointly continuous in $\mathbf{L}(\mathcal{H})$.*

In the study of compact Riemann surfaces we often encounter, sometimes in an unexpected way, results from functional analysis needed to prove that some underlying vector spaces (for instance cohomology spaces) are finite dimensional. One such example is presented in the sequel in Question 5.3.3. Another such result is given in the following question, taken from Gunning's book [168] on Riemann surfaces.

Question 4.2.27 (see [168, Lemma 9, p. 66]). *Let S and K be linear continuous operators from the Hilbert space $\mathcal{H}_1$ into the Hilbert space $\mathcal{H}_2$. Assume that S is onto and K is compact. Then the operator $S - K$ has closed range of finite codimension.*

The notion of positive matrices was recalled in Definition 1.6.1. More generally in the setting of Hilbert spaces, one has:

Definition 4.2.28. A bounded operator Γ from the Hilbert space $\mathcal{H}$ into itself is called positive if
$$\langle \Gamma h, h \rangle_{\mathcal{H}} \geq 0, \quad \forall h \in \mathcal{H}. \tag{4.2.5}$$

4.2. Operators on normed spaces

The notion can be extended further to operators from a topological vector space into its anti-dual. See Remark 4.2.37 and Definition 5.5.1. One can also consider unbounded operators. See Definition 4.3.10.

Exercise 4.2.29. Let Γ be a everywhere positive operator in the Hilbert space $\mathcal{H}$. Show that Γ is Hermitian and bounded, and that Γ^n is positive for every $n \in \mathbb{N}$.

Exercise 4.2.30. Let $\mathcal{H}$ be a Hilbert space and let A be a bounded positive operator from $\mathcal{H}$ into itself.

(1) Show that

$$\|Af\| \leq \sqrt{\|A\|}\sqrt{\langle Af, f \rangle}, \quad \forall f \in \mathcal{H}. \tag{4.2.6}$$

(2) Show that a weakly convergent increasing sequence of positive operators converges strongly.
(3) Let $(A_n)_{n \in \mathbb{N}}$ be an increasing family of bounded positive operators such that

$$\lim_{n \to \infty} \langle A_n f, f \rangle < \infty, \quad \forall f \in \mathcal{H}. \tag{4.2.7}$$

Show that $(A_n)_{n \in \mathbb{N}}$ converges strongly to a positive operator.

As a corollary of the previous exercise we have:

Exercise 4.2.31. Let Γ be a bounded positive operator. Show that there exists a positive operator X such that $X^2 = \Gamma$.

One way is to proceed as in [170, Problem 121, p. 64]. Assume that Γ has norm less or equal to 1 and define a sequence of operators $X_0, X_1, \ldots$ by $X_0 = 0$ and

$$X_{n+1} = \frac{1}{2}((I - \Gamma) + X_n^2), \quad n = 0, 1, \ldots,$$

and apply Exercise 4.2.30. We here propose in the hint below another way, used in [38, Lemma 2.2] to tackle the quaternionic setting.

Hint: Consider the power series expansion

$$\sqrt{1 - z} = 1 - \sum_{n=1}^{\infty} a_n z^n, \quad |z| < 1,$$

and let

$$D_N = I - \sum_{n=1}^{N} a_n (I - \Gamma)^n, \quad N \in \mathbb{N}. \tag{4.2.8}$$

Assuming Γ strictly contractive, show that $(D_N)_{N \in \mathbb{N}}$ is a decreasing sequence of positive operators converging in the operator norm to an operator X satisfying $X^2 = \Gamma$.

The operator X in the previous exercise is uniquely defined and denoted by $\sqrt{\Gamma}$, or $\Gamma^{1/2}$. It is called the (positive) squareroot of Γ. The existence of the squareroot is part of the various equivalent characterizations of a positive operator, as given in the following theorem. For the finite-dimensional version see [CABP, Proposition 14.3.2, p. 485].

Theorem 4.2.32. *Let* Γ *be a bounded linear operator from the Hilbert space* $\mathcal{H}$ *into itself. Then the following are equivalent:*

(1) $\Gamma \geq 0$.

(2) Γ *is self-adjoint, and its spectrum (which is real) lies inside a bounded set of* $[0, \infty)$.

(3) *There is a positive operator* X *from* $\mathcal{H}$ *into itself such that* $\Gamma = X^2$.

(4) *There exist a Hilbert space* $\mathcal{C}$ *and a linear bounded operator* C *from* $\mathcal{H}$ *into* $\mathcal{C}$ *such that* $\ker C^* = \{0\}$ *and*

$$\Gamma = C^*C. \tag{4.2.9}$$

For matrices, yet another characterization is:

Theorem 4.2.33. *The matrix* $A \in \mathbb{C}^{N \times N}$ *is positive if and only if there exist a Hilbert space* $\mathcal{H}$ *and elements* $f_1, \ldots, f_N \in \mathcal{H}$ *such that*

$$a_{jk} = \langle f_k, f_j \rangle_{\mathcal{H}}$$

The matrix is strictly positive if and only if the vectors $f_1, \ldots, f_N$ *are linearly independent.*

As an application we have:

Exercise 4.2.34. *The matrix* (1.6.5)

$$\begin{pmatrix} \frac{1}{1} & \frac{1}{2} & \cdots & \frac{1}{1+N} \\ \frac{1}{2} & \frac{1}{3} & \cdots & \frac{1}{2+N} \\ & & & \\ \frac{1}{N+1} & \frac{1}{N+2} & \cdots & \frac{1}{2N+1} \end{pmatrix}$$

is strictly positive.

Definition 4.2.35. Given a Banach space $\mathcal{B}$, its anti-dual $\mathcal{B}^*$ is the space of its anti-linear continuous functionals.

Question 4.2.36. *The anti-dual of a Banach space is a vector space. It is a Banach space when endowed with the norm*

$$\|f\|_{\mathcal{B}^*} = \sup_{b \in \mathcal{B}, \|b\| \leq 1} |f(b)|.$$

Note that the following remark requires notions which appear in the sequel of the book.

Remark 4.2.37. It is a fascinating fact that the notion of positivity can be extended to operators Γ from a Banach space into its anti-dual, or more generally, from a topological vector space V into its anti-dual V^*, by replacing (4.2.5) by

$$\langle \Gamma v, v \rangle_{V,V^*} \geq 0, \tag{4.2.10}$$

where $\langle \cdot, \cdot \rangle_{V,V^*}$ denotes the duality between V and V^*. Factorizations of the kind (4.2.9) do not exist in every topological vector space. The class of spaces for which factorizations always exist has been characterized in [162]. It includes Banach spaces and nuclear spaces.

Exercise 4.2.38. *Let Γ be a positive operator in the Hilbert space $\mathcal{H}$. Show that*

$$\ker \Gamma = \ker \sqrt{\Gamma}.$$

For the next exercise, see for instance [120, Theorem 10, p. 604]. The result itself is called the spectral mapping theorem. We will use it in Exercise 6.1.31 to compute the spectrum of the Fourier transform.

Exercise 4.2.39. *Let T be a bounded operator in the Hilbert space $\mathcal{H}$, and let p be a polynomial. Then,*

$$p(\sigma(T)) = \sigma(p(T)).$$

Operator ranges play an important role in the theory of reproducing kernel Hilbert spaces. We review in the next exercise some of the relevant aspects of this notion. For a follow-up exercise when the space is a reproducing kernel Hilbert space, see Exercise 7.5.17. The case of unbounded operators is considered in Question 4.3.14.

Exercise 4.2.40. *Let $\mathcal{H}$ be a Hilbert space and let Γ be a positive bounded operator in $\mathcal{H}$. Let $\sqrt{\Gamma}$ denote the unique positive squareroot of Γ (see Exercise 4.2.31 for the latter) and let π denote the orthogonal projection from $\mathcal{H}$ onto $\mathrm{Ker}\,\pi$. The space $\mathrm{ran}\,\sqrt{\Gamma}$ with the norm*

$$\|\sqrt{\Gamma}u\|_{\Gamma} = \|(I - \pi)u\|_{\mathcal{H}} \tag{4.2.11}$$

is a Hilbert space. Moreover, for $u, v \in \mathcal{H}$

$$\langle \Gamma u, \Gamma v \rangle_{\Gamma} = \langle \Gamma u, v \rangle_{\mathcal{H}}. \tag{4.2.12}$$

In fact, as proved in [283, p. 150], any Hilbert subspace of the Hilbert space $\mathcal{H}$ is of this form. See [283, formulas (4.19), (4.20), p. 151].

Exercise 4.2.40 is a particular case of the following result in [283, §8, p. 174]. Take $\mathcal{E}$ and $\mathcal{F}$ two locally convex spaces with appropriate properties and let u be a weakly continuous map form $\mathcal{E}$ into $\mathcal{F}$. Then $u(\mathcal{E})$ is a Hilbertian subspace of $\mathcal{F}$. We refer to the discussion in [283, §8, p. 174] for more details. We also refer to [283, §10, p. 208] where the case where $\mathcal{E}'$ is a subspace of $\mathcal{E}$ is considered; it includes in particular the case where $\mathcal{E}$ is a Hilbert space.

If $\mathcal{H}_1$ is a closed subspace of a Hilbert space $\mathcal{H}$ it is well known that there exists an orthogonal complement to $\mathcal{H}_1$ in $\mathcal{H}$. This space, denoted $\mathcal{H} \ominus \mathcal{H}_1$, or $\mathcal{H}_1^\perp$, has the following property:

Every element $h \in \mathcal{H}$ admits a unique decomposition

$$h = h_1 + h_2, \quad h_1 \in \mathcal{H}_1, \quad h_2 \in \mathcal{H}_1^\perp$$

such that

$$\|h\|^2 = \|h_1\|^2 + \|h_2\|^2.$$

The theory of complementation is an extension of this result when the space $\mathcal{H}_1$ is contractively included in $\mathcal{H}$.

Exercise 4.2.41. *Let $(\mathcal{H}_1, \langle \cdot, \cdot \rangle_1)$ be a Hilbert space contractively included in a Hilbert space $(\mathcal{H}, \langle \cdot, \cdot \rangle)$. There exists a uniquely defined Hilbert space $(\mathcal{H}_2, \langle \cdot, \cdot \rangle_2)$ contractively included in $(\mathcal{H}, \langle \cdot, \cdot \rangle)$ and with the following property: Every element h in $\mathcal{H}$ admits decompositions of the form $h = h_1 + h_2$ with $h_i \in \mathcal{H}_i$, $i = 1, 2$. For every such decomposition*

$$\|h\|^2 \leq \|h_1\|_1^2 + \|h_2\|_2^2, \tag{4.2.13}$$

and there is a unique decomposition for which equality holds. More precisely, let i_1 denote the inclusion map from $\mathcal{H}_1$ into $\mathcal{H}$. Then

$$\mathcal{H}_1 = \operatorname{ran} \sqrt{i_1 i_1^*}, \quad \mathcal{H}_2 = \operatorname{ran} \sqrt{I_\mathcal{H} - i_1 i_1^*}, \tag{4.2.14}$$

with the corresponding lifted norms and

$$h = (i_1 i_1^*)h + (I_\mathcal{H} - i_1 i_1^*)h \tag{4.2.15}$$

is the decomposition of h for which equality holds in (4.2.13).

The space $\mathcal{H}_2$ is called the complementary space of $\mathcal{H}_1$ in $\mathcal{H}$. We will denote $\mathcal{H}_2 = \mathcal{H}_1^c$. From the uniqueness of the complementary space we have

$$\mathcal{H}_1^{cc} = \mathcal{H}_1.$$

Complementation plays an important role in the theory of reproducing kernel Hilbert spaces, and in particular in the theory of de Branges–Rovnyak spaces; see [85, 86, 104].

In the setting of operator ranges, Exercise 4.2.41 gives the following generalization of Exercise 1.6.11 (see [133]). See also Theorem 7.5.22.

Question 4.2.42. *In the notation of Exercise 4.2.40, let Γ_1 and Γ_2 be two continuous positive operators in the Hilbert space $\mathcal{H}$. Then*

$$\operatorname{ran} \sqrt{\Gamma_1 + \Gamma_2} = \operatorname{ran} \sqrt{\Gamma_1} + \operatorname{ran} \sqrt{\Gamma_2},$$

and the sum (endowed with the lifted norm) is orthogonal if and only if

$$\operatorname{ran} \sqrt{\Gamma_1} \cap \operatorname{ran} \sqrt{\Gamma_2} = \{0\}.$$

Hint: We outline part of proof following the arguments in Exercise 1.6.11. In view of the positivity of the operators Γ_1 and Γ_2, the map j_1 defined by

$$j_1((\Gamma_1 + \Gamma_2)h) = \Gamma_1 h, \quad h \in \mathcal{H},$$

extends to a contraction from $\operatorname{ran} \sqrt{\Gamma_1 + \Gamma_2}$ into $\operatorname{ran} \sqrt{\Gamma_1}$. Its adjoint is given by

$$j_1^*(\sqrt{\Gamma_1}h) = (\sqrt{\Gamma_1}h),$$

and so $\operatorname{ran} \sqrt{\Gamma_1} \subset \operatorname{ran} \sqrt{\Gamma_1 + \Gamma_2}$. Defining similarly a map j_2 we get

$$\operatorname{ran} \sqrt{\Gamma_1} + \operatorname{ran} \sqrt{\Gamma_2} \subset \operatorname{ran} \sqrt{\Gamma_1 + \Gamma_2}.$$

Define now a densely defined map

$$T : \operatorname{ran} \sqrt{\Gamma_1 + \Gamma_2} \longrightarrow \operatorname{ran} \sqrt{\Gamma_1} + \operatorname{ran} \sqrt{\Gamma_2}$$

by

$$T((\Gamma_1 + \Gamma_2)h) = (\Gamma_1 h, \Gamma_2 h).$$

Then T extends to an isometry, with adjoint

$$T^*(\sqrt{\Gamma_1}h, \sqrt{\Gamma_1}g) = \sqrt{\Gamma_1}h + \sqrt{\Gamma_1}g, \quad h, g \in \mathcal{H}.$$

The result follows since

$$\operatorname{ran} T^* \dotplus \ker T =, \sqrt{\Gamma_1 + \Gamma_2},$$

and $\ker T = \{0\}$. The last claim follows also the arguments in Exercise 1.6.11.

We now present a simple, but illustrative, instance of Exercise 4.2.41.

Example 4.2.43. *Let $\mathcal{H} = \mathbb{C}$ with the usual norm (the absolute value) and let $\mathcal{H}_1$ denote also the complex numbers, but with the norm:*

$$\|x\|_1 = \frac{|x|}{\epsilon}$$

where $0 < \epsilon < 1$. Then $\mathcal{H}_1$ is contractively included in $\mathcal{H}$ and the complementary space $\mathcal{H}_2$ is equal to $\mathbb{C}$ endowed with the norm

$$\|x\|_2 = \frac{|x|}{\sqrt{1 - \epsilon^2}}.$$

The decomposition

$$x = \epsilon^2 x + (1 - \epsilon^2)x$$

is the minimal decomposition of x since

$$\|\epsilon^2 x\|_1^2 = \frac{\epsilon^4}{\epsilon^2}|x|^2 = \epsilon^2 |x|^2 \quad and \quad \|(1 - \epsilon^2)x\|_2^2 = \frac{(1 - \epsilon^2)^2}{1 - \epsilon^2}|x|^2 = (1 - \epsilon^2)|x|^2.$$

The limiting case $\epsilon = 1$ corresponds to $\mathcal{H} = \mathcal{H}_1 = \mathbb{C}$ as Hilbert spaces and $\mathcal{H}_2 = \{0\}$.

For an application of the following exercise see Exercise 7.1.7.

Exercise 4.2.44. Let $(\mathcal{H}, \langle \cdot, \cdot \rangle_{\mathcal{H}})$ be a Hilbert space, and for $k \in \mathcal{H}$, set M_k to be the multiplication operator:

$$M_k(c) = ck, \quad c \in \mathbb{C}.$$

Compute M_k^* and show that

$$M_k^* M_h = \langle h, k \rangle_{\mathcal{H}}. \tag{4.2.16}$$

Remark 4.2.45. The function (4.2.16) is an instance of a positive definite kernel (see Definition 7.1.1). This seemingly trivial function is a key player in two important fields. In white noise space analysis, one builds for any *real* Hilbert space $\mathcal{H}$ a probability space $(\Omega, \mathcal{B}, P)$ and a map $h \mapsto Q_h$ from $\mathcal{H}$ into the Lebesgue space $\mathbf{L}_2(\Omega, \mathcal{B}, P)$ such that

$$\langle h, k \rangle = \langle Q_h, Q_k \rangle_{dP}.$$

See [72, 188]. In free analysis one builds a type II_1 von Neumann algebra with trace τ and a map $h \mapsto X_h$ from $\mathcal{H}$ into $\mathcal{M}$ such that

$$\langle h, k \rangle = \tau(X_h^* X_k).$$

See [317].

We now meet one of the important connections between operator theory and analytic functions of one complex variable, namely (one instance of) the characteristic operator function. The underlying problem was to develop a theory for non self-adjoint operators, and this led to the definition of a new object in complex analysis, namely J-contractive functions, see Definition 4.2.47 after the exercise. These functions appear in a wide class of problems in analysis and their structure was given by V. Potapov in [253]. A central question is to relate the properties of the function Θ and of the operator T. For more information on characteristic operator functions and operator models we refer in particular to the works of Sz.-Nagy and Foias (see [240] and the enlarged edition [300] together with Berkovici and Kérchy), of Gohberg and Kreĭn [158] and de Branges and Rovnyak [85].

Before stating the question we recall that signature matrices were defined in (1.5.4) as matrices which are both self-adjoint and unitary.

Question 4.2.46. Let $\mathcal{H}$ be a Hilbert space and let T be a bounded linear operator from $\mathcal{H}$ into itself. We assume that $\frac{T - T^*}{2i}$ is a finite-dimensional operator, say of rank m.

(1) Show that there exist a signature matrix $J \in \mathbb{C}^{m \times m}$ and a bounded operator C from $\mathcal{H}$ into $\mathbb{C}^m$ such that

$$T - T^* = iC^* J C. \tag{4.2.17}$$

(2) *Let*

$$\Theta(z) = I_{\mathcal{H}} + izC(I_{\mathcal{H}} - zT)^{-1}C^*J. \tag{4.2.18}$$

Show that

$$C(I_m - zT)^{-1}(I_m - wT)^{-*}C^* = \frac{J - \Theta(z)J\Theta(w)^*}{-i(z - \overline{w})}.$$

The finite-dimensional version of the above question is studied in Exercise 7.7.3.

Definition 4.2.47. Let $J \in \mathbb{C}^{m \times m}$-valued signature matrix. The $\mathbb{C}^{m \times m}$-valued function Θ is called J-contractive if it is analytic in an open subset Ω of the open upper half-plane and

$$\Theta(z)^*J\Theta(z) \le J, \quad \forall z \in \Omega.$$

Exercise 4.2.48. *Let $\mathcal{H}_1$ and $\mathcal{H}_2$ be two Hilbert spaces, and let T denote a contraction from $\mathcal{H}_1$ into itself. Let C be a bounded operator from $\mathcal{H}_1$ into $\mathcal{H}_2$. Show that the operator matrix*

$$\begin{pmatrix} \frac{1}{2}(I_{\mathcal{H}_1} + T) & C^* \\ CT & CC^* \end{pmatrix} : \quad \mathcal{H}_1 \oplus \mathcal{H}_2 \longrightarrow \mathcal{H}_1 \oplus \mathcal{H}_2$$

has a real positive part.

We conclude this section with an important notion, which plays a special role in the case of indefinite inner product spaces (then the norms in (4.2.19) are replaced by self inner products). Let $\mathcal{H}$ and $\mathcal{G}$ be two Hilbert spaces and let T be a linear operator from Dom $D(T) \subset \mathcal{H}$ into $\mathcal{G}$. The graph R_T of the operator T is defined by:

$$R_T = \{(h, Th) \, ; \, h \in \text{Dom} \, D(T)\}.$$

It is a linear subspace of $\mathcal{H} \times \mathcal{G}$. More generally:

Definition 4.2.49. Let $\mathcal{H}$ and $\mathcal{G}$ be two Hilbert spaces. A linear subspace of $\mathcal{H} \times \mathcal{G}$ is called a linear relation. The domain of the relation R is defined by

$$\text{Dom}\,(R) = \{h \in \mathcal{H} \, ; \, \exists g \in \mathcal{G} \text{ such that } (f, g) \in R\}.$$

The relation is called contractive if

$$(f, g) \in R \implies \|g\|_{\mathcal{G}} \le \|f\|_{\mathcal{H}}. \tag{4.2.19}$$

Exercise 4.2.50.
(1) *Show that not every linear relation is an operator graph.*
(2) *Show that a densely defined contractive relation extends uniquely to the graph of a contraction.*

4.3 Unbounded operators

Definition 4.3.1. Let $\mathcal{H}_1$ and $\mathcal{H}_2$ be two Hilbert spaces.

(1) A linear operator T with domain $\text{Dom}\,(T) \subset \mathcal{H}_1$ and range inside $\mathcal{H}_2$ is closed if its graph $\mathscr{G}(T)$

$$\mathscr{G}(T) = \{(u, Tu)\,;\, u \in \text{Dom}\,(T)\} \qquad (4.3.1)$$

is a closed subspace of $\mathcal{H}_1 \times \mathcal{H}_2$.

(2) The linear operator T is closable if it has a closed extension.

Question 4.3.2.

(1) *Give an example of an everywhere defined linear operator which is not closed.*
(2) *Give an example of a non closable operator.*

Hints: For (1) define an operator in an appropriate way on a countable set of linearly independent vectors, complete this set to a Hamel basis and define then the operator on the whole of the space using the basis (this construction in fact builds an everywhere defined unbounded operator; see also Exercise 7.5.14 below).

For (2) take for instance the example from [6, p. 120], that is

$$(Tf)(x) = xf(1)$$

defined on an appropriate subspace of $\mathbf{L}_2(0,1)$.

Question 4.3.3. *Let $\mathcal{H}_1$ and $\mathcal{H}_2$ be two Hilbert spaces, and let T be a densely defined operator with domain $\text{Dom}\,(T) \subset \mathcal{H}_1$ and closed range inside $\mathcal{H}_2$. Assume that T has an adjoint T^*. Show that T^* is closed.*

The following exercise is important in connection with quantum mechanics.

Exercise 4.3.4. *Find a linear subspace of $\mathbf{L}_2(\mathbb{R}, \mathcal{B}, dx)$ on which both operators*

$$(Qf)(x) = -if'(x) \quad and \quad (Pf)(x) = xf(x)$$

are self-adjoint.

Hint: (see [6, pp. 156–157]) Consider the space of functions in $\mathbf{L}_2(\mathbb{R}, \mathcal{B}, dx)$, which are absolutely continuous, and with derivative in $\mathbf{L}_2(\mathbb{R}, \mathcal{B}, dx)$. For such a function, compute

$$\int_a^b f'(x)\overline{f(x)}dx, \quad \text{with } a, b \in \mathbb{R},$$

and conclude that

$$\lim_{x \to \pm\infty} f(x) = 0.$$

These operators satisfy

$$(-iQ)P - P(-iQ) = I. \qquad (4.3.2)$$

Exercise 4.3.5. *There do not exist continuous operators P and Q in a normed space which satisfy (4.3.2).*

We note that there exist linear bounded operators in a Hilbert space such that

$$QP - qPQ = I$$

for every choice of $q \in [-1, 1)$. A trivial example is given by

$$P = Q = \frac{1}{\sqrt{1-q}} I.$$

A more interesting example is given by the annihilation and creation operators in the corresponding q-Fock spaces.

Question 4.3.6. *Consider the operator*

$$(Af)(x) = -f''(x) + V(x)f(x),$$

where V is a continuous positive function.

(1) *Find a condition on the function b such that the factorization*

$$A = XX^*$$

holds with $(Xf)(x) = -f'(x) + b(x)f(x)$. Here X^* denotes the formal (also called Lagrange) adjoint of X, that is

$$(X^*f)(x) = f'(x) + b(x)f(x).$$

(2) *Find $b(x)$ when $V(x) = x^2 + 1$.*

To solve the following question the reader should first do the various exercises on Hermite functions and Hermite polynomials and in particular Exercise 6.1.16.

Question 4.3.7.

(1) *Find the domain on which the differential operator*

$$Hf(x) = -f''(x) + (x^2 + 1)f(x) \tag{4.3.3}$$

is self-adjoint in $\mathbf{L}_2(\mathbb{R}, \mathcal{B}, dx)$.

(2) *Show that it is boundedly invertible and that its inverse is a Hilbert–Schmidt operator (see Question 4.2.22 for the latter).*

Hint: Following Naĭmark's book [241, p. 60] we define $\mathscr{D}$ to be the space of functions f absolutely continuous and with absolutely continuous derivative and such that f, f' and Hf all belong to $\mathbf{L}_2(\mathbb{R})$. Then (see [241, p. 70]), H is self-adjoint on the space of functions y such that

$$\lim_{a \to -\infty} [y, f](a) = \lim_{b \to +\infty} [y, f](b), \quad \forall f \in \mathscr{D},$$

where (see [241, p. 50])

$$[y, f](x) = y(x)f'(x) - y'(x)f(x).$$

Conditions for the spectrum of a self-adjoint operator defined by an expression

$$Hf(x) = -f''(x) + q(x)f(x), \quad x \in \mathbb{R} \tag{4.3.4}$$

to be purely discrete are given in [241, pp. 239–245]. We mention in particular the following result of Molchanov (see [241, Theorem 13, p. 245] and [237]):

Theorem 4.3.8. *Assume in* (4.3.4) *that q is bounded from below. The spectrum of the self-adjoint operator defined by H is discrete if and only if*

$$\lim_{x \to \pm\infty} \int_x^{x+a} q(u)du = \infty$$

for every $a \in \mathbb{R}$.

The following easy consequence of the previous theorem is closely related to the Fourier transform; see Exercise 6.1.33. The (discrete) spectrum of B is in fact computed in Exercise 6.1.16.

Question 4.3.9. *Let*

$$Bf(x) = \frac{-f''(x) + (x^2 - 1)f(x)}{2}. \tag{4.3.5}$$

Find the domain of B on which it is self-adjoint and show that the spectrum of B is discrete.

Definition 4.3.10. The densely defined operator T with domain $\mathrm{Dom}\,(T)$ and range inside the Hilbert space $\mathcal{H}$ is positive if

$$\langle Tf, f \rangle_{\mathcal{H}} \geq 0, \quad \forall f \in \mathrm{Dom}\,(T). \tag{4.3.6}$$

Question 4.3.11. *Show that the operator* (4.3.3) *is positive.*

Question 4.3.12.

(1) *Show that the operator $Hf(x) = -f''(x)$ (with appropriate domain) is self-adjoint and positive.*
(2) *Show that for every $a > 0$ the operator $H + a^2 I$ is boundedly invertible.*
(3) *Show that the operator $(H + a^2 I)^{-1}$ has a "kernel" in the sense that there exists a function $K_a(x, y)$ such that*

$$((H + a^2 I)^{-1}f)(x) = \int_{\mathbb{R}} K_a(x, y)f(y)dy, \quad \forall f \in \mathbf{L}_2(\mathbb{R}). \tag{4.3.7}$$

(4) *Show that the function $K_a(x, y)$ is positive definite in the sense that*

$$\iint_{\mathbb{R}^2} K_a(x, y)f(x)\overline{f(y)}dxdy \geq 0, \quad \forall f \in \mathbf{L}_2(\mathbb{R}). \tag{4.3.8}$$

Hint: Try

$$K_a(x, y) = \frac{1}{2a} e^{-a|x-y|}.$$

Remarks 4.3.13.

(1) The function $K_a(x, y)$ is the Green function of the one-dimensional Schrödinger operator. See the papers [185, 186] by H. Rabitz and his collaborators for applications.

(2) Of course not every linear bounded operator T from $\mathbf{L}_2(\mathbb{R})$ into itself has a kernel, in the sense that it cannot always be written in the integral form

$$(Tf)(x) = \int_{\mathbb{R}} k(x, y) f(y) dy$$

for some function $k(x, y)$ (called its kernel), with appropriate conditions. See also item (1) in Remarks 7.8.2 and the discussion at the beginning of Section 7.6. Here there is a kernel because the operator $(H + a^2 I)^{-1}$ is of Hilbert–Schmidt class. Schwartz' kernel[1] theorem gives a characterization of such operators between topological vector spaces (with appropriate properties).

(3) Equation (4.3.8) means that $K_a(x, y)$ is positive definite in the sense of Mercer; see the discussion in Aronszajn's paper [52, p. 339].

Question 4.3.14. *Solve the counterpart of Exercises* 4.2.40 *and* 7.5.17 *when the operator* Γ, *still self-adjoint, is unbounded.*

For an illustration of the following result, see [236, p. 454].

Question 4.3.15. *Let* T *be a Hermitian densely defined operator in the Hilbert space* $\mathcal{H}$. *The following are equivalent:*

(1) *The operator* T *is self-adjoint.*
(2) *The domain of* T *is complete when endowed with the norm*

$$\|f\|_T^2 = \|Tf\|^2 + \|f\|^2.$$

(3) *The operator* $T + iI$ *is one-to-one onto from its domain to* $\mathcal{H}$.

4.4 Indefinite inner product spaces

An indefinite inner product space is a pair $(\mathcal{V}, [\cdot, \cdot]_\mathcal{V})$, where $\mathcal{V}$ is a vector space and $[\cdot, \cdot]_\mathcal{V}$ is a Hermitian form (see Definition 1.3.12) on $\mathcal{V}$. The theory of these spaces has a long history, which can be traced to the structure of Hermitian forms for the finite-dimensional case and to the work of Pontryagin for the infinite-dimensional case. We refer to Bognár's book [76, pp. 118–119 and p. 207] for historical background. In the circle of ideas presented in this book, indefinite inner product spaces appear in the study of close to self-adjoint or close to unitary operators.

---

[1] To quote Jean Dieudonné in 1978 (see [113, p. 3]), *peut-être le plus important de toute l'Analyse fonctionnelle linéaire moderne.*

Definition 4.4.1. The vector space $\mathcal{V}$ endowed with the Hermitian form $[\cdot, \cdot]$ is called a Kreĭn space if it can be written as

$$\mathcal{V} = \mathcal{V}_+ + \mathcal{V}_- \tag{4.4.1}$$

where:

(1) The space $\mathcal{V}_+$ endowed with $[\cdot, \cdot]$ is a Hilbert space.
(2) The space $\mathcal{V}_-$ endowed with $-[\cdot, \cdot]$ is a Hilbert space.
(3) The spaces $\mathcal{V}_-$ and $\mathcal{V}_+$ have a zero intersection, $\mathcal{V}_+ \cap \mathcal{V}_- = \{0\}$, and are orthogonal in the sense that

$$[v_+, v_-] = 0, \quad \forall v_+ \in \mathcal{V}_+ \text{ and } v_- \in \mathcal{V}_-.$$

The decomposition $\mathcal{V} = \mathcal{V}_+ + \mathcal{V}_-$ is then called a fundamental decomposition. It will not be unique, unless $\mathcal{V}$ is a Hilbert space or an anti-Hilbert space.

The proof of item (2) in the exercise below requires some notions from general topological spaces. The result itself is a key result in the theory of Kreĭn spaces.

Question 4.4.2. *Let* $(\mathcal{V}, [\cdot, \cdot])$ *be a Kreĭn space, with fundamental decomposition* $\mathcal{V} = \mathcal{V}_+ + \mathcal{V}_-$.

(1) *Show that* $\mathcal{V}$ *endowed with the Hermitian form:*

$$\langle v, w \rangle_{\mathcal{V}} = [v_+, w_+] - [v_-, w_-] \tag{4.4.2}$$

is a Hilbert space.

(2) *Let* $\mathcal{V} = \mathcal{W}_+ + \mathcal{W}_-$ *denote another fundamental decomposition. Show that the two associated norms* (4.4.2) *are equivalent.*

For a given decomposition we define

$$P(v_+ + v_-) = v_+ \quad \text{and} \quad Q(v_+ + v_-) = v_-. \tag{4.4.3}$$

Question 4.4.3. *Let* $\mathcal{V}$ *be a Kreĭn space and let* T *be a bi-contraction:*

$$[Tv, Tv] \le [v, v] \quad \text{and} \quad [T^*v, T^*v] \le [v, v].$$

Show that $P - TQ$ *is invertible (where* P *and* Q *are defined by* (4.4.3)*) and that*

$$(P - TQ)^{-1}(Q - PT)$$

is a contraction from the Hilbert space $(\mathcal{V}, \langle \cdot, \cdot \rangle_{\mathcal{V}})$ *into itself.*

The Kreĭn space is called a Pontryagin space if the space $\mathcal{V}_-$ in the direct and orthogonal decomposition (4.4.1) is finite dimensional. In a number of works in the former Soviet Union, see in particular the fundamental monograph [190], the other convention dim $\mathcal{V}_+ < \infty$ is taken.

We conclude with an important definition.

Definition 4.4.4. Let $h_1, \ldots, h_N$ be elements of the indefinite inner product vector space $(\mathcal{V}, [\cdot, \cdot]_{\mathcal{V}})$. Then, the $N \times N$ Hermitian matrix with (ℓ, j) entry equal to $[h_j, h_\ell]_{\mathcal{V}}$ is called the Gram matrix of $h_1, \ldots, h_N$.

4.5   Solutions of the exercises

Solution of Exercise 4.1.5: The examples $\mathcal{V} = \mathbb{R}$ and $A = [-1/2, 1]$ and $\mathcal{V} = \mathbb{C}$ and $A = [-1, 1] \cup [-i, i]$ answer respectively the first and second question.   □

Solution of Exercise 4.1.8: The application p_K is bounded since K is compact. It is a semi-norm, as follows from the properties of the absolute value. Let now $f \in \mathcal{C}(\Omega)$. The condition $p_K(f) = 0$ will not force $f = 0$, and so p_K is only a semi-norm. On the other hand, if f is analytic in Ω and such that $p_K(f) = 0$, we have that $f(z) = 0$ in a set with an accumulation point (since the interior of K is assumed non-empty). Since Ω is assumed connected, the zero theorem forces f to be identically vanishing in Ω, and hence p_K is a norm.   □

We note that the claim in Exercise 4.1.8 is false if we do not assume Ω connected.

Solution of Exercise 4.1.10: If $X(\varphi)(v) = \varphi(v) = 0$ for all $v \in \mathcal{V}$, then $\varphi = 0$, and the map X is indeed one-to-one.

Let now U be an open subset of $X(\mathcal{V}')$. By definition, $U = X(\mathcal{V}') \cap \Omega$, where Ω is an open subset of $\mathbb{C}^{\mathcal{V}}$ in the product topology. By Question 3.5.2 it is enough to consider Ω of the form

$$\Omega = \{(z_v)_{v \in \mathcal{V}} \; ; \; |z_{v_j} - z_j| < \epsilon_j, \; j = 1, 2, \dots, N\},$$

where $N \in \mathbb{N}$, $z_1, \dots, z_N \in \mathbb{C}$, $\epsilon_1, \dots, \epsilon_N \in (0, \infty)$. Since $U \subset X(\mathcal{V}')$ we have

$$z_j = \varphi_0(v_j), \quad j = 1, \dots, N$$

for some $\varphi_0 \in \mathcal{V}'$ and $v_1, \dots, v_N \in \mathcal{V}$. Hence

$$U = \{(\varphi(v))_{v \in \mathcal{V}} \; ; \; |\varphi(v_j) - \varphi_0(v_j)| < \epsilon_j, \; j = 1, 2, \dots, N\},$$

and

$$\begin{aligned} X^{-1}(U) &= \{\varphi \in \mathcal{V}' \; ; \; X(\varphi) \in U\} \\ &= \{\varphi \in \mathcal{V}' \; ; \; |\varphi(v_j) - \varphi_0(v_j)| < \epsilon_j, \; j = 1, 2, \dots, N\}, \end{aligned}$$

which is open in the $\sigma(\mathcal{V}', \mathcal{V})$ topology.

Conversely, let (with the same notation as above)

$$E = \{\varphi \in \mathcal{V}' \; ; \; |\varphi(v_j) - \varphi_0(v_j)| < \epsilon_j, \; j = 1, 2, \dots, N\}$$

be an open set in the $\sigma(\mathcal{V}', \mathcal{V})$ topology. Then,

$$\begin{aligned} (X^{-1})^{-1}(E) = X(E) \\ = \{(z_v)_{v \in \mathcal{V}}, \, z_v = \varphi(v) \text{ for some (uniquely defined)} \\ \varphi \in \mathcal{V}' \text{ and } |\varphi(v_j) - \varphi_0(v_j)| < \epsilon_j, \; j = 1, 2, \dots\} \end{aligned}$$

is open in the topology induced by $\mathbb{C}^{\mathcal{V}}$.   □

Solution of Exercise 4.1.11:

(1) Each of the functions

$$\varphi \mapsto |\varphi(v)|$$

is continuous with respect to the $\sigma(\mathcal{V}', \mathcal{V})$ topology (by definition of the latter). The first claim follows since the supremum of continuous functions is lower semi-continuous. See the discussion above Definition 3.4.7.

(2) Denote by C the closed ball (4.1.9):

$$C = \{\varphi \in \mathcal{V}; \ \|\varphi - \varphi_0\|' \leq r\}.$$

Since the norm $\|\cdot\|'$ is lower semi-continuous (see (1) of the exercise), C is closed in the $\sigma(\mathcal{V}', \mathcal{V})$ topology. We now assume that $\mathcal{V}$ is separable, and we use the hint and proceed as follows to prove compactness. Let $\overline{B(0,r)}$ denote the closed disk centered at the origin and with radius r. Using the map

$$Y(\varphi) = ((\varphi - \varphi_0)(v_n))_{n \in \mathbb{N}}$$

which sends C into $\overline{B(0,r)}^{\mathbb{N}}$, we will show that C is a closed subset of a compact set. We first check that C is metrizable.

C is metrizable: Let $v_1, v_2, \ldots$ be a countable dense subset of the closed unit ball of $\mathcal{V}$, and let $\mathcal{T}$ denote the topology generated by the maps

$$\varphi \mapsto |\varphi(v_n)|, \quad n = 1, 2, \ldots$$

By definition of $\sigma(\mathcal{V}', \mathcal{V})$ we have $\mathcal{T} \subset \sigma(\mathcal{V}', \mathcal{V})$. To prove the converse inclusion, consider $O \subset C$ open in the $\sigma(\mathcal{V}', \mathcal{V})$ topology. Every $\varphi' \in O$ is included in a set of the form

$$\{\psi \in \mathcal{V}', \ |(\psi - \varphi)(b_j)| < \rho, \ j = 1, \ldots, M\} \tag{4.5.1}$$

included in O, where $M \in \mathbb{N}$ and $b_1, \ldots, b_M \in \mathcal{V}$ are of norm less or equal to 1. To show that O is a union of open sets of $\mathcal{T}$ and hence is an open set of $\mathcal{T}$ we show that any set of the form (4.5.1) contains a set of the same form, but with now the b_j chosen among the set $v_1, v_2, \ldots$. To prove this last claim, take v_j from the above sequence such that

$$\|v_j - b_j\| < \rho_1,$$

and consider the set

$$\{\psi \in \mathcal{V}', \ |(\psi - \varphi)(v_j)| < \rho_2, \ j = 1, \ldots, M\} \tag{4.5.2}$$

where ρ_1 and ρ_2 are strictly positive numbers. We have

$$|(\psi - \varphi)(b_j)| \leq |(\psi - \varphi)(b_j - v_j)| + |(\psi - \varphi)(v_j)| \leq 2r\rho_1 + \rho_2 < \rho$$

for ρ_1 and ρ_2 small enough.

4.5. Solutions of the exercises

The topology defined by the countable family of semi-metrics

$$d_n(\varphi, \psi) = |(\varphi - \psi)(v_n)|, \quad n = 1, 2, \dots$$

is metrizable, and one can choose the metric

$$d(\varphi, \psi) = \sum_{n=1}^{\infty} \frac{1}{2^n} \frac{d_n(\varphi, \psi)}{1 + d_n(\varphi, \psi)}$$

to define the topology. It follows that C is metrizable in the $\sigma(\mathcal{V}', \mathcal{V})$ topology.

$Y(C)$ **lies inside a compact set:** The space $\overline{B(0, r)}^{\mathbb{N}}$ endowed with the metric

$$D((z_n), (w_n)) = \sum_{n=1}^{\infty} \frac{1}{2^n} \frac{|z_n - w_n|}{1 + |z_n - w_n|} \tag{4.5.3}$$

is compact and we have

$$d(\varphi, \psi) = D(Y(\varphi), Y(\psi)).$$

Thus $Y(C)$ is closed, and is compact as a closed subset of a compact set.

C **is compact:** The map Y is one-to-one and the map Y^{-1} is continuous from $Y(C)$ onto C. So $C = Y^{-1}(Y(C))$ is compact. □

Solution of Exercise 4.1.12: Let $C \subset \mathcal{V}'$ be bounded and closed. Using the map X defined by (4.1.8) we consider the image $X(C) \subset \mathbb{C}^{\mathcal{V}}$. Since C is bounded, there exists $K > 0$ such that

$$|\varphi(v)| \le K\|v\|, \quad \forall \varphi \in C.$$

By Tychonoff's theorem, the set $\prod_{v \in \mathcal{V}} \overline{B(0, K\|v\|)}$ is compact. But (as we recalled in Theorem 3.2.4) a closed subset of a compact set is compact, and this ends the proof. □

Solution of Exercise 4.1.14: We follow [114, (12.16.3), p. 84]. Let us denote by p the semi-norm in the normed space $\mathcal{V}$, and assume that p is lower semi-continuous. Then the sets

$$p^{-1}(n, \infty], \quad n = 1, 2, \dots,$$

are open (see Definition 3.4.7) and therefore the sets

$$C_n = p^{-1}\{\mathbb{R} \setminus (n, \infty]\} = p^{-1}[0, n], \quad n = 1, 2, \dots,$$

are closed. We have $\cup_{n \in \mathbb{N}} C_n = \mathcal{V}$. The space $\mathcal{V}$ need not be complete, but every point $v_0 \in \mathcal{V}$ has a neighborhood which is a complete metric space (for instance, $\{v \in \mathcal{V}; \|v - v_0\| \le 1\}$). We can thus apply Theorem 3.9.13. We have that C_{n_0} has a non-empty interior for some $n_0 \in \mathbb{N}$ (and in fact for all $n \geq n_0$, since the family $(C_n)_{n\in\mathbb{N}}$ is increasing). So there are $v_0 \in C_{n_0}$ and $r_0 > 0$ such that

$$B(v_0, r_0) \subset C_{n_0}.$$

Thus

$$\|v - v_0\| < r_0 \implies p(v) \leq n_0,$$

and so for $w \in \mathcal{V}$ such that $\|w\| < r_0$ we have

$$p(w) = p(w+v_0-v_0) \leq p(w+v_0)+p(v_0) \leq n_0+p(v_0) \quad \text{(since } \|w + v_0 - v_0\| < r_0).$$

Thus p is bounded on an open neighborhood of the origin. Let now $z \neq 0 \in \mathcal{V}$. Then

$$p(z) = p\left(\frac{r_0 z}{\|z\|}\frac{\|z\|}{r_0}\right) = \frac{\|z\|}{r_0}p\left(\frac{r_0 z}{\|z\|}\right) \leq \frac{\|z\|}{r_0}(n_0 + p(v_0)),$$

and so p is continuous. $\qquad\square$

Solution of Exercise 4.1.17: The space $\mathcal{B}_p(\Omega)$ is isometrically included in

$$\mathbf{L}_p(\Omega, \mathcal{B}, dxdy),$$

where $\mathcal{B}$ denote the sigma-algebra of Borel subsets of Ω and $dxdy$ denotes the two-dimensional Lebesgue measure. Let $z_0 \in \Omega$ and let r_0 be such that $\overline{B(z_0, r_0)} \subset \Omega$. Cauchy's formula gives, for $r \in (0, r_0]$,

$$f(z_0) = \frac{1}{2\pi}\int_0^{2\pi} f(z_0 + re^{it})dt,$$

so that

$$|f(z_0)| \leq \frac{1}{2\pi}\int_0^{2\pi} |f(z_0 + re^{it})|dt.$$

Thus we have (where we use Minkowski's inequality to go from the third line to the fourth one):

$$\frac{r_0^2}{2}|f(z_0)| = \int_0^{r_0} |f(z_0)|r\,dr$$

$$\leq \frac{1}{2\pi}\int_0^{r_0}\left(\int_0^{2\pi} |f(z_0 + re^{it})|dt\right)r\,dr$$

$$= \frac{1}{2\pi}\int_{B(z_0,r_0)} |f(z)|dxdy$$

$$\leq \frac{1}{2\pi}\left(\int_{B(z_0,r_0)} |f(z)|^p dxdy\right)^{1/p}\left(\int_{B(z_0,r_0)} dxdy\right)^{1/q}.$$

4.5. *Solutions of the exercises*

Thus
$$|f(z_0)| \le c_{z_0} \|f\|_p, \quad \text{where} \quad c_{z_0} = \frac{1}{(\pi r_0^2)^q}. \tag{4.5.4}$$

Let now $w \in \Omega$ and $r < R$ be such that
$$\overline{B(w,r)} \subset \overline{B(w,R)} \subset \Omega.$$

Then, for every $z_0 \in \overline{B(w,r)}$ there exists r_0, such that
$$\overline{B(z_0,r_0)} \subset \overline{B(w,R)}.$$

The numbers r_0 can be chosen to be uniformly bounded from below by
$$\rho = \min\{R - r, r\}.$$

Thus for $z_0 \in \overline{B(w,r)}$ we have the uniform bound
$$|f(z_0)| \le \frac{1}{(\pi \rho^2)^q} \|f\|_p.$$

It follows that if $(f_n)_{n \in \mathbb{N}}$ is a Cauchy sequence in $\mathcal{B}_p(\Omega)$, the sequence also converges pointwise uniformly on sets of the form $\overline{B(w,r)}$, and hence uniformly on compact sets. The limit
$$f(z) = \lim_{n \to \infty} f_n(z), \quad z \in \Omega,$$

is analytic in Ω. We now show that it belongs to $\mathcal{B}_p$ and that $\lim_{n \to \infty} \|f_n - f\|_p = 0$. Since $(f_n)_{n \in \mathbb{N}}$ is a Cauchy sequence, there exists N such that, for $n, m \ge N$,
$$\int_\Omega |f_n(z) - f_m(z)|^p dx dy \le \epsilon^p.$$

Let $m \to \infty$. Fatou's lemma gives
$$\int_\Omega \liminf_{m \to \infty} |f_n(z) - f_m(z)|^p dx dy \le \liminf_{m \to \infty} \int_\Omega |f_n(z) - f_m(z)|^p dx dy,$$

and so $\|f_n - f\|_p \le \epsilon^p$ for $n \ge N$. Hence $f = f - f_n + f_n$ belongs to $\mathbf{L}_p(\Omega, \mathcal{B}, dx dy)$ in view of Minkovski's inequality. Since ϵ is arbitrary we have
$$\lim_{n \to \infty} \|f - f_n\|_p = 0$$

which ends the proof of the completeness of $\mathcal{B}_p(\Omega)$.

The boundedness of the point evaluations was proved in (4.5.4). $\square$

Solution of Exercise 4.1.21:

(1) It suffices to take a nonzero vector h_{N+1} orthogonal to $h_1, \ldots, h_N$. The vector $x h_{N+1}$ belong to the set (4.1.17) for every $x > 0$.

(2) It suffices to take the unit sphere $\mathbb{S} = \{h \in \mathcal{H} \,;\, \|h\|_{\mathcal{H}} = 1\}$. Since $\mathcal{H}$ is infinite dimensional it has a countable family $(e_n)_{n \in \mathbb{N}}$ of orthonormal vectors (if $\mathcal{H}$ is not separable, this family will not be a Hilbert basis). Then

$$\sum_{n=1}^{\infty} |\langle h, e_n \rangle_{\mathcal{H}}|^2 \leq \|h\|_{\mathcal{H}}^2, \quad \forall h \in \mathcal{H},$$

and so

$$\lim_{n \to \infty} \langle h, e_n \rangle_{\mathcal{H}} = 0.$$

Thus, the sequence $(e_n)_{n \in \mathbb{N}}$ converges weakly to 0, which does not belong to $\mathbb{S}$.

(3) Let C be closed and convex in $\mathcal{H}$, and let $h \notin \mathcal{H}$. There is $\epsilon > 0$ such that $B(h, \epsilon) \cap C = \emptyset$, and h is an internal point (see Definition 1.3.2) of $B(h, \epsilon)$. By the basic separation theorem (see Theorem 1.3.3) there is a (*a priori* non continuous) non identically vanishing functional on $\mathcal{H}$ and $c \in \mathbb{R}$ such that

$$\begin{cases} \operatorname{Re} f(u) \leq c, & u \in B(h, \epsilon), \\ \operatorname{Re} f(u) \geq c, & u \in C. \end{cases}$$

By Proposition 4.1.23, f can be chosen continuous, and is therefore an inner product by Riesz theorem (see Theorem 4.1.19 for the latter):

$$f(u) = \langle u, h_f \rangle$$

for a uniquely defined $h_f \in \mathcal{H}$. The set

$$\left\{ u \in \mathcal{H} \,;\, |\langle u - h, h_f \rangle| < \frac{c}{2} \right\}$$

is then a weakly open neighborhood of h which does not intersect C, and so $\mathcal{H} \setminus C$ is weakly open and C is weakly closed.

$\square$

Solution of Exercise 4.2.3: Let ι denote the inclusion map from X_1 into X_2, and let $O_2 \subset X_2$ be open. Then $\iota^{-1}(O_2)$ is open in X_1. To conclude, remark that

$$\iota^{-1}(O_2) = X_1 \cap O_2.$$

This set is open in X_1 since τ is continuous.

$\square$

Remark 4.5.1. The preceding exercise holds in fact for general topological spaces.

Solution of Exercise 4.2.7: As mentioned before the exercise, we focus on (2). In view of (1.1.3) it is enough to show that the image of an open ball is open. Let $b_0 \in \mathcal{B}$ and $\epsilon_0 > 0$, and let $b \in B(b_0, \epsilon_0)$. By definition of the quotient norm, we have

$$\|\widetilde{b_0 - b}\|_{\mathcal{M}} < \epsilon_0.$$

Thus $\widetilde{b} \in B(\widetilde{b_0}, \epsilon_0)$, and so $B(\widetilde{b_0}, \epsilon_0) \subset q(B(b_0, \epsilon_0))$, so that the latter is open in the quotient space.

$\square$

4.5. Solutions of the exercises

Solution of Exercise 4.2.13: We follow the discussion after the statement of Exercise 1.5.5. Let C and D be as in that discussion. We have (where we use the resolvent equation (1.7.4) to go from the first line to the second line; see also Question 4.2.10):

$$P^2 = \left(\frac{1}{2\pi i}\right)^2 \int_C \int_D (zI_{\mathcal{B}} - T)^{-1}(wI_{\mathcal{B}} - T)^{-1} dzdw$$

$$= \left(\frac{1}{2\pi i}\right)^2 \int_C \int_D \frac{1}{w-z}\left((zI_{\mathcal{B}} - T)^{-1} - (wI_{\mathcal{B}} - T)^{-1}\right) dzdw$$

$$= \left(\frac{1}{2\pi i}\right)^2 \int_C \int_D \frac{1}{w-z}(zI_{\mathcal{B}} - T)^{-1} dzdw -$$

$$- \left(\frac{1}{2\pi i}\right)^2 \int_C \int_D \frac{1}{w-z}(wI_{\mathcal{B}} - T)^{-1} dzdw.$$

Each of these integrals can be computed using (1.5.3) and we get $P^2 = P$. $\qquad\square$

Solution of Exercise 4.2.17: Denote by B_1 the closed unit ball of $\mathcal{B}_1$. When T is one-to-one it follows from the open mapping theorem (see Theorem 4.2.15) that T^{-1} is continuous, and it follows that the image $T(B_1) = (T^{-1})^{-1}(B_1)$ is closed. Thus $T(B_1)$ is compact since T is compact. Therefore $B_1 = T^{-1}(T(B_1))$ is compact as the continuous image of a compact map. B_1 is thus finite dimensional, and hence $\mathcal{B}_2$ is also finite dimensional since T is one-to-one.

When T is not one-to-one, we adapt the previous argument as follows. Since T is continuous the kernel $\ker T$ is a closed subspace of $\mathcal{B}_1$ and so the quotient space $\mathcal{B}_1/\ker T$ is a Banach space (see Exercise 4.2.7). By definition of the quotient topology, the map

$$\widetilde{T}(\widetilde{b}) = Tb, \quad b \in \mathcal{B}_1$$

is continuous, where $\widetilde{b}$ denotes the equivalence class of $b \in \mathcal{B}_1$. It is also one-to-one. It is onto since T is onto. As in the previous paragraph the open mapping theorem asserts that $\widetilde{T}^{-1}$ is continuous and so

$$\widetilde{T}^{-1}(T(B_1)) = q(B_1) \tag{4.5.5}$$

is compact in $\mathcal{B}_1/\ker T$, where q denotes the quotient map. We now remark that (4.5.5) is the closed unit ball $\widetilde{B}_1$ of the quotient space. Indeed, if $b \in B_1$, then

$$\|q(b)\|_{\mathcal{B}_1/\ker T} = \inf_{u \in \ker T} \|b + u\|_{\mathcal{B}_1} \leq \|b\|_{\mathcal{B}_1} \leq 1.$$

Thus $q(B_1)$ contains the closed unit ball of the quotient space. Conversely, if $b \in \mathcal{B}_1$ is such that $\|q(b)\|_{\mathcal{B}_1/\ker T} \leq 1$. Then there exists a sequence $u_1, u_2, \ldots$ of elements in $\ker T$ such that

$$\|b + u_n\|_{\mathcal{B}_1} \leq 1 + \frac{1}{n}, \quad n = 1, 2, \ldots$$

Thus the sequence $u_1, u_2, \ldots$ is bounded in norm, and has a weakly-$*$ convergent subsequence, with weak-$*$ limit u (see Exercise 4.1.11). Since T is continuous, and in particular weak-$*$ continuous, we have $u \in \ker T$. The element $b + u$ has norm less than or equal to 1 (as is seen by writing $b + u = b + u_n + u - u_n$ and using the Hahn–Banach theorem to compute $\|b + u\|$) and so $q(b) = q(b + u) \in q(B_1)$, that is, the closed unit ball of the quotient space contains $q(B_1)$. Thus $\overline{B_1} = q(B_1)$. We can now proceed as in the previous paragraph to show that $\mathcal{B}_1 / \ker T$, and so $\mathcal{B}_2$, is finite dimensional.

$\square$

Solution of Exercise 4.2.18: (See also [200, p. 9]).

(1) By assumption, the function $t \mapsto \|X(t)b\|$ is continuous for every $b \in \mathcal{B}$. It is therefore bounded since $[0, 1]$ is compact. There exists $k_b < \infty$ such that

$$\|X(t)b\| \le k_b, \quad \forall t \in [0, 1].$$

The result follows from the uniform boundedness principle.

(2) The lower semi-continuity of the function

$$t \mapsto \|X(t)\| = \sup_{\substack{b \in \mathcal{B} \\ \|b\| = 1}} \|X(t)b\|$$

follows from item (4) in Exercise 3.4.9 since, by hypothesis, the functions $t \mapsto \|X(t)b\|$ are continuous.

$\square$

Solution of Exercise 4.2.19: We follow the arguments in [114, p. 87].

(1) Let $g \in \mathcal{H}$. By hypothesis the map

$$h \mapsto \langle Ah, g \rangle$$

is continuous. By Riesz' representation theorem (see Theorem 4.1.19 above) we have

$$\langle Ah, g \rangle = \langle h, A^* g \rangle.$$

So

$$\|A^* g\| = \sup_{\substack{h \in \mathcal{H} \\ \|h\| \le 1}} |\langle Ah, g \rangle| = \sup_{\substack{h \in \mathcal{H} \\ \|h\| \le 1}} |\langle h, A^* g \rangle|.$$

Each of the maps $g \mapsto |\langle Ah, g \rangle|$ is continuous and so the map (4.2.2) is lower semi-continuous. We can now apply Exercise 4.1.14 to conclude that (4.2.2) is continuous. So the map $g \mapsto \|A^* g\|$ is continuous and so A^* is bounded, and so continuous (see Exercise 4.2.2). It follows that $A = A^{**}$ is continuous.

(2) Let $(h_n)_{n \in \mathbb{N}}$ be a sequence of elements in $\mathcal{H}$ which tends weakly to 0. Then, for every $h \in \mathbf{H}$,

$$\lim_{n \to \infty} \langle Ah, h_n \rangle = 0,$$

and so, since $A = A^*$,

$$\lim_{n \to \infty} \langle h, Ah_n \rangle = 0.$$

Thus $(Ah_n)_{n \in \mathbb{N}}$ tends weakly to 0, and A is weakly continuous. One can now apply item (1) to conclude the proof. $\qquad\square$

Solution of Exercise 4.2.20: One direction is clear. If A is bounded (and so, equivalently, continuous), the maps

$$h \mapsto \langle Ah, g \rangle \tag{4.5.6}$$

are continuous and so the adjoint exists and is everywhere defined. The equation

$$\langle Ah, g \rangle = \langle h, A^* g \rangle$$

leads then easily to the fact that A^* is bounded and has the same norm as A.

To prove the converse it suffices to apply the uniform boundedness principle (see Theorem 4.2.5) to the maps

$$T_g(f) = \langle Af, g \rangle = \langle f, A^* g \rangle,$$

where $\|g\| \le 1$. We have

$$\sup_{\|g\| \le 1} \|T_g(f)\| = \|Af\| < \infty, \quad \forall f \in \mathcal{H}.$$

It follows that $\sup_{\|g\| \le 1} \|T_g\| < \infty$. This ends the proof since $\|T_g\| = \|A^* g\|$. We leave the last claim to the reader. $\qquad\square$

Solution of Exercise 4.2.23: We just outline the proof. A set of the form

$$\{T \in \mathbf{L}(\mathcal{H}) \text{ such that } |\langle Tf, g \rangle_{\mathcal{H}}| < \epsilon\} \tag{4.5.7}$$

where $f, g \in \mathcal{H} \setminus \{0\}$ contains elements such that $\|Tf\|$ is arbitrarily large (take for instance $T_n = nT_0$, where $g \in \ker T_0^*$ and $T_0 f \ne 0$), and so cannot be included in a set of the form

$$\{M \in \mathbf{L}(\mathcal{H}) \text{ such that } \|Mh\|_{\mathcal{H}} < \eta\} \tag{4.5.8}$$

where $h \in \mathcal{H}$ and $\eta > 0$ are pre-assigned. The same conclusion holds, in the infinite-dimensional case, for finite intersections of sets of the form (4.5.7).

Similarly, a set of the form (4.5.8) contains elements such that $\|Tf\|$ is arbitrarily large (take for instance $T_n = nT_0$, where $f \in \ker T_0$ and $Tf \ne 0$), and so cannot be included in a set of the form

$$\{N \in \mathbf{L}(\mathcal{H}) \text{ such that } \|N\| < \nu\}$$

where $\nu > 0$ is pre-assigned.

This shows inclusions. Strict inclusions will be true only in the infinite-dimensional case. $\qquad\square$

Solution of Exercise 4.2.25: Let $e_0, e_1, \ldots$ be an orthonormal basis of $\mathcal{H}$, and set

$$T_n(f) = \langle f, e_n \rangle_{\mathcal{H}} e_0.$$

Then, $\|T_n\| = 1$ for all n, and

$$T_n^*(f) = \langle f, e_0 \rangle_{\mathcal{H}} e_n.$$

The sequence $(T_n)_{n \in \mathbb{N}_0}$ tends strongly (and hence weakly) to 0 since the Fourier coefficient $\langle f, e_n \rangle_{\mathcal{H}}$ tends to 0. For the same reason, the sequence of adjoints tends weakly (but not strongly) to 0. But

$$T_n T_n^* f = \langle f, e_0 \rangle_{\mathcal{H}} e_0$$

does not go to 0 weakly. So the product is not jointly weakly continuous when the original space is endowed with the strong operator topology.

We now prove that the product is not jointly continuous neither in the strong topology nor in the weak topology, following the hints given at the end of the exercise. See [116, Exercise 2, pp. 45–46] and [67, Exercise 7, p. 116] from which these hints are taken).

(a) Take $m \in \mathcal{H}$ of norm 1 and orthogonal to $h_1, \ldots, h_N$, and define an operator T_μ by

$$T_\mu h_j = 0, \quad j = 1, \ldots, N \quad \text{and} \quad T_\mu m = \mu m, \quad \mu > 0.$$

The operator T_μ has a bounded extension to all of $\mathcal{H}$, which we still denote by T_μ. By construction $T_\mu \in \mathcal{W}$ and $\|T_\mu\| \geq \mu$. Thus, (4.2.4) is in force.

(b) Assume first that h does not belong to the linear span of the vectors $h_1, \ldots, h_N$. By direct construction (or using the Hahn–Banach theorem) we define a bounded operator T_1 on $\mathcal{H}$ of norm less than 1 and such that

$$T_1 h_j = 0, \quad j = 1, \ldots, N \quad \text{and} \quad T_1 h = m.$$

By construction $T_1 \in \mathcal{W}$. Assume now that h belongs to the linear span of $h_1, \ldots, h_N$, say $h = \sum_{j=1}^{J} z_j h_{i_j}$, where $h_{i_1}, \ldots, h_{i_J}$ are linearly independent and $z_1, \ldots, z_J \in \mathbb{C}$. Once more using the Hahn–Banach theorem we define a bounded operator T_1 of norm less than 1 on $\mathcal{H}$ such that

$$T_1 h_j = 0, \quad j \notin \{i_1, \ldots, i_J\} \quad \text{and} \quad T_1 h = m.$$

(c) Such an operator exists in view of (a).

(d) By definition of the operator norm there is $k_0 \in \mathcal{H}$ such that

$$|\langle T_2 T_1 h, k_0 \rangle| > 1.$$

We build a contraction in $\mathcal{H}$ such that $T_3^* k_0 = k$.

The previous steps show the lack of continuity of the product when the domain is endowed with the strong operator topology and the range with either the strong operator topology or the weak operator Indeed, starting from the weak operator topology neighborhood $\mathcal{U}$ of the origin, any strong operator topology neighborhood $\mathcal{V}$ of the origin contains operators $T_4 = T_3T_2$ and T_1 such that $T_4T_1 \notin \mathcal{U}$. This shows that multiplication is not continuous when the domain is endowed with the strong operator topology and the range with the weak operator topology. The same conclusion holds when the range is endowed with the strong operator topology since the latter is finer than the weak operator topology. The same conclusion can be also be obtained directly from (c). $\qquad\square$

Remark 4.5.2. The previous example shows in particular that the map $T \mapsto T^*$ is not continuous in the strong topology.

Solution of Exercise 4.2.26:

(1) Let $\mathcal{B}_1$ denote the closed unit ball of $\mathbf{L}(\mathcal{H})$. Let $h \neq 0 \in \mathcal{H}$ and consider the open covering of $\mathcal{B}_1$ by the sets

$$\{T \in \mathcal{B}_1 \,;\, \|(T - V)h\| < 1\}, \quad V \in \mathcal{B}_1.$$

Assume that there exists a finite covering of $\mathcal{B}_1$,

$$\{T \in \mathcal{B}_1 \,;\, \|(T - V_n)h\| < 1\}, \quad n = 1, 2, \dots, N,$$

by such sets. Let $u \in \mathcal{H}$ be of norm 1 orthogonal to the vectors $V_n h, n = 1, \dots, N$. We define an operator T by

$$Te = \begin{cases} \epsilon u, & \text{if } e = u, \\ 0 & \text{if } \langle e, u \rangle = 0, \end{cases}$$

where $\epsilon \in (0, 1]$ satisfies

$$\epsilon \geq \min_{n=1,2,\dots,N} \|V_n h\|.$$

Then, $T \in \mathcal{B}_1$. On the other hand,

$$\|(T - V_n)h\|^2 = \epsilon^2 + \|V_n h\|^2 \geq 1,$$

and T does not belong to the given sub-covering and so $\mathcal{B}_1$ is not compact in the strong topology. That $\mathcal{B}_1$ is compact in the weak operator topology is a special case of item (2) of Exercise 4.1.11.

(2) That $\mathcal{B}_1$ is metrizable in the weak operator topology also follows from Exercise 4.1.11. We here give a more detailed argument. We need to show that this topology can be defined by a countable family of semi-norms on this set. For $u, v \in \mathcal{H}$ and $\epsilon \in (0, 1)$ we define:

$$V(u, v, \epsilon) = \{T \in \mathcal{B}_1 \,;\, |\langle Tu, v \rangle| < \epsilon\}.$$

Let $M = (u_n)_{n\in\mathbb{N}}$ denote a dense and countable subset of $\mathcal{H}$. We show that for every choice of $(u, v, \epsilon) \in \mathcal{H} \times \mathcal{H} \times (0,1)$ there exist $(u_1, v_1) \in M \times M$ and $\epsilon_1, \epsilon_2 \in (0,1)$ such that

$$V(u_1, v_1, \epsilon_1) \subset V(u, v, \epsilon) \subset V(u_1, v_1, \epsilon_2).$$

Indeed, let $\eta \in (0,1)$ and let $u_1, v_1 \in M$ be such that

$$\|u - u_1\| < \eta \quad \text{and} \quad \|v - v_1\| < \eta.$$

Then

$$\begin{aligned}
|\langle Tu, v\rangle| &\le |\langle T(u - u_1), v - v_1\rangle| + |\langle T(u - u_1), v_1\rangle| \\
&\quad + |\langle Tu_1, v_1\rangle| + |\langle Tu_1, v - v_1\rangle| \\
&\le |\langle Tu_1, v_1\rangle| + \eta^2 + \eta(\|u\| + \|v\| + 2),
\end{aligned}$$

where we have used the fact that $\|T\| \le 1$ and

$$\|u_1\| \le \|u\| + 1 \quad \text{and} \quad \|v_1\| \le \|v\| + 1.$$

Thus (and here we write out the dependence of u_1 and v_1 on η) we have, for $T \in V(u_1(\eta), v_1(\eta), \eta)$,

$$\begin{aligned}
|\langle Tu, v\rangle| &\le |\langle Tu_1, v_1\rangle| + \eta^2 + \eta(\|u\| + \|v\| + 2) \\
&\le \eta + \eta^2 + \eta(\|u\| + \|v\| + 2) \\
&= \eta^2 + \eta(\|u\| + \|v\| + 3),
\end{aligned}$$

and so:

$$V(u_1(\eta), v_1(\eta), \eta) \subset V(u, v, \eta^2 + \eta(\|u\| + \|v\| + 3)).$$

By symmetry

$$V(u, v, \eta) \subset V(u_1(\eta), v_1(\eta), \eta^2 + \eta(\|u\| + \|v\| + 3)).$$

Let $\varphi(\eta) = \eta^2 + \eta(\|u\| + \|v\| + 3)$ defined on $(0, k)$, where $\varphi(k) = 1$. We have

$$V(u_1(\eta), v_1(\eta), \varphi^{-1}(\eta)) \subset V(u, v, \eta) \subset V(u_1(\eta), v_1(\eta), \varphi(\eta)).$$

(3) The claim follows from the uniform boundedness theorem. See [170, Solution of Exercise 113, p. 250].

$\square$

Remark 4.5.3. The set $\mathcal{B}_1$ is also metrizable in the strong operator topology.

4.5. Solutions of the exercises

Solution of Exercise 4.2.29: The fact that Γ is Hermitian is a direct consequence of the polarization formula (1.3.8): For $f, g \in \mathcal{H}$ we have:

$$\langle \Gamma h, g \rangle_{\mathcal{H}} = \frac{1}{4} \sum_{k=0}^{3} i^k \langle \Gamma(h + i^k g), h + i^k g \rangle_{\mathcal{H}}$$

$$= \frac{1}{4} \sum_{k=0}^{3} i^k \overline{\langle \Gamma(h + i^k g), h + i^k g \rangle_{\mathcal{H}}} \quad \text{(since } \Gamma \text{ is positive)}$$

$$= \frac{1}{4} \sum_{k=0}^{3} i^k \langle h + i^k g, \Gamma(h + i^k g) \rangle_{\mathcal{H}}$$

$$= \frac{1}{4} \sum_{k=0}^{3} i^k \langle \Gamma^*(h + i^k g), h + i^k g \rangle_{\mathcal{H}}$$

$$= \langle \Gamma^* h, g \rangle_{\mathcal{H}}.$$

Now if $n = 2p$ (with $p \in \mathbb{N}$) we have

$$\langle \Gamma^{2p} h, h \rangle_{\mathcal{H}} = \langle \Gamma^p h, \Gamma^p h \rangle_{\mathcal{H}} \geq 0,$$

while

$$\langle \Gamma^{2p+1} h, h \rangle_{\mathcal{H}} = \langle \Gamma(\Gamma^p h), (\Gamma^p h) \rangle_{\mathcal{H}} \geq 0$$

if $n = 2p + 1$. $\qquad\square$

Solution of Exercise 4.2.30: The proof follows the arguments in [6, §33]. In the arguments f and g denote elements of $\mathcal{H}$.

(1) The formula

$$[f, g] = \langle Af, g \rangle$$

defines a possibly degenerate positive Hermitian form, and so the Cauchy–Schwarz inequality holds for it and we can write

$$|\langle Af, g \rangle|^2 \leq \langle Af, f \rangle \langle Ag, g \rangle \leq \langle Af, f \rangle \|A\| \|g\|^2.$$

Setting $g = Af$ we get

$$\|Af\|^4 \leq \langle Af, f \rangle \|A\| \|Af\|^2,$$

and so we obtain (4.2.6) by dividing by $\|Af\|^2$ both sides of this inequality if $Af \neq 0$. Note that (4.2.6) holds trivially if $Af = 0$.

(2) From

$$\lim_{n \to \infty} \langle A_n f, g \rangle = \langle Af, g \rangle,$$

we have in particular, for every $f, g \in \mathcal{H}$,

$$\sup_{n \in \mathbb{N}} |\langle A_n f, g \rangle| < \infty,$$

and the uniform boundedness theorem implies that

$$\sup_{n \in \mathbb{N}} \|A_n f\| < \infty.$$

Yet another application of this theorem leads to $\sup_{n \in \mathbb{N}} \|A_n\| < \infty$. By hypothesis, the operator $A - A_n$ is positive. We can write (with $K = \sup_{n \in \mathbb{N}} \|A_n\| < \infty$)

$$\|(A - A_n)f\| \le \sqrt{\|A - A_n\|} \sqrt{\langle (A - A_n)f, f \rangle} \le \sqrt{\|A\| + K} \sqrt{\langle (A - A_n)f, f \rangle},$$

and hence the result since

$$\lim_{n \to \infty} \langle (A - A_n)f, f \rangle = 0, \quad \forall f \in \mathcal{H}.$$

(3) We follow [6, pp. 98–99]. The polarization formula (1.3.8) gives that the limit

$$\lim_{n \to \infty} \langle A_n f, g \rangle$$

exists for all $f, g \in \mathcal{H}$. The uniform boundedness theorem applied to the maps

$$g \mapsto \langle g, A_n f \rangle$$

implies that $\sup_{n \in \mathbb{N}} \|A_n f\| < \infty$. That same theorem now applied to the maps

$$f \mapsto \|A_n f\|$$

implies that $K = \sup_{n \in \mathbb{N}} \|A_n\| < \infty$. Applying inequality (4.2.6) to $A_m - A_n$ with $m \ge n$ we have

$$
\begin{aligned}
\|A_m f - A_n f\| &\le \sqrt{\|A_m - A_n\|} \cdot \sqrt{\langle A_m f - A_n f, f \rangle} \\
&\le \sqrt{\|A_m\|} \cdot \sqrt{\langle A_m f - A_n f, f \rangle} \\
&\le \sqrt{K} \sqrt{\langle A_m f - A_n f, f \rangle},
\end{aligned}
$$

which shows that A_m converges strongly to some everywhere defined operator A. It remains to show that A is linear and bounded. Linearity is clear from the weak convergence, and A is positive and hence self-adjoint from (4.2.7). By Exercise 4.2.19, the operator A is then bounded. $\qquad\square$

Solution of Exercise 4.2.31: Recall (see for instance [CAPB, Exercise 4.4.4, p. 158]) that $\sqrt{1 - z}$ corresponds to setting $\alpha = \frac{1}{2}$ and replacing z by $-z$ in the power series expansion

$$f_\alpha(z) = 1 + \sum_{n=1}^{\infty} \frac{\alpha(\alpha - 1) \cdots (\alpha - n + 1)}{n!} z^n. \tag{4.5.9}$$

It follows that

$$a_n = \frac{(-1)^{n-1}}{n!} \prod_{j=0}^{n-1} \left(\frac{1}{2} - j \right) > 0.$$

4.5. Solutions of the exercises

From the equality $(f_{\frac{1}{2}}(z))^2 = 1 - z$ and comparing coefficients in the power series expansions, we obtain

$$2a_n = \sum_{m=1}^{n} a_m a_{n-m}, \quad n = 2, 3, \ldots \tag{4.5.10}$$

Furthermore $\sum_{n=1}^{\infty} a_n = 1$.

The operators D_N are selfadjoint in view of Exercise 4.2.29. They are positive since $I - \Gamma$ is a contraction and

$$\sum_{n=1}^{N} \langle a_n \langle (I - \Gamma)^n h, h \rangle_{\mathcal{H}} \le \sum_{n=1}^{N} a_n \le 1$$

for $h \in \mathcal{H}$ of unit norm, and converge in norm since

$$\|D_{N+P} - D_N\| = \left\| \sum_{n=N+1}^{N+P} a_n (I - \Gamma)^n \right\| \le \sum_{n=N+1}^{N+P} a_n, \quad N, P \in \mathbb{N}.$$

Taking into account equalities (4.5.10), (and since $2a_1 = 1$) we have

$$\Gamma - D_N^2 = \Gamma - (I - 2\sum_{n=1}^{N} a_n (I - \Gamma)^n + \sum_{u=1}^{N} \sum_{v=1}^{N} a_u a_v (I - \Gamma)^{u+v})$$

$$= \Gamma - (I - (I - \Gamma)) + 2\sum_{n=2}^{N} a_n (I - \Gamma)^n - \sum_{n=2}^{N} \left(\sum_{m=1}^{n} a_m a_{n-m} \right)(I - \Gamma)^n$$

$$- \sum_{n=N+1}^{2N} \left(\sum_{\substack{u+v=n \\ 1 \le u, v \le N}} a_u a_v \right)(I - \Gamma)^n$$

$$= - \sum_{n=N+1}^{2N} \left(\sum_{\substack{u+v=n \\ 1 \le u, v \le N}} a_u a_v \right)(I - \Gamma)^n.$$

From (4.5.10) (and since $a_1 = \frac{1}{2}$) we have

$$\sum_{\substack{u+v=n \\ 1 \le u, v \le N}} a_u a_v \le 2a_n \le 1,$$

and we have

$$\|\Gamma - D_N^2\| \le \sum_{n=N+1}^{\infty} \|I - \Gamma\|^n.$$

This last expression goes to 0 as $N \to \infty$ since the series $\sum_{n=1}^{\infty} \|I - \Gamma\|^n$ converges. Hence $X = \lim_{N \to \infty} D_N$ satisfies $X^2 = \Gamma$. $\square$

Solution of Exercise 4.2.34: It suffices to write

$$\frac{1}{j+k+1} = \int_0^\infty e^{-jx}e^{-kx}e^{-x}dx,$$

and to note that the functions $x \mapsto e^{-jx}$, $j = 0, \ldots, N$, are linearly independent in $\mathbf{L}_2(\mathbb{R}_+, e^{-x}dx)$.

$\square$

Solution of Exercise 4.2.38: We have

$$\langle \Gamma u, u \rangle_{\mathcal{H}} = \|\sqrt{\Gamma}u\|_{\mathcal{H}}^2.$$

Thus $\ker \Gamma \subset \ker \sqrt{\Gamma}$. Conversely, if $\sqrt{\Gamma}u = 0$, then $\Gamma u = \sqrt{\Gamma}(\sqrt{\Gamma}u) = 0$, so that $\ker \Gamma \supset \ker \sqrt{\Gamma}$.

$\square$

Solution of Exercise 4.2.39: Let $\lambda \notin \sigma(p(T))$ (that is, $\lambda \in \rho(p(T))$). There exists a bounded operator B such that

$$B(p(T) - \lambda I) = (p(T) - \lambda I)B = I. \tag{4.5.11}$$

By the fundamental theorem of algebra (see [CAPB, §13.6, p. 470] for an elementary proof not using Liouville's theorem), there exists $u \in \mathbb{C}$ such that $\lambda = p(u)$. Let $p(z) = \sum_{k=0}^n a_k z^k$. Writing (see for instance [CAPB, p. 53])

$$p(T) - p(u) = (T - uI)V = V(T - uI) \tag{4.5.12}$$

where

$$V = \sum_{k=1}^n a_k \left(\sum_{\ell=0}^{k-1} T^k u^{k-1-\ell} \right),$$

we obtain from (4.5.11) and (4.5.12) that

$$BV(T - uI) = (T - uI)VB = I.$$

But $BV = VB$ since

$$BV = BV(T - uI)VB = (BV(T - uI))VB = VB$$

and hence $(T - uI)$ is invertible. Since $\lambda = p(u) \in \rho(p(T))$ we have $u \in p^{-1}(\rho(p(T)))$ and so:

$$p^{-1}(\rho(p(T))) \subset \rho(T).$$

But $\rho(p(T)) = \mathbb{C} \setminus \sigma(p(T))$, and by (1.1.2) we have

$$p^{-1}(\rho(p(T))) = \mathbb{C} \setminus p^{-1}(\sigma(p(T))),$$

and so:

$$\mathbb{C} \setminus p^{-1}(\sigma(p(T))) \subset \mathbb{C} \setminus \sigma(T),$$

4.5. *Solutions of the exercises*

and

$$\sigma(T) \subset p^{-1}(\sigma(p(T))).$$

By (1.1.5) with $U = \sigma(T)$, we have

$$p(\sigma(T)) \subset \sigma(p(T)). \tag{4.5.13}$$

To prove the converse inclusion, let $\lambda \in \mathbb{C} \setminus p(\sigma(T))$ and let u be such that $p(u) = \lambda$. Then $p(u) \in \mathbb{C} \setminus p(\sigma(T))$. Thus $p(u) \notin p(\sigma(T))$ and $u \notin \sigma(T)$. So u belongs to $\rho(T)$. Let $u_1, \ldots, u_N$ (counting multiplicities) be such that $p(u) = \lambda$. Then

$$p(x) - \lambda = c \cdot \prod_{k=1}^{N} (x - u_k)$$

for some $c \in \mathbb{C} \setminus \{0\}$. Thus,

$$p(T) - \lambda = c \cdot \prod_{k=1}^{N} (T - u_k I)$$

is invertible since every operator $T - u_k I$ is invertible. So $\lambda \in \rho(p(T))$. So we have

$$\mathbb{C} \setminus p(\sigma(T)) \subset \rho(p(T)), \quad \text{that is,} \quad \sigma(p(T)) \subset p(\sigma(T)),$$

which, together with (4.5.13), ends the proof. $\qquad\square$

Solution of Exercise 4.2.40: We first check that (4.2.11) defines a norm, induced by the inner product (4.2.12). We have that

$$\|(I - \pi)u\|_{\mathcal{H}} = 0$$

if and only if $u \in \ker \Gamma$, that is, if and only if $\Gamma u = 0$. This is equivalent to $\sqrt{\Gamma} u = 0$ in view of Exercise 4.2.38. Thus, condition (3) in Definition 4.1.1 holds. In particular, (4.2.11) is well defined, and is the same for $u, v \in \mathcal{H}$ such that $\sqrt{\Gamma} u = \sqrt{\Gamma} v$. The other conditions in Definition 4.1.1 follow from the fact that $\| \cdot \|_{\mathcal{H}}$ is a norm. That (4.2.12) defines an inner product follows from the fact that the norm $\| \cdot \|_{\mathcal{H}}$ is itself induced from an inner product.

We now check the completeness of ran $\sqrt{\Gamma}$ with the associated norm $\| \cdot \|_{\Gamma}$. Let $(\sqrt{\Gamma} u_n)_{n \in \mathbb{N}}$ be a Cauchy sequence in this norm. By definition, $((I - \pi)u_n)_{n \in \mathbb{N}}$ is a Cauchy sequence in $\mathcal{H}$. Since $\mathcal{H}$ is complete, there is $v \in \mathcal{H}$ such that

$$\lim_{n \to \infty} (I - \pi)u_n = v$$

in the topology of $\mathcal{H}$. By continuity of $(I - \pi)$ we have

$$(I - \pi)v = (I - \pi) \lim_{n \to \infty} (I - \pi)u_n = \lim_{n \to \infty} (I - \pi)(I - \pi)u_n = \lim_{n \to \infty} (I - \pi)u_n = v.$$

Hence

$$\lim_{n\to\infty} \|\sqrt{\Gamma}u_n - \sqrt{\Gamma}v\| = \lim_{n\to\infty} \|(I - \pi)u_n - (I - \pi)v\|_{\mathcal{H}}$$
$$= \lim_{n\to\infty} \|(I - \pi)u_n - v\|_{\mathcal{H}} = 0,$$

which proves the completeness of ran $\sqrt{\Gamma}$ in the norm $\| \cdot \|_{\Gamma}$.   □

Solution of Exercise 4.2.41: The proof proceeds in a number of steps which we will not discuss in detail:

STEP 1: *Let i_1 denote the injection map from $\mathcal{H}_1$ into $\mathcal{H}$. Then $i_1(\mathcal{H}_1)$ endowed with the norm $\|i_1(h_1)\| = \|h_1\|_{\mathcal{H}_1}$ is equal to ran $\sqrt{i_1 i_1^*}$.*

STEP 2: *The space ran $\sqrt{I_{\mathcal{H}} - i_1 i_1^*}$ answers the requirements of the exercise.*

STEP 3: *There is only one space which answers the requirements of the exercise.*

□

Solution of Exercise 4.2.44: Let $h, k \in \mathcal{H}$. We have

$$\langle M_k^* h, 1 \rangle_{\mathbb{C}} = \langle h, M_k 1 \rangle_{\mathcal{H}}$$
$$= \langle h, k \rangle_{\mathcal{H}},$$

so that

$$M_k^* h = \langle h, k \rangle_{\mathcal{H}}.$$

This last equation can also be rewritten as (4.2.16).   □

Solution of Exercise 4.2.48: Writing

$$\begin{pmatrix} \frac{1}{2}(I_{\mathcal{H}_1} + T) & C^* \\ CT & CC^* \end{pmatrix} = \begin{pmatrix} I_{\mathcal{H}_1} & 0 \\ 0 & C \end{pmatrix} \begin{pmatrix} \frac{1}{2}(I_{\mathcal{H}_1} + T) & I_{\mathcal{H}_1} \\ T & I_{\mathcal{H}_1} \end{pmatrix} \begin{pmatrix} I_{\mathcal{H}_1} & 0 \\ 0 & C^* \end{pmatrix},$$

we see that it is enough to check that the operator

$$\begin{pmatrix} \frac{1}{2}(I_{\mathcal{H}_1} + T) & I_{\mathcal{H}_1} \\ T & I_{\mathcal{H}_1} \end{pmatrix}$$

has a positive real part. But

$$\frac{1}{2}\left\{ \begin{pmatrix} \frac{1}{2}(I_{\mathcal{H}_1} + T) & I_{\mathcal{H}_1} \\ T & I_{\mathcal{H}_1} \end{pmatrix} + \begin{pmatrix} \frac{1}{2}(I_{\mathcal{H}_1} + T) & I_{\mathcal{H}_1} \\ T & I_{\mathcal{H}_1} \end{pmatrix}^* \right\}$$
$$= \begin{pmatrix} \frac{1}{4}(2I_{\mathcal{H}_1} + T + T^*) & \frac{1}{2}(I_{\mathcal{H}_1} + T^*) \\ \frac{1}{2}(I_{\mathcal{H}_1} + T) & I_{\mathcal{H}_1} \end{pmatrix}$$
$$= \begin{pmatrix} \frac{1}{2}(I_{\mathcal{H}_1} + T^*) \\ I_{\mathcal{H}_1} \end{pmatrix} \begin{pmatrix} \frac{1}{2}(I_{\mathcal{H}_1} + T^*) \\ I_{\mathcal{H}_1} \end{pmatrix}^* + \begin{pmatrix} \spadesuit & 0 \\ 0 & 0 \end{pmatrix},$$

where

$$\spadesuit = \frac{1}{4}\left(2I_{\mathcal{H}_1} + T + T^* - (I_{\mathcal{H}_1} + T^*)(I_{\mathcal{H}_1} + T)\right) = \frac{1}{4}\left(I_{\mathcal{H}_1} - T^*T\right) \geq 0.$$

The result follows. □

Solution of Exercise 4.2.50:

(1) It suffices to consider the space

$$\{0\} \times \mathcal{H} = \{(0, h) \in \mathcal{H} \times \mathcal{H}, \, h \in \mathcal{H}\},$$

where $\mathcal{H}$ is some Hilbert space different from $\{0\}$.

(2) Let $R \subset \mathcal{H} \times \mathcal{G}$ be a linear contractive densely defined relation, where $\mathcal{H}$ and $\mathcal{G}$ are two Hilbert spaces. In view of (4.2.19),

$$(0, g) \in R \implies g = 0$$

and we can define a (possibly non linear) densely defined operator T by $Tf = g$ for $(f, g) \in R$. Since $(\lambda_1 f_1 + \lambda_2 f_2, \lambda_1 g_1 + \lambda_2 g_2) \in R$ for every $\lambda_1, \lambda_2 \in \mathbb{C}$ and $(f_1, g_1), (f_2, g_2) \in R$ we have

$$T(\lambda_1 f_1 + \lambda_2 f_2) = \lambda_1 g_1 + \lambda_2 g_2 = \lambda_1 T f_1 + \lambda_2 T f_2,$$

and thus T is linear, with dense domain.

It remains to show that T extends to an everywhere defined contraction. This is done by setting $Th = \lim_{n \to \infty} T h_n$, where for $h \in \mathcal{H}$ the elements $h_1, h_2 \ldots$ in the domain of T are such that $\lim_{n \to \infty} h_n = h$. It is easily shown that the limit does not depend on the given sequence and that the operator so extended is a linear contraction. □

Solution of Exercise 4.3.4: (see [6, pp. 156–157]) We consider only the operator Q and leave the case of P to the reader. Set $\mathrm{Dom}\,(Q)$ to be the space of absolutely continuous functions in $\mathbf{L}_2(\mathbb{R}, \mathcal{B}, dx)$ with derivative in $\mathbf{L}_2(\mathbb{R}, \mathcal{B}, dx)$. Let $f \in \mathrm{Dom}\,(Q)$ and let $a, b \in \mathbb{R}$. Integrating by parts we have

$$\int_a^b f'(t)\overline{f(t)}dt = |f(b)|^2 - |f(a)|^2 - \int_a^b f(t)\overline{f'(t)}dt.$$

By the Cauchy–Schwarz inequality both $f'\overline{f}$ and its conjugate $f\overline{f'}$ belong to $\mathbf{L}_1(\mathbb{R}, \mathcal{B}, dx)$. Thus letting $b \to +\infty$ we see that $\lim_{b \to +\infty} |f(b)|$ exists, and similarly $\lim_{a \to -\infty} |f(a)|$ exists. Since f is in $\mathbf{L}_2(\mathbb{R}, \mathcal{B}, dx)$, these limits are equal to 0 and we have

$$\int_{\mathbb{R}} f'(t)\overline{f(t)}dt = -\int_{\mathbb{R}} f(t)\overline{f'(t)}dt, \tag{4.5.14}$$

and so Q is Hermitian on Dom (Q). We now show that Q is self-adjoint. Let $g \in \mathrm{Dom}\,(Q^*)$ and let $h = Q^*g$. We have

$$\int_{\mathbb{R}} if'(t)\overline{g(t)}dt = \int_{\mathbb{R}} f(t)\overline{h(t)}dt, \quad \forall f \in \mathrm{Dom}\,(Q). \tag{4.5.15}$$

Take now f to be with compact support (say inside $[a,b]$ for some real numbers a and b, with $a < b$). Since $f(a) = f(b) = 0$, integrating by part the right-hand side of (4.5.15) leads to

$$\int_a^b f(t)\overline{h(t)}dt = \left[f(t)\left(\int_a^t h(u)du + K\right)\right]_{t=a}^{t=b} - \int_a^b f'(t)\left(\int_a^t h(u)du + K\right)dt$$

$$= -\int_a^b f'(t)\left(\int_a^t h(u)du + K\right)dt$$

where K is an arbitrary constant. Hence (4.5.15) becomes

$$\int_a^b if'(t)\overline{g(t)}dt = -\int_a^b f'(t)\left(\int_a^t h(u)du + K\right)dt \tag{4.5.16}$$

for all such functions f, and so

$$\int_a^b f'(t)\left(-ig(t) + \int_a^t h(u)du + K\right)dt = 0, \quad \forall K \in \mathbb{C}. \tag{4.5.17}$$

Take now

$$f(t) = \left(\int_a^t \left(-ig(v) + \int_a^v h(u)du\right)dv\right)$$
$$- (t-a)\frac{\int_a^b \left(-ig(v) + \int_a^v h(u)du\right)dv}{b-a}, \quad t \in [a,b],$$

and

$$K = K_0 = -\frac{\int_a^b \left(-ig(v) + \int_a^v h(u)du\right)dv}{b-a}.$$

Then, $f(a) = f(b) = 0$ and plugging this f in (4.5.17) gives:

$$\int_a^b |-ig(t) + \int_a^t h(u)du + K_0|^2 dt = 0.$$

Hence,

$$-ig(t) + \int_a^t h(u)du + K_0 = 0, \quad \forall t \in [a,b].$$

Thus g is absolutely continuous, and $h = ig'$ (and $K_0 = ig(a)$). This ends the proof since the interval $[a,b]$ is arbitrary. $\qquad\square$

4.5. Solutions of the exercises

Remark 4.5.4. With f as above, set $n(t) = -ig(t) + \int_a^t h(u)du$, so that

$$f'(t) = n(t) - K_0 \quad \text{and} \quad K_0 = \frac{\int_a^b n(v)dv}{b-a}.$$

If we set $K = 0$ in (4.5.16) we obtain

$$\int_a^b |n(t)|^2 dt = \frac{1}{b-a} \left| \int_a^b n(t)dt \right|^2.$$

The equality case in the Cauchy–Schwarz inequality leads to the conclusion that n is a constant.

Solution of Exercise 4.3.5: In the finite-dimensional case P and Q can be seen as matrices in $\mathbb{C}^{N \times N}$ for some $N \in \mathbb{N}$, and the result follows from the fact that

$$\mathrm{Tr}\,(QP) = \mathrm{Tr}\,(PQ),$$

so that (4.3.2) would imply

$$0 = \mathrm{Tr}\,(QP - PQ) = \mathrm{Tr}\,I_N = N,$$

where N is the dimension of the space where the operators P and Q act. This argument will not work in the case of a general Banach space, and one can proceed as follows: Assume that there exist bounded linear operators P and Q in the Banach space $\mathscr{B}$ (with norm $\|\cdot\|$) such that (4.3.2) holds. Then an easy induction shows that

$$Q^{n+1}P - PQ^{n+1} = (n+1)Q^n, \quad n = 0, 1, \ldots \tag{4.5.18}$$

and so,

$$(n+1)\|Q^n\| \le 2\|Q^n\| \cdot \|P\| \cdot \|Q\|, \quad n = 0, 1, \ldots \tag{4.5.19}$$

where we have used $\|Q^{n+1}P\| \le \|Q^n\| \cdot \|P\| \cdot \|Q\|$ and similarly for $\|PQ^{n+1}\|$. Equation (4.5.18) implies in particular that there does not exist $n \in \mathbb{N}$ such that $Q^n \neq 0$ and $Q^{n+1} = 0$, and therefore $Q^n \neq 0$ for all n. Thus (4.5.19) implies that

$$(n+1) \le 2\|P\| \cdot \|Q\|, \quad n = 0, 1, \ldots$$

$\square$

which is impossible.

Remark 4.5.5. We note that the proof shows in fact that equality (4.3.2) is impossible in a normed algebra.

Chapter 5

Locally Convex Topological Vector Spaces

Locally convex topological vector spaces form an important class of topological spaces. We have already seen some examples in the previous chapter. This chapter is motivated in part by the study of the space of functions analytic in a given open set and of its dual, and by the study of spaces of test functions and of distributions (for instance the Schwartz space and the space of tempered distributions). For general references on topological vector spaces we refer to [146, 251, 307], and to the book [202] for counterexamples in the theory.

5.1 Topological vector spaces

Let $\mathcal{V}$ be a vector space (over the complex numbers, or the real numbers, endowed with the usual topology), endowed with a topology. It is called a topological vector space if the maps

$$(\lambda, v) \mapsto \lambda v, \quad \mathbb{C} \times \mathcal{V} \longrightarrow \mathcal{V}, \tag{5.1.1}$$

(or from $\mathbb{R} \times \mathcal{V}$ into $\mathcal{V}$), and

$$(v, w) \mapsto v + w, \quad \mathcal{V} \times \mathcal{V} \longrightarrow \mathcal{V}, \tag{5.1.2}$$

are continuous with respect to this topology, and when the domain spaces are endowed with the product topology.

In a vector space one can define the counterpart of Cauchy sequences, namely Cauchy filters, as follows (see for instance [307, Definition 5.1, p. 37]). For filters see Definition 3.1.18.

Definition 5.1.1. A filter in a topological vector space is called a Cauchy filter if for every neighborhood N of the origin there is a subset M of the filter such that $M - M \subset N$.

We have recalled the notion of convergent filters in Definition 3.1.18 (see [307, p. 10]). A subset of a topological vector space is complete if every Cauchy filter converges. Important for the sequel is the following fact (see, e.g., [307, Proposition 5.3, p. 38]):

Proposition 5.1.2. *A complete subset of a Hausdorff topological vector space is closed.*

The following question is related to the notion of completion. It is of particular importance in the setting of signal processing. See [131] and the book [132, p. 176] of Feintuch and Saeks for more information on this exercise and on the notion of extended resolution space and its applications to feedback systems.

Question 5.1.3. *Let $t \in \mathbb{R}$. The function*

$$\left(\int_{-\infty}^{t} |f(u)|^2 du \right)^{1/2} \tag{5.1.3}$$

is a semi-norm on the Lebesgue space $\mathbf{L}_2(\mathbb{R}, \mathcal{B}, dx)$. Characterize the completion of $\mathbf{L}_2(\mathbb{R}, \mathcal{B}, dx)$ with respect to the topology defined by the family of semi-norms (5.1.3).

Remark 5.1.4. The topology defined by the above family of semi-norms is called the resolution topology, and is in fact metrizable (see [130, Chapter 5]). The extended resolution space is hence a Fréchet space (see Definition 5.2.8 below).

Remark 5.1.5. The following result seems trivial, but its proof takes full advantage of the continuity of the operations in a topological vector space. A (maybe too short) proof is given in [183]. Another proof follows from [307, Theorem 9.1, p. 79] in Treves' book. See Corollary p. 80 there. We here follow the proof given in Narasimhan's book; see [242, p. 41]. In that book, the fact itself is used in the proof of a result needed in the study of the closedness of a certain space of differential forms. See Question 5.3.3. This closedness property in turn is used there in the proof of Serre's duality theorem (see [242, p. 47]).

Exercise 5.1.6. *A finite-dimensional subspace (say $\mathcal{V}$) of a Hausdorff topological vector space (say, $\mathcal{W}$) is closed.*

Hint: (following [242, p. 41]) Define a natural algebraic isomorphism T between $\mathbb{C}^N$, where $N = \dim V$, and $\mathcal{V}$, which is moreover continuous (here enters the hypothesis that $\mathcal{W}$ is a topological vector space). For every $\epsilon > 0$, the image of the sphere of radius ϵ in $\mathbb{C}^N$ is compact, and hence closed, in $\mathcal{W}$. Use this fact to show that T is a topological isomorphism, and to conclude.

Remark 5.1.7. It is proved in [307, Theorem 9.1, p. 79] that any finite-dimensional topological vector space is topologically isomorphic to $\mathbb{C}^N$ for some $N \in \mathbb{N}$, and Exercise 5.1.6 is then presented as a corollary of this result. Here we give a direct proof.

The continuity of the vector space operations forces special properties of the neighborhoods. Recall that absorbing and balanced sets were defined in Definition 4.1.2. We only prove part of the first claim in the next exercise, and send the reader to [307, pp. 21–25] for more details for the other items.

Exercise 5.1.8 (see [307, p. 21, and Proposition 3.1 and Corollary, p. 24]). *Let $\mathcal{V}$ be a topological vector space. Then:*

(1) *Every neighborhood of the origin is absorbing, and contains a balanced neighborhood.*
(2) *Every neighborhood contains a closed neighborhood.*
(3) *There is a basis of neighborhoods of the origin made of closed balanced sets.*

Although convexity is not mentioned in the previous exercise we already give the following definition:

Definition 5.1.9. (see for instance [307, Definition 7.1, p. 58]) Let $\mathcal{V}$ be a topological vector space. An absorbing balanced closed convex set is called a barrel.

One is interested in topological vector spaces in which the origin has a basis of neighborhoods made of barrels (these will be the locally convex topological vector spaces). A smaller and important class is the one for which every barrel is a neighborhood of the origin. These are called barreled spaces. These spaces are particularly important in the context of the Banach–Steinhaus theorem. See [307, Chapter 33].

In connection with Question 4.1.6 we mention (see, e.g., [251, pp. 6–7]).

Theorem 5.1.10. *A locally convex topology can always be defined by a family of semi-norms, namely the semi-norms continuous with respect to this topology.*

Not every vector space endowed with a topology is a topological vector space, as is illustrated by the following counterexample, also taken from Trèves' book; see [307, p. 25].

Question 5.1.11. *Consider the set of formal power series with complex coefficients $\mathbb{C}[[X]]$ and the family of spaces*

$$\mathcal{M}_n = X^n \mathbb{C}[[X]], \quad n = 0, 1, \dots$$

Then:

(1) *The family $\mathcal{M}_n$ defines a basis of neighborhood of the origin for a given topology.*
(2) *The space $\mathbb{C}[[X]]$ is not a topological vector space in this topology.*

Hint: The sets $\mathcal{M}_n$ are not absorbing for $n \geq 1$.

The definition of a topological vector space is too general, in the sense that the topological dual of the space may be reduced to the zero functional.

Normed spaces and Fréchet spaces (see Definition 5.2.8 below) are particular instances of topological vector spaces. The topology of a normed space (resp. of a Fréchet space) is defined by a norm (resp. by a countable family of semi-norms). More generally:

Definition 5.1.12. The topology defined by a family (countable or not) of semi-norms in a given vector space $\mathcal{V}$ is the coarsest topology (that is, the smallest topology) on $\mathcal{V}$ with respect to which all the semi-norms of the given family are continuous.

Definition 5.1.13. A vector space whose topology is Hausdorff and is defined by a family of semi-norms is called locally convex.

Equivalently, a vector space is locally convex if it admits a basis of neighborhoods of the origin which are convex. In fact, it has a basis of neighborhoods consisting of sets which are much more structured than convex sets, namely barrels (see Definition 5.1.9 for the latter).

Not every topological vector space is locally convex, as is illustrated by the following example, taken from [183, Example 2, p. 86]. See also Question 3.1.21.

Exercise 5.1.14. *The space $C[0,1]$ of all continuous real-valued functions defined on $[0,1]$ endowed with the topology defined in Question 3.1.21 is a topological vector space which is not locally convex.*

Hint: Following [183], show that no convex set lies between two sets of the form $V_{\epsilon,\eta}(0)$. To that purpose, consider open subsets I and J of $[0,1]$ such that $J \subsetneq I$ and build functions which are in modulus less than ϵ on $I \setminus J$ and greater than $n\epsilon$ on J, where n is the number of connected components of I.

The following question illustrates the fact that barrels appear in a natural way in the theory of locally convex topological vector spaces.

Question 5.1.15. *Let p be a continuous semi-norm in a topological vector space $\mathcal{V}$. Show that the set*

$$\{v \in \mathcal{V} : p(v) \leq 1\}$$

is a barrel. Conversely, any given barrel is defined as the closed unit ball of a uniquely defined semi-norm. This semi-norm is continuous if and only if the given barrel is a neighborhood of the origin.

In the previous question, the semi-norm associated with the barrel B is called its Minkowski gauge (also called Minkowski functional)

$$p_B(v) = \inf_{t>0} \{t \text{ such that } v \in tB\}. \tag{5.1.4}$$

See for instance [257, p. 128], [277, p. 733] for the latter.

In view of Question 5.1.16 we mention (see Exercise 5.4.2) that the dual of a (non-trivial) locally convex space is never trivial.

Question 5.1.16 (see, e.g., [120, p. 422]). *A convex subset of a locally convex topological space is $\mathcal{V}'$-closed if and only if it is closed.*

Definition 5.1.17. A subset A of a topological vector space is called bounded if for every neighborhood V of the origin there exists a positive number $M > 0$ such that

$$A \subset MV. \qquad (5.1.5)$$

The following exercise exhibits an important difference between normed spaces and general locally convex topological vector spaces (see, e.g., [183, Proposition 2.6.1]). See Exercise 5.2.12 for an illustration.

Exercise 5.1.18. *In a locally convex topological vector space whose topology is defined by an infinite number of norms, neighborhoods of the origin are not bounded.*

Definition 5.1.19. A Montel space is a locally convex topological vector space which is Hausdorff and barreled, and in which every bounded set is relatively compact.

See Remark 5.2.14 in connection with the previous definition.

Definition 5.1.20. A locally convex Hausdorff space $\mathcal{V}$ is a Schwartz space if for every balanced closed convex neighborhood U of 0, there exists a neighborhood V of the origin with image in $\mathcal{V}_U$ precompact (and $\mathcal{V}_U$ defined from the semi-norm associated with U).

5.2 Countably normed spaces and Fréchet spaces

The topologies of Banach and Hilbert spaces are defined by a single norm. We now consider the case of vector spaces whose topology is defined by a countable family of norms or, in general, of semi-norms. See Definition 4.1.1 for the latter.

Exercise 5.2.1. *Let $\mathcal{V}$ be a vector space and let $(p_n)_{n\in\mathbb{N}}$ be a countable family of increasing norms on $\mathcal{V}$:*

$$n \leq m \implies p_n(v) \leq p_m(v), \quad v \in \mathcal{V}.$$

Describe the smallest topology with respect to which all the p_n are continuous. Show that $\mathcal{V}$ endowed with this topology is metrizable and locally convex. When is $\mathcal{V}$ complete?

Completing a given space endowed with a number of norms requires a condition on the norms, called compatibility. The following definition and the ensuing example are taken from [166, p. 13].

Definition 5.2.2. Let $\mathcal{V}$ be a linear vector space. Two norms on $\mathcal{V}$ are called compatible if, given a sequence $(v_n)_{n \in \mathbb{N}}$ which is a Cauchy sequence with respect to the two norms, it converges to 0 with respect to one of the norms if and only if it converges to 0 with respect to the second one.

Such norms are also called pairwise coordinated; see [183, p. 156].

Exercise 5.2.3. *Show that the norms in Exercise 3.9.26 are not compatible.*

Hint: (see [166, p. 13]) It is enough to construct a sequence of functions converging uniformly to 0 but such that the values $f'_n(0)$ all coincide with some given nonzero number.

The following question also follows [166, p. 13]. It exhibits the importance of compatible norms, and arises in the construction of a complete space from a countable set of (say increasing) norms $p_1, p_2, \ldots$ given on a pre-assigned vector space $\mathcal{V}$.

Question 5.2.4. *Let p_1, p_2 be two norms given on the vector space $\mathcal{V}$, and assume that $p_1 \leq p_2$ and that p_1 and p_2 are compatible. Let $\mathcal{V}_1$ and $\mathcal{V}_2$ be the respective completion of $\mathcal{V}$ with respect to p_1 and p_2. Let $(v_n)_{n \in \mathbb{N}}$ be a Cauchy sequence with respect to p_2, with respective limits $\ell_2 \in \mathcal{V}_2$, and $\ell_1 \in \mathcal{V}_1$. Then, the map $\ell_2 \mapsto \ell_1$ is linear, continuous and one-to-one.*

The following example is due to J. Górniak, and is an example of a topological vector space which does not have the factorization property, that is of a topological vector space in which not every positive operator (see Definition 5.5.1 for the latter) from $\mathcal{V}$ into its anti-dual can be factorized via a Hilbert space. See the papers [162, 161, 160] of Górniak and Weron. In the statement recall that a subset K of the integers has density zero if

$$\lim_{n \to \infty} \frac{\mathrm{Card}\,(K \cap \{1, \ldots, n\})}{n} = 0.$$

The density zero hypothesis is not needed in the exercise itself, but in its follow-up which exhibits a continuous operator from $\mathcal{V}$ into its anti-dual with no factorization. See Exercise 5.5.6.

Exercise 5.2.5 (see [160, p. 70]). *Let $\mathcal{V}$ denote the vector space of sequences $z = (z_n)_{n \in \mathbb{N}}$ of complex numbers indexed by the integers, and with at most a finite number of nonzero elements, and let $p_{K,M}$ denote the semi-norm*

$$p_{K,M}(z) = \sum_{n \in K} M_n |z_n|, \tag{5.2.1}$$

where $K \subset \mathbb{N}$ has density zero and where $M = (M_n)_{n \in \mathbb{N}}$ is a sequence of non-negative numbers. Characterize the bounded sets for the topology defined by these semi-norms.

Hint: Let A be a bounded set. Show first that there exists $n_A \in \mathbb{N}$ such that

$$a \in A \implies a_n = 0 \text{ for } n \geq n_A.$$

Remark 5.2.6. We note that a finite sum of semi-norms of the form (5.2.1) is of the same form. Thus for every finite family $q_1, \ldots, q_M$ of semi-norms of this form the semi-norm $q = \sum_{j=1}^{M} q_j$ is a semi-norm of the same family such that

$$q_j(z) \leq q(z), \quad j = 1, \ldots, M.$$

By (for instance) [251, p. 7] a basis of neighborhoods of the topology defined by the semi-norms (5.2.1) is made of the open balls defined by the semi-norms, that is of the sets of the form

$$\{z \in \mathcal{V}; \, p_{K,M}(z) < \epsilon\}$$

where $\epsilon > 0$.

Question 5.2.7. *Let $\mathcal{F}$ be a countably normed space, with topology defined by the semi-norms $(p_n)_{n \in \mathbb{N}}$.*

(1) *Show that $A \subset \mathcal{F}$ is bounded if and only if the following condition hold:*

$$\forall n \in \mathbb{N}, \, \exists M_n > 0, \, \forall f \in A, \, p_n(f) \leq M_n.$$

(2) *Assume that the topology of $\mathcal{F}$ cannot be defined by a single norm. Show that any bounded set is nowhere dense.*

(3) *Find a condition for $\mathcal{F}$ (endowed with the above topology) to be a normed space.*

See for instance [167, p. 56] for the preceding question.

Definition 5.2.8 (see, e.g., [307, Chapter 10, p. 85]). A locally convex topological vector space which is metrizable and complete is called a Fréchet space.

Remark 5.2.9. Let $\mathcal{F}$ be a Fréchet space. Since the space is metrizable, it is enough to check completeness using sequences. Furthermore, assume that d_1 and d_2 are two metrics which define the topology and for which $\mathcal{F}$ is complete. Since a continuous and onto linear map between metrizable (not necessarily locally convex) vector spaces is an isomorphism (see for instance [79, Corollaire 1, p. EVT I.19]), the corresponding metric space structures are isomorphic.

In the definition of a metric space, E need not be a vector space. The case when E is a vector space over the field of complex numbers (and then we use the notation $\mathcal{V}$ rather than E) plays an important role. An important example in this book, $\mathcal{V}$ is the space $\mathcal{H}(\Omega)$ of the functions holomorphic in the open set $\Omega \subset \mathbb{C}$, endowed with the metric

$$d(f,g) = \sum_{n=0}^{\infty} \frac{1}{2^n} \frac{p_{K_n}(f-g)}{1 + p_{K_n}(f-g)}, \quad f, g \in \mathcal{H}(\Omega), \tag{5.2.2}$$

where $K_0 \subset K_1 \subset \cdots$ is an increasing family of compact sets which cover Ω. We will also sometimes consider the space $\mathcal{C}(\Omega)$ of functions continuous in the open subset Ω. Note that (5.2.2) is still a metric on $\mathcal{C}(\Omega)$ in view of (4.1.5). This topology is independent of the given choice of sequence K_n and will be the topology associated with $\mathcal{C}(\Omega)$ and $\mathcal{H}(\Omega)$. In the next exercise we illustrate this point for $\Omega = \mathbb{D}$, the open unit disk, and $K_n = \{z \in \mathbb{C} \; ; |z| \le r_n\}$, where $(r_n)_{n \in \mathbb{N}_0}$ is a strictly increasing sequence of positive numbers in $(0, 1)$ with limit 1. The result holds for every open (not necessarily connected) set Ω, when one considers an increasing sequence of compact sets (with non-empty interiors) covering Ω.

Exercise 5.2.10. *Let $(r_n)_{n \in \mathbb{N}_0}$ be a sequence of numbers in $(0, 1)$ converging to 1, and let $d_n(f, g) = \max_{|z| \le r_n} |f(z) - g(z)|$, where $f, g \in \mathcal{C}(\mathbb{D})$. Let*

$$d(f, g) = \sum_{n=0}^{\infty} \frac{1}{2^n} \frac{d_n(f, g)}{1 + d_n(f, g)}, \quad f, g \in \mathcal{C}(\mathbb{D}).$$

Show that the sequence $(f_p)_{p \in \mathbb{N}_0}$ converges uniformly on compact subsets of $\mathbb{D}$ to the function f if and only if

$$\lim_{p \to \infty} d(f_p, f) = 0. \tag{5.2.3}$$

Another important vector space whose topology is defined by a countable family of norms is:

Definition 5.2.11. The Schwartz space is the space of C^∞ functions f such that, for all $p, q \in \mathbb{N}_0$,

$$\lim_{x \to \pm\infty} x^p f^{(q)}(x) = 0,$$

endowed with the norms

$$\|f\|_{p,q} = \max_{x \in \mathbb{R}} |x^p f^{(q)}(x)|.$$

We will denote by $\mathscr{S}$ the space of Schwartz functions and by $\mathscr{S}_\mathbb{R}$ the space of real-valued Schwartz functions. The strong dual of $\mathscr{S}_\mathbb{R}$, that is, the space of (real-valued) tempered distributions, will be denoted by $\mathscr{S}_\mathbb{R}'$. See Exercise 7.2.19 for a related exercise.

We now give an illustration of Exercise 5.1.18 using the Schwartz space.

Exercise 5.2.12. *Show that the set*

$$\left\{ s \in \mathscr{S} \; ; \max_{x \in \mathbb{R}} |s(x)| < 1 \right\}$$

is not bounded in the Schwartz space.

Question 5.2.13. *A Fréchet space is barreled.*

Hint: One uses Theorem 3.9.12 and proceeds as follows. The given space, say X, has a basis of neighborhoods made of barrels. Let B be such a barrel. Each of the sets nB is closed and $X = \cup_{n=1}^{\infty}(nB)$. At this stage one uses the fact that X is a Baire space (see Theorem 3.9.12) to insure that one of the closed sets nB has a non-empty interior. See [307, Proposition 33.2, p. 346] for more details if need be.

Remark 5.2.14. In the setting of Fréchet spaces, Gelfand and Shilov called Montel spaces *perfect*.

Theorem 5.2.15. *Assume that V is a complete countably normed Hilbert space, $V = \cap_{n=1}^{\infty} V_n$, and that for every n there exists $m \geq n$ such that the injection is compact. Then, V is a Montel space (in fact, V is a particular case of a Schwartz space).*

Remark 5.2.16. The terminology is a bit unfortunate, but the Schwartz space of rapidly decreasing functions is an instance of a Schwartz space.

5.3 Operators in countably normed spaces

We prove only (2) of the following exercise.

Exercise 5.3.1.

(1) *Show that a linear operator T from the countably normed space V (whose topology is defined by the semi-norms $p_1, p_2, \ldots$) into the countably normed space W (whose topology is defined by the semi-norms $q_1, q_2, \ldots$) is continuous if and only if for every norm q_n there exist a finite number of norms $p_{k_1}, \ldots, p_{k_n}$ and a number $K > 0$ such that*

$$q_n(Tv) \leq K \max\{p_{k_1}(v), \ldots, p_{k_n}(v)\}, \quad \forall v \in V. \tag{5.3.1}$$

(2) *Assume that the norms are increasing:*

$$p_n(v) \leq p_{n+1}(v), \quad and \quad q_n(w) \leq q_{n+1}(w), \quad n = 1, 2, \ldots,$$

for all $v \in V$ and $w \in W$. Express the fact that a linear operator T is continuous from V into W.

(3) *Show that a linear operator in a countably normed space is continuous if and only if it is bounded.*

Exercise 5.3.2. *Show that the operators*

$$Pf(x) = xf(x) \quad and \quad Qf(x) = f'(x)$$

are continuous from $\mathscr{S}$ into itself.

We note that P and Q satisfy the identity (4.3.2)

$$QP - PQ = I$$

on $\mathscr{S}$. In Exercise 5.6.4 below one is asked to prove that the operators P and Q, this time acting on the space $\mathcal{H}(\Omega)$ of functions analytic in the open set Ω (with the topology defined as in Exercise 5.6.3), are continuous. This is another example, of particular importance in the setting of the present book, of a Fréchet space where there exist continuous operators satisfying (4.3.2). On the other hand, recall that such operators do not exist in a normed space. See Exercise 4.3.5 above.

The following result is used in particular in [242, §8, p. 41] in the study of spaces of differential forms. See also Remark 5.1.5 in that respect.

Question 5.3.3. *Let T be a continuous map from the Fréchet space $\mathcal{F}_1$ into the Fréchet space $\mathcal{F}_2$. Assume that the quotient space $\mathcal{F}_2/T(\mathcal{F}_1)$ is finite dimensional. Show that the image of T is closed.*

We conclude with an example of a countably normed space of entire functions, which appears in [41]. See Remark 7.8.2. The space was first introduced in the paper [314, §2] of van Eijndhoven and Meyers, and is used in the theory of coherent states; see [8]. The union of the $\mathscr{G}_p$ was introduced by Gelfand and Shilov [150] in an equivalent way, as the space of entire functions f for which there exist positive numbers a, b and c (depending on f) such that

$$|f(z)| \le c e^{-ax^2+by^2}, \quad z \in \mathbb{C}.$$

See [314, §4] for more information. The intersection of the $\mathscr{G}_p$ was studied in [41] and its dual is an example of an algebra with a certain inequality; see Exercise 5.4.8.

Question 5.3.4. *Let $p \in \mathbb{N}$ and denote by $\mathscr{G}_p$ the Hilbert space of entire functions f such that*

$$C_p \iint_{\mathbb{C}} |f(z)|^2 e^{\frac{1-2^{-p}}{1+2^{-p}}x^2 - \frac{1+2^{-p}}{1-2^{-p}}y^2}\, dx dy < \infty,$$

where

$$C_p = \frac{2^{1-p}}{\sqrt{\pi(1-2^{-2p})}}.$$

Show that $\mathscr{G}_p$ contains non-trivial functions and that $\cap_{p=1}^{\infty}\mathscr{G}_p$ is a Fréchet space.

The following result is used in [242, p. 41] in the study of spaces of differential forms. See also Exercise 5.3.3.

Question 5.3.5. *The quotient of a Fréchet space by a finite-dimensional space is a Fréchet space when endowed with the quotient topology.*

The following result is taken from [114, (12.16.3), p. 84] (but of course has a much longer history). See Exercise 4.1.14 and the discussion preceding it. The notion of lower semi-continuous functions has been recalled in Definition 3.4.7.

Question 5.3.6. *Every lower semi-continuous semi-norm in a Fréchet space is continuous.*

5.4 Dual of a Fréchet space

We first recall that the dual of a topological vector space may be trivial, as is illustrated from the following example taken from Khaleelulla's book [202, 6, p. 14].

Question 5.4.1 (see [202, 6, p. 14]). *Show that the space $\mathbf{L}_{1/2}[0,1]$ (endowed with an appropriate topology) is a topological vector space, with trivial dual.*

Hint: The topology is defined by $\|f\|_{1/2} = \int_0^1 \sqrt{|f(t)|}dt$. The latter is not a norm.

Question 5.4.2. *Show that the dual of a locally convex topological vector space is not trivial when the space is itself not trivial.*

Question 5.4.3 (see [150, p. 34]). *Let $\mathcal{F}$ be a Fréchet space, with topology defined by the semi-norms $p_1, \dots$. The linear map φ belongs to the dual $\mathcal{F}'$ if and only if there exist $n \in \mathbb{N}$ and $C_n > 0$ such that*

$$|\varphi(v)| \le C_n \|v\|_n, \quad \forall v \in \mathcal{F}.$$

For the following question, see for instance [257, p. 134]. It is set in one variable, but the result holds for functions and distributions of n real variables.

Question 5.4.4. *Show that the Schwartz space $\mathscr{S}$ is continuously included and dense in $\mathscr{S}'$ in the $\sigma(\mathscr{S}', \mathscr{S})$ topology.*

Question 5.4.5 ([312, Chapter 1]). *Show that the space $\mathbb{R}^{\mathbb{N}}$ endowed with the metric (3.9.3) is a Fréchet space and that its dual is $\mathbb{R}_0^{\mathbb{N}}$, where $\mathbb{R}_0$ denotes the space of sequences of real numbers indexed by $\mathbb{N}$ and with only a finite number of nonzero entries.*

The following exercises are taken from the joint work of the author with Guy Salomon, see [41, 42, 43], where a new class of topological algebras is defined and studied. These algebras, called *strong algebras*, are a generalization of a space of stochastic distributions originally introduced by Yuri Kondratiev. See [182, p. 28] for the latter.

The proofs are based on iterated use of the Cauchy–Schwarz inequality. We give some hints, but it seems difficult to be more helpful without giving away the whole argument.

Exercise 5.4.6. *Let (where the c_n are complex numbers)*

$$\mathcal{K}_{-p} = \left\{ c = (c_n)_{n \in \mathbb{N}_0} \; ; \; \sum_{n=0}^{\infty} (n+1)^{-2p}|c_n|^2 < \infty \right\}, \quad p \in \mathbb{N}, \qquad (5.4.1)$$

and

$$\|c\|_{-p} = \left(\sum_{n=0}^{\infty} (n+1)^{-2p}|c_n|^2 \right)^{\frac{1}{2}}.$$

Let $a \in \mathcal{K}_{-q}$ and $b \in \mathcal{K}_{-p}$ with $p, q \in \mathbb{N}$ such that $p - q \geq 2$. Let $a \star b$ denote the convolution of the sequences a and b:

$$(a \star b)_n = \sum_{u=0}^{n} a_u b_{n-u}, \quad n = 0, 1, \dots$$

Show that $a \star b \in \mathcal{K}_{-2p}$ and that

$$\|a \star b\|_{-2p} \leq A(p - q)\|a\|_{-q}\|b\|_{-p}, \tag{5.4.2}$$

where

$$A(p - q) = \left(\sum_{n=0}^{\infty} (n + 1)^{2(q-p)} \right)^{\frac{1}{2}}. \tag{5.4.3}$$

Hint: When computing $\|a \star b\|_{-2p}$, use the inequality

$$(u + 1)^p (v + 1)^p (n - u + 1)^p (n - v + 1)^p \leq (n + 1)^{4p} \tag{5.4.4}$$

valid for integers $u, v \in \{0, \dots, n\}$ to bound terms of the form

$$|a_u| \cdot |a_v| \cdot |b_{n-u}| \cdot |b_{n-v}|.$$

The Cauchy–Schwarz inequality is also applied a number of times to get estimates.

Remark 5.4.7. The space $\cup_{p=0}^{\infty} \mathcal{K}_{-p}$, with $\mathcal{K}_{-p}$ defined by (5.4.1), is the space of tempered distributions.

Exercise 5.4.8. Let $\alpha = (\alpha_n)_{n \in \mathbb{N}_0}$ be a sequence of strictly positive numbers such that

$$\alpha_{n+m} \geq \alpha_n \alpha_m, \quad \forall n, m \in \mathbb{N}_0, \tag{5.4.5}$$

and such that

$$\sum_{n=0}^{\infty} \alpha_n^{-d} < \infty \tag{5.4.6}$$

for some $d \in \mathbb{N}$. Let (where the c_n are complex numbers)

$$\mathcal{K}_{-p}(\alpha) = \left\{ c = (c_n)_{n \in \mathbb{N}_0} ; \sum_{n=0}^{\infty} \alpha_n^{-2p} |c_n|^2 < \infty \right\}, \quad p \in \mathbb{N}, \tag{5.4.7}$$

and

$$\|c\|_{-p} = \left(\sum_{n=0}^{\infty} \alpha_n^{-2p} |c_n|^2 \right)^{\frac{1}{2}}. \tag{5.4.8}$$

Let $a \in \mathcal{K}_{-q}(\alpha)$ and $b \in \mathcal{K}_{-p}(\alpha)$ with $p, q \in \mathbb{N}$ such that $p - q \geq d$. Let $a \star b$ denote the convolution of the sequences a and b. Show that $a \star b \in \mathcal{K}_{-p}(\alpha)$ and that

$$\|a \star b\|_{-p} \leq A(p - q)\|a\|_{-q}\|b\|_{-p}, \tag{5.4.9}$$

where

$$A(p-q) = \left(\sum_{n=0}^{\infty} \alpha_n^{2(q-p)} \right)^{\frac{1}{2}} \tag{5.4.10}$$

(which is finite in view of (5.4.6)).

Hint: Now (5.4.4) is replaced by

$$\alpha_u^p \alpha_v^p \alpha_{n-u}^p \alpha_{n-v}^p \leq \alpha_n^{2p}. \tag{5.4.11}$$

Note the difference of the power ($2p$ instead of $4p$) with the right-hand side of (5.4.4).

Remarks 5.4.9.

(1) A sequence of positive numbers satisfying (5.4.5) is called superexponential. The case $\alpha_n = 2^n$ corresponds to the (dual) of the spaces considered in Question 5.3.4. Inequality (5.4.9) was proved in [41].

(2) The sequence $\alpha_n = (n+1)$, $n = 0, 1, 2, \ldots$ is not superexponential, but rather subexponential since

$$(n+1)(m+1) = nm + n + m + 1 \geq (n+m+1).$$

Note that to get (5.4.4) (and so to prove (5.4.2)) one uses in fact the property

$$(n+1)(m+1) \leq (n+m+1)^2.$$

(3) Inequalities (5.4.2) and (5.4.9) are of a totally different nature. Using the first one, one cannot obtain useful bounds for the norms of the elements $a^{\star n}$. On the other hand, inequality (5.4.9) implies in particular that $\forall n \in \mathbb{N}$, $a^{\star n} \in \mathcal{K}_{-p-d}(\alpha)$. Moreover,

$$\|a^{\star n}\|_{p+d} \leq A(d)^n \|a\|_p^n.$$

This last inequality suggests that one can define a functional calculus in the space $\cup_{p=0}^{\infty} \mathcal{K}_{-p}(\alpha)$. Inequalities of this kind seem to have been first found by Våge in the setting of a space of stochastic distributions. See [311].

In the following question we describe the set of stochastic distributions just alluded to, and introduced by Yuri Kondratiev. See [182]. We denote by ℓ the set of sequences of elements in $\mathbb{N}_0$, indexed by $\mathbb{N}$, and with at most a finite number of nonzero entries (thus, $\ell = \ell_J$ with $J = \mathbb{N}$ in the notation defined in the proof of item (2) of Exercise 7.1.17). We set (see [182]) for $\alpha \in \ell$:

$$(2\mathbb{N})^\alpha = 2^{\alpha_1} \cdot (2 \cdot 2)^{\alpha_2} \cdot (2 \cdot 3)^{\alpha_3} \cdots$$

Question 5.4.10. *Define for $p = 1, 2, \ldots$*

$$K_{-p} = \left\{ (f_\alpha)_{\alpha \in \ell} \; ; \; \|f\|_p^2 = \sum_{\alpha \in \ell} |f_\alpha|^2 (2\mathbb{N})^{-p\alpha} < \infty \right\}.$$

Show that for $p \geq q + 2$, there exists a finite constant $A(p, q)$ such that

$$\|f \star g\|_p \leq A(p, q) \|f\|_p \cdot \|g\|_q, \quad \forall f \in K_{-p} \text{ and } g \in K_{-q}. \tag{5.4.12}$$

Remark 5.4.11. (5.4.12) is called Våge's inequality.

5.5   Positive operators

We met earlier the notion of a positive matrix (see Definition 1.6.1) and of a positive operator in a Hilbert space (see Definition 4.2.28) and in Banach space (see Remark 4.2.37). The latter definition still makes sense for general topological vector spaces.

Definition 5.5.1. Let $\mathcal{V}$ be a topological vector space and let $\mathcal{V}^*$ denote its anti-dual, that is the space of continuous anti-linear functionals. The linear operator M from $\mathcal{V}$ into $\mathcal{V}^*$ is called positive if

$$\langle Mv, v \rangle_{\mathcal{V}, \mathcal{V}^*} \geq 0, \quad \forall v \in \mathcal{V}.$$

Exercise 5.5.2. *Let $\mathcal{H}$ be a Hilbert space, let $\mathcal{V}$ be a topological vector space (with anti-dual denoted by $\mathcal{V}^*$) and let T be a continuous linear from $\mathcal{V}$ into $\mathcal{H}$. Let T^* be defined from $\mathcal{H}$ into $\mathcal{V}^*$ by*

$$\langle T^* h, b \rangle_{\mathcal{V}^*, \mathcal{V}} = \langle h, Tb \rangle_{\mathcal{H}}.$$

Show that the operator $M = T^*T$ is linear and positive.

Remark 5.5.3 (see also (7.4.4)). The adjoint map T' from $\mathcal{H}'$ into $\mathcal{V}'$ is defined by

$$\langle T'(\varphi_h), b \rangle_{\mathcal{V}', \mathcal{V}} = \varphi_h(Tb) = \langle Tb, h \rangle_{\mathcal{H}}, \quad \forall h \in \mathcal{H} \text{ and } b \in \mathcal{V},$$

where $\varphi_h \in \mathcal{H}'$ is uniquely defined by $h \in \mathcal{H}$ using Riesz' theorem:

$$\langle \varphi_h, u \rangle_{\mathcal{H}', \mathcal{H}} = \langle u, h \rangle_{\mathcal{H}}.$$

Hence the operators T' and T^* are related by

$$\langle T^* h, b \rangle_{\mathcal{V}^*, \mathcal{V}} = \overline{\langle T'(\varphi_h), b \rangle_{\mathcal{V}', \mathcal{V}}} = \langle h, Tb \rangle_{\mathcal{H}}. \tag{5.5.1}$$

The converse to Exercise 5.5.2 will not hold in general. A continuous positive operator need not admit a factorization via a Hilbert space. Such a factorization will always hold in spaces with the factorization property. To motivate the definition, we begin with an easy question.

5.5. Positive operators

Question 5.5.4. *Let V be a topological vector space (with antidual V^*), and let M be a positive operator from V into V^*. Assume that $M = T^*T$, where T is a continuous operator from V into a Hilbert space $\mathcal{H}$. Then the map*

$$v \mapsto \langle Mv, v \rangle \qquad (5.5.2)$$

is continuous.

The continuity of the application (5.5.2) was singled out by J. Górniak to characterize locally convex topological vector spaces V in which all continuous positive operators from V into V^* can be factored via a Hilbert space; see [161, Theorem (1.7), p. 71]. Such spaces are called spaces with the factorization property. Górniak in fact considered antilinear operators from V into V', but the situation is equivalent. For real spaces (which form an important family of examples), the two approaches trivially coincide. An example of a space which does not have the factorization property (also due to Górniak) is given in Exercise 5.5.6. We first mention that important families of spaces, such as Banach and nuclear spaces, do have the factorization property. It is interesting to note that the intersection of these two families consists of finite-dimensional spaces.

Question 5.5.5. *Let $\mathcal{B}$ be a Banach space and let M be a linear positive operator from $\mathcal{B}$ into $\mathcal{B}^*$.*

(1) *Show that M can be factorized via a Hilbert space.*
(2) *Show that the result still holds when $\mathcal{B}$ is replaced by a space having the factorization property (that is, a space for which (5.5.2) holds).*

Hint: Show that the range of M endowed with the Hermitian form

$$\langle Mu, Mv \rangle = \langle Mu, v \rangle_{\mathcal{B}^*, \mathcal{B}}$$

is a pre-Hilbert space. Let $\mathcal{H}$ denote its closure and let i denote the map $i(u) = Mu$ from $\mathcal{B}$ into $\mathcal{H}$. Using the factorization property in the second case (and directly in the first case), the equality

$$\|i(u)\|^2_{\mathcal{H}} = \langle Mu, u \rangle_{\mathcal{B}^*, \mathcal{B}}$$

shows that i is continuous. Compute its adjoint to conclude.

Exercise 5.5.6. *Let V be as in Exercise 5.2.5. Show that the natural injection map from V into V^* (the latter endowed with the strong topology) is continuous and positive, but that it cannot be factored via a Hilbert space.*

5.6   Topology of the space of analytic functions

Banach spaces and Hilbert spaces of analytic functions play an important role in functional analysis; see for instance [84, 121, 126, 180, 206]. Banach spaces are complete topological vector spaces whose topology is defined by a single norm. When one considers the space $\mathcal{H}(\Omega)$ of *all* functions analytic in a given open set Ω, the appropriate structure is more complicated. The topology is defined by a countable number of norms, and we have a Fréchet space (that is, a metrizable, complete, locally convex, topological vector space; see Definition 5.2.8 and [307, p. 95]). As shown in Exercise 5.2.10 for the case of $\mathbb{D}$, the metric space structure is such that convergence is equivalent to uniform convergence on compact sets. In particular, the underlying topology can be characterized by convergence of sequences.

Exercise 5.6.1. *Let Ω be an open subset of $\mathbb{C}$, and let $(f_n)_{n\in\mathbb{N}}$ denote a sequence of functions analytic in Ω and converging uniformly on compact subsets of Ω to a function f. Show that f is analytic in Ω.*

Question 5.6.2.

(a) *Show that an open path-connected subset of the complex plane is connected.*
(b) *Show that a connected open subset of the complex plane can be written as a union of increasing compact connected sets $K_1, K_2, \ldots$ such that*

$$K_n \subset \overset{\circ}{K_{n+1}}, \quad n = 1, 2, \ldots.$$

For (b) above see for instance [144, Lemma A.20, p. 372].

Let Ω be open and connected, and let $(K_n)_{n\in\mathbb{N}}$ be a sequence of compact sets as in (b) of Exercise 5.6.2. The maps

$$\|f\|_n = \max_{z\in K_n} |f(z)|, \quad n = 1, 2, \ldots$$

define norms in $\mathcal{H}(\Omega)$.

Question 5.6.3. *Describe the smallest topology for which the norms $\|\cdot\|_n$ are continuous. Show that this topology is metrizable, and hence that $\mathcal{H}(\Omega)$ endowed with this topology is a Fréchet space. Show that a sequence of elements in $\mathcal{H}(\Omega)$ converges in this topology if and only if it converges uniformly on compact subsets of Ω.*

Exercise 5.6.4. *Let Ω be open and connected. Show that the operators*

$$Pf(z) = zf(z) \quad \text{and} \quad Qf(z) = f'(z) \tag{5.6.1}$$

are continuous on $\mathcal{H}(\Omega)$.

The continuity of the operator $f \mapsto f'$ on $\mathcal{H}(\Omega)$ is used in particular in the proof of Riemann's mapping theorem; see for instance [95].

Other important examples of Fréchet spaces include the Schwartz space $\mathscr{S}$ of rapidly vanishing smooth functions, see the definition above Exercise 5.3.2, and various other spaces which play the role of test functions in the theory of distributions. The duals of these spaces and their topologies (in particular the strong and weak topologies) play a key role in various problems.

Since $\mathcal{H}(\Omega)$ is a complete metric space, Baire's category theorem asserts that it is a second category space. For the following question, where a result of Kierst and Szpilrajn is presented, see the discussion after (3.2.1).

Question 5.6.5. *The set of functions analytic in the open unit disk and which have a radial tangential limit (possibly infinite) at at least one point on the unit circle is a first category set.*

Now that we have a topology structure on $\mathcal{H}(\Omega)$ we can ask various questions relative to this topology. For instance:

(1) What are the bounded subsets (see Definition 5.1.17 for the latter) of $\mathcal{H}(\Omega)$?
(2) What are the compact subsets of $\mathcal{H}(\Omega)$?
(3) What is the dual of $\mathcal{H}(\Omega)$?

Surprisingly, the characterization of compact subsets of $\mathcal{H}(\Omega)$ drifts us away from classical paradigms of functional analysis, and plays a key role in a proof of Riemann's mapping theorem.

Exercise 5.6.6. *Characterize the bounded subsets of $\mathcal{H}(\Omega)$.*

Question 5.6.7. *Let as above $\Omega \subset \mathbb{C}$ be open and connected, and consider now a countable set $\{z_0, z_1, \ldots\}$ dense in Ω, and let $K_n = \{z_0, \ldots, z_n\}$. Note that the interior of K_n is empty, and that the union of the K_n is different from Ω. Show that the corresponding semi-metric d is a metric, and characterize the convergence in this metric in the spaces $\mathcal{C}(\Omega)$ and $\mathscr{H}(\Omega)$.*

Consider a sequence of functions $(f_n)_{n \in \mathbb{N}}$ which converges uniformly on compact subsets of $\mathbb{D}$ to a function f. Then, f is also analytic in $\mathbb{D}$ (use for instance Morera's theorem to see that; see Exercise 5.6.1 if need be), and Cauchy's formula applied to $f_n - f$,

$$\frac{f_n^{(p)}(0) - f^{(p)}(0)}{p!} = \frac{1}{2\pi i} \int_{|\zeta| = r} \frac{f_n(\zeta) - f(\zeta)}{\zeta^{p+1}} d\zeta$$

where r is any fixed number in $(0, 1)$ implies that

$$\lim_{n \to \infty} f_n^{(p)}(0) = f^{(p)}(0), \quad p = 0, 1, \ldots \tag{5.6.2}$$

The following exercise is taken from Cartan's book, see [95, Lemma 2, p. 167], and concerns the converse statement: When does (5.6.2) imply uniform convergence on compact sets? The original statement speaks of a bounded set in $\mathcal{H}(\Omega)$. Since $\mathcal{H}(\Omega)$ is a metric space, item (2) shows that a bounded and closed subspace of $\mathcal{H}(\Omega)$ is compact.

Exercise 5.6.8. *Let $(f_n)_{n\in\mathbb{N}}$ be a sequence of functions analytic in the open unit disk, and bounded by 1 in modulus there.*

(1) *Show that $(f_n)_{n\in\mathbb{N}}$ converges to f uniformly on compact subsets of the open unit disk if and only if (5.6.2) holds.*
(2) *Show that $(f_n)_{n\in\mathbb{N}}$ admits a subsequence converging uniformly on compact subsets of the open unit disk.*

Question 5.6.9. *Using the previous result prove that*

$$\lim_{n\to\infty}\left(1-\frac{z}{n}\right)^n = e^z, \quad z\in\mathbb{C}. \tag{5.6.3}$$

Exercise 5.6.10. *Let $q\in\mathbb{D}\setminus\{0\}$. Show that the series*

$$f(z) = \sum_{\substack{n\in\mathbb{Z}\\ n\neq 0}}\frac{z^n}{1-q^n} \tag{5.6.4}$$

converges in the open ring $|q| < |z| < 1$.

The derivative of the Laurent series (5.6.4) is equal to

$$\sum_{\substack{n\in\mathbb{Z}\\ n\neq 0}}\frac{nz^{n-1}}{1-q^n},$$

and converges uniformly on every closed subring of $|q| < |z| < 1$. Let now $r\in(0,1)$. Series of the form

$$\sum_{\substack{n\in\mathbb{Z}\\ n\neq 0}}\frac{nz^n}{1-r^{2n}},$$

appear in particular in the computation of the Bergman kernel for the annulus $r < |z| < 1$. More precisely, see [70, p. 10], the Bergman kernel of the annulus $r < |z| < 1$ is given by the formula

$$-\frac{1}{2\pi z\overline{w}\ln r} + \frac{1}{\pi z\overline{w}}\sum_{\substack{n\in\mathbb{Z}\\ n\neq 0}}\frac{n(z\overline{w})^n}{1-r^{2n}}. \tag{5.6.5}$$

This sum can be written in terms of the Weierstrass function:

$$\wp(\ln z) = -\frac{\eta_1}{\pi i} + zf'(z),$$

where $\wp(z)$ is a Weierstrass function with periods $T_1 = i\pi$ and $T_2 = \ln r$, and where η_1 is some complex constant.

In relation to the above, and in relation to elliptic functions, we mention the following (see [315, p. 66]). We set for fixed $r \in (0,1)$:

$$f(a,z) = \sum_{n \in \mathbb{Z}} \frac{z^n}{1 - ar^{2n}}$$

to be the Jordan–Kronecker function.

Question 5.6.11. *Show that*

$$f(a,z)f(b,z) = zf(ab,z)' + f(ab,z)(\rho_1(a) + \rho_1(b))$$

where

$$\rho_1(t) = \frac{1}{2} + \sum_{\substack{n \in \mathbb{Z} \\ n \neq 0}} \frac{z^n}{1 - r^{2n}}.$$

5.7 Normal families

The completeness property of the space $\mathcal{H}(\Omega)$ of functions analytic in an open set Ω endowed with the topology defined by the norms (4.1.4) is referred to also as the normal family theorem (for analytic functions). We refer for instance to [4, p. 220] for more on the normal family.

Question 5.7.1. *Let Ω be an open subset of $\mathbb{C}$. Show that the unit ball of $\mathcal{H}(\Omega)$ is compact.*

Exercise 5.7.2. *Let Ω_1 and Ω_2 be two open subsets of $\mathbb{C}$ and assume that $\overline{\Omega_1} \subset \Omega_2$. Show that the restriction map from $\mathcal{H}^\infty(\Omega_2)$ into $\mathcal{H}^\infty(\Omega_1)$ is compact. Is it onto?*

Question 5.7.3. *Let $R \in (0,1)$ and let $\mathbb{D}_R$ denote the open disk centered at the origin and of radius R. Show that the natural injection from $\mathcal{H}(\mathbb{D})$ into $\mathcal{H}(\mathbb{D}_R)$ is compact.*

The following result is one of the keys to a proof of Riemann's theorem.

Question 5.7.4. *The sequence of functions $(f_k)_{k \in \mathbb{N}}$ of $\mathcal{H}(\mathbb{D})$ converges in the topology of $\mathcal{H}(\mathbb{D})$ if and only if*

$$\forall n \in \mathbb{N}_0, \exists \lim_{k \to \infty} f_k^{(n)}(0).$$

5.8   The dual of the space of analytic functions

Exercise 5.8.1. *Let φ be a continuous linear map from $\mathcal{H}(\mathbb{D})$ into $\mathbb{C}$. Show that there exist $R \in (0,1)$ and $C > 0$ such that*

$$|\varphi(z^n)| \le CR^n, \quad n = 0, 1, \ldots \tag{5.8.1}$$

Conversely, given a sequence of numbers $(b_n)_{n \in \mathbb{N}_0}$ such that

$$|b_n| \le CR^n, \quad n = 0, 1, \ldots \tag{5.8.2}$$

with $R \in (0,1)$ and $C > 0$, show that the formula

$$\varphi(f) = \sum_{n=0}^{\infty} a_n b_n, \tag{5.8.3}$$

where $f(z) = \sum_{n=0}^{\infty} a_n z^n$, defines a continuous functional from $\mathcal{H}(\mathbb{D})$ into $\mathbb{C}$.

Question 5.8.2. *Characterize the topological dual of $\mathcal{H}(A_R)$, where A_R is the annulus defined by (2.1.14):*

$$A_R = \{z \in \mathbb{C} \,; 1 < |z| < R\}.$$

Question 5.8.3. *Characterize the topological dual of the space of entire functions.*

Hint: Consider the space of germs at the origin. To that purpose, recall the following: Let $a \in \mathbb{C}$. Consider the set of pairs (V, f) where V is an open subset of the complex plane which contains a, and f is analytic in V. Say that $(V_1, f_1) \sim (V_2, f_2)$ if there is an open set V_3 containing a and such that

$$V_3 \subset V_1 \cap V_2 \quad \text{and} \quad f_1(x) = f_2(x), \quad x \in V_3.$$

This defines indeed an equivalence relation, and the equivalence classes are called germs at a.

5.9   Solutions of the exercises

Solution of Exercise 5.1.6: Let $v_1, \ldots, v_N$ be a basis of $\mathcal{V}$. Since $\mathcal{W}$ is a topological vector space the vector space operations are continuous and so the algebraic isomorphism

$$T(z_1, \ldots, z_N) = \sum_{n=1}^{N} z_n v_n$$

from $\mathbb{C}^N$ onto $\mathcal{V}$ is continuous. We now show that T^{-1} is also continuous. Let $\epsilon > 0$. The sphere

$$\mathbb{S}_\epsilon = \left\{ (z_1, \ldots, z_N) \in \mathbb{C}^N \,; \sum_{n=1}^{N} |z_n|^2 = \epsilon^2 \right\}$$

is compact and so is its image under T in $\mathcal{W}$. Furthermore, $0 \notin T(\mathbb{S}_\epsilon)$. Recall that a compact set is closed and so $T(\mathbb{S}_\epsilon)$ is closed in $\mathcal{W}$, and there is a neighborhood of the origin, say N_ϵ, such that

$$T(\mathbb{S}_\epsilon) \cap N_\epsilon = \emptyset.$$

Thus $N_\epsilon \subset T(\mathbb{C}^N \setminus \mathbb{S}_\epsilon)$. The latter has two connected components, the one which contains the origin being $T(\mathbb{B}_\epsilon)$, with

$$\mathbb{B}_\epsilon = \left\{ (z_1, \dots, z_N) \in \mathbb{C}^N ; \sum_{n=1}^{N} |z_n|^2 < \epsilon^2 \right\}$$

(recall that the continuous image of a connected set is connected; see Theorem 3.4.11). By taking the connected component of N_ϵ which contains the origin we may assume N_ϵ connected. Furthermore we note that

$$N_\epsilon \subset T(\mathbb{B}_\epsilon),$$

since 0 belongs to both sets and both sets are connected. It follows that

$$T^{-1}(N_\epsilon) \subset \left\{ (z_1, \dots, z_N) \in \mathbb{C}^N ; \sum_{n=1}^{N} |z_n|^2 < \epsilon^2 \right\}$$

and so T^{-1} is continuous. Thus $\mathbb{C}^N$ and $\mathcal{V}$ are topologically isomorphic, and hence $\mathcal{V}$ is complete and so closed in $\mathcal{W}$. $\qquad\square$

Solution of Exercise 5.1.8: We denote by $\mathcal{N}(0)$ the set of neighborhoods of the origin in $\mathcal{V}$, and by V a given element of $\mathcal{N}(0)$.

(1) The map $(\lambda, v) \mapsto \lambda v$ is jointly continuous. In particular it is separately continuous, and for any given $v \in \mathcal{V}$ the map $\lambda \mapsto \lambda v$ is continuous in the variable λ. Thus for any neighborhood V of the origin there is $\epsilon > 0$ such that

$$|\lambda| < \epsilon \implies \lambda v \in V.$$

Taking any $\alpha \in (0, \epsilon)$ we have

$$|\lambda| \leq \alpha \implies \lambda v \in V,$$

and so V is absorbing.

Recall that a basis of neighborhoods of the origin for the product topology of $\mathbb{C} \times \mathcal{V}$ consists of sets of the form $B(0, \rho) \times W$, where W is a neighborhood of the origin in $\mathcal{V}$. Since the product is continuous at the point $(0, x)$ with $x \in V$ there exist $\rho_x > 0$ and a neighborhood of the origin W_x such that

$$z \in B(0, \rho_x) \quad \text{and} \quad w \in W_x \implies zw \in V.$$

Each of the sets $\rho_x W_x \in \mathcal{N}(0)$ and is included in V. Thus, the set $\cup_{x \in V} \rho_x W_x$ answers the question. $\qquad\square$

Solution of Exercise 5.1.14: The joint continuity of the addition at the point 0 follows from the inclusion

$$V_{\epsilon/2,\eta/2}(0) + V_{\epsilon/2,\eta/2}(0) \subset V_{\epsilon,\eta}(0).$$

The joint continuity of $(\lambda, f) \mapsto \lambda f$ is proved similarly.

We now show that $C[0,1]$ endowed with the given topology is not locally convex (and in particular the sets $V_{\epsilon/2,\eta/2}(0)$ are not convex). We follow the argument in [183] (outlined in the hint to the exercise) to show that no convex set X lies between two sets of the form $V_{\epsilon,\eta}(0)$.

Let $V_{\epsilon_1,\eta_1}(0) \subset X \subset V_{\epsilon_2,\eta_2}(0)$, and let

$$I = \cup_{i=1}^{n} I_i \subset [0,1],$$

where the I_i are open intervals, each of them of length less than η_1, and of total length strictly greater than η_2. For $i = 1, \ldots, n$, let J_i be an open interval such that $\overline{J_i} \subsetneq I_i$. In particular the length of J_i is less than η_1. There exist continuous functions $f_1, \ldots, f_n \in V_{\epsilon_1,\eta_1}$ such that

$$f_i(t) = \begin{cases} 0, & t \notin J_i, \\ 2n\epsilon_2, & t \in J_i. \end{cases}$$

The function $\sum_{i=1}^{n} \frac{f_i}{n}$ is greater in modulus than ϵ_2 on an open set of length greater than η_2 and in particular will not belong to $V_{\epsilon_2,\eta_2}(0)$. Thus $\sum_{i=1}^{n} \frac{f_i}{n} \notin X$. □

Solution of Exercise 5.1.18: Let $\mathcal{V}$ be the underlying space, and let V be a neighborhood of the origin. It contains a set of the form

$$\cap_{u=1}^{m} \{v \in \mathcal{V} ; \ p_u(v) < \epsilon_u\},$$

where $p_1, \ldots, p_m$ belong to the set of semi-norms defining the topology of $\mathcal{V}$. Let now q be another such norm and let $\epsilon > 0$. Assume that V is bounded. Then there exists $k > 0$ such that

$$V \subset k\{v \in \mathcal{V} ; \ q(v) < \epsilon\},$$

and so

$$\cap_{u=1}^{m} \left\{v \in \mathcal{V} ; \ p_u(v) < \frac{\epsilon_u}{k}\right\} \subset \{v \in \mathcal{V} ; \ q(v) < \epsilon\}.$$

The topology of $\mathcal{V}$ is therefore defined by a finite number of semi-norms, and hence by one semi-norm. □

Solution of Exercise 5.2.1: A basis of open neighborhoods of the origin is given by the sets of the form

$$\cap_{i=1}^{m} \{v \in \mathcal{V}; p_{n_i}(v) < \epsilon_i\}, \tag{5.9.1}$$

5.9. *Solutions of the exercises*

where m varies in $\mathbb{N}$, $n_1, \ldots, n_m \in \mathbb{N}$, and $\epsilon_1, \ldots, \epsilon_m$ are all strictly positive numbers. Any such neighborhood contains the open neighborhood

$$\{v \in \mathcal{V}; p_{n_0}(v) < \epsilon_0\},$$

where $n_0 = \max\{n_1, \ldots, n_m\}$ and $\epsilon = \min\{\epsilon_1, \ldots, \epsilon_m\}$. Thus the sets

$$\{v \in \mathcal{V}; p_n(v) < \epsilon\}, \quad n \in \mathbb{N}, \quad \epsilon > 0,$$

also form a basis of neighborhoods of the topology.

We now show that the above topology is metrizable, with the metric

$$d(v, w) = \sum_{n=1}^{\infty} \frac{1}{2^n} \frac{p_n(v - w)}{1 + p_n(v - w)}. \tag{5.9.2}$$

We set for $r > 0$

$$B_d(0, r) = \{v \in \mathcal{V}; d(0, v) < r\}.$$

It is enough to show that

(a) for every $M \in \mathbb{N}$ and $\epsilon > 0$ there exists $r > 0$ such that

$$B_d(0, r) \subset \{v \in \mathcal{V}; p_M(v) < \epsilon\},$$

and that, conversely,

(b) for every $r > 0$ there exist M and ϵ as above such that

$$\{v \in \mathcal{V}; p_M(v) < \epsilon\} \subset B_d(0, r). \tag{5.9.3}$$

We first prove (a): Let us take r such that $2^M r < 1$. The condition

$$\sum_{n=1}^{\infty} \frac{1}{2^n} \frac{p_n(v)}{1 + p_n(v)} < r$$

implies in particular that

$$\frac{1}{2^M} \frac{p_M(v)}{1 + p_M(v)} < r,$$

or equivalently

$$p_M(v) < \frac{2^M r}{1 - 2^M r}.$$

This last expression will be less than ϵ as soon as

$$2^M r < \frac{\epsilon}{1 + \epsilon}.$$

We now prove (b). Consider a ball $B_d(0, r)$. Since the distance d (defined by (5.9.2)) is bounded by 1, it is enough to consider the case $r < 1$. Let $M \in \mathbb{N}$ be such that

$$\sum_{n=M+1}^{\infty} \frac{1}{2^n} < \frac{r}{2}.$$

Since the norms are increasing and since the function $u \mapsto \frac{u}{1+u}$ is increasing, we moreover have that

$$\sum_{n=1}^{M} \frac{1}{2^n} \frac{p_n(v)}{1+p_n(v)} \leq \frac{p_M(v)}{1+p_M(v)} \times \left(\frac{1}{2} + \cdots + \frac{1}{2^M} \right).$$

Chose $\epsilon > 0$ such that

$$\frac{\epsilon}{1+\epsilon} \times \left(\frac{1}{2} + \cdots + \frac{1}{2^M} \right) < \frac{r}{2}.$$

Then, for $p_M(v) < \epsilon$ we have that

$$\frac{p_M(v)}{1+p_M(v)} \times \left(\frac{1}{2} + \cdots + \frac{1}{2^M} \right) < \frac{r}{2}.$$

For such ϵ, (5.9.3) is in force. $\square$

Solution of Exercise 5.2.3: Following the hint, take

$$f_n(t) = \frac{1}{n} e^{-tn}, \quad t \in [0,1].$$

Then

$$f_n'(0) = -e^{-1},$$

and so $\|f_n - f_m\|_1 = \|f_n - f_m\|_2$, and $(f_n)_{n \in \mathbb{N}}$ is a Cauchy sequence with respect to the two norms. It converges to 0 with respect to the first norm. It does not converge to 0 with respect to the second norm since

$$\|f_n\|_2 \geq e^{-1}.$$ $\square$

Solution of Exercise 5.2.5: Let $A \subset V$ be a bounded set. We follow the hint, and assume by contradiction that there exists a strictly increasing sequence of integers $n_1, n_2, \ldots$ (which we can suppose of density zero) such that

$$\forall m \in \mathbb{N}, \ \exists a^{(m)} \in A, \text{ such that } a_{n_m}^{(m)} \neq 0.$$

Consider the semi-norm

$$p(z) = \sum_{m=0}^{\infty} n_m \frac{|z_{n_m}|}{|a_{n_m}^{(m)}|},$$

and the corresponding neighborhood $V = \{z \in V \ ; \ p(z) < 1\}$. Since

$$p(a^{(m)}) \geq n_m,$$

the boundedness condition (5.1.5) cannot hold.

Consider now the semi-norms $p_n(z) = |z_n|$. Since A is assumed bounded there exists for every $n \in \mathbb{N}$ a number $M_n > 0$ such that

$$A \subset M_n \cdot p_n^{-1}[0,1).$$

Hence

$$C_n = \{|z_n|, \; z \in A\}$$

is bounded. Combining with the beginning of the argument we see that there exists an integer N_A and a constant K such that

$$a \in A \implies |a_n| \leq K \quad \text{and} \quad a_n = 0 \text{ for } n > N_A.$$

Conversely, any such set is bounded. □

Solution of Exercise 5.2.10: We note that d is indeed a metric. Each of the d_n is only a semi-norm on $\mathcal{C}(\mathbb{D})$, but $d(f,g) = 0$ if and only if $f = g$ since the K_n cover all the open unit disk.

To prove the second claim, we first assume that (5.2.3) is in force. Let $\epsilon > 0$ and $N \in \mathbb{N}_0$ preassigned. There exists $\eta > 0$ such that

$$2^N \eta < 1 \quad \text{and} \quad \frac{2^N \eta}{1 - 2^N \eta} \leq \epsilon.$$

Given η as above, there exists $P \in \mathbb{N}$ such that:

$$p \geq P \implies d(f, f_p) < \eta,$$

and in particular

$$\frac{1}{2^N} \frac{d_N(f, f_p)}{1 + d_N(f, f_p)} < \eta,$$

which implies $d_N(f, f_p) < \epsilon$.

Conversely, assume that the sequence $(f_p)_{p \in \mathbb{N}_0}$ converges to f uniformly on compact subsets of $\mathbb{D}$. Fix $\epsilon > 0$. There exists $N \in \mathbb{N}_0$ such that

$$\sum_{n=N+1}^{\infty} \frac{1}{2^n} < \frac{\epsilon}{2},$$

and in particular

$$\sum_{n=N+1}^{\infty} \frac{1}{2^n} \frac{d_n(f, f_p)}{1 + d_n(f, f_p)} < \frac{\epsilon}{2}.$$

Moreover, there exists $P \in \mathbb{N}_0$ such that:

$$p > P \implies d_n(f, f_p) < \frac{\epsilon}{2(N+1)}, \quad n = 0, 1, \ldots, N.$$

Thus, for $p > P$, we have that $d(f, f_p) < \epsilon$. □

Solution of Exercise 5.2.12: The set

$$\left\{ s \in \mathscr{S} \; ; \; \max_{x \in \mathbb{R}} |s'(x)| < 1 \right\}$$

is an open neighborhood of the origin. Should the set

$$V = \left\{ s \in \mathscr{S} \; ; \; \max_{x \in \mathbb{R}} |s(x)| < 1 \right\}$$

be bounded there would exist an $M > 0$ such that

$$\left\{ s \in \mathscr{S} \; ; \; \max_{x \in \mathbb{R}} |s(x)| < 1 \right\} \subset M \left\{ s \in \mathscr{S} \; ; \; \max_{x \in \mathbb{R}} |s'(x)| < 1 \right\}.$$

This is not possible since the set V contains functions with first derivative arbitrarily large in absolute value at the origin, as is seen by taking the functions $x \mapsto f_m(x) = \frac{1}{2} e^{-mx^2}$, which belong to V for every $m > 0$. We detail the computation for the convenience of the reader. We have

$$f_m'(x) = -mx e^{-mx^2} \quad \text{and} \quad f_m''(x) = me^{-mx^2}(-1 + 2x^2 m).$$

Hence the extrema of f_m' is at the points $\pm\sqrt{\frac{1}{2m}}$ and are both equal in absolute value to

$$\sqrt{\frac{m}{2}} e^{-1/2}. \qquad \qquad \square$$

Solution of Exercise 5.3.1: As mentioned before the exercise, we prove only the second item. Since the semi-norms are ordered, a basis of neighborhoods of the origin in $\mathcal{V}$ is given by the sets

$$\{ v \in \mathcal{V} \; ; \; p_n(v) < \epsilon \}, \quad \epsilon > 0,$$

and similarly for $\mathcal{W}$. Thus, a necessary and sufficient condition of continuity for the linear operator T is:

$$\forall n \in \mathbb{N} \quad \text{and} \quad \forall \epsilon > 0, \; \exists m \in \mathbb{N} \quad \text{and} \quad \eta > 0, \text{ such that}$$
$$T \{ v \in \mathcal{V} \; ; \; p_m(v) < \eta \} \subset \{ w \in \mathcal{W} \; ; \; q_n(w) < \epsilon \}. \qquad \square$$

Solution of Exercise 5.3.2: From

$$(xf(x))^{(q)} = xf^{(q)}(x) + qf(x),$$

we obtain

$$\|Pf\|_{p,q} \le \|f\|_{p+1,q} + q\|f\|_{p,0}. \qquad (5.9.4)$$

Let now

$$\mathcal{W} = \{ f \in \mathscr{S} \; ; \; \|f\|_{p_i, q_i} < \epsilon_i, \; i = 1, 2, \ldots, N \},$$

5.9. Solutions of the exercises

with $N \in \mathbb{N}$, $p_i, q_i \in \mathbb{N}_0$ and $\epsilon_i > 0$ for $i = 1, \ldots, N$, be an open neighborhood of the origin. Then, with

$$V = \left\{ f \in \mathscr{S} \, ; \, \|f\|_{p_i+1,q_i} < \frac{\epsilon_i}{2} \text{ and } \|f\|_{p_i,0} < \frac{\epsilon_i}{2q_i}, \, i = 1, 2, \ldots, N \right\},$$

we have $PV \subset W$, and so P is continuous.

The continuity of Q is proved in a similar way, using

$$\|Qf\|_{p,q} = \|f\|_{p,q+1}.$$

$\square$

Solution to Exercise 5.4.6: Let $r \in \mathbb{N}$. We have

$$
\begin{aligned}
\|a \star b\|_{-r}^2 &= \sum_{n=0}^{\infty} \left| \sum_{u=0}^{n} a_u b_{n-u} \right|^2 (n+1)^{-2r} \\
&\leq \sum_{n=0}^{\infty} \left(\sum_{u=0}^{n} |a_u| \cdot |b_{n-u}| \right)^2 (n+1)^{-2r} \\
&= \sum_{n=0}^{\infty} \left(\sum_{u,v=0}^{n} |a_u| \cdot |a_v| \cdot |b_{n-u}| \cdot |b_{n-v}| \right) (n+1)^{-2r}.
\end{aligned}
$$
(5.9.5)

Since the quantities $u+1, v+1, n-u+1$ and $n-v+1$ are all greater or equal to 0 and less or equal to $n+1$ for $u, v \in \{0, \ldots, n\}$ we have:

$$(u+1)^p (v+1)^p (n-u+1)^p (n-v+1)^p \leq (n+1)^{4p}. \tag{5.9.6}$$

Hence,

$$
\begin{aligned}
(n+1)^{-4p} &= (u+1)^{-p}(v+1)^{-p}(n-u+1)^{-p}(n-v+1)^{-p} \\
&\quad \times (u+1)^p (v+1)^p (n-u+1)^p (n-v+1)^p (n+1)^{-4p} \\
&\leq (u+1)^{-p}(v+1)^{-p}(n-u+1)^{-p}(n-v+1)^{-p}.
\end{aligned}
$$

Thus, (5.9.5) becomes with $r = 2p$:

$$
\begin{aligned}
\|a \star b\|_{-2p}^2 &\leq \sum_{n=0}^{\infty} \left(\sum_{u,v=0}^{n} |a_u|(u+1)^{-p} \cdot |a_v|(v+1)^{-p} \cdot \right. \\
&\qquad\qquad \left. \cdot |b_{n-u}|(n-u+1)^{-p} \cdot |b_{n-v}|(n-v+1)^{-p} \right) \\
&= \sum_{u,v=0}^{\infty} |a_u|(u+1)^{-p} \cdot |a_v|(v+1)^{-p} \times \\
&\qquad \times \left(\sum_{n \geq u,v} |b_{n-u}|(n-u+1)^{-p} \cdot |b_{n-v}|(n-v+1)^{-p} \right).
\end{aligned}
$$
(5.9.7)

The Cauchy–Schwarz inequality gives

$$\sum_{n \geq u,v} |b_{n-u}|(n - u + 1)^{-p} \cdot |b_{n-v}|(n - v + 1)^{-p}$$

$$\leq \left(\sum_{n \geq u,v} |b_{n-u}|^2(n - u + 1)^{-2p} \right)^{\frac{1}{2}} \left(\sum_{n \geq u,v} |b_{n-v}|^2(n - v + 1)^{-2p} \right)^{\frac{1}{2}}$$

$$\leq \|b\|^2_{-p}.$$

Thus (5.9.7) becomes

$$\|a \star b\|^2_{-2p} \leq \left(\sum_{u,v=0}^{\infty} |a_u|(u + 1)^{-p} \cdot |a_v|(v + 1)^{-p} \right) \cdot \|b\|^2_{-p}. \tag{5.9.8}$$

But

$$\sum_{u,v=0}^{\infty} |a_u| \cdot |a_v|(u + 1)^{-p}(v + 1)^{-p} = \left(\sum_{u=0}^{\infty} |a_u|(u + 1)^{-p} \right)^2$$

$$= \left(\sum_{u=0}^{\infty} |a_u|(u + 1)^{-q}(u + 1)^{q-p} \right)^2$$

$$\leq \left(\sum_{u=0}^{\infty} |a_u|^2(u + 1)^{-2q} \right) \left(\sum_{u=0}^{\infty} (u + 1)^{2(q-p)} \right)$$

$$= \|a\|^2_{-q}(A(p - q))^2,$$

where $A(p-q)$ is given by (5.4.3), which is finite for $p-q \geq 2$. Hence the result. $\qquad\square$

Solution to Exercise 5.4.8: The proof goes along the lines of the arguments in the previous exercise, but this time (5.9.6) is replaced by an inequality taking into account (5.4.5). More precisely we have for $r \in \mathbb{N}$:

$$\|a \star b\|^2_{-r} = \sum_{n=0}^{\infty} \left| \sum_{u=0}^{n} a_u b_{n-u} \right|^2 \alpha_n^{-2r}$$

$$\leq \sum_{n=0}^{\infty} \left(\sum_{u=0}^{n} |a_u| \cdot |b_{n-u}| \right)^2 \alpha_n^{-2r} \tag{5.9.9}$$

$$= \sum_{n=0}^{\infty} \left(\sum_{u,v=0}^{n} |a_u| \cdot |a_v| \cdot |b_{n-u}| \cdot |b_{n-v}| \right) \alpha_n^{-2r}.$$

From (5.4.5) we have

$$\alpha_u^p \alpha_v^p \alpha_{n-u}^p \alpha_{n-v}^p \leq \alpha_n^{2p}, \tag{5.9.10}$$

and thus:

$$\alpha_n^{-2p} = \alpha_u^{-p}\alpha_v^{-p}\alpha_{n-u}^{-p}\alpha_{n-v}^{-p}$$
$$\times (\alpha_u^p\alpha_v^p\alpha_{n-u}^p\alpha_{n-v}^p) \times \alpha_n^{-2p}$$
$$\leq \alpha_u^{-p}\alpha_v^{-p}\alpha_{n-u}^{-p}\alpha_{n-v}^{-p}.$$

Hence, (5.9.9) becomes (with $r = p$; recall that in the previous exercise one had to choose $r = 2p$):

$$\|a \star b\|_{-p}^2 \leq \sum_{n=0}^{\infty} \left(\sum_{u,v=0}^{n} |a_u|\alpha_u^{-p} \cdot |a_v|\alpha_v^{-p} \right.$$
$$\left. \times |b_{n-u}|\alpha_{n-u}^{-p} \cdot |b_{n-v}|\alpha_{n-v}^{-p} \right)$$

$$= \sum_{u,v=0}^{\infty} |a_u|\alpha_u^{-p} \cdot |a_v|\alpha_v^{-p}$$
$$\times \left(\sum_{n \geq u,v} |b_{n-u}|\alpha_{n-u}^{-p} \cdot |b_{n-v}|\alpha_{n-v}^{-p} \right).$$

$$(5.9.11)$$

The Cauchy–Schwarz inequality gives

$$\sum_{n \geq u,v} |b_{n-u}|\alpha_{n-u}^{-p} \cdot |b_{n-v}|\alpha_{n-v}^{-p} \leq \left(\sum_{n \geq u,v} |b_{n-u}|^2\alpha_{n-u}^{-2p} \right)^{\frac{1}{2}} \left(\sum_{n \geq u,v} |b_{n-v}|^2\alpha_{n-v}^{-2p} \right)^{\frac{1}{2}}$$
$$\leq \|b\|_{-p}^2.$$

Thus (5.9.11) becomes

$$\|a \star b\|_{-p}^2 \leq \left(\sum_{u,v=0}^{\infty} |a_u|\alpha_u^{-p} \cdot |a_v|\alpha_v^{-p} \right) \cdot \|b\|_{-p}^2. \qquad (5.9.12)$$

But, as in the previous exercise:

$$\sum_{u,v=0}^{\infty} |a_u| \cdot |a_v|\alpha_u^{-p}\alpha_v^{-p} = \left(\sum_{u=0}^{\infty} |a_u|\alpha_u^{-p} \right)^2$$
$$= \left(\sum_{u=0}^{\infty} |a_u|\alpha_u^{-q}\alpha_u^{q-p} \right)^2$$
$$\leq \left(\sum_{u=0}^{\infty} |a_u|^2\alpha_u^{-2q} \right) \left(\sum_{u=0}^{\infty} \alpha_u^{2(q-p)} \right)$$
$$= \|a\|_{-q}^2 \cdot (A(p-q))^2,$$

where $A(p-q)$ is defined by (5.4.10), and is finite in view of (5.4.6) since $p-q \geq d$. $\qquad \square$
The result follows.

Solution of Exercise 5.5.2: The claim follows from the very definition of T^*.   □

Solution of Exercise 5.5.6: The strong topology in $\mathcal{V}'$ is defined by the bounded sets in $\mathcal{V}$. More precisely, a basis of open sets is given by sets of the form

$$\left\{ u \in \mathcal{V}' ; \sup_{z \in B} |\langle u, z \rangle_{\mathcal{V}', \mathcal{V}}| < \epsilon \right\}, \tag{5.9.13}$$

where B is a bounded set in $\mathcal{V}$. Using Exercise 5.2.5 we know that B is included in a set of the form

$$\{z \in \mathcal{V},\ |z_n| < c \text{ for } n = 1, \ldots, n_0 \text{ and } z_n = 0 \text{ for } n > n_0\}, \tag{5.9.14}$$

where $c > 0$ and $n_0 \in \mathbb{N}$.

With these definitions out of the way we prove that the natural map (the canonical inclusion)

$$(I(v))(w) = \sum_{n \in \mathbb{N}} w_n \overline{v_n}, \quad v, w \in \mathcal{V}, \tag{5.9.15}$$

indeed sends $\mathcal{V}$ into $\mathcal{V}^*$, that it is continuous and positive, but that I cannot be factored via a Hilbert space.

STEP 1: *I sends $\mathcal{V}$ into $\mathcal{V}^*$.*

To prove this claim, let $v \in \mathcal{V}$ and let Λ denote the support of v. We have

$$|I(v)(w)| = |\sum_{n \in \Lambda} w_n \overline{v_n}| \leq M p(w)$$

where $M = \sum_{n \in \Lambda} |v_n|$ and $p(w) = \sum_{n \in \Lambda} |w_n|$. The semi-norm p belongs to the family of norms defining the topology, and the continuity of $I(v)$ follows.

STEP 2: *I is continuous $\mathcal{V}$ into $\mathcal{V}^*$.*

Let N' be a neighborhood of $0 \in \mathcal{V}^*$, which we will take of the form

$$N' = \{\varphi \in \mathcal{V}^* :\ |\varphi(v_n)| < \epsilon \text{ for } n = 1, \ldots, N_0, \text{and } |v_n| < c\},$$

for some preassigned $\epsilon > 0$, $N_0 \in \mathbb{N}$ and $c > 0$. Let now

$$N = \left\{ (v_n)_{n \in \mathbb{N}} \in \mathcal{V} :\ \sum_{n=1}^{N_0} |v_n| < \frac{\epsilon}{c} \right\}.$$

Then $I(N) \subset N'$, and hence the continuity of the map I.

STEP 3: *The map I is positive.*

This follows from (note that we replace $\mathcal{V}'$ by $\mathcal{V}^*$)

$$\langle I(v), v \rangle_{\mathcal{V}^*, \mathcal{V}} = \sum_{n \in \mathbb{N}} |v_n|^2.$$

5.9. *Solutions of the exercises*

Assume now by contradiction that $I = J^*J$, where J is a linear continuous operator from $\mathcal{V}$ into some Hilbert space $\mathcal{H}$. Then, the equality

$$\sum_{n=1}^{\infty} |v_n|^2 = \langle Iv, v \rangle_{\mathcal{V}^*, \mathcal{V}} = \langle J^*Jv, v \rangle_{\mathcal{V}^*, \mathcal{V}} = \|Jv\|_{\mathcal{H}}^2$$

would imply that the map $v \mapsto \sum_{n=1}^{\infty} |v_n|^2$ is continuous from $\mathcal{V}$ into $\mathbb{R}$. But this is not the case. To see this take any semi-norm $p_{K,M}$ and any number $t > 0$. Choose $z \in \mathcal{V}$ such that $z_{n_0} = 1$ for some $n_0 \notin K$ and $z_n = 0$ otherwise (such n_0 exists since K has density 0). Then

$$\sum_{n=1}^{\infty} |z_n|^2 = 1 > t \cdot p_{K,M}(z) = 0.$$

In view of Remark 5.2.6, this shows that the semi-norm $\sqrt{\sum_{n=1}^{\infty} |z_n|^2}$ is not continuous in $\mathcal{V}$. See also [251, p. 7]. $\qquad \square$

Remark 5.9.1. We note that, in the previous exercise, not every set of integers can be chosen. The case $K = \mathbb{N}$ (which, of course, has not density 0) leads to $p(z) = \sum_{n=1}^{\infty} |z_n|$ and the function $\sqrt{\sum_{n=1}^{\infty} |z_n|^2}$ is clearly continuous with respect to this norm. We note that sequences of zero density rather than sequences with finite support are used in the definition of the norms to get a topology as small as possible on s_0. Finally we remark that the space $\mathcal{V}$ in the previous exercise plays a key role in the theory of sequences spaces. See Ruckle's book [262, Chapter 3, Section 3].

Solution of Exercise 5.6.1: The limit f is continuous since the convergence is uniform on compact subsets of Ω. Since analyticity is a local property, it is enough to show that f is analytic in any open convex subset, say E, of Ω. Let Δ be a triangle inside E, with oriented boundary $\partial\Delta$. For every $n \in \mathbb{N}$, Cauchy's theorem implies that

$$\int_{\partial\Delta} f_n(z)dz = 0. \qquad (5.9.16)$$

Since $\partial\Delta$ is a compact subset of Ω, the uniform convergence of the sequence $(f_n)_{n\in\mathbb{N}}$ on $\partial\Delta$ allows us to interchange limit and integral in (5.9.16) and write

$$0 = \lim_{n\to\infty} \int_{\partial\Delta} f_n(z)dz = \int_{\partial\Delta} (\lim_{n\to\infty} f_n(z))dz = \int_{\partial\Delta} f(z)dz.$$

Morera's theorem allows us to conclude that f is analytic in E, and hence in all of Ω. $\qquad \square$

Solution of Exercise 5.6.4: The space $\mathcal{H}(\Omega)$ is metrizable. So, in view of Exercise 3.9.22 it is enough to consider sequences to prove continuity. Furthermore, we use the characterization of the topology of $\mathcal{H}(\Omega)$ given in Question 5.6.3. Let thus $(f_n)_{n \in \mathbb{N}}$ be a sequence of functions analytic in Ω and converging uniformly on compact subsets of Ω to the function f (which is analytic, as is well known and follows for instance from Morera's theorem; see the previous exercise). Let $K \subset \Omega$ be a given compact set and let K_1 be another compact set such that

$$K \subset \mathring{K_1} \subset K_1 \subset \Omega.$$

(See Question 5.6.2 for the latter.) Then $K \subset \cup_{u=1}^N B(z_u, \epsilon_u) \subset K_1$ for some $N \in \mathbb{N}, z_1, \ldots, z_N \in K$ and $\epsilon_1, \ldots, \epsilon_N > 0$. We have

$$f_n'(z) - f'(z) = \frac{1}{2\pi i} \int_{|z_u - \zeta| = \epsilon_u} \frac{f_n(\zeta) - f(\zeta)}{(\zeta - z)^2} d\zeta, \quad z \in B(z_u, \epsilon_u),$$

and thus, with $\epsilon = \min \{\epsilon_1, \ldots, \epsilon_N\}$,

$$\max_{z \in K} |f_n'(z) - f'(z)| \le \frac{1}{\epsilon} \max_{z \in K_1} |f_n(z) - f(z)|,$$

and hence the continuity of the differentiation. Continuity of multiplication by z is proved similarly by writing

$$z f_n(z) - z f(z) = \frac{1}{2\pi i} \int_{|z_u - \zeta| = \epsilon_u} \frac{\zeta(f_n(\zeta) - f(\zeta))}{\zeta - z} d\zeta, \quad z \in B(z_u, \epsilon_u). \qquad \Box$$

Solution of Exercise 5.6.6: (see [95, p. 165]) Since a basis of neighborhoods of the origin consists of sets of functions uniformly bounded on a given compact, it follows that a set is bounded if and only if its elements are uniformly bounded on compact subsets of Ω.

$\Box$

Remark 5.9.2 (see [95, Proposition 1.1, p. 165]). Cauchy's formula implies that the operator of differentiation sends bounded sets into bounded sets. This is of course not surprising since $\mathcal{H}(\Omega)$ is a metric space and $f \mapsto f'$ is continuous.

Solution of Exercise 5.6.8:

(1) We follow [95, p. 168]. Conditions (5.6.2) follow from the assumed uniform convergence on compact subsets and from Cauchy's formula, and do not require the condition that the functions be bounded in modulus by 1 in the open unit disk. Under this condition we now prove that the sequence $(f_n)_{n \in \mathbb{N}}$ converges uniformly on compact subsets to f. Let $r \in (0, 1)$. Cauchy's formula

$$\frac{f_n^{(p)}(0)}{p!} = \frac{1}{2\pi i} \int_{|\zeta| = r} \frac{f_n(\zeta)}{\zeta^{p+1}} d\zeta, \quad p = 0, 1, \ldots$$

5.9. *Solutions of the exercises*

implies that the numbers $\frac{f_n^{(p)}(0)}{p!}$ are uniformly bounded by $\frac{1}{r^n}$ in modulus, and so are the limits $\frac{f^{(p)}(0)}{p!}$. Since r is arbitrary in $(0,1)$, these various numbers are in fact bounded by 1 in modulus:

$$\left|\frac{f_n^{(p)}(0)}{p!}\right| \le 1, \quad p = 0, 1, \ldots, \tag{5.9.17}$$

and

$$\left|\frac{f^{(p)}(0)}{p!}\right| \le 1, \quad p = 0, 1, \ldots. \tag{5.9.18}$$

Fix now $r \in (0,1)$. Then (5.9.17) and (5.9.18) allow us to use the dominated convergence theorem (certainly too heavy a tool here), or an $\epsilon/3$ argument (which we leave to the reader), to show that

$$\lim_{n\to\infty} \sum_{p=0}^{\infty} \frac{f_n^{(p)}(0)}{p!} z^p = \sum_{p=0}^{\infty} \frac{f^{(p)}(0)}{p!} z^p, \quad |z| \le r,$$

which ends the proof of (1).

(2) Inequalities (5.9.17) are still in force. The diagonal process will provide a subsequence of integer numbers for which all the limits $\lim_{n\to\infty} \frac{f_n^{(p)}(0)}{p!}$ exist. We refer the reader to Cartan's book [95] for more details. $\quad\square$

Solution of Exercise 5.6.10: Since $|q| < 1$, there exists $n_0 \in \mathbb{N}_0$ such that

$$n \ge n_0 \implies |q|^n \le \frac{1}{2}, \tag{5.9.19}$$

and therefore, for $n \ge n_0$ we have

$$\left|\frac{z^n}{1 - q^n}\right| \le \frac{|z|^n}{1 - |q|^n} \le \frac{|z|^n}{1 - \frac{1}{2}} = 2|z|^n. \tag{5.9.20}$$

Let now $n \in \{\ldots, -2, -1\}$, and write $m = -n$. Since $z \ne 0$ we can write:

$$\frac{z^n}{1 - q^n} = \frac{(qz^{-1})^m}{q^m - 1}$$

and for n_0 as in (5.9.19) and $m \ge n_0$ we have

$$\left|\frac{z^n}{1 - q^n}\right| = \left|\frac{(qz^{-1})^m}{q^m - 1}\right| \le 2|qz^{-1}|^m = 2|qz^{-1}|^{-n}. \tag{5.9.21}$$

The sum for positive n's converges in $|z| < 1$ in view of (5.9.20), while the sum for negative n's converges in $|q| < |z|$ in view of (5.9.21). These same bounds show that the convergence is uniform in every closed ring $r_1 \le |z| \le r_2$ with $|q| < r_1 < r_2 < 1$. $\quad\square$

Solution of Exercise 5.7.2: Let ι denote the restriction operator from $\mathcal{H}^\infty(\Omega_2)$ into $\mathcal{H}^\infty(\Omega_1)$. To show that ι is compact we will show that the image of the closed unit ball in $\mathcal{H}^\infty(\Omega_2)$ is relatively compact in $\mathcal{H}^\infty(\Omega_1)$. It is enough to show sequential compactness since $\mathcal{H}^\infty(\Omega_1)$ is a metric space. Let $(\iota(f_n))_{n\in\mathbb{N}}$ be a sequence of elements in the image $\iota(\mathcal{H}^\infty(\Omega_2))$, where the functions f_n belong to the closed unit ball of $\mathcal{H}^\infty(\Omega_2)$. By Montel's theorem, there exists a subsequence $(f_{n_k})_{k\in\mathbb{N}}$ of the sequence $(f_n)_{\in\mathbb{N}}$ which converges uniformly on compact subsets of Ω_2, and in particular in $\overline{\Omega_1}$, to a function $f \in \mathcal{H}^\infty(\Omega_2)$. Thus

$$\lim_{k\to\infty} \sup_{z\in\overline{\Omega_1}} |f_{n_k}(z) - f(z)| = 0.$$

But

$$\sup_{z\in\overline{\Omega_1}} |f_{n_k}(z) - f(z)| = \sup_{z\in\overline{\Omega_2}} |\iota(f_{n_k})(z) - \iota(f)(z)|.$$

It follows that $\iota(f)$ belongs to the closed unit ball of $\mathcal{H}^\infty(\Omega_2)$ and is the limit of the functions $\iota(f_{n_k})$ in the norm of this space.

The restriction map is not onto since there exist functions analytic in Ω_1 which do not have an analytic extension to Ω_2. $\square$

Remark 5.9.3. When Ω_2 is connected, ι is one-to-one and one can also resort to the open mapping theorem (see Theorem 4.2.15) to show that ι is not onto. If it was onto, the open mapping theorem would imply that ι^{-1} is continuous and so the identity map from $\mathcal{H}^\infty(\Omega_2)$ into itself would be compact. This cannot be since the space is infinite dimensional.

Solution of Exercise 5.8.1: By definition of a continuous linear map between Fréchet spaces there exist $R \in (0,1)$ and $C > 0$ such that

$$|\varphi(f)| \le C \max_{|z|=R} |f(z)|.$$

We obtain condition (5.8.1) by taking $f(z) = z^n$, $n = 0, 1, \dots$.

Conversely, let $(b_n)_{n\in\mathbb{N}_0}$ be a sequence of complex numbers subject to (5.8.2). We note that (5.8.3) makes sense for every $f \in \mathcal{H}(\mathbb{D})$ when (5.8.1) is in force. Indeed, let

$$f(z) = \sum_{n=0}^{\infty} a_n z^n$$

be the Taylor series of f at the origin. It has radius of convergence at least one and so the series

$$\sum_{n=0}^{\infty} a_n \varphi(z^n)$$

converges absolutely since

$$\left|\sum_{n=0}^{\infty} a_n \varphi(z^n)\right| \le \sum_{n=0}^{\infty} |a_n| \cdot |\varphi(z^n)| \le C \sum_{n=0}^{\infty} |a_n| R^n < \infty \quad \text{for } R \in (0,1).$$

5.9. *Solutions of the exercises*

As in the solution of item (2) of Exercise 2.3.4 we define a function g by (2.8.10):

$$g(z) = \sum_{m=0}^{\infty} \frac{b_m}{z^m}, \quad |z| > R.$$

It holds that

$$\frac{1}{2\pi i} \int_{|z|=R_1} \frac{f(z)g(z)}{z} dz = \sum_{n=0}^{\infty} a_n \left(\frac{1}{2\pi i} \int_{|z|=R_1} z^{n-1} g(z) dz \right).$$

It follows that

$$\varphi(f) = \frac{1}{2\pi i} \int_{|z|=R_1} \frac{f(z)g(z)}{z} dz$$

since

$$\frac{1}{2\pi i} \int_{|z|=R_1} z^{n-1} g(z) dz = b_n. \tag{5.9.22}$$

(Use once more the dominated convergence theorem if need be to prove (5.9.22), but this is just the formula for the negative coefficients in a Laurent series; see, e.g., [CAPB, (7.1.1), p. 318]). Thus

$$|\varphi(f)| \le C \max_{|z|=R_1} |f(z)|$$

$\square$

with $C = \max_{|z|=R_1} |g(z)|$, and φ is continuous.

Chapter 6

Some Functional Analysis

Functional analysis and complex variables interact in numerous places. In this chapter we picked out some instances of these connections. We first discuss the Fourier transform and the Stieltjes integral. Then, we focus on some density results. We also discuss the Hankel transform and realization of analytic functions.

6.1 Fourier transform

Recall that the Fourier transform of a function $f \in \mathbf{L}_1(\mathbb{R}, \mathcal{B}, du)$ is given by

$$\widehat{f}(t) = \int_{\mathbb{R}} e^{-iut} f(u) du. \tag{6.1.1}$$

Note that other definitions are possible, such as

$$\int_{\mathbb{R}} e^{-2\pi i u t} f(u) du, \tag{6.1.2}$$

$$\frac{1}{\sqrt{2\pi}} \int_{\mathbb{R}} e^{-iut} f(u) du, \tag{6.1.3}$$

(see Remark 6.4.2 in connection with this definition), or

$$\frac{1}{2\pi} \int_{\mathbb{R}} e^{-iut} f(u) du.$$

Each of these definitions has its own advantages and disadvantages in computations. A very interesting characterization of the Fourier transform is given in [54, Theorem 1.3, p. 36] by Artstein-Avidan, Faifman and Milman. Their result is given in $\mathbb{R}^N$. For $N = 1$ it reads as follows (note that neither continuity nor linearity are assumed in the statement):

Theorem 6.1.1 (S. Artstein-Avidan, D. Faifman and V. Milman). *Let F be a bijection from the Schwartz space into itself such that*

$$F(f \star g) = F(f)F(g),$$

where ⋆ denotes convolution. Then, there is a diffeomorphism x of the real line such that

$$(F(f))(t) = \int_{\mathbb{R}} e^{-2\pi i u x(t)} f(u)du \quad or \quad (F(f))(t) = \overline{\int_{\mathbb{R}} e^{-2\pi i u x(t)} f(u)du}.$$

The Fourier transform extends (up to a factor $\frac{1}{\sqrt{2\pi}}$) to a unitary map from $\mathbf{L}_2(\mathbb{R}, \mathcal{B}, du)$ into itself,

$$\|f\|_2 = \frac{1}{\sqrt{2\pi}} \|\widehat{f}\|_2, \tag{6.1.4}$$

where we have denoted by $\| \cdot \|_2$ the norm in $\mathbf{L}_2(\mathbb{R}, \mathcal{B}, du)$, with inverse given by

$$\check{f}(u) = \frac{1}{2\pi} \int_{\mathbb{R}} e^{iut} f(t)dt, \tag{6.1.5}$$

when f is summable. It holds that

$$\|f\|_2 = \sqrt{2\pi} \|\check{f}\|_2, \tag{6.1.6}$$

and this allows us to extend the inverse Fourier transform to all of $\mathbf{L}_2(\mathbb{R}, \mathcal{B}, du)$. See item (3) of Exercise 6.1.22 for an explicit expression for the extension of the Fourier transform to the whole of $\mathbf{L}_2(\mathbb{R}, \mathcal{B}, du)$.

A first and classical example of a Fourier transform has been given in Exercise 2.1.26. More precisely, with $f(u) = e^{-\frac{u^2}{2}}$, we have

$$\widehat{f} = \sqrt{2\pi} f \quad and \quad \check{f} = \frac{1}{\sqrt{2\pi}} f,$$

as follows from (2.1.30). See (6.1.23) for the counterpart of this last equality to arbitrary elements in $\mathbf{L}_2(\mathbb{R}, \mathcal{B}, dx)$.

We note the following. When choosing definition (6.1.2) for the Fourier transform of the function f, then the inverse Fourier transform is $H = F \circ \phi$ with $\phi(t) = -t$ and

$$\|F\|_2 = \|H\|_2 = \|f\|_2.$$

See for instance [163, Theorem 2.2.14].

In the next example the function $f \notin \mathbf{L}_1(\mathbb{R}, \mathcal{B}, dx)$. Therefore one cannot use directly formula (6.1.5) to compute the inverse Fourier transform.

6.1. Fourier transform

Exercise 6.1.2. *Compute the Fourier transform of the function* $f(u) = \frac{1}{u+i}$.

Hint: Check first that

$$\frac{1}{u+i} = \frac{1}{2\pi} \int_0^\infty e^{itu} e^{-t}(-2\pi i)dt, \qquad (6.1.7)$$

where the integral is absolutely convergent, and use the formula for the inverse Fourier transform.

More generally:

Exercise 6.1.3.

(1) *Compute the inverse Fourier transform of the functions*

$$\frac{1}{(u+i)^n} \quad and \quad \frac{1}{(u-i)^n}, \qquad n = 1, 2, \cdots$$

(2) *Show that the functions* f_n *defined by*

$$\widehat{f_n}(u) = \frac{\sqrt{2}}{u+i} \left(\frac{u-i}{u+i} \right)^n, \qquad n \in \mathbb{Z} \qquad (6.1.8)$$

are orthonormal in $\mathbf{L}_2(\mathbb{R}, \mathcal{B}, du)$.
(3) *What is the support of* f_n *for* $n \geq 0$ *and* $n < 0$?

Hint: Start from (6.1.7) and, using the dominated convergence theorem, differentiate with respect to t both sides of the equation.

Exercise 6.1.4. *Compute the Fourier transforms of*

$$\frac{1}{(t^2 + 1)^n} \quad and \quad \frac{1}{(t^{2n} + 1)}$$

respectively.

We refer to [71, p. 277] for a discussion related to the second kernel. More generally, the computation of the Fourier transform of a rational function without poles on the real line, and analytic at infinity, plays an important role in various questions in stochastic processes and linear system theory. For instance Gelfand and Yaglom use such computations to evaluate the amount of information of a given stationary Gaussian process $y(t)$ inside another such process. See [149]. In the next exercise, we present an approach based on the notion of realization of rational functions. For the definition of the Riesz projection in the exercise, see Exercise 4.2.13 above. For more on, and applications of, formula (6.1.9) see for instance the paper [65] of Bart, Kaashoek and Gohberg for applications to convolution equations and linear systems.

Exercise 6.1.5 (Fourier transform of a rational function). *Let $A \in \mathbb{C}^{N \times N}$ without spectrum on the real line, and let B, C be matrices of appropriate sizes. Show that*

$$-\frac{1}{2\pi} \int_{\mathbb{R}} e^{-i\lambda t} C(\lambda I_N - A)^{-1} B d\lambda = \begin{cases} -iCe^{-itA} PB, & t < 0 \\ iCe^{-itA}(I_N - P)B, & t > 0 \end{cases}, \quad (6.1.9)$$

where P denotes the Riesz projection corresponding to the eigenvalues of A in the open upper half-plane.

Hint: Check that (see [65, p. 284]):

$$-C(\lambda I_N - A)^{-1} B = \int_{\mathbb{R}} e^{i\lambda t} k(t) dt, \quad \lambda \in \rho(A), \quad (6.1.10)$$

where the function $k(t)$ is given by

$$k(t) = \begin{cases} -iCe^{-itA} PB, & t < 0, \\ iCe^{-itA}(I_N - P)B, & t > 0, \end{cases} \quad (6.1.11)$$

and use formula (6.1.5) for the inverse Fourier transform.

We note that (6.1.7) is a very simple instance of (6.1.10) with

$$A = -i, \quad P = 0, \quad \text{and} \quad B = C = 1.$$

When $f \in \mathbf{L}_1(\mathbb{R}, \mathcal{B}, dx)$ the Fourier transform is easily seen to be continuous; see [CAPB, Exercise 15.6.3, p. 503]. Another important property of the Fourier transform in this case is given in the next exercise, and called Riemann's lemma.

Exercise 6.1.6. *Let $f \in \mathbf{L}_1(\mathbb{R}, \mathcal{B}, dx)$. Show that*

$$\lim_{t \to \pm\infty} \widehat{f}(t) = 0.$$

The Lebesgue space $\mathbf{L}_2(\mathbb{R}, \mathcal{B}, dx)$ is invariant under the Fourier transform. Here is another key example of a space invariant under this transform.

Exercise 6.1.7. *Show that the Schwartz space (see Definition 5.2.11) is invariant under the Fourier transform.*

Hint: Use Exercise 6.1.6.

Remark 6.1.8. For a proof of this classical result, see, e.g., [326, pp. 183–184]. Furthermore, if one knows (1) that the Hermite functions form an orthonormal basis of $\mathbf{L}_2(\mathbb{R}, \mathcal{B}, dx)$, and (2) that they are eigenvectors of the Fourier transform, and finally, (3) how to characterize Schwartz functions in terms of their expansion along the Hermite functions, the above exercise has a very quick solution.

The following example will be used in Section 7.2 to illustrate Bochner's theorem. See Exercise 7.2.10. It is taken from the paper [81] of Marek Bozejko and Takahiro Hasebe. See Exercise 7.2.10 there. This integral and some related

6.1. Fourier transform

ones appear in the note[1] [66] of Jean Bass and Paul Lévy. The authors trace the example to the 1937 thesis of G. Kunetz. We also mention the work [73] where the functions considered in the next exercise and question appear in connection with Riemann's function and the theta functions. See also Remark 7.2.16 and Exercise 6.1.31.

Exercise 6.1.9. *Compute the Fourier transform of the function* $\frac{1}{\cosh t}$.

Following [66] we ask:

Question 6.1.10. *Compute the Fourier transforms of the functions* $\frac{1}{\cosh^2 t}$ *and* $\frac{t}{\sinh t}$.

Exercise 6.1.11. *Solve equation* (2.1.34)

$$z^n = \int_{\mathbb{R}} x_n(u)\overline{h_z(u)}du, \quad n = 0, 1, \ldots$$

and show that

$$x_n(u) = \frac{1}{\sqrt[4]{\pi}2^{\frac{n}{2}}} e^{\frac{u^2}{2}} \left(e^{-u^2}\right)^{(n)}, \quad n = 0, 1, \ldots \tag{6.1.12}$$

Hint: Rewrite (2.1.34) as a Fourier transform and use the inverse Fourier transform. To find the corresponding inverse transform differentiate (2.1.30) n-times.

Definition 6.1.12. The functions

$$\xi_n(z) = (-1)^n \frac{1}{\sqrt[4]{\pi}2^{\frac{n}{2}}} e^{\frac{z^2}{2}} \left(e^{-z^2}\right)^{(n)}$$

$$= \frac{1}{\sqrt[4]{\pi}2^{\frac{n}{2}}} e^{-\frac{z^2}{2}} H_n(z), \quad n = 0, 1, \ldots \tag{6.1.13}$$

are the (unnormalized) Hermite functions, and the polynomials

$$H_n(z) = (-1)^n e^{z^2} \left(e^{-z^2}\right)^{(n)}, \quad n = 0, 1, \ldots$$

(already appearing in Exercise 2.1.16; see (2.1.19)) are the Hermite polynomials.

In relation with the following exercise, see also Exercise 6.1.31.

Exercise 6.1.13. *Compute the Fourier transform of the Hermite functions.*

Hint: Let

$$g_n(z) = (-1)^n \sqrt[4]{\pi}2^{\frac{n}{2}} \xi_n(z) = e^{\frac{z^2}{2}} \left(e^{-z^2}\right)^{(n)}, \quad n = 0, 1, \ldots$$

---

[1]We mention to the reader that this note can be obtained at

http://gallica.bnf.fr/ark:/12148/bpt6k3182n/f815.image.

Using formula (2.1.30), compute the sum

$$\sum_{n=0}^{\infty} \frac{i^n s^n}{n!} \widehat{\mathcal{G}_n}(t), \quad s \in \mathbb{C}.$$

We here have followed the definitions of Hille; see [177, (3), p. 431 and (14), p. 432]. Other variations appear in the literature. See for instance [182, p. 208], [188, p. 36]. We note that (see, e.g., [177, (31), p. 436])

$$\int_{\mathbb{R}} e^{-u^2} H_n(u) H_m(u) du = \sqrt{\pi} 2^n n! \delta_{m,n}. \tag{6.1.14}$$

Thus the functions $(\sqrt{\pi} 2^n n!)^{-1/2} H_n(u)$, $n = 0, 1, 2, \ldots$ can be obtained from the functions $1, u, u^2, \ldots$ by the Gram–Schmidt orthogonalization process, and are the orthogonal polynomials (defined, *a priori*, up to a unitary constant) associated with the weight e^{-u^2}.

Remark 6.1.14. It holds that

$$\langle \xi_n, \xi_m \rangle_2 = \delta_{m,n} n! \tag{6.1.15}$$

This follows from the fact that in the reproducing kernel Hilbert space $\mathcal{F}$ with reproducing kernel $e^{z\overline{w}}$ (the Fock space;[2] see Chapter 11) we have

$$\langle z^n, z^m \rangle_{\mathcal{F}} = \delta_{n,m} n!.$$

It can also be verified from formula (6.1.14). The functions

$$\eta_n(u) = \frac{1}{\sqrt[4]{\pi} 2^{\frac{n}{2}} \sqrt{n!}} e^{\frac{u^2}{2}} \left(e^{-u^2} \right)^{(n)}, \quad n = 0, 1, \ldots \tag{6.1.16}$$

are the normalized Hermite functions.

Exercise 6.1.15. *Show that the linear span of the Hermite functions is dense in* $L_2(\mathbb{R}, \mathcal{B}, dx)$.

Hint: Use formula (2.1.20) with $z = -\frac{it}{2}$, t real.

Exercise 6.1.16. *Let ξ_n be as in Exercise 6.1.11. Show that*

$$\xi_n''(z) = (z^2 - 2n - 1)\xi_n(z), \tag{6.1.17}$$

Hint: Express ξ_n in terms of the Hermite polynomial H_n (see (2.1.19) or (6.1.13)) and use (2.1.22).

---

[2]More precisely, the symmetric Fock space associated with $\mathbb{C}$, or the boson Fock space with one degree of freedom; see for instance [246, p. 227] for the latter.

Remark 6.1.17. The classical differential equation (6.1.17) can be found for instance in [177, (15), p. 432]. It expresses the fact that the Hermite functions are the eigenvalues of the positive (densely defined unbounded) self-adjoint operator (4.3.3):

$$Hf(x) = -f''(x) + (x^2 + 1)f(x),$$

that is

$$H\xi_n = 2(n+1)\xi_n, \quad n = 0, 1, \dots \tag{6.1.18}$$

See for instance [7, Theorem 5, p. 499 and p. 543] for a discussion of the self-adjointness of the minimal operator defined by H, that is, of the operator defined by H and with domain $C^2(\mathbb{R})$ functions f such that f, f' and Hf are in $\mathbf{L}_2(\mathbb{R}, \mathcal{B}, dx)$.

The closely related operator (4.3.5):

$$Bf(x) = \frac{-f''(x) + (x^2 - 1)f(x)}{2}$$

defined in Exercise 4.3.9 is such that (see Exercise 6.1.33)

$$(e^{i\pi B})f = \frac{1}{\sqrt{2\pi}}\widehat{f}, \quad f \in \mathbf{L}_2(\mathbb{R}, \mathcal{B}, dx). \tag{6.1.19}$$

Remark 6.1.18. We note that the Schwartz space $\mathscr{S}$ is equal to

$$\mathscr{S} = \cap_{p=0}^{\infty} \mathcal{K}_p,$$

where the Hilbert space $\mathcal{K}_p$ is given by

$$\mathcal{K}_p = \left\{ f = \sum_{n=0}^{\infty} a_n \eta_n ; \sum_{n=0}^{\infty} (n+1)^{2p} |a_n|^2 < \infty \right\}, \tag{6.1.20}$$

that is, in terms of the operator H defined by (4.3.3) and appearing in Remark 6.1.17,

$$\mathscr{S} = \cap_{p=1}^{\infty} \mathrm{Dom}\, H^p.$$

We take this opportunity to recall the following formula, due in particular to Mehler, see [231], [177, (39), p. 439]:

$$\sum_{n=0}^{\infty} t^n \frac{H_n(z)H_n(w)}{2^n n!} = \frac{1}{\sqrt{1-t^2}} e^{\frac{2zwt - t^2(z^2 + w^2)}{1 - t^2}}, \quad t \in \mathbb{D}, \tag{6.1.21}$$

or, in terms of the Hermite functions,

$$\sum_{n=0}^{\infty} t^n \frac{\xi_n(z)\xi_n(w)}{n!} = \frac{1}{\sqrt{\pi(1-t^2)}} e^{\frac{4zwt - (z^2 + w^2)(1 + t^2)}{2(1 - t^2)}}, \quad t \in \mathbb{D}. \tag{6.1.22}$$

Three proofs of this formula, including one due to Hardy, can be found in Watson's paper [319].

Remark 6.1.19. Replacing the function e^{-u^2} by $|u|^{2\mu}e^{-u^2}$ with $\mu > -\frac{1}{2}$ the Gram-Schmidt orthogonalization process leads to the generalized Hermite polynomials, and most (if not all) of the previous analysis can be extended to this case. We refer to Rosenblum's paper [260] for a fascinating account and references. We will only mention two points. The Lebesgue space $\mathbf{L}_2(\mathbb{R}, \mathcal{B}, du)$ is now replaced by $\mathbf{L}_2(\mathbb{R}, \mathcal{B}, |u|^{2\mu}du)$, and the exponential function e^u is replaced by the function (see [260, (2.2.4), p. 371])

$$E_\mu(u) = \sum_{n=0}^{\infty} \frac{u^n}{\gamma_\mu(n)},$$

where

$$\gamma_\mu(2n) = \frac{2^{2n}n!\Gamma(n + \mu + 1/2)}{\Gamma(\mu + 1/2)},$$

$$\gamma_\mu(2n + 1) = \frac{2^{2n+1}n!\Gamma(n + \mu + 3/2)}{\Gamma(\mu + 1/2)}.$$

The function

$$E_\mu(z\overline{w}) = \sum_{n=0}^{\infty} \frac{z^n\overline{w}^n}{\gamma_\mu(n)}$$

is called the Dunkl kernel. See the article [290, (3), p. 93] of Sifi and Soltani for this and for a generalization of the Fock space to the present setting.

Question 6.1.20. *Prove that, for $\mu > -\frac{1}{2}$,*

$$\frac{2^{2n}n!\Gamma(n + \mu + 1/2)}{\Gamma(\mu + 1/2)} = (2n)!\frac{\Gamma(n + \mu + 1/2)\Gamma(1/2)}{\Gamma(n + 1/2)\Gamma(\mu + 1/2)},$$

$$\frac{2^{2n+1}n!\Gamma(n + \mu + 3/2)}{\Gamma(\mu + 1/2)} = (2n + 1)!\frac{\Gamma(n + \mu + 3/2)\Gamma(1/2)}{\Gamma(n + 3/2)\Gamma(\mu + 1/2)}.$$

Exercise 6.1.21. *Let $\phi(t) = -t$. Prove the formulas, valid for $f, g \in \mathbf{L}_2(\mathbb{R}, \mathcal{B}, dx)$,*

$$\widehat{f \circ \phi} = 2\pi f, \tag{6.1.23}$$

$$\langle \widehat{f}, g \rangle_2 = \langle f, \widehat{g \circ \phi} \rangle_2, \tag{6.1.24}$$

where we have denoted by $\langle \cdot, \cdot \rangle_2$ the inner product in $\mathbf{L}_2(\mathbb{R}, \mathcal{B}, dx)$.

We also refer the reader to the remark that appears after the proof of the exercise.

Formulas (6.1.1) and (6.1.5) are to be understood in general in the $\mathbf{L}_2$ sense, and the following result, due to Paul Lévy, is therefore of special interest. See [308, (32), p. 123], [220]. It provides an extension of the Fourier transform to the whole of $\mathbf{L}_2(\mathbb{R}, \mathcal{B}, dx)$. See item (3) in the exercise. For further discussion of this latter point, see for instance [6, p. 116].

6.1. Fourier transform

Exercise 6.1.22.

(1) Let $f \in \mathbf{L}_2(\mathbb{R}, \mathcal{B}, dx)$ with Fourier transform $\widehat{f}$. Then, for every $t \in \mathbb{R}$,

$$\int_0^t \widehat{f}(u)du = \int_{\mathbb{R}} \frac{1 - e^{-itu}}{iu} f(u)du. \qquad (6.1.25)$$

(2) Show that the functions

$$\gamma_t(u) = \begin{cases} \frac{e^{itu}-1}{iu}, & u \neq 0, \\ t & u = 0, \end{cases} \qquad (6.1.26)$$

are dense in $\mathbf{L}_2(\mathbb{R}, \mathcal{B}, dx)$ when t runs through $\mathbb{R}$.

(3) Show that for $f \in \mathbf{L}_2(\mathbb{R}, \mathcal{B}, dx)$ it holds that

$$\widehat{f}(t) = \frac{d}{dt} \int_{\mathbb{R}} \frac{1 - e^{-itu}}{iu} f(u)du, \quad a.e.$$

Hint: Use (6.1.24) to prove the first item.

More generally, let μ be a positive Borel measure on $\mathbb{R}$ such that (3.7.1) holds,

$$\int_{\mathbb{R}} \frac{d\mu(t)}{t^2 + 1} < \infty.$$

Then the functions γ_t belong to $\mathbf{L}_2(\mathbb{R}, \mathcal{B}, d\mu)$. The description of the closed linear span of the functions γ_t when t is in some possibly semi-infinite interval of $\mathbb{R}$ is an important and difficult problem, which involves Hilbert spaces of entire functions introduced by de Branges, and is related to the prediction theory of stationary second-order stochastic processes. See [84, 126] for more information.

In view of the (solution of the) following exercise, we define $\mathbf{1}_t$ by

$$\mathbf{1}_t = \begin{cases} \mathbf{1}_{[0,t]}, & \text{for } t \geq 0 \\ \mathbf{1}_{[t,0]}, & \text{for } t \leq 0. \end{cases} \qquad (6.1.27)$$

Exercise 6.1.23. *Show that*

$$\frac{|t| + |s| - |t - s|}{2} = \frac{1}{2\pi} \int_{\mathbb{R}} \frac{(1 - e^{itu})(1 - e^{-isu})}{u^2} du, \quad t, s \in \mathbb{R}. \qquad (6.1.28)$$

Hint: Use (6.1.25) and formula (6.1.5) for an appropriate choice of f.

See Exercise 7.2.6 for a general family of functions which includes as a special case (6.1.28).

Formally differentiating (6.1.25) on both sides, one gets back to the definition (6.1.1) for the Fourier transform. Such a differentiation is not justified in general for $f \in \mathbf{L}_2(\mathbb{R}, \mathcal{B}, dx)$. But the Fourier transform of such a function f is also in $\mathbf{L}_2(\mathbb{R}, \mathcal{B}, dx)$, and so is locally in $\mathbf{L}_1(\mathbb{R}, \mathcal{B}, dx)$. The fundamental theorem of calculus (see, e.g., [264, Théorème 8.17, p. 158]) allows us to claim that the function $t \mapsto \int_0^t \widehat{f}(u)du$ is absolutely continuous, and differentiable almost everywhere, and:

$$\frac{\mathrm{d}}{\mathrm{d}t} \int_0^t \widehat{f}(u)du = \widehat{f}(t), \quad \text{a.e.}$$

This last equality can also be understood in the sense of distributions, as explained in the next exercise:

Exercise 6.1.24. *Let $f \in \mathbf{L}_2(\mathbb{R}, \mathcal{B}, du)$ and let g be a Schwartz function (see Definition 5.2.11). Show that*

$$\int_{\mathbb{R}} \left(\int_{\mathbb{R}} \frac{1 - e^{-itu}}{iu} f(u)du \right) g'(t)dt = - \int_{\mathbb{R}} \widehat{f}(u)g(u)du. \tag{6.1.29}$$

Exercise 6.1.25. *Assume that both f and $t \mapsto tf(t)$ are in $\mathbf{L}_2(\mathbb{R}, \mathcal{B}, dx)$.*

(1) *Show that $\widehat{f}$ is continuous.*

(2) *Let $g(t) = itf(t)$. Show that for every Schwartz function s (see Definition 5.2.11) it holds that*

$$\langle s', \widehat{f} \rangle = \langle s, \widehat{g} \rangle. \tag{6.1.30}$$

(3) *Compute $\|f\|^2 + \|\widehat{g}\|^2$.*

Hint for (1): Show that $f \in \mathbf{L}_1(\mathbb{R}, \mathcal{B}, dx)$ and use [Exercise 15.6.3, p. 503, CABP].

Remark 6.1.26. Equation (6.1.30) means that the derivative of $\widehat{f}$ in the sense of distributions is $-\widehat{g}$. The function $\widehat{f}$ belongs to the Sobolev space of functions in $\mathbf{L}_2(\mathbb{R}, \mathcal{B}, dx)$ with distributional derivative of the Fourier transform also in $\mathbf{L}_2(\mathbb{R}, \mathcal{B}, dx)$.

Item (2) in the following question can be used to show that the operator $i\frac{\mathrm{d}}{\mathrm{d}t}$ with domain the functions f in $\mathbf{L}_2(\mathbb{R}, \mathcal{B}, dx)$ which are absolutely continuous and with derivative also in $\mathbf{L}_2(\mathbb{R}, \mathcal{B}, dx)$ is self-adjoint.

Question 6.1.27.

(1) *Show that both f and f' are in $\mathbf{L}_2(\mathbb{R}, \mathcal{B}, dx)$ if and only if both $\widehat{f}$ and $t \mapsto t\widehat{f}(t)$ are in $\mathbf{L}_2(\mathbb{R}, \mathcal{B}, dx)$ and that*

$$\widehat{f'}(t) = it\widehat{f}(t).$$

(2) *Let g and h in $\mathbf{L}_2(\mathbb{R}, \mathcal{B}, dx)$ be such that*

$$\int_{\mathbb{R}} f'(u)\overline{g(u)}du = \int_{\mathbb{R}} f(u)\overline{h(u)}du$$

for all Schwartz functions. Relate g and h.

The following exercise plays an important role in the proof of Polya's theorem which gives sufficient conditions for a function to be positive definite; see Exercise 7.2.23. We here consider a simpler case. The arguments are taken from [224, pp. 83–85]. See also after the proof of the exercise the discussion on the conditions Polya assumed.

Exercise 6.1.28. *Let f be an even continuously differentiable function from $\mathbb{R}$ into itself, and assume that:*

1. *f belongs to $\mathbf{L}_1(\mathbb{R}, du)$,*
2. *$f'(u) \leq 0$ for $u \geq 0$ and $-f'$ is decreasing to 0 on $[0, \infty)$,*
3. *$\lim_{u \to \infty} f(u) = 0$.*

Show that $\widehat{f}$ is real and positive on the real line.

Hint: (see [224, p. 84]) Show that, for $t > 0$,

$$\widehat{f}(t) = -\frac{2}{t} \int_0^{\frac{\pi}{t}} \left(\sum_{n=0}^{\infty} (-1)^n f'(u + \frac{n\pi}{t}) \right) \sin(ut) du, \quad t > 0 \qquad (6.1.31)$$

and use Abel's lemma on alternating series.

Remark 6.1.29. Polya's conditions require f to be convex on $(0, \infty)$ rather than continuously differentiable. It then has a right derivative everywhere. See [224, §4.3, p. 83].

The Fourier transform (6.1.1) maps $\mathbf{L}_2(\mathbb{R}_+, du)$ onto an important subspace of $\mathbf{L}_2(\mathbb{R}, du)$, the elements of which are boundary values of a Hilbert space of functions analytic in the open right half-plane $\mathbb{C}_r$. This space, called the Hardy space of the open right half-plane, and denoted by $\mathbf{H}_2(\mathbb{C}_r)$, is studied in Section 8.6 below. More generally, the space $\mathbf{H}_2^\nu(\mathbb{C}_r)$ defined in item (2) in the following question is called the fractional Hilbert space of order ν. It plays an important role in the study of self-similar systems; see [230]. See Section 8.7 for more details.

Question 6.1.30. *Let $\nu > -1$.*

(1) *Let $f \in \mathbf{L}_2(\mathbb{R}_+, du)$ be such that*

$$F(z) = \int_0^\infty t^{\frac{\nu}{2}} f(u) e^{-zu} du = 0, \quad \forall z \in \mathbb{C}_r. \qquad (6.1.32)$$

Show that $f = 0$.

(2) *Show that the space $\mathbf{H}_2^\nu(\mathbb{C}_r)$ of functions of the form (6.1.32), endowed with the norm*

$$\|F\|_2 = \|f\|_2$$

is a Hilbert space.

(3) *Let $\mathbf{H}_2^\nu(\mathbb{C}_r)$ denote the space in (2), and let $a > 0$. Show that the map which to F associates the function $a^{\frac{\nu+1}{2}} F(az)$ is unitary from $\mathbf{H}_2^\nu(\mathbb{C}_r)$ into itself.*

Related to the following exercise, see also Exercise 6.1.13 and Remark 6.4.2.

Exercise 6.1.31.

(1) *Let*

$$Ff = \frac{1}{\sqrt{2\pi}}\widehat{f}.$$

Show that

$$F^4 = I.$$

(2) *Compute the spectrum of the Fourier transform (as an operator from $\mathbf{L}_2(\mathbb{R}, \mathcal{B}, dx)$ into itself) using Exercise 4.2.39.*

Question 6.1.32. *Show that $\sigma(F)$ is made of eigenvalues and compute the associated eigenvectors.*

The Fourier transform is part of a semi-group of operators, called the Fourier rotation operators. It is a pleasure to thank Prof. J. Snygg for pointing out the reference [292], where these operators are defined and studied in the context of quantum mechanics.

Question 6.1.33.

(1) *Prove (6.1.19) (with B defined by (4.3.5)):*

$$(e^{i\pi B})f = \frac{1}{\sqrt{2\pi}}\widehat{f}, \quad f \in \mathbf{L}_2(\mathbb{R}, \mathcal{B}, dx).$$

(2) *Let $f \in \mathbf{L}_2(\mathbb{R}, \mathcal{B}, dx)$ with Hermite expansion $f(x) = \sum_{n=0}^{\infty} \eta_n(x)f_n$ (where the η_n are the normalized Hermite functions, see (6.1.16) for the latter, and $(f_n)_{n\in\mathbb{N}_0} \in \ell_2(\mathbb{N}_0)$) and for $t \in \mathbb{R}$,*

$$T_t f = \sum_{n=0}^{\infty} e^{itn} \eta_n f_n. \tag{6.1.33}$$

Show that T_t is unitary and that

$$T_{t+s} = T_t T_s, \quad t, s \in \mathbb{R}. \tag{6.1.34}$$

(3) *(see [314, Theorem 4.2, p. 97]) Compute the domain of e^{uB} for $u > 0$.*

Hints: In principle, one needs the spectral theorem to give formula (6.1.19) a precise meaning. Here we can, at least informally, avoid it and define $e^{i\pi B}$ (and more generally e^{itB} with $t \in \mathbb{R}$) first on the Hermite functions via

$$e^{i\pi B}\eta_n = e^{i\pi n}\eta_n, \quad \text{and more generally,} \quad e^{itB}\eta_n = e^{itn}\eta_n, \quad n = 0, 1, 2, \ldots$$

Claim (1) then follows by using (6.4.15). Furthermore one then defines

$$e^{itB}f = \sum_{n=0}^{\infty} e^{itn} f_n\eta_n, \quad \text{with } f = \sum_{n=0}^{\infty} f_n\eta_n.$$

6.1. *Fourier transform*

We conclude this section with the proof that the Bargmann transform is injective. See Exercise 2.1.27.

Remark 6.1.34. Let $f \in \mathbf{L}_2(\mathbb{R}, \mathcal{B}, dx)$ be such that the function

$$\frac{1}{\pi^{\frac{1}{4}}} \int_{\mathbb{R}} f(u) e^{\left\{-\frac{1}{2}(z^2+u^2)+\sqrt{2}zu\right\}} du \equiv 0.$$

Setting $z = it$ we obtain that

$$\int_{\mathbb{R}} f(u) e^{\left\{-\frac{u^2}{2}+\sqrt{2}itu\right\}} du \equiv 0,$$

and so $f(u)e^{-\frac{u^2}{2}} \equiv 0$ since $f(u)e^{-\frac{u^2}{2}} \in \mathbf{L}_2(\mathbb{R}, \mathcal{B}, dx)$, and using the fact that the Fourier transform is one-to-one on this space.

Remark 6.1.35. The Fourier transform defines a unitary mapping from $\mathbf{L}_2(\mathbb{R}, \mathcal{B}, dx)$ onto itself. We mention that another important transform, namely the Hankel transform, defines a unitary mapping from $\mathbf{L}_2(\mathbb{R}_+, \mathcal{B}, dx)$ onto itself.

Let J_ν denote the Bessel function

$$J_\nu(z) = \sum_{n=0}^\infty \frac{(-1)^n \left(\frac{z}{2}\right)^{2n+\nu}}{n!(\nu+1)(\nu+2)\cdots(\nu+n)}.$$

The Hankel (or Fourier–Bessel) transform of f is defined as

$$g(y) = \int_0^\infty x J_\nu(xy) f(x) dx, \tag{6.1.35}$$

first for functions f measurable on $[0, \infty)$ and such that

$$\int_0^\infty \sqrt{x}|f(x)| dx < \infty.$$

Hankel's integral theorem (see [215, (5.14.11), p. 130]) holds:

$$f(x) = \int_0^\infty y J_\nu(xy) \left(\int_0^\infty z J_\nu(yz) f(z) dz\right) dy, \tag{6.1.36}$$

from which follows that

$$\int_0^\infty x|f(x)|^2 dx = \int_0^\infty y|g(y)|^2 dy.$$

We note that the map (6.1.35), or more precisely, the closely related map

$$\tilde{f}(y) = \int_0^\infty J_\nu(xy) f(x) \sqrt{xy} dx$$

is used in de Branges' book [84, §50, p. 185] in his study of homogeneous spaces.

The study of the range of the Hankel transform when f has support inside a preassigned interval leads to reproducing kernel formulas involving the Bessel functions, for instance (see [215, (5.14.3), p. 128] in Lebedev's book on special functions)

$$\frac{1}{T}\int_0^T rJ_\nu(ar)J_\nu(br)dr = \frac{bJ_\nu(aT)J_\nu'(bT) - aJ_\nu(bT)J_\nu'(aT)}{a^2 - b^2}.$$

Similar formulas are used in [127] in the prediction problem for the fractional Brownian motion.

6.2   Stieltjes integral

In complex analysis, Stieltjes integrals appear in a number of topics, and in particular for the integral formula for functions analytic and with a positive real part in the open unit disk, or, equivalently, for the integral formula for functions which are positive and harmonic in the open unit disk.

First the definition: Let f and g be real-valued functions defined on the interval $[a, b]$. The Stieltjes integral is defined to be the limit, if it exists, of the sums of the form

$$\sum_{k=0}^n f(s_k)(g(t_k) - g(t_{k-1})),$$

where

$$t_0 = a < t_1 < t_2 < \cdots < t_{n-1} < t_n = b,$$

and

$$s_k \in [t_k, t_{k+1}], \quad k = 0, \ldots, n - 1,$$

as the mesh $\max_{k=1,\ldots,n} |t_k - t_{k-1}|$ goes to 0.

The case $g(t) = t$ corresponds to Riemann's integral. As in the case of Riemann's integral, the definition is extended via limits to intervals of infinite length. When g is defined on $\mathbb{R}$, increasing and right continuous, it defines a Borel measure on $\mathbb{R}$. See Exercise 6.2.3 below. In the corresponding Lebesgue space $\mathbf{L}_1(\mathbb{R}, \mathcal{B}, m_g)$ not every equivalent class contains a Stieltjes function integrable with respect to g (as is seen by considering the case $g(t) = t$), but, if f is Stieltjes integrable with respect to g and is in $\mathbf{L}_1(\mathbb{R}, \mathcal{B}, m_g)$ it holds that

$$\int_{\mathbb{R}} f(t)dg(t) = \int_{\mathbb{R}} f(t)dm_g(t). \tag{6.2.1}$$

For the next exercise, see for instance [74, Lemma 5.9, p. 143]. The result, of importance here, is a direct application of the characterization of the extension of an additive set function to a sigma-additive set function. Recall first that, given a set E, a family of sets $\mathcal{A} \subset \mathcal{P}(E)$ is called an algebra if it is closed under complementation and finite union, and contains E. We also recall the following definition (see [120, Definition 11, p. 137]; here we consider the finite and positive case):

Definition 6.2.1. Let E be a topological space, and let $\mathcal{A} \subset \mathcal{P}(E)$ be an algebra. The bounded positive additive function μ on $\mathcal{A}$ is called regular if for every $A \in \mathcal{A}$ and $\epsilon > 0$ there exist F and O in $\mathcal{A}$ such that

$$\overline{F} \subset A \subset \overset{\circ}{O}, \tag{6.2.2}$$

and $\mu(O \setminus F) \leq \epsilon$.

The following extension result can be found for instance in [120, Theorem 14, p. 138].

Theorem 6.2.2. *Let E be locally compact and let μ be a bounded regular additive set function on an algebra $\mathcal{A} \subset \mathcal{P}(E)$. Then μ has a unique sigma-additive extension to the sigma-algebra generated by $\mathcal{A}$.*

Exercise 6.2.3. *Let g be an increasing function from $\mathbb{R}$ into itself, and define*

$$m_g(s,t] = g(t) - g(s), \quad s, t \in \mathbb{R}. \tag{6.2.3}$$

Then, m_g extends to a measure if and only if g is right-continuous.

More generally, see [263, Examples 11.6, p. 308], the function m_g defined by

$$m_g(s,t) = g(t_-) - g(s_+)$$
$$m_g\{s\} = g(s_+) - g(s_-)$$

extends to a measure on the Borel sets of $\mathbb{R}$.

In view of the following exercise, recall that 1_t has been defined in (6.1.27). Recall also that a function f from $[a,b]$ into $\mathbb{C}$ is of bounded variation if there exists an $M > 0$ such that

$$\sum_{j=0}^{N-1} |f(t_{j+1}) - f(t_j)| \leq M$$

for every choice N and of partition $t_0 = a < t_1 < \cdots < t_{N-1} < t_N = b$ of the given interval. The function f is called absolutely continuous if the following condition holds: For every $\epsilon > 0$ there exists $\delta > 0$ such that for any $N \in \mathbb{N}$ and any choice of non-intersecting open intervals $I_1 = (s_1, t_1), \ldots, I_N = (s_N, t_N)$ inside $[a,b]$ of total length less or equal to δ,

$$\sum_{j=1}^{N} |f(t_j) - f(s_j)| \leq \epsilon.$$

We will present only the proof of (1) of the following exercise.

Exercise 6.2.4. *Let $f \in \mathbf{L}_2(\mathbb{R}, \mathcal{B}, dt)$.*

(1) *Show that the function*

$$\sigma(t) = \int_{1_t} f(u) du$$

is absolutely continuous on every bounded interval.

(2) *Let $g \in \mathbf{L}_2(\mathbb{R}, \mathcal{B}, dt)$ be continuous. Show that*

$$\underbrace{\int_{\mathbb{R}} g(u) d\sigma(u)}_{\text{Stieltjes integral}} = \underbrace{\int_{\mathbb{R}} g(u) f(u) du}_{\text{Lebesgue integral}} \ .$$

In relation with Exercises 6.2.3 and 6.2.4 it is well to recall the following result.

Theorem 6.2.5. *(see for instance [264, Théorème 1.29, p. 23] for the case of general topological spaces). Let $f \in \mathbf{L}_2(\mathbb{R}, \mathcal{B}, dt)$ with values in $[0, \infty)$, and let for $B \in \mathcal{B}$*

$$\sigma_f(B) = \int_B f(u) du. \tag{6.2.4}$$

Then, σ_f is a measure, and it holds that

$$\int_{\mathbb{R}} m(u) d\sigma(u) = \int_{\mathbb{R}} m(u) f(u) du$$

for all positive measurable functions m.

For the next exercise, see for instance [243, p. 229].

Exercise 6.2.6. *Prove the integration by parts formula for the Stieltjes integral: Let f and g be real-valued functions defined on the interval $[a, b]$. If the Stieltjes integral of f with respect to g,*

$$\int_a^b f dg$$

exists so does the integral $\int_a^b g df$ and one has

$$\int_a^b g(t) df(t) = f(b)g(b) - f(a)g(a) - \int_a^b f(t) dg(t). \tag{6.2.5}$$

As remarked in Natanson's book, see [243, p. 231], this formula implies that an increasing function (or, more generally, a bounded variation function) is always Stieltjes integrable with respect to a continuous function.

We end this section with the Stieltjes–Perron inversion formula.

Proposition 6.2.7 (see [5, p. 155]). *Let σ be a real function of a real variable, of bounded variation on every finite interval, and assume that*

$$\int_{\mathbb{R}} \frac{|d\sigma|(u)}{1 + |u|^3} < \infty.$$

Let

$$u(x, y) = \int_{\mathbb{R}} \left\{ \frac{y}{(u-x)^2 + y^2} - \frac{y}{1+u^2} \right\} d\sigma(u).$$

Then, for every $a, b \in \mathbb{R}$,

$$\frac{\sigma(b_+) + \sigma(b_-)}{2} - \frac{\sigma(a_+) + \sigma(a_-)}{2} = \lim_{y \to 0} \int_a^b u(x, y) dx. \tag{6.2.6}$$

When the stronger requirement

$$\int_{\mathbb{R}} \frac{|d\sigma|(u)}{1 + u^2} < \infty \tag{6.2.7}$$

is made on $d\sigma$, formula (6.2.6) takes the simpler form

$$\frac{\sigma(b_+) + \sigma(b_-)}{2} - \frac{\sigma(a_+) + \sigma(a_-)}{2} = \lim_{y \to 0} \int_a^b \left(\int_{\mathbb{R}} \frac{y}{(u-x)^2 + y^2} d\sigma(u) \right) dx. \tag{6.2.8}$$

Note that the left-hand side of (6.2.8) can be rewritten as

$$\int_{(a_+, b_-)} d\sigma(u) + \frac{\sigma(b_+) - \sigma(b_-)}{2} + \frac{\sigma(a_+) - \sigma(a_-)}{2}$$

$$= \lim_{y \to 0} \int_a^b \left(\int_{\mathbb{R}} \frac{y}{|u - z|^2} d\sigma(u) \right) dx, \tag{6.2.9}$$

with $z = x + iy$.

Exercise 6.2.8. *Let $f \in \mathbf{L}_2(\mathbb{R}, \mathcal{B}, du)$ and suppose that*

$$\int_{[a,b]} f(u) du = 0, \quad \forall a, b \in \mathbb{R} \, (a < b).$$

Show that $f = 0$ (a.e.)

Hint: Check that formula (6.2.4) now defines a complex measure on every finite interval $(-n, n)$. Show that it vanishes on the Borel sets.

Exercise 6.2.9. *Let $f \in \mathbf{L}_2(\mathbb{R}, \mathcal{B}, du)$ be such that*

$$\int_{\mathbb{R}} \frac{f(u)}{u - z} du = 0, \quad \forall z \in \mathbb{C} \setminus \mathbb{R}. \tag{6.2.10}$$

Using the Stieltjes–Perron inversion formula, prove that $f = 0$ (a.e.).

The preceding exercise in fact shows that rational functions which vanish at infinity and without poles on the real line are dense in $\mathbf{L}_2(\mathbb{R}, \mathcal{B}, du)$ (see Exercise 6.3.1 below) and is therefore a good link to Section 6.3, devoted to various density results.

6.3   Density results

Continuous functions with compact support are dense in $\mathbf{L}_2(\mathbb{R}, \mathcal{B}, du)$. A proof of this well-known fact is based on Lusin's theorem. See for instance [264, Théorème 3.14, p. 66]. We pause to mention that this last theorem is itself proved using Urysohn's lemma (see [264, Théorème 2.23, p. 52] and [264, p. 38] respectively). Here we consider two other families of functions, of importance in engineering, namely rational functions without poles on the real line and vanishing at infinity, and the functions

$$\chi_t(u) = \int_0^t e^{iau} da = \begin{cases} \frac{e^{itu}-1}{iu}, & u \neq 0, \\ t, & u = 0. \end{cases} \tag{6.3.1}$$

That the space of rational functions without poles on the real line and vanishing at infinity is dense in $\mathbf{L}_2(\mathbb{R}, \mathcal{B}, dx)$ is clear if you know that the functions f_n defined in (6.1.8) above are an orthonormal basis of $\mathbf{L}_2(\mathbb{R}, \mathcal{B}, dx)$, or equivalently, that the Laguerre functions span a dense linear subspace of $\mathbf{L}_2(\mathbb{R}_+, \mathcal{B}, dx)$ (and then, their reflections span a dense linear space in $\mathbf{L}_2(\mathbb{R}_-, \mathcal{B}, dx)$). Here we follow another path, the key being to use the Perron–Stieltjes inversion formula.

Exercise 6.3.1. *The rational functions without poles on the real line and vanishing at infinity are dense in $\mathbf{L}_2(\mathbb{R}, \mathcal{B}, dx)$.*

Hint: Show that a function f orthogonal to all the given set of rational functions satisfies (6.2.10) and apply Exercise 6.2.9.

    More generally:

Exercise 6.3.2. *Let σ be an increasing function from $\mathbb{R}$ into itself, such that*

$$\int_{\mathbb{R}} \frac{d\sigma(u)}{u^2 + 1} < \infty \tag{6.3.2}$$

holds.

(1) *Show that the rational functions without poles on the real line and vanishing at infinity are dense in $\mathbf{L}_2(\mathbb{R}, \mathcal{B}, d\sigma)$.*
(2) *Assume that $\int_{\mathbb{R}} d\sigma(u) < \infty$. Show that the linear span of the functions $u \mapsto e^{iut}$ $(t \in \mathbb{R})$ is dense in $\mathbf{L}_2(\mathbb{R}, \mathcal{B}, d\sigma)$.*
(3) *Show that the linear span of the functions $\chi_t, t \in \mathbb{R}$ is dense in $\mathbf{L}_2(\mathbb{R}, \mathcal{B}, d\sigma)$.*

Hints: For (1), use formula (6.2.8) with measure $f d\sigma$. For (2), consider for $z \in \mathbb{C}_+$ the function $\varphi(t) = e^{izt} 1_{[0,\infty)}(t)$ and for $z \in \mathbb{C}_-$ the function $\varphi(t) = e^{izt} 1_{(-\infty,0]}$, and compute

$$\int_{\mathbb{R}} \varphi(t) \left(\int_{\mathbb{R}} e^{iut} f(u) d\sigma(u) \right) dt, \tag{6.3.3}$$

using Fubini's theorem and item (1). The argument for (3) is similar.

6.3. Density results

The closed linear span of the functions χ_t for $|t| \leq T$ plays an important role in the theory of prediction of stationary processes. See [126].

The criteria for completeness presented in Exercise 6.3.5 below is a special case of Vitali's completeness theorem, see [270, Theorem 26, p. 25], [309, p. 93], for the latter. It is first useful to prove a preliminary result, used in the proof:

Exercise 6.3.3. *Let* $g \neq 0 \in \mathbf{L}_2(\mathbb{R}, \mathcal{B}, dx)$. *Then,*

$$\int_0^t g(u)du \neq 0, \quad a.e.$$

Remark 6.3.4. The previous claim means that the linear span of the functions $1_{[0,t]}$ $(t \in \mathbb{R})$ is dense in $\mathbf{L}_2(\mathbb{R}, \mathcal{B}, dx)$.

Exercise 6.3.5. *The orthonormal family* $(g_n)_{n \in \mathbb{N}_0}$ *in* $\mathbf{L}_2(\mathbb{R}, \mathcal{B}, dx)$ *is complete if and only if*

$$|t| = \sum_{n=0}^{\infty} \left| \int_0^t g_n(u)du \right|^2, \quad \forall t \in \mathbb{R}. \tag{6.3.4}$$

In the following exercises we denote by $d\mu_g$ the Gaussian probability measure

$$d\mu_g(x) = \frac{1}{\sqrt{\pi}} e^{-x^2} dx. \tag{6.3.5}$$

Exercise 6.3.6.

(1) *Prove that the polynomials are dense in* $\mathbf{L}_2(\mathbb{R}, d\mu_g)$.

(2) *Show that an orthonormal basis for* $\mathbf{L}_2(\mathbb{R}^2, \frac{e^{-(x^2+y^2)}}{\pi} dxdy)$ *is given by the functions*

$$1 \quad and \quad \frac{z^n}{\sqrt{n!}}, \quad \frac{\bar{z}^n}{\sqrt{n!}}, \quad n \in \mathbb{N}.$$

Hint: For (1) compute the Fourier transform of the function $f(x)e^{-x^2}$ when f is orthogonal to all the polynomials. The proof makes uses of Fubini's theorem, and we here quote the relevant version needed, specialized from Brezis' book [90, Théorème IV.5, p. 55].

Theorem 6.3.7. *Let* $f \in \mathbf{L}_1(\mathbb{R}^2, dxdy)$. *Then,*

$$\int_{\mathbb{R}} |f(x,y)|dx < \infty \ (a.e. \ y) \quad and \quad \int_{\mathbb{R}} |f(x,y)|dy < \infty \ (a.e. \ x), \tag{6.3.6}$$

and

$$\int_{\mathbb{R}} \left(\int_{\mathbb{R}} f(x,y)dy \right) dx = \int_{\mathbb{R}} \left(\int_{\mathbb{R}} f(x,y)dx \right) dy = \iint_{\mathbb{R}^2} f(x,y)dxdy. \tag{6.3.7}$$

We conclude this section with some exercises on orthogonal polynomials with respect to the weight e^{-x^2}. Consider the space $\mathcal{P}_N$ of polynomials of degree less than or equal to N, endowed with the inner product of $\mathbf{L}_2(\mathbb{R}, d\mu_g)$. Then,

$$\langle z^n, z^m \rangle_{d\mu_g} = \frac{1}{\sqrt{\pi}} \int_{\mathbb{R}} x^{n+m} e^{-x^2} dx.$$

Thus the Gram matrix (see Definition 4.4.4 for the latter) of $1, z, \ldots, z^N$ is given by

$$(H_N)_{n,m} = \frac{1}{\sqrt{\pi}} \int_{\mathbb{R}} x^{n+m} e^{-x^2} dx$$

$$= \begin{cases} 0, & \text{if } n+m \text{ is odd,} \\ \Gamma(p + \frac{1}{2}), & \text{if } n+m = 2p \text{ is even,} \end{cases}$$

where Γ denotes the Gamma function. It is constant on the off diagonals, and is in particular a Hankel matrix. Let L_N denote a lower triangular matrix such that

$$L_N H_N L_N^* = I_{N+1},$$

and let $p_0, \ldots, p_N$ denote the polynomials defined by

$$\begin{pmatrix} p_0(z) \\ p_1(z) \\ \vdots \\ p_N(z) \end{pmatrix} = L_N \begin{pmatrix} 1 \\ z \\ \vdots \\ z^N \end{pmatrix}.$$

Then, p_n has degree n and it holds that

$$\langle p_n, p_m \rangle = \delta_{nm}, \quad n, m = 0, \ldots, N. \tag{6.3.8}$$

Question 6.3.8. *Prove that L_N exists and is unique when requiring that the diagonal elements are positive. Prove* (6.3.8).

Up to a normalization factor, the p_n are the Hermite polynomials (see Definition 6.1.12). More precisely, assuming L_N with positive diagonal elements, set (see [177, (31), p. 436]:

$$h_n(z) = \sqrt{\sqrt{\pi} 2^n n!}\, p_n(z).$$

We note the following formula, see [177, (41), p. 440]:

$$\sum_{n=0}^{N} \frac{h_n(z)\overline{h_n(w)}}{2^n n!} = \frac{h_{N+1}(z)\overline{h_N(w)} - h_N(z)\overline{h_{N+1}(w)}}{2^N N!(z - \overline{w})}, \tag{6.3.9}$$

where $z \neq \overline{w} \in \mathbb{C}$.

6.3. *Density results*

Hille calls this formula a Christoffel formula. It is a particular instance of a general formula associated with Hankel matrices, and which is the counterpart of the Christoffel–Darboux formula associated with Toeplitz matrices (recall that these are square matrices constant on the diagonals). These are particular cases of reproducing kernel formulas for finite-dimensional reproducing kernel spaces when the Gram matrix satisfies a certain Stein or Lyapunov equation.

Question 6.3.9. *Let $a_0, a_1, \ldots$ be a sequence of positive numbers, and define a (possibly indefinite and degenerate) Hermitian form on polynomials via*

$$\langle z^n, z^m \rangle = a_{n+m}, \quad n, m = 0, 1, \ldots$$

Show that π_n defined by

$$\pi_n(z) = \begin{vmatrix} 1 & z & \cdots & z^n \\ a_0 & a_1 & \cdots & a_n \\ \vdots & \vdots & & \vdots \\ a_{n-1} & h_n & \cdots & a_{2n-1} \end{vmatrix}, \quad n = 0, 1, \ldots,$$

is a polynomial of degree n and that

$$\langle \pi_n, \pi_m \rangle = 0, \quad for \quad n \neq m.$$

Related to the next question, see also Exercise 6.3.13 at the end of the section.

Question 6.3.10. *In the notation of the previous exercise, let H_n denote the Hankel matrix with (ℓ, k) entry equal to $a_{\ell+k}$, where $\ell, k \in \{0, \ldots, n\}$. Assume that H_n is invertible, and let, for $n \in \mathbb{N}_0$,*

$$t_n(z) = \begin{pmatrix} 1 & z & \cdots & z^n \end{pmatrix} H_n^{-1} \begin{pmatrix} 0 \\ 0 \\ \vdots \\ 1 \end{pmatrix}.$$

Show that

$$\langle t_n, z^j \rangle = 0, \quad j = 0, \ldots, n-1.$$

The following question relates to another important class of polynomials, namely the Laguerre polynomials.

Exercise 6.3.11. *Show that the polynomials are dense in $\mathbf{L}_2(\mathbb{R}_+, e^{-x}dx)$.*

Hint: Compute

$$\int_0^\infty e^{ixz} f(x) e^{-x} dx$$

for $z \in \mathbb{D}$, for $f \in \mathbf{L}_2(\mathbb{R}_+, e^{-x}dx)$ orthogonal to all the polynomials.

Remark 6.3.12. Here too, the Gram matrix associated with $1, z, \ldots, z^N$ is a Hankel matrix.

Let us mention another density result, related to Bergman spaces. Polynomials are dense in the Bergman spaces associated with a bounded open set Ω whose complement is connected (thus, an annulus will not fit, for instance). The proof presented in Nehari's book [244] uses Riemann's mapping theorem, and has an essential complex variable flavor.

To conclude this section we give the following generalization of Question 6.3.10.

Exercise 6.3.13. *Let* $f_1, \ldots, f_N$ *be elements of a complex vector space endowed with a Hermitian form* $\langle \cdot, \cdot \rangle$*. Let* P *denote the associated Gram matrix, that is* P *is the* $N \times N$ *Hermitian matrix with* (ℓ, k) *entry* $P_{\ell,k} = \langle f_k, f_\ell \rangle$*, and assume that* P *is invertible. Finally let*

$$e_N = \begin{pmatrix} f_1 & f_2 & \cdots & f_N \end{pmatrix} P^{-1} \begin{pmatrix} 0 \\ \vdots \\ 0 \\ 1 \end{pmatrix}. \tag{6.3.10}$$

(1) *Show that*

$$\langle e_N, f_j \rangle = 0, \quad j = 1, \ldots, N - 1.$$

(2) *Show that* $\langle e_N, e_N \rangle \neq 0$*.*

6.4   Solutions of the exercises

Solution of Exercise 6.1.2: The function $f(u) = \frac{1}{u+i}$ is in the Hardy space $\mathbf{H}_2(\mathbb{C}_+)$, and is therefore the inverse Fourier transform of a function g with support in $[0, \infty)$, which is given in the hint, namely

$$g(t) = -2\pi i e^{-t} \mathbf{1}_{\mathbb{R}_+}(t).$$

We will not follow this path, but go the pedestrian way. Of course the Fourier transform cannot be computed via formula (6.1.1) since f does not belong to $\mathbf{L}_1(\mathbb{R}, dx)$, and we follow the original definition of the Fourier transform via approximation. Consider for $\epsilon > 0$ the function

$$f_\epsilon(u) = \frac{i}{(u + i)(\epsilon u + i)}.$$

Then,

$$\int_{\mathbb{R}} |f(u) - f_\epsilon(u)|^2 du = \int_{\mathbb{R}} \frac{1}{1 + u^2} \frac{\epsilon^2 u^2}{\epsilon^2 u^2 + 1} du,$$

and the dominated convergence theorem implies that $\lim_{\epsilon \to 0} \|f - f_\epsilon\|_2 = 0$. The residue theorem allows us to compute the Fourier transform $\widehat{f_\epsilon}$. Indeed, for $t \le 0$ we have

$$|e^{-itz}| \le 1, \quad z \in \mathbb{C}_+,$$

and the function $e^{-itz} f_\epsilon(z)$ is analytic in the closed upper half-plane. Hence,

$$\widehat{f_\epsilon}(t) = 0, \quad \text{for} \quad t \le 0, \tag{6.4.1}$$

and similarly (for $\epsilon \ne 1$)

$$\widehat{f_\epsilon}(t) = -2\pi i \left\{ \operatorname{Res}\left(e^{-itu} f_\epsilon(u), -i\right) + \operatorname{Res}\left(e^{-itu} f_\epsilon(u), -i/\epsilon\right) \right\}$$

$$= -2\pi i \left\{ \frac{e^{-t}}{1 - \epsilon} + \frac{e^{-t/\epsilon}}{1 - 1/\epsilon} \right\}, \tag{6.4.2}$$

$$= -\frac{2\pi i}{1 - \epsilon} \left\{ e^{-t} - \epsilon e^{-t/\epsilon} \right\}, \quad \text{for} \quad t > 0.$$

Let

$$g(t) = \begin{cases} 0, & \text{for } t \le 0, \\ -2\pi i e^{-t}, & \text{for } t > 0. \end{cases}$$

Then $g \in \mathbf{L}_2(\mathbb{R}, dt)$ and the equality (recall that $\epsilon > 0$)

$$\|\widehat{f_\epsilon} - g\|_2^2 = \frac{4\pi^2 \epsilon^2}{(1 - \epsilon)^2} \int_0^\infty \left| e^{-t} - e^{-t/\epsilon} \right|^2 dt$$

shows that

$$\lim_{\substack{\epsilon \to 0 \\ \epsilon > 0}} \|\widehat{f_\epsilon} - g\|_2 = 0. \qquad \square$$

Remark 6.4.1. D. Sarason in [273, p. 30] calculates the above Fourier transform by applying the residue theorem to compute the limit

$$\lim_{R \to \infty} \int_{-R}^{R} \frac{e^{iut}}{u + i} du.$$

Solution of Exercise 6.1.3:

(1) We switch the symbols u and t in (6.1.7). Differentiating then n times (6.1.7) with respect to t we have that, for $n = 0, 1, \cdots$

$$\frac{1}{(t + i)^{n+1}} = \frac{1}{2\pi} \int_0^\infty e^{itu} g_n(u) du, \tag{6.4.3}$$

with

$$g_n(u) = \frac{2\pi(-i)^{n+1}}{n!} e^{-u} u^n \mathbf{1}_{\mathbb{R}_+}(u). \tag{6.4.4}$$

Taking adjoints on both sides of (6.4.3) we have

$$\frac{1}{(t-i)^{n+1}} = \frac{1}{2\pi} \int_{-\infty}^{0} e^{itu}\overline{g_n(-u)}du. \tag{6.4.5}$$

(2) This follows from the fact that the functions b_n defined by

$$u \mapsto b_n(u) = \frac{1}{\sqrt{\pi}} \frac{1}{u+i} \left(\frac{u-i}{u+i}\right)^n, \quad n \in \mathbb{Z},$$

are orthonormal in $\mathbf{L}_2(\mathbb{R}, \mathcal{B}, du)$ and from (6.1.4). Indeed, we have:

$$\begin{aligned} 2\pi\langle f_n, f_m\rangle_2 &= \langle \widehat{f_n}, \widehat{f_m}\rangle_2 \\ &= 2\pi\langle b_n, b_m\rangle_2 \\ &= 2\pi\delta_{m,n}, \quad n, m \in \mathbb{Z}. \end{aligned}$$

(3) In view of (1) the support of f_n is $[0,\infty)$ for $n \geq 0$ and $(-\infty,0]$ for $n < 0$ since the functions b_u are linear combinations of the functions considered in the first item (use the partial fraction expansion of a rational function, if need be, to check this last fact). $\qquad\square$

We note the following. The isometry property (6.1.6) applied to (6.4.3) leads to

$$\frac{1}{2\pi} \int_{\mathbb{R}} |g_n(u)|^2 du = \int_{\mathbb{R}} \frac{dt}{(t^2+1)^{n+1}}. \tag{6.4.6}$$

Since

$$\int_0^\infty e^{-2u}u^{2n}du = \frac{\Gamma(2n+1)}{2^{n+1}} = \frac{(2n)!}{2^{2n+1}},$$

(6.4.6) is equivalent to

$$2\pi\frac{(2n)!}{2^{n+1}(n!)^2} = \int_{\mathbb{R}} \frac{dt}{(t^2+1)^{n+1}}.$$

This in turn is a well-known integral; see for instance [CAPB, p. 378].

Solution of Exercise 6.1.9: We consider the closed contour Γ_R made of the rectangle with vertices

$$A_R = (-R, 0), \quad B_R = (R, 0), \quad C_R = (R, i\pi), \quad D_R = (-R, i\pi).$$

The zeros of the function $\cosh z$ are the solutions of the equation $e^{2z} = -1$, and the only one inside the above contour is $z = i\frac{\pi}{2}$. By the residue theorem,

$$\int_{\Gamma_R} \frac{e^{-itz}dz}{\cosh z} = 2\pi i \text{Res}\left(\frac{e^{-itz}}{\cosh z}, i\frac{\pi}{2}\right) = 2\pi e^{\frac{\pi t}{2}},$$

where we have used formula (2.8.12) (see for instance [CAPB, (7.3.2), p. 325]) to compute the residue. On the other hand, and since $\cosh(z + i\pi) = -\cosh z$, we have:

$$\int_{[C_R,D_R]} \frac{e^{-itz}dz}{\cosh z} = -e^{\pi t}\int_{[A_R,B_R]} \frac{e^{-itz}dz}{\cosh z}.$$

Furthermore, for $t < 0$,

$$\lim_{R\to\infty}\int_{[B_R,C_R]} \frac{e^{-itz}dz}{\cosh z} = \lim_{R\to\infty}\int_{[D_R,A_R]} \frac{e^{-itz}dz}{\cosh z} = 0.$$

It follows that

$$\int_{\mathbb{R}} \frac{e^{-iut}du}{\cosh u} = \frac{\pi}{\cosh\left(\frac{\pi t}{2}\right)}, \tag{6.4.7}$$

first for negative t, and by symmetry for all real t. $\qquad\square$

Remark 6.4.2. As remarked in the paper of Bass and Lévy [66], the above formula gives that the function

$$t \mapsto \frac{1}{\cosh\left(t\sqrt{\frac{\pi}{2}}\right)}$$

is an eigenfunction, with eigenvalue 1, of the Fourier transform when this one is defined as in (6.1.3), that is,

$$\frac{1}{\sqrt{2\pi}}\int_{\mathbb{R}} e^{-itu} f(u)du.$$

Solution of Exercise 6.1.4: By the residue theorem, we have for $t \geq 0$,

$$\int_{\mathbb{R}} \frac{e^{itx}}{(x^2+1)^n}dx = 2\pi i \mathrm{Res}\left(\frac{e^{itz}}{(z^2+1)^n}, z = i\right)$$

$$= 2\pi i \frac{\left(\frac{e^{itz}}{(z+i)^n}\right)^{(n-1)}\Big|_{z=i}}{(n-1)!}.$$

But using the formula for the Nth derivative of a product with $N = n-1$ (see for instance [CAPB, Exercise 4.2.1, p. 148]), and with the understanding that

$$\prod_{\ell=0}^{k-1}(n+\ell) = 1, \quad \text{for} \quad k = 1,$$

we have:

$$\left(\frac{e^{itz}}{(z+i)^n}\right)^{(n-1)} = \sum_{k=0}^{n-1}\binom{n-1}{k}\frac{(-1)^k\prod_{\ell=0}^{k-1}(n+\ell)}{(z+i)^{n+k}}(it)^{n-1-k}e^{itz}.$$

Therefore,

$$\int_{\mathbb{R}} \frac{e^{itx}}{(x^2+1)^n} dx = \frac{2\pi i e^{-t}}{2^n} \left(\sum_{k=0}^{n-1} t^{n-1-k} \frac{(-1)^k \prod_{\ell=0}^{k-1}(n+\ell)}{k!(n-1-k)!2^k i^{n+k}} i^{n-1-k} \right)$$

$$= \frac{\pi e^{-t}}{2^{n-1}} \left(\sum_{k=0}^{n-1} t^{n-1-k} c_k \right),$$

where

$$c_k = \frac{\prod_{\ell=0}^{k-1}(n+\ell)}{k!(n-1-k)!2^k}, \quad k = 0, 1, \ldots, n-1.$$

Thus the corresponding positive definite function is:

$$\frac{\pi e^{-|t-s|}}{2^{n-1}} \left(\sum_{k=0}^{n-1} c_k |t-s|^{n-1-k} \right). \tag{6.4.8}$$

We now turn to the case $\sigma'(u) = \frac{1}{u^{2n}+1}$. Let

$$z_k = e^{i\left(\frac{\pi}{2n} + \frac{\pi k}{n}\right)}, \quad k = 0, 1, \ldots, n-1,$$

be the roots of order n of -1 which lie in the open upper half-plane. By the residue theorem, we have for $t \geq 0$:

$$\int_{\mathbb{R}} \frac{e^{itu} du}{u^{2n}+1} = 2\pi i \sum_{k=0}^{n-1} \frac{e^{itz_k}}{2nz_k^{2n-1}}$$

$$= -\frac{i\pi}{n} \sum_{k=0}^{n-1} z_k e^{itz_k},$$

where we have used that $z_k^{2n-1} = -z_k^{-1}$ to go from the first to the second line. Let

$$f(t) = \sum_{k=0}^{n-1} e^{itz_k}, \tag{6.4.9}$$

so that the above sum reads

$$\int_{\mathbb{R}} \frac{e^{itu} du}{u^{2n}+1} = -\frac{\pi}{n} f'(t).$$

To compute f, we distinguish the cases n even and n odd. Let first $n = 2p$ with $p \in \mathbb{N}$. Then,

$$z_k = -\overline{z_{2p-k-1}}, \quad k = 0, \ldots, p-1,$$

6.4. Solutions of the exercises

and, with $z_k = \alpha_k + i\beta_k$, where

$$\alpha_k = \cos\left(\frac{\pi}{2n} + \frac{\pi k}{n}\right) \quad \text{and} \quad \beta_k = \sin\left(\frac{\pi}{2n} + \frac{\pi k}{n}\right),$$

we have

$$f(t) = \sum_{k=0}^{p-1} \left(e^{itz_k} + e^{-it\overline{z_k}}\right) = 2\sum_{k=0}^{p-1} e^{-\beta_k t} \cos(\alpha_k t).$$

Thus

$$\int_{\mathbb{R}} \frac{e^{itu}\,du}{u^{2n}+1} = \frac{2\pi}{n}\sum_{k=0}^{p} e^{-\beta_k t}\left(\beta_k \cos(\alpha_k t) + \alpha_k \sin(\alpha_k t)\right)$$

$$= \frac{2\pi}{n}\sum_{k=0}^{p} e^{-\beta_k t}\sin\left(\alpha_k t + \frac{\pi}{2n} + \frac{\pi k}{n}\right)$$

and the corresponding positive definite function is:

$$\frac{2\pi}{n}\sum_{k=0}^{p} e^{-\beta_k |t-s|}\sin\left(\alpha_k |t-s| + \frac{\pi}{2n} + \frac{\pi k}{n}\right).$$

When n is odd, $n = 2p+1$, then there are $2p+1$ roots of -1 in the open upper half-plane, $2p$ of them in pairs and the root $z = i$. The function (6.4.9) is now equal to

$$f(t) = e^{-t} + \underbrace{\sum_{k=0}^{p-1}(e^{itz_k} + e^{-it\overline{z_k}})}_{\text{equal } 0 \text{ if } p = 0},$$

and the corresponding positive definite function is now

$$\frac{\pi}{n}\left(e^{-|t-s|} + \underbrace{2\sum_{k=0}^{p-1} e^{-\beta_k |t-s|}\sin\left(\alpha_k |t-s| + \frac{\pi}{2n} + \frac{\pi k}{n}\right)}_{\text{equal } 0 \text{ if } p = 0}\right). \qquad \square$$

We note that in (6.4.8) the function

$$K(t,s) = \sum_{k=0}^{n-1} c_k |t-s|^{n-1-k}$$

is not positive definite. To verify this, consider the matrix

$$\begin{pmatrix} K(t,t) & K(t,0) \\ K(0,t) & K(0,0) \end{pmatrix}$$

and let t be large enough.

Solution of Exercise 6.1.5: The spectrum of PA is in the open upper half-plane and so the integral

$$\int_{-\infty}^{0} e^{it(\lambda I_N - PA)|_{\mathrm{ran}\,P})}\,dt \tag{6.4.10}$$

converges for real λ. To see this, write A as

$$PA|_{\mathrm{ran}\,P} = T\mathrm{diag}\,(J_1,\dots)T^{-1}$$

where each of the J_j is a Jordan block, and where $T \in \mathbb{C}^{N\times N}$ is invertible. Let $J_j \in \mathbb{C}^{n_j \times n_j}$,

$$J_j = \lambda_j I_{n_j} + N_j, \quad \text{with} \quad N_j = \begin{pmatrix} 0 & 1 & 0 & \cdots & 0 & 0 \\ 0 & 0 & 1 & 0 & \cdots & 0 \\ & & & & & \\ 0 & 0 & 0 & 0 & 0 & 1 \\ 0 & 0 & 0 & 0 & 0 & 0 \end{pmatrix},$$

and where $\mathrm{Im}\,\lambda_j > 0$. We have

$$e^{-itJ_j} = e^{-it\lambda_j}\left(\sum_{n=0}^{n_j-1}\frac{(-it)^n}{n!}N_j^n\right) = e^{-it(\mathrm{Re}\,\lambda_j)}e^{t(\mathrm{Im}\,\lambda_j)}\left(\sum_{n=0}^{n_j-1}\frac{(-it)^n}{n!}N_j^n\right),$$

and so the integral

$$\int_{-\infty}^{0} e^{it(\lambda I_{n_j} - J_j)}\,dt$$

converges for real λ, and therefore so does the integral (6.4.10). Thus for real λ we have

$$\int_{-\infty}^{0} e^{it(\lambda I_N - PA)|_{\mathrm{ran}\,P})}\,dt = \left[e^{it(\lambda I_N - PA)|_{\mathrm{ran}\,P})}(i(\lambda I_N - PA)|_{\mathrm{ran}\,P}))^{-1}\right]_{t=-\infty}^{t=0}$$

$$= -i((\lambda I_N - PA)|_{\mathrm{ran}\,P}))^{-1}$$

$$= -i(\lambda I_N - A)^{-1}|_{\mathrm{ran}\,P},$$

and so

$$\int_{-\infty}^{0} -ie^{it\lambda}e^{-it(\lambda I_N - A)|_{\mathrm{ran}\,P})}\,dt = -(\lambda I_N - A)^{-1}|_{\mathrm{ran}\,P}.$$

Checking that

$$\int_{0}^{\infty} ie^{it\lambda}e^{-it(A|_{\mathrm{ran}\,(I_N-P)})}\,dt = -(\lambda I_N - A)^{-1}|_{\mathrm{ran}\,(I_N-P)}$$

is proved in the same way. Since

$$(\lambda I_N - A)^{-1} = (\lambda I_N - A)^{-1}|_{\mathrm{ran}\,P} + (\lambda I_N - A)^{-1}|_{\mathrm{ran}\,(I_N-P)}$$

we have

$$\int_{\mathbb{R}} e^{it\lambda} l(t) dt = -(\lambda I_N - A)^{-1}$$

with

$$l(t) = \begin{cases} -ie^{-itA}|_{\text{ran}(P)}, & t < 0, \\ ie^{-itA}|_{\text{ran}(I_N - P)}, & t > 0. \end{cases}$$

The claim follows by multiplying this equality by C on the left and B on the right and applying the formula for the inverse Fourier transform. $\qquad\square$

Solution of Exercise 6.1.6: For a fixed $\epsilon > 0$ choose first $A > 0$ such that $\int_{|u|>A} |f(u)| du \le \epsilon$. The functions $e_n(u) = e^{\frac{in\pi u}{A}}, n \in \mathbb{Z}$ are dense in $\mathbf{L}_2[-A, A]$. Consider then a finite linear combination $p(u)$ of the functions e_n such that $\|f - p\|_2 \le \epsilon$. By the Cauchy–Schwarz inequality,

$$\int_{[-A,A]} |f(u) - p(u)| du \le \sqrt{2A}\epsilon.$$

For a given e_n we have

$$\int_{[-A,A]} e_n(u) e^{-iut} du = \int_{[-A,A]} e^{iu(\frac{n\pi}{A} - t)} du = \frac{e^{i(n\pi - At)} - e^{-i(n\pi - At)}}{i(\frac{n\pi}{A} - t)}$$

and so

$$\lim_{t \to \pm\infty} \int_{[-A,A]} e_n(u) e^{-iut} du = 0.$$

This allows us to conclude the proof. $\qquad\square$

Solution of Exercise 6.1.7: (see also the remark after the statement of the exercise). We follow the hint. For $p, q \in \mathbb{N}_0$ we have

$$\widehat{s}^{(q)}(t) = \int_{\mathbb{R}} s(u)(-iu)^q e^{-iut} du,$$

where the interchange of integration and differentiation is justified using the dominated convergence theorem. Furthermore successive integration by parts allows us to write $t^p \widehat{s}^{(q)}(t)$ as the Fourier transform of a Schwartz function. The result follows then from the previous exercise since Schwartz functions belong to $\mathbf{L}_1(\mathbb{R})$. $\qquad\square$

Solution of Exercise 6.1.11: Let $f \in \mathbf{L}_2(\mathbb{R}, \mathcal{B}, dx)$. For $z = -\frac{it}{\sqrt{2}}$, the Bargmann transform (2.1.33) can be rewritten as

$$e^{-\frac{t^2}{4}} F\left(-\frac{it}{\sqrt{2}}\right) = \frac{1}{\sqrt[4]{\pi}} \int_{\mathbb{R}} f(u) e^{-\frac{u^2}{2}} e^{-itu} du. \qquad (6.4.11)$$

Using formula (6.1.5) for the inverse Fourier transform we have

$$e^{-\frac{u^2}{2}} f(u) = \frac{\sqrt[4]{\pi}}{2\pi} \int_{\mathbb{R}} e^{itu} F\left(-\frac{it}{\sqrt{2}}\right) e^{-\frac{t^2}{4}} dt. \tag{6.4.12}$$

When $F(z) = z^n$, the corresponding function f, which we will denote by x_n, satisfies

$$\int_{\mathbb{R}} e^{itu} t^n e^{-\frac{t^2}{4}} dt = (-i)^{-n} 2^{\frac{n}{2}} e^{-\frac{u^2}{2}} \frac{2\pi}{\sqrt[4]{\pi}} x_n(u). \tag{6.4.13}$$

Consider now equation (2.1.30), which, after a change of notation, can be rewritten as:

$$\sqrt{2\pi} e^{-\frac{v^2}{2}} = \int_{\mathbb{R}} e^{-\frac{s^2}{2}} e^{-isv} ds.$$

With $v = \sqrt{2}u$, and the change of variable $s = \frac{t}{\sqrt{2}}$ we have

$$\sqrt{2\pi} e^{-u^2} = \frac{1}{\sqrt{2}} \int_{\mathbb{R}} e^{-\frac{t^2}{4}} e^{-iut} dt.$$

Differentiating n times we get

$$\sqrt{2\pi} \left(e^{-u^2}\right)^{(n)} = \int_{\mathbb{R}} e^{-\frac{t^2}{4}} t^n e^{-itu} (-i)^n \frac{1}{\sqrt{2}} dt. \tag{6.4.14}$$

Comparing (6.4.13) and (6.4.14) we get

$$2^{\frac{n}{2}} e^{-\frac{u^2}{2}} \frac{2\pi}{\sqrt[4]{\pi}} x_n(u) = \sqrt{2}\sqrt{2\pi} \left(e^{-u^2}\right)^{(n)},$$

and hence we obtain (6.1.12):

$$x_n(u) = \frac{1}{\sqrt[4]{\pi} 2^{\frac{n}{2}}} e^{\frac{u^2}{2}} \left(e^{-u^2}\right)^{(n)}. \qquad \square$$

Remark 6.4.3. Equations (6.4.11) and (6.4.12) give the relations between the Bargmann transform and the Fourier transform.

Solution of Exercise 6.1.13: Using the hint we write for $s \in \mathbb{C}$:

$$\sum_{n=0}^{\infty} \frac{(is)^n}{n!} \widehat{g_n}(t) = \int_{\mathbb{R}} e^{-itu} e^{\frac{u^2}{2}} \left(\underbrace{\sum_{n=0}^{\infty} \frac{(is)^n}{n!} \left(e^{-u^2}\right)^{(n)}}_{\text{equal to } e^{-(u+is)^2} \text{ by Taylor's formula}} \right) du$$

$$= \int_{\mathbb{R}} e^{-itu} e^{\frac{u^2}{2}} e^{-(u+is)^2} du$$

$$= e^{s^2} \int_{\mathbb{R}} e^{-i(t+2s)u} e^{-\frac{u^2}{2}} du$$

$$= e^{s^2} \sqrt{2\pi} e^{-\frac{(t+2s)^2}{2}} \quad \text{(where we have used (2.1.30))}$$

$$= \sqrt{2\pi} e^{\frac{t^2}{2}} e^{-(t+s)^2}$$

$$= \sqrt{2\pi} \sum_{n=0}^{\infty} \frac{s^n}{n!} e^{\frac{t^2}{2}} \left(e^{-t^2}\right)^{(n)}.$$

It follows that

$$\widehat{g_n}(t) = \sqrt{2\pi}(-i)^n g_n(t), \quad n = 0, 1, \ldots \tag{6.4.15}$$

and similarly for the Hermite functions ξ_n since these are scalar multiples of the functions g_n. $\qquad\square$

Solution of Exercise 6.1.15: A function f orthogonal to all the Hermite functions is orthogonal to the functions $e^{-u^2} u^n$, $n = 0, 1, \ldots$. Thus, by (2.1.20) we have

$$\int_{\mathbb{R}} f(u) e^{2uz - u^2} du = \sum_{n=0}^{\infty} \frac{2^n z^n}{n!} \int_{\mathbb{R}} f(u) e^{-u^2} u^n du \equiv 0,$$

where the interchange of summation and integral is made using the dominated convergence theorem. Setting $z = -\frac{it}{2}$ we get

$$\int_{\mathbb{R}} f(u) e^{-u^2} e^{-iut} du \equiv 0,$$

$\qquad\square$

and so $f(u) e^{-u^2} \equiv 0$.

Solution of Exercise 6.1.16: Let H_n be the Hermite polynomial defined by (2.1.19). Then $\xi_n(z) = e^{-\frac{z^2}{2}} H_n(z)$. Hence

$$\xi_n'(z) = -z\xi_n(z) + e^{-\frac{z^2}{2}} H_n'(z), \tag{6.4.16}$$

$$\xi_n''(z) = -z\xi_n'(z) - \xi_n(z) - z e^{-\frac{z^2}{2}} H_n'(z) + e^{-\frac{z^2}{2}} H_n''(z). \tag{6.4.17}$$

Using (6.4.16) and then (2.1.22), equation (6.4.17) becomes

$$\xi_n''(z) = -z(-z\xi_n(z) + e^{-\frac{z^2}{2}} H_n'(z)) - \xi_n(z) - z e^{-\frac{z^2}{2}} H_n'(z) + e^{-\frac{z^2}{2}} H_n''(z)$$

$$= (z^2 - 1)\xi_n(z) + \underbrace{(-2zH_n'(z) + H_n''(z))}_{\substack{\text{equal to } -2nH_n(z) \\ \text{in view of (2.1.22)}}} e^{-\frac{z^2}{2}}$$

$$= (z^2 - 1 - 2n)\xi_n(z).$$

$\qquad\square$

Solution of Exercise 6.1.21: From the definition of the inverse Fourier transform (see (6.1.5)), and with $\phi(t) = -t$, we have:

$$\check{f} = \frac{1}{2\pi}\widehat{f \circ \phi},$$

and hence we obtain (6.1.23) by taking the Fourier transform of both sides. To obtain the second identity we note that (6.1.4) is equivalent to

$$2\pi\langle f, h\rangle_2 = \langle \widehat{f}, \widehat{h}\rangle_2, \quad f, h \in \mathbf{L}_2(\mathbb{R}, \mathcal{B}, dx).$$

Take now $h = \widehat{g \circ \phi}$ where $g \in \mathbf{L}_2(\mathbb{R}, \mathcal{B}, dx)$, and apply (6.1.23) to get (6.1.24):

$$2\pi\langle f, \widehat{g \circ \phi}\rangle = \langle \widehat{f}, \widehat{\widehat{g \circ \phi}}\rangle = 2\pi\langle \widehat{f}, g\rangle.$$

$\square$

Remark 6.4.4. When one assumes f and g to be in $\mathbf{L}_1(\mathbb{R}, \mathcal{B}, dx)$ one can also apply Fubini's theorem to the function

$$f(x)e^{-ixy}g(y)$$

to obtain

$$\iint_{\mathbb{R}^2} f(x)e^{-ixy}g(y)dxdy = \int_{\mathbb{R}} f(x)\left(\int_{\mathbb{R}} e^{-ixy}g(y)dy\right)dx$$

$$= \int_{\mathbb{R}}\left(\int_{\mathbb{R}} f(x)e^{-ixy}dx\right)g(y)dy,$$

so that

$$\int_{\mathbb{R}} f(x)\widehat{g}(x)dx = \int_{\mathbb{R}} \widehat{f}(x)g(x)dx.$$

See for instance [235, (3.2), p. 7]. Replacing g by $\overline{g}$ we obtain then (6.1.24).

Solution of Exercise 6.1.22:

(1) Let $g(v) = 1_{[a,b]}(v)$, with $a < b$. Then

$$\widehat{g \circ \phi}(u) = \int_{-b}^{-a} e^{-iuv}dv = \frac{e^{iua} - e^{iub}}{-iu},$$

and (6.1.24) leads to

$$\int_a^b \widehat{f}(u)du = \int_{\mathbb{R}} f(u)\frac{e^{-iua} - e^{-iub}}{iu}du.$$

The cases $a = 0$, $b = t > 0$ and $a = t < 0$, $b = 0$ lead to (6.1.25).

(2) Let $f \in \mathbf{L}_2(\mathbb{R}, \mathcal{B}, dx)$ be orthogonal to all the γ_t. Then, (6.1.25) leads to

$$\int_0^t \widehat{f}(u)du = 0, \quad \forall t \in \mathbb{R}.$$

It follows from the fundamental theorem of calculus (see, e.g., [264, Théorème 8.17, p. 158]) that $\widehat{f} = 0$ almost everywhere and so $f = 0$. □

Solution of Exercise 6.1.23: Let f be such that $\widehat{f} = \mathbf{1}_s$, where $s \in \mathbb{R}$. Then, by formula (6.1.5)

$$f(u) = \begin{cases} \begin{cases} \frac{1}{2\pi} \frac{e^{ius}-1}{iu}, & u \neq 0, \\ s, & u = 0, \end{cases} & \text{when } s \geq 0, \\ \begin{cases} \frac{1}{2\pi} \frac{1-e^{ius}}{iu}, & u \neq 0, \\ s, & u = 0, \end{cases} & \text{when } s \leq 0, \end{cases}$$

and (6.1.25) leads then to

$$\min(t,s) = \frac{1}{2\pi} \int_{\mathbb{R}} \frac{(1-e^{-itu})(1-e^{isu})}{u^2} du, \quad t, s \in \mathbb{R}.$$

To conclude we note that

$$\min(t,s) = \frac{|t| + |s| - |t - s|}{2}, \quad t, s \in \mathbb{R}.$$ □

Solution of Exercise 6.1.24: Let g be a Schwartz function. Then, the function

$$\frac{1-e^{-iut}}{iu} f(u)g'(t)$$

is absolutely summable with respect to the Lebesgue measure of $\mathbb{R}^2$, and Fubini's theorem together with an integration by parts leads to:

$$\int_{\mathbb{R}} \left(\int_{\mathbb{R}} \frac{1-e^{-iut}}{iu} f(u)du \right) g'(t)dt = \int_{\mathbb{R}} f(u) \left(\int_{\mathbb{R}} \frac{1-e^{-iut}}{iu} g'(t)dt \right) du$$

$$= -\int_{\mathbb{R}} f(u) \left(\int_{\mathbb{R}} e^{-iut} g(t)dt \right) du$$

$$= -\int_{\mathbb{R}} f(u)\widehat{g}(u)du.$$

On the other hand, (6.1.24) with $\overline{g}$ instead of g leads to

$$\int_{\mathbb{R}} \widehat{f}(t)g(t)dt = \int_{\mathbb{R}} f(u) \left(\int_{\mathbb{R}} e^{-iuv}\overline{g}(-v)dv \right) du$$

$$= \int_{\mathbb{R}} f(u) \left(\int_{\mathbb{R}} e^{iuv} g(-v) \right) dv$$

$$= \int_{\mathbb{R}} f(u)\widehat{g}(u)du,$$

and hence we obtain (6.1.29). □

Solution of Exercise 6.1.25:

(1) Following the hint, we have

$$\int_{\mathbb{R}} |f(u)|du \leq \int_{[-1,1]} |f(u)|du + \int_{\mathbb{R}\setminus[-1,1]} \frac{1}{|u|}|u||f(u)|du$$

$$\leq \int_{[-1,1]} |f(u)|du + \left(\int_{\mathbb{R}\setminus[-1,1]} \frac{1}{u^2}\right)^{1/2} \left(\int_{\mathbb{R}\setminus[-1,1]} u^2|f(u)|^2 du\right)^{1/2} < \infty.$$

The result is then the content of [Exercise 15.6.3, p. 503, CABP].

(2) Since $f \in \mathbf{L}_1(\mathbb{R}, \mathcal{B}, dx)$ we can apply Fubini's theorem in computing

$$\langle s', \hat{f} \rangle = \int_{\mathbb{R}} s'(t) \left(\int_{\mathbb{R}} e^{itu} \overline{f(u)} du \right) dt$$

$$= \int_{\mathbb{R}} \left(\int_{\mathbb{R}} s'(t) e^{iut} dt \right) \overline{f(u)} du$$

$$= \int_{\mathbb{R}} \left([s(t)e^{iut}]_{-\infty}^{\infty} - \int_{\mathbb{R}} ius(t)e^{iut} dt \right) \overline{f(u)} du$$

$$= -\int_{\mathbb{R}} \left(\int_{\mathbb{R}} s(t)e^{iut} dt \right) iu\overline{f(u)} du$$

$$= \int_{\mathbb{R}} s(t) \left(\int_{\mathbb{R}} (-iu)e^{itu} \overline{f(u)} du \right) dt$$

$$= \langle s, \hat{g} \rangle$$

and this leads to (6.1.30).

(3) In view of (6.1.4) we have

$$\|f\|^2 + \|\hat{g}\|^2 = \int_{\mathbb{R}} (1 + 2\pi u^2)|f(u)|^2 du. \qquad \Box$$

Solution of Exercise 6.1.28: (see [224, p. 84]) Since f is even and tends to 0 at infinity we can write for $t > 0$,

$$\hat{f}(t) = 2 \int_0^\infty \cos(ut)f(u)du$$

$$= -\frac{2}{t} \int_0^\infty \sin(ut)f'(u)du$$

$$= -\frac{2}{t} \sum_{n=0}^\infty \int_{\frac{n\pi}{t}}^{\frac{(n+1)\pi}{t}} \sin(ut)f'(u)du$$

$$= -\frac{2}{t} \sum_{n=0}^\infty (-1)^n \int_0^{\frac{\pi}{t}} \sin(ut)f'(u + \frac{n\pi}{t})du$$

$$= -\frac{2}{t} \int_0^{\frac{\pi}{t}} \left(\sum_{n=0}^\infty (-1)^n f'(u + \frac{n\pi}{t}) \right) \sin(ut)du,$$

where the interchange of summation and integration is justified as follows. For fixed u (and t), and in view of the hypothesis on f', the sequence

$$n \mapsto (-f'(u + \frac{n\pi}{t}))$$

is decreasing to 0. Thus, by Abel's theorem on alternating series (see for instance [CAPB, Theorem 3.3.5, p. 95]), the alternating series

$$\sum_{n=0}^{\infty} (-1)^n (-f'(u + \frac{n\pi}{t}))$$

converges to a positive number. Thus

$$\widehat{f}(t) = -\frac{2}{t} \int_0^{\frac{\pi}{t}} \left(\sum_{n=0}^{\infty} (-1)^n (-f'(u + \frac{n\pi}{t})) \right) \sin(ut) du \geq 0$$

$\square$

since $\sin(ut) \geq 0$ on $[0, \frac{\pi}{t}]$.

Remark 6.4.5. In fact, rather than assuming the existence of a derivative, it is enough to assume that f, besides being even continuous, real-valued and going to 0 at infinity, is convex. See [224, pp. 83–85].

Solution of Exercise 6.1.31:

(1) Equation (6.1.23) can be rewritten as

$$F^2(f \circ \phi) = f, \quad \text{that is,} \quad F^2(f) = f \circ \phi.$$

Thus

$$F^4 f = F^2(f \circ \phi) = f.$$

(2) By the spectral mapping theorem (see Exercise 4.2.39), we have

$$(\sigma(F))^4 = \sigma(F^4) = 1,$$

$\square$

and hence $\sigma(F) \subset \{1, -1, i, -i\}$.

Solution of Exercise 6.2.3: We first consider an interval of the form $E = (a, b]$, with the topology induced by the usual topology of $\mathbb{R}$, and consider the algebra $\mathcal{A}$ generated by intervals of the form $(c, d]$, where $a \leq c < d \leq b$. We note that

$$E \setminus (c, d] = \begin{cases} (a, c] \cup (d, b] & if\,a<c<d<b, \\ (d, b] & if\,a=c<d<b, \\ (a, c] & if\,a<c<d=b. \end{cases}$$

Furthermore, if $(c_1, d_1] \cap (c_2, d_2] \neq \emptyset$ we have $\max\{c_1, c_2\} < \min\{d_1, d_2\}$ and

$$(c_1, d_1] \cup (c_2, d_2] = (\max\{c_1, c_2\}, \min\{d_1, d_2\}]$$

we see that every element of $\mathcal{A}$ can be written as a finite disjoint union of sets of the form $(c, d]$. So m_g will be regular if and only if it is regular on each set of this form. Assume first that g is right continuous, let $(c, d] \subsetneq (a, b]$, and take $\epsilon > 0$. Take $F = (c, d]$ and $O = (c, d_n]$ where $d < d_n < b$ is such that $g(d_n) - g(d) < \epsilon$. Such a d_n exists since g is assumed right-continuous. Then (6.2.2) is in force and $m_g(O \setminus F) = g(d_n) - g(d) < \epsilon$. Therefore m_g is regular and Theorem 6.2.2 insures the existence of the extension of m_g to a sigma-additive function on $(a, b]$. Conversely, the same theorem asserts that if m_g admits such an extension it is regular. We can assume $F = E$ and O to be of the form $(c, h]$. The condition $m_g(G \setminus F) \le \epsilon$ gives $g(h) - g(b) < \epsilon$ and so g is right-continuous. The case $E = \mathbb{R}$ is done by writing $\mathbb{R} = \cup_{n \in \mathbb{N}}(-n, n]$. $\square$

Solution of Exercise 6.2.4:

(1) The function f is in $\mathbf{L}_1([a, b], \mathcal{B}, dx)$. Thus we have, for $a < s < t < b$,

$$\left| \int_{(s,t)} f(u)du \right| \le \int_{(s,t)} |f(u)|du < \infty,$$

and thus

$$\sum_{j=1}^{N} |\sigma(t_j) - \sigma(s_j)| = \sum_{j=1}^{N} \left| \int_{(s_j,t_j)} f(u)du \right|$$

$$\le \sum_{j=1}^{N} \left| \int_{(s_j,t_j)} |f(u)|du \right|$$

$$= \int_{\cup_{j=1}^{N}(s_j,t_j)} |f(u)|du$$

and, using the Cauchy–Schwarz inequality,

$$\le \sqrt{\sum_{j=1}^{N}(t_j - s_j) \left(\int_{\mathbb{R}} |f(u)|^2 du \right)} \le \sqrt{\delta \left(\int_{\mathbb{R}} |f(u)|^2 du \right)}$$

and it suffices to take $\delta = \frac{\epsilon^2}{\int_{\mathbb{R}} |f(u)|^2 du}$ to conclude the proof. $\square$

Solution of Exercise 6.2.6: (See [243, p. 249]). Let $a_1 < a_2 < \cdots < a_{n-1} < a_n$ and $b_1 \le b_2 \le \cdots \le b_{n-1} \le b_n$. We first note that:

$$\sum_{k=0}^{n-1} a_k(b_{k+1} - b_k) = a_n b_n - a_0 b_0 - \sum_{k=1}^{n}(a_k - a_{k-1})b_k. \qquad (6.4.18)$$

Indeed we have

$$\sum_{k=0}^{n-1} a_k(b_{k+1} - b_k) + \sum_{k=1}^{n}(a_k - a_{k-1})b_k$$

$$= a_0(b_1 - b_0) + (a_n - a_{n-1})b_n + \sum_{k=1}^{n-1}(a_k(b_{k+1} - b_k) + (a_k - a_{k-1})b_k)$$

$$= a_n b_n - a_0 b_0 + a_0 b_1 - a_{n-1}b_n + \sum_{k=1}^{n-1}(a_k b_{k+1} - a_{k-1}b_k))$$

$$= a_n b_n - a_0 b_0,$$

which proves (6.4.18). Formula (6.2.5) follows by taking limits. □

Solution of Exercise 6.2.8: Let $n \in \mathbb{N}$. By the Cauchy–Schwarz inequality, the restriction of σ_f to $(-n, n)$ is absolutely continuous, and (6.2.4) now defines a complex measure on $(-n, n)$. The set of Borel sets where it vanishes is a sigma-algebra, which contains the open intervals in $(-n, n)$, and hence contains all the Borel sets since the Borel sets form the minimal sigma-algebra generated by the open intervals. Taking real and imaginary parts we see that

$$\int_B \operatorname{Re} f(u)du = 0 \quad \text{and} \quad \int_B \operatorname{Im} f(u)du = 0,$$

for every Borel subset B of $(-n, n)$. Take now $\epsilon > 0$ and

$$B = B_\epsilon = \{u \in (-n, n) : \operatorname{Re} f(u) > \epsilon\}.$$

We have

$$0 = \int_{B_\epsilon} \operatorname{Re} f(u)du \ge \epsilon\lambda(B_\epsilon), \quad \text{(where } \lambda \text{ denotes the Lebesgue measure)},$$

and so $\lambda(B_\epsilon) = 0$. Taking now

$$B = C_\epsilon = \{u \in (-n, n) : \operatorname{Re} f(u) < -\epsilon\}$$

we have

$$0 = \int_{C_\epsilon} \operatorname{Re} f(u)du \le -\epsilon\lambda(C_\epsilon),$$

and $\lambda(C_\epsilon) = 0$. It follows that $\operatorname{Re} f = 0$ a.e. The case of $\operatorname{Im} f$ is treated in the same way. Thus $f = 0$ (a.e.) on every interval $(-n, n)$ and so on $\mathbb{R}$. □

Solution of Exercise 6.2.9: We have for $\omega \ne \bar{\omega}$,

$$\int_{\mathbb{R}} f(u)\left\{\frac{1}{u - \omega} - \frac{1}{u - \bar{\omega}}\right\} du = 0,$$

so that

$$\int_{\mathbb{R}} \frac{f(u)}{|u-\omega|^2} du = 0, \quad \forall \omega \in \mathbb{C} \setminus \mathbb{R}.$$

Define a signed measure $d\mu$ by

$$d\mu(u) = f(u)du.$$

Then, condition (6.2.7) is in force and since μ carries no jumps, the Stieltjes–Perron formula (6.2.9) gives

$$\int_{(a,b)} d\mu(u), \quad \forall a, b \in \mathbb{R}.$$

So,

$$\mu(a,b) = 0, \quad \forall a, b \in \mathbb{R},$$

and so μ vanishes on the sigma-algebra generated by the open intervals, and so vanishes identically. $\qquad\square$

Solution of Exercise 6.3.1: Let $f \in \mathbf{L}_2(\mathbb{R}, \mathcal{B}, du)$ be orthogonal to all rational functions belonging to that space. Then, for $\omega \neq \overline{\omega}$,

$$\int_{\mathbb{R}} f(u) \left\{ \frac{1}{u-\omega} - \frac{1}{u-\overline{\omega}} \right\} du = 0,$$

and the previous exercise allows us to conclude. $\qquad\square$

Solution of Exercise 6.3.2:

(1) Let $f \in \mathbf{L}_2(\mathbb{R}, \mathcal{B}, d\sigma)$ be such that

$$\int_{\mathbb{R}} \frac{f(u)d\sigma(u)}{u-z} = 0, \quad \forall z \in \mathbb{C} \setminus \mathbb{R}.$$

Then, as in the proof of Exercise 6.3.1 we have

$$\frac{1}{z-\overline{z}} \left(\int_{\mathbb{R}} \frac{f(u)d\sigma(u)}{u-z} - \int_{\mathbb{R}} \frac{f(u)d\sigma(u)}{u-\overline{z}} \right) = \int_{\mathbb{R}} \frac{f(u)d\sigma(u)}{|u-z|^2} = 0, \quad \forall z \in \mathbb{C} \setminus \mathbb{R}.$$

In a way similar to the preceding exercise, we define a signed measure $d\mu$ by

$$d\mu(u) = f(u)d\sigma(u).$$

Then, condition (6.2.7) is in force and we can use formula (6.2.9) to obtain

$$\int_{(a,b)} d\mu(u) + \frac{\mu(b_+) - \mu(b_-)}{2} + \frac{\mu(a_+) - \mu(a_-)}{2} = 0, \quad \forall a, b \in \mathbb{R}.$$

6.4. Solutions of the exercises

Write $\mu = \mu_1 + \mu_2$, where μ_2 carries the jump of μ. Thus

$$\int_{(a,b)} d\mu_1(u) + \frac{\mu_2(b_+) - \mu_2(b_-)}{2} + \frac{\mu_2(a_+) - \mu_2(a_-)}{2} = 0, \quad \forall a, b \in \mathbb{R}.$$

Thus the measure $\mu_1 + \frac{\mu_2}{2}$ vanishes on the intervals $[a, b]$, and hence on all the Borel sets, that is $\mu = 0$. In particular

$$0 = \int_{\mathbb{R}} \overline{f(u)} d\mu(u) = \int_{\mathbb{R}} |f(u)|^2 d\sigma(u),$$

and so $f = 0$ (σ a.e.).

(2) Let $f \in \mathbf{L}_2(\mathbb{R}, \mathcal{B}, d\sigma)$ be such that

$$\int_{\mathbb{R}} e^{iut} f(u) d\sigma(u) = 0, \quad \forall t \in \mathbb{R}.$$

We follow the hint. With $z \in \mathbb{C}_+$ and $\varphi(t) = e^{izt} 1_{[0,\infty)}(t)$ we have:

$$0 = -i \int_{\mathbb{R}} \varphi(t) \left(\int_{\mathbb{R}} e^{iut} f(u) d\sigma(u) \right) dt$$

$$= -i \int_0^\infty e^{izt} \left(\int_{\mathbb{R}} e^{iut} f(u) d\sigma(u) \right) dt$$

$$= -i \int_{\mathbb{R}} \left(\int_0^\infty e^{itz} e^{iut} dt \right) f(u) d\sigma(u)$$

$$= \int_{\mathbb{R}} \frac{f(u)}{u + z} d\sigma(u).$$

Similarly

$$\int_{\mathbb{R}} \frac{f(u)}{u + z} d\sigma(u) = 0, \quad \forall z \in \mathbb{C}_-.$$

Exercise 6.2.9 allows us to conclude.

(3) Let $f \in \mathbf{L}_2(\mathbb{R}, \mathcal{B}, d\sigma)$ be such that

$$\int_{\mathbb{R}} \chi_t(u) f(u) d\sigma(u) = 0, \quad \forall t \in \mathbb{R}.$$

Then, with $z \in \mathbb{C}_+$ we have

$$0 = \int_{\mathbb{R}} \varphi(t) \left(\int_{\mathbb{R}} \left(\int_0^t e^{iua} da \right) f(u) d\sigma(u) \right) dt$$

$$= \int_{\mathbb{R}} \left(\int_0^\infty e^{itz} \left(\int_0^t e^{iua} da \right) dt \right) f(u) d\sigma(u).$$

But, for $T > 0$, integration by parts gives:

$$\int_0^T e^{itz} \left(\int_0^t e^{iua} da \right) dt = \frac{1}{iz} e^{iTz} \int_0^T e^{iua} da - \frac{1}{iz} \int_0^T e^{itz} e^{itu} dt.$$

Thus,

$$\int_0^\infty e^{itz} \left(\int_0^t e^{iua} da \right) dt = -\frac{1}{iz} \int_0^\infty e^{izt} e^{itu} dt = -\frac{1}{z(z+u)}$$

and the proof proceeds then as in (2).

$\square$

Solution of Exercise 6.3.3: Let $T > 0$. The function $g \in \mathbf{L}_1([0,T], dx)$ and therefore the map

$$\mu(B) = \int_B g(t) dt$$

defines a (signed) measure on $[0,T]$ (see [264, Théorème 1.29, p. 23] for the positive case). Since μ vanishes on open intervals in $[0,T]$, it vanishes on the sigma-algebra they generate, and thus vanishes identically. See for instance [264, Théorème 1.39 (b), p. 29].

$\square$

Solution of Exercise 6.3.5: The proof follows closely the argument in [270, pp. 25–26]. Assume first that $(g_n)_{n \in \mathbb{N}_0}$ is an orthonormal basis. Then, Parseval's equality applied to the function $f(u) = \mathbf{1}_t(u)$ (where $\mathbf{1}_t$ has been defined in (6.1.27)) leads to

$$\|f\|_2^2 = \sum_{n=0}^\infty |\langle f, g_n \rangle|^2,$$

that is,

$$|t| = \sum_{n=0}^\infty \left| \int_0^t g_n(u) du \right|^2,$$

and (6.3.4) holds. Assume now that $(g_n)_{n \in \mathbb{N}}$ is not complete, and let $\mathcal{M}$ denote the closed linear span of the functions g_n. Then, there exists a function $g \not\equiv 0$, of unit norm, and orthogonal to all of the g_n. Parseval's inequality applied to the function $f(u) = \mathbf{1}_t(u)$ leads to

$$|t| \geq \sum_{n=0}^\infty \left| \int_0^t g_n(u) du \right|^2 + \left| \int_0^t g(u) du \right|^2.$$

By Exercise 6.3.3,

$$\int_0^t g(u) du \not\equiv 0,$$

and so there is a strict inequality in (6.3.4) for at least one t.

$\square$

Solution of Exercise 6.3.6:

(1) Since $d\mu_g$ (defined in (6.3.5)) is a probability measure (and in particular finite), it holds that

$$\mathbf{L}_2(\mathbb{R}, d\mu_g) \subset \mathbf{L}_1(\mathbb{R}, d\mu_g),$$

and so $\int_{\mathbb{R}} e^{itx} f(x) e^{-x^2} dx$ converges absolutely for all real t (this could also be checked directly using the Cauchy–Schwarz inequality in $\mathbf{L}_2(\mathbb{R}, dx)$ with the functions $f(x)e^{-\frac{x^2}{2}}$ and $e^{-\frac{x^2}{2}}$). Assume now f orthogonal to all the polynomials in $\mathbf{L}_2(\mathbb{R}, d\mu_g)$. Thus

$$\int_{\mathbb{R}} x^n f(x) e^{-x^2} dx = 0, \quad n = 0, 1, 2, \cdots$$

Let for a fixed $t \in \mathbb{R}$,

$$f_N(x) = \left(\sum_{n=0}^{N} \frac{(itx)^n}{n!} \right) f(x) e^{-x^2} \quad \text{and} \quad g(x) = |f(x)| e^{|tx| - x^2}.$$

Then, $g \in \mathbf{L}_1(\mathbb{R}, dx)$ and

$$|f_N(x)| \leq g(x), \quad \text{a.e.}$$

It follows from the dominated convergence theorem in $\mathbf{L}_1(\mathbb{R}, dx)$ that

$$\lim_{N \to \infty} \int_{\mathbb{R}} f_N(x) dx = \int_{\mathbb{R}} (\lim_{N \to \infty} f_N(x)) dx = \int_{\mathbb{R}} e^{itx} f(x) e^{-x^2} dx.$$

Hence

$$\int_{\mathbb{R}} e^{itx} f(x) e^{-x^2} dx = 0, \quad \forall t \in \mathbb{R},$$

and $f(x) = 0$ as is seen by taking the inverse Fourier transform.

(2) We first show that the linear span of the monomials $x^n y^m$ is dense in the space. Let $f \in \mathbf{L}_2(\mathbb{R}^2, d\mu_g(x) d\mu_g(y))$ and let $n, m \in \mathbb{N}$. The Cauchy–Schwarz inequality applied to $f(x,y) e^{-\frac{x^2+y^2}{2}}$ and $e^{-\frac{x^2+y^2}{2}} x^n y^m$ will show that the function

$$f(x,y) e^{-\frac{x^2+y^2}{2}} x^n y^m \in \mathbf{L}_1(\mathbb{R}, dxdy).$$

Let f be orthogonal to the linear span of the monomials $x^n y^m$. By Theorem 6.3.7 we have

$$\int_{\mathbb{R}} x^n e^{-x^2} \left(\int_{\mathbb{R}} f(x,y) e^{-\frac{y^2}{2}} y^m dy \right) dx = 0, \quad m, n \in \mathbb{N}.$$

Using (1) we see that

$$b(x) = \int_{\mathbb{R}} f(x,y) e^{-\frac{y^2}{2}} y^m dy = 0, \quad a.e. \, x, \quad m \in \mathbb{N}. \tag{6.4.19}$$

The argument of (1) will show that

$$\int_{\mathbb{R}} f(x,y) e^{ity} dy, \quad a.e. \, x, \quad a.e. \, y,$$

and so $f = 0$ a.e.

The span of the polynomials is equal to the span of the functions z^n and $\overline{z}^m$. Furthermore (and this fact will be used in particular in Exercise 11.1.4 on the Bargmann–Segal–Fock space)

$$\frac{1}{\pi} \iint_{\mathbb{R}^2} z^n \overline{z}^m e^{-(x^2+y^2)} \, dx dy = \begin{cases} 0, & n \neq m, \\ n!, & n = m. \end{cases} \tag{6.4.20}$$

See [CAPB, Exercise 5.6.13, p. 214] for (6.4.20).

$\square$

Remarks 6.4.6. In the proof of (2) there is a delicate point. The function b is defined a.e. One needs to take a representative of b in its equivalence class in $\mathbf{L}_1(\mathbb{R}, dx)$ and apply (1) to this representative. We have then that for almost every x the corresponding element $f(x, y)$ is a.e. equal to 0 in y.

Still related to the proof of (2) in the preceding exercise, one could apply the dominated convergence theorem as in (1) directly to the functions

$$f_N(x, y) = f(x, y) \left(\sum_{n=0}^{N} \frac{(itx)^n}{n!} \right) \left(\sum_{n=0}^{N} \frac{(isy)^n}{n!} \right) e^{-x^2 - y^2},$$

and

$$g(x, y) = |f(x, y)| e^{|tx| + |sy| - x^2 - y^2}.$$

One then obtains that

$$\iint_{\mathbb{R}^2} f(x, y) e^{i(tx + sy)} e^{-x^2 - y^2} \, dx dy \equiv 0,$$

and uses the uniqueness theorem for the Fourier transform in two variables.

Solution of Exercise 6.3.11: We follow the hint and take $z \in \mathbb{D}$. Then, applying the dominated convergence theorem in $\mathbf{L}_2(\mathbb{R}, dx)$ to

$$F_N(x) = f(x) e^{-x} \left(\sum_{n=0}^{N} \frac{(izx)^n}{n!} \right) \quad \text{and} \quad G(x) = |f(x)| e^{(|z| - 1)x},$$

we see that

$$\int_0^\infty e^{izx} f(x) e^{-x} \, dx = \sum_{n=0}^{\infty} \frac{(iz)^n}{n!} \int_0^\infty f(x) x^n e^{-x} \, dx = 0, \quad z \in \mathbb{D}. \tag{6.4.21}$$

Furthermore, the function

$$F(z) = \int_0^\infty f(x) e^{-x} e^{izx} \, dx$$

is analytic in Im $z > -1$. Indeed $x \geq 0$, and so we have

$$\left| \frac{e^{ihx} - 1}{h} \right| = \left| \int_0^x e^{ihu} du \right| \leq x e^{|h|x}.$$

Take now $r_0 < 1$ and, let $(h_n)_{n \in \mathbb{N}}$ be a sequence of complex numbers going to 0, and all of modulus less than r_0. Applying the dominated convergence theorem to

$$F_n(x) = f(x) e^{-x} e^{izx} \frac{e^{ih_n} - 1}{h_n} \quad \text{and} \quad G(x) = x |f(x)| e^{x(r_0 - 1)},$$

we see that
$$\lim_{n \to \infty} \frac{F(z + h_n) - F(z)}{h_n} = \int_0^\infty ix f(x) e^{-x} e^{ixz} dx.$$

Since it is enough to compute the limit on sequences, we conclude that F is analytic in Im $z > -1$. In view of (6.4.21) F vanishes in the open unit disk, and hence identically in Im $z > -1$, and in particular on the real line. It follows that $f(x) = 0$ a.e. on $\mathbb{R}_+$, as is seen using the inverse Fourier transform for instance. $\square$

Solution of Exercise 6.3.13:

(1) Let
$$(P^{-1})_{i,k} = \gamma_{i,k}^{(N)}, \quad i, k = 1, \ldots, N.$$

Then, $e_N = \sum_{i=1}^N f_i \gamma_{i,N}^{(N)}$, and

$$\langle e_N, f_j \rangle = \sum_{i=1}^N \langle f_i, f_j \rangle \gamma_{i,N}^{(N)}$$

$$= \sum_{i=1}^N P_{j,i} \gamma_{i,N}^{(N)} = \begin{cases} 0, & j = 1, \ldots, N-1 \\ 1, & j = N. \end{cases}$$

(2) Should $\langle e_N, e_N \rangle = 0$ we would have $\langle e_N, f \rangle = 0$ for all f in the given vector space, and this contradicts the assumed invertibility of P. $\square$

Part III

Hilbert Spaces of Analytic Functions

Chapter 7

Positive Definite Functions and Kernels, and Reproducing Kernel Hilbert Spaces

> *... afin de pouvoir mieux mettre en lumière quelques idées simples, fondamentales dans l'Analyse harmonique non commutative. Une de ces notions, celle de fonction de type positif, apparaît comme fortuite et quelque peu artificielle tant qu'on ne s'est pas rendu compte qu'elle est la contrepartie, pour les représentations linéaires des groupes, de la notion de repésentation monogène.*
>
> Jean Dieudonné, [112, p. 2]

Positive definite kernels (and the associated reproducing kernel Hilbert spaces) play an important role in various fields in mathematics, and Dieudonné's judgment is very harsh, and somewhat unjustified. Besides representation theory and harmonic analysis, they appear in function theory (the kernel function associated with a domain), in stochastic processes (every positive definite function is a correlation function, and vice versa; see Michel Loève's book [223]), in infinite-dimensional analysis, in learning theory (see for instance [276, 232, 291, 181]), and in linear system theory (positivity translates into dissipativity of some underlying linear system), to name a few. In this chapter we present exercises which reflect some of this diversity. We refer to the books [267, 268] of Saitoh and to the papers [179] by Hille and [302, 301] by Szafraniec for background information and applications.

7.1   Positive definite kernels

We first recall the definition of a positive definite kernel.

Definition 7.1.1. Let Ω be a set. The function $K(z,w)$ from $\Omega \times \Omega$ into $\mathbb{C}$ (one uses usually the term *kernel* in this framework) is called a *positive definite kernel*[1] if the following condition holds: For every $N \in \mathbb{N}$, every choice of $w_1, \ldots, w_N \in \Omega$, and every choice of $c_1, \ldots, c_N \in \mathbb{C}$,

$$\sum_{j,k=1}^{N} \overline{c_j} K(w_j, w_k) c_k \geq 0. \tag{7.1.1}$$

We already met a number of examples earlier in the book. In particular:

Question 7.1.2. *Show that the kernel* (1.6.14) *is positive definite in the open unit disk.*

Hint: Use Exercise 2.1.36.

Condition (7.1.1) is equivalent to saying that the $N \times N$ matrix with (j,k) entry $K(w_j, w_k)$ is positive. More generally, there is the following fundamental notion, due to Kreĭn:

Definition 7.1.3. Let Ω be a set. The function $K(z,w)$ from $\Omega \times \Omega$ into $\mathbb{C}$ has a *finite number of negative squares*, say κ, if the following condition holds: It is Hermitian,

$$K(z,w) = \overline{K(w,z)}, \quad \forall z, w \in \Omega,$$

and for every $N \in \mathbb{N}$, every choice of $w_1, \ldots, w_N \in \Omega$, the $N \times N$ matrix with (j,k) entry $K(w_j, w_k)$ has at most κ strictly negative eigenvalues, and exactly κ strictly negative eigenvalues for some choice of N and $w_1, \ldots, w_N$.

Assume now that Ω has a group structure with operation denoted by $\circ$ and where the inverse of $a \in \Omega$ is denoted by a^{-1}. When Ω is Abelian, the function f from Ω into $\mathbb{C}$ is called a positive definite function if the kernel $f(z \circ w^{-1})$ is positive definite in Ω. For instance:

Definition 7.1.4. The function $f(t)$ of the real variable t is called positive definite if the associated kernel $f(t-s)$ is positive definite in the sense of Definition 7.1.1.

When Ω is not Abelian, one can consider this latter kernel, but also the kernel $f(w^{-1} \circ z)$. The two kernels need not be simultaneously positive definite. When Ω is a finite-dimensional real vector space, the function f has a special structure. Setting $\Omega = \mathbb{R}^N$, Bochner's theorem states that, provided f is continuous, there

---

[1]Note that the terminology *positive definite* is a bit misleading since the inequality in (7.1.1) is not strict.

exists a uniquely defined positive Borel measure μ on $\mathbb{R}^N$ such that

$$f(t) = \int \cdots \int_{\mathbb{R}^N} e^{i\langle u,t\rangle_{\mathbb{R}^N}} d\mu(u), \quad \forall t \in \mathbb{R}^N. \tag{7.1.2}$$

The continuity can be dispensed with. Precise results are given in [275].

Question 7.1.5.

(1) *Show that the kernel*

$$K(x,y) = 1_{\mathbb{Q}}(x-y)$$

is positive definite on $\mathbb{R}$.

(2) *Can it written as the Fourier transform of a finite positive Borel measure?*

A representation (7.1.2) will not hold in general in the case of an arbitrary topological (real) vector space V (with the measure defined on the Borel sets of the dual of V). See Question 7.2.18 for a counterexample, where V is an infinite-dimensional real Hilbert space.

It is sometimes easy to check that (7.1.1) holds, but sometimes the positivity property (7.1.1) is quite below the surface for an outsider, as the examples

$$K(t,s) = e^{-|t-s|}, \quad t,s \in \mathbb{R},$$

(see (7.2.11) below) and (0.0.5)

$$\frac{1}{2}\left(|t|^{2H} + |s|^{2H} - |t-s|^{2H}\right), \quad t,s \in \mathbb{R},$$

(where $H \in (0,1)$ is fixed) illustrate. See Exercise 7.2.6 for a family of positive definite functions which include (7.2.11) as a special case, and Exercise 7.2.9 for the kernel (0.0.5). Yet another example is the apparently simple function (11.3.7),

$$K(z,w) = e^{z\overline{w}}(1 - |z-w|^2),$$

which is positive definite in $\mathbb{C}$. An elementary proof of this fact is given in Exercise 7.2.5 (where the function is denoted by $F_2(z,w)$). See also Exercise 11.2.1 and Remark 11.3.4 at the end of the book.

As we already mentioned, positive definite functions play an important role in complex analysis, in the theory of stochastic processes and in infinite-dimensional analysis, and in other surprising fields such as learning theory. In Section 7.2 we present various examples related to some of these topics. We begin here with some general properties.

Exercise 7.1.6.

(1) *Show that* (7.1.1) *implies that* $K(z,w)$ *is Hermitian:*

$$K(z,w) = \overline{K(w,z)} \quad \text{for all} \quad z,w \in \Omega.$$

(2) *The sum of two positive definite kernels is still positive definite.*
(3) *The product of two complex-valued positive definite kernels is still positive definite.*
(4) *If $K(z, w)$ is positive definite so is $L(z, w) = \overline{K(z, w)}$.*

We note that the first and second claims (this last one with adjoint instead of conjugate) still hold in the vector-valued case. On the other hand, (4) does not hold in that case. See Exercise 7.4.2 for a counterexample. As for (3), the product has to be replaced by tensor product in the vector-valued case; see Exercise 7.4.3.

One important way to check that a given function is positive definite is to express it as an inner product:

Exercise 7.1.7. *Let Ω be some set and $(\mathscr{H}, \langle \cdot, \cdot \rangle_{\mathscr{H}})$ be a Hilbert space. Let $z \mapsto h_z$ be a function from Ω into $\mathscr{H}$.*

(1) *Show that the function*

$$K(z, w) = \langle h_w, h_z \rangle_{\mathscr{H}} \tag{7.1.3}$$

is positive definite in Ω.

(2) *The representation (7.1.3) is called minimal if the linear span of the functions h_z, $z \in \Omega$, is dense in $\mathscr{H}$. Show that a minimal representation is unique up to an isomorphism of Hilbert spaces.*

(3) *Under the minimality hypothesis, show that the set $\mathcal{H}(K)$ of functions of the form*

$$F(z) = \langle f, h_z \rangle_{\mathscr{H}}, \quad f \in \mathscr{H}, \tag{7.1.4}$$

endowed with the norm

$$\|F\|_{\mathcal{H}(K)} = \|f\|_{\mathscr{H}} \tag{7.1.5}$$

is a Hilbert space. Show that the function $K_w : z \mapsto K(z, w)$ belongs to $\mathcal{H}(K)$ and that for every $w \in \Omega$ and $F \in \mathcal{H}(K)$,

$$\langle F, K_w \rangle_{\mathcal{H}(K)} = F(w). \tag{7.1.6}$$

(4) *Using Exercise 4.2.44 show that (7.1.3) can be rewritten as*

$$\langle h_w, h_z \rangle_{\mathscr{H}} = T_z^* T_w, \tag{7.1.7}$$

where for all $z \in \Omega$ we have $T_z \in \mathbf{L}(\mathbb{C}, \mathscr{H})$.

As an example of the previous exercise we mention Exercise 2.1.27, and in particular equations (2.1.32) and (2.1.33) there. The following exercises also present illustrations of Exercise 7.1.7.

Exercise 7.1.8. *Show that the functions*

$$K(z, w) = \sum_{n=1}^{N} z_n \overline{w_n}, \quad z, w \in \mathbb{C}^N$$

(where $N \in \mathbb{N}$) and

$$K(z,w) = \sum_{n=1}^{\infty} z_n \overline{w_n}, \quad z, w \in \ell^2(\mathbb{N})$$

are of the form (7.1.3) and characterize the corresponding space $\mathcal{H}(K)$ as defined in (3) in Exercise 7.1.7.

The following exercise pertains to the white noise. The case $\alpha \equiv 1$ corresponds to the covariance function of the Brownian motion.

Exercise 7.1.9. Let α be a continuous positive function defined on $[0, \infty)$. Show that the function

$$K(t,s) = \int_0^{t \wedge s} \alpha(u) du \qquad (7.1.8)$$

is positive definite on $[0, \infty)$.

Remark 7.1.10. Differentiating (7.1.8) in the sense of distributions (see the following exercise) we get

$$\frac{\partial^2}{\partial t \partial s} K(t,s) = \alpha(t)\delta(t - s).$$

The case $\alpha \equiv 1$ corresponds to the white noise.

Remark 7.1.11. More generally, one can consider positive definite functions of the form

$$a(t)b(t \wedge s)a(s)^*, \quad t, s \in \mathbb{R}. \qquad (7.1.9)$$

rather than (7.1.8). In this expression, the functions a and b are matrix-valued and b is differentiable with derivative being nonnegative (as a matrix) for every t. The corresponding stochastic processes are those which admit a finite-dimensional Markovian realization. See the book [71, Example 2, p. 58] of Berlinet and Thomas-Agnan and Kalman's paper [197] for a survey.

Exercise 7.1.12. Let K be of the form (7.1.9). Show that for any pair of Schwartz functions φ, ψ,

$$\iint_{[0,\infty)^2} \varphi'(t)\psi'(s)K(t,s)dtds = \int_0^{\infty} \alpha(u)\varphi(u)\psi(u)du.$$

The space $\mathcal{H}(K)$ defined in item (3) in Exercise 7.1.7 above is called the reproducing kernel Hilbert space associated with the positive definite function $K(z,w)$. Exercises related to such spaces are given in Section 7.5 below.

Remark 7.1.13. In fact, every positive definite function can be written in the form (7.1.3), or, equivalently, as a sum of "squares" $f_j(z)\overline{f_j(w)}$,

$$K(z,w) = \sum_{j \in J} f_j(z)\overline{f_j(w)}, \quad z, w \in \Omega, \qquad (7.1.10)$$

where the index J need not be countable.

These facts will be proved in Exercise 7.5.4 using the reproducing kernel Hilbert space associated with a positive definite function. An example of a function of the form (7.1.10) is given by Mehler's formula (6.1.22):

$$\sum_{n=0}^{\infty} t^n \frac{\xi_n(z)\xi_n(w)}{n!} = \frac{1}{\sqrt{\pi(1-t^2)}} e^{\frac{4zwt-(z^2+w^2)(1+t^2)}{2(1-t^2)}},$$

when $t \in (0,1)$.

Another example is related to the fractional Hardy space of the open right half-plane.

Exercise 7.1.14. *Show that the function*

$$\frac{\Gamma(1+\nu)}{(z+\overline{w})^{1+\nu}} \tag{7.1.11}$$

is positive definite in $\mathbb{C}_r$.

Hint: Use Exercise 2.1.33.

The following question focuses on the converse: What can we say about a sum of the form (7.1.10) in general?

Question 7.1.15. *Let* Ω *be a set and let* $(\varphi_n)_{n\in\mathbb{N}}$ *be a family of (say complex-valued, and not necessarily linearly independent) functions defined on* Ω. *Let*

$$\Omega_0 = \left\{ x \in \Omega \ \Big| \ \sum_{n\in\mathbb{N}} |\varphi_n(x)|^2 < \infty \right\}.$$

The function $K(x,y) = \sum_{n\in\mathbb{N}} \varphi_n(x)\overline{\varphi_n(y)}$ *is positive on* Ω_0 *and the corresponding reproducing kernel Hilbert space is the set of functions of the form*

$$f(x) = \sum_{n\in\mathbb{N}} a_n\varphi_n(x), \quad a_n \in \mathbb{C}, \quad \sum_{n\in\mathbb{N}} |a_n|^2 < \infty, \tag{7.1.12}$$

with norm the minimum of the sums $\sqrt{\sum_{n\in\mathbb{N}} |a_n|^2}$.

The functions φ_n in Question 7.1.15 may belong to some underlying Hilbert space. In this Hilbert space they need not be orthogonal. Even when orthogonal the associated set Ω_0 may be empty, or reduced to a point. When linearly independent, they are orthogonal in the Hilbert space with reproducing kernel K. An example where Ω_0 is empty is provided by the Hermite functions (see (6.1.16) for the latter). Another example of interest appears in Meschkowski's book [234, pp. 3–4] with the Rademacher functions. Then Ω_0 is a countable dense subset of $[0,1]$.

Question 7.1.16. *For every $N \in \mathbb{N}$ divide the interval $[0,1]$ into 2^N sub-intervals of equal length, and define the Rademacher function φ_N to be equal alternatively to 1 and -1 in the corresponding open sub-intervals, and 0 at the boundary points. Characterize Ω_0.*

In (7.1.13) in the following exercise, we denote by ℓ the set of sequences of elements in $\mathbb{N}_0$, indexed by $\mathbb{N}$, and with at most a finite number of nonzero entries.

Exercise 7.1.17.

(1) *Let $(a_n)_{n \in \mathbb{N}}$ be an absolutely convergent series of complex numbers. Show that*

$$\left(\sum_{n=1}^{\infty} a_n \right)^N = \sum_{\substack{k \in \ell \\ |k| = N}} a^k \frac{N!}{k!} \tag{7.1.13}$$

where

$$k! = \prod_{j=1}^{\infty} k_j!, \quad |k| = \sum_{j=1}^{\infty} k_j, \quad \text{and} \quad a^k = \prod_{n=1}^{\infty} a_j^{k_j}$$

for $a = (a_n)_{n \in \mathbb{N}}$. Note that the various expressions make sense since at most a finite number of k_j are different from 0.

(2) *Let K be positive definite on Ω, and of the form (7.1.3). Show that $e^{K(z,w)}$ is positive definite in Ω, and find a representation of the same form for this function.*

(3) *Show that the function*

$$K(z,w) = e^{\langle z,w \rangle_{\ell_2}} \tag{7.1.14}$$

is positive definite on $\Omega = \ell_2$, and represent it in the form (7.1.3).

We send the reader to Remark 11.1.10 in connection to the kernel (7.1.14). This kernel is in the background in Question 5.4.10.

We also note the following: When $z_{N+1} = z_{N+2} = \cdots = 0$, we have

$$e^{\langle z,w \rangle_{\ell_2}} = e^{\sum_{j=1}^{N} z_j \overline{w_j}},$$

and an important factorization of the function

$$e^{\sum_{j=1}^{N} z_j \overline{w_j}}$$

is given in [61, p. 20], and is related to the Bargmann transform. For the case where $z_2 = z_3 = \cdots = 0$, see (2.1.32).

We conclude this section with an exercise where the function is defined on a group. We consider the case of $\mathbb{C}$ but the underlying idea is much more general.

Exercise 7.1.18. *Assume in (7.1.3) that $\Omega = \mathbb{C}$ and that $K(z, w)$ is a function of the difference $z - w$, that is $K(z, w) = k(z - w)$. Assume furthermore given a representation (7.1.3) (assumed to exist; see Remark 7.1.13 if need be). Show that there is a group of unitary operators $(U_z)_{z \in \mathbb{C}}$ such that*

$$U_z(h_v) = h_{z+v}, \quad z, v \in \mathbb{C}. \tag{7.1.15}$$

Hint: Use item (2) of Exercise 7.1.7 to justify the existence of U_z.

7.2 Examples of positive definite functions and kernels

In the first example, $\mathcal{P}(\Omega)$ denotes the set of subsets of a given set Ω, and Δ denotes the symmetric difference of sets:

$$A \Delta B = (A \cap (\Omega \setminus B)) \cup (B \cap (\Omega \setminus A)) = (A \cup B) \setminus (A \cap B).$$

Furthermore we denote by $\#(A)$ the number of elements of the finite set A.

Exercise 7.2.1. *Let $q \in (0, 1)$ and let Ω be a finite set. Show that the function*

$$K(A, B) = q^{\#(A \Delta B)}, \quad A, B \in \mathcal{P}(\Omega), \tag{7.2.1}$$

is positive definite on $\mathcal{P}(\Omega)$.

Hint: Write (with 1_A denoting the characteristic function of the set A)

$$1_{A \Delta B}(x) = 1_A(x) + 1_B(x) - 2(1_{A \cap B}(x)), \tag{7.2.2}$$

and note that

$$\#(A) = \int_\Omega 1_A(x) d\mu(x),$$

where $d\mu$ is the counting measure on Ω and that the function

$$\#(A \cap B)$$

is positive definite on $\mathcal{P}(\Omega)$.

Positive definite functions on a group play a key role in harmonic analysis; see for instance [114]. We now give an example of functions positive definite on the group S_n of permutations of $\{1, 2, \dots, n\}$. The result is used in particular in the study of the q-Fock space. See [248]. We first need a definition.

Definition 7.2.2. *The parity (denoted by $i(\sigma)$) of a permutation σ is the numbers of pairs $(u, v) \in \{1, 2, \dots, n\} \times \{1, 2, \dots, n\}$ such that*

$$u < v \quad \text{and} \quad \sigma(u) > \sigma(v).$$

Question 7.2.3. *Let $q \in [-1, 1]$. Show that the kernels*

$$q^{i(\sigma_1^{-1}\sigma_2)} \quad and \quad q^{i(\sigma_2^{-1}\sigma_1)}$$

are positive definite on S_n.

Hint: Assume first $q > 0$ and write $q = e^r$ for some $r < 0$. Show that, with $\Omega = \{1, 2, \ldots, n\}$, it holds that

$$i(\sigma_1^{-1}\sigma_2) = \sharp(\sigma_1(\Omega)\Delta\sigma_2(\Omega)),$$

and use the previous exercise.

Exercise 7.2.4. *Show that the function*

$$K(m, n) = \binom{m + n}{n} \tag{7.2.3}$$

is positive definite on $\mathbb{N}_0$.

Hint: Use the Chu–Vandermonde formula (1.6.3).

See [33] for a study of the harmonic analysis associated with the function $K(m, n)$.

The functions (0.0.7)

$$F_N(z, w) = e^{z\overline{w}} \sum_{k=0}^{N-1} (-1)^k \binom{N}{k+1} \frac{1}{k!} |z - w|^{2k}$$

are positive definite in the whole complex plane for $N = 1, 2, \ldots$. See Exercise 11.2.1 and Remark 11.3.3. In the next exercise one asks for an elementary proof of this fact for $N = 2$.

Exercise 7.2.5. *Show that the function*

$$F_2(z, w) = e^{z\overline{w}}(2 - |z - w|^2)$$

is positive definite in $\mathbb{C}$.

Hint: Using the power series expansion of $e^{z\overline{w}}$ and writing

$$|z - w|^2 = |z|^2 + |w|^2 - z\overline{w} - \overline{z}w,$$

express the kernel as a sum of "squares", as in (7.1.10).

Exercise 7.2.6.

(1) *Let σ be an increasing function from $\mathbb{R}$ into $\mathbb{R}$ such*

$$\sigma(+\infty) - \sigma(-\infty) < \infty. \tag{7.2.4}$$

Show that the kernel

$$k(t, s) = \int_{\mathbb{R}} e^{iu(t-s)} d\sigma(u) \tag{7.2.5}$$

is positive definite on $\mathbb{R}$.

(2) *Assume now that σ satisfies (6.3.2):*

$$\int_{\mathbb{R}} \frac{d\sigma(u)}{u^2 + 1} < \infty.$$

Show that the kernel

$$K(t, s) = \int_{\mathbb{R}} \frac{e^{iut} - 1}{u} \frac{e^{-ius} - 1}{u} d\sigma(u) \tag{7.2.6}$$

is positive definite on $\mathbb{R}$.

(3) *Assume that σ satisfies (7.2.4) and that its support is inside $[0, \infty)$. Then, the kernel*

$$K(z, \omega) = \int_0^{\infty} e^{iu(z-\overline{\omega})} d\sigma(u)$$

is positive definite in the open upper half-plane $\mathbb{C}_+$.

Remark 7.2.7. The kernels (7.2.5) and (7.2.6) are related by

$$\frac{\partial^2}{\partial t \partial s} K(t, s) = k(t, s)$$

in the sense of distributions. See [11].

Exercise 7.2.8. *Show that (7.2.6) can be rewritten as*

$$K(t, s) = r(t) + \overline{r(s)} - r(t - s), \tag{7.2.7}$$

where

$$r(t) = -\int_{\mathbb{R}} \left(e^{itu} - 1 - \frac{itu}{u^2 + 1} \right) \frac{d\sigma(u)}{u^2}. \tag{7.2.8}$$

The case $r(t) = \frac{|t|}{2}$ corresponds to the Lebesgue measure. See Exercise 6.1.23 above.

Exercise 7.2.9. *Let $d\sigma(u) = |u|^{1-2H} du$ with $H \in (0, 1)$. Compute $r(t)$ and deduce that the kernel (0.0.5) is positive definite on the real line.*

Positive definite kernels of the forms (7.2.5) and (7.2.6) play an important role in the theory of Gaussian stochastic processes. As mentioned earlier in the book, Bochner's theorem states that any positive function f continuous at the origin is of the form

$$f(t) = \int_{\mathbb{R}} e^{itu} d\sigma(u), \qquad (7.2.9)$$

where σ is as in (1) in the previous exercise, that is, f is such that the associated kernel $f(t-s)$ is of the form (7.2.5). Kernels of the form (7.2.5) can take a variety of forms. The following example is a follow-up to Exercise 6.1.9. See also Remark 7.2.16 below.

Exercise 7.2.10. *Show that the kernel*

$$k(t,s) = \frac{1}{\cosh(t-s)} \qquad (7.2.10)$$

is positive definite on the real line and find its representation (7.2.5).

Hint: Use Exercise 6.1.9.

Another example of importance of such kernel is presented in the following exercise, and is related to Shannon's sampling theorem.

Exercise 7.2.11. *Show that the kernel*

$$k(t,s) = \begin{cases} \frac{\sin(t-s)}{t-s}, & t \neq s, \\ 1, & t = s, \end{cases}$$

is positive definite on the real line.

Hint: Write the kernel in the form (7.2.5) with $d\mu$ is absolutely continuous with respect to Lebesgue measure, with derivative

$$\mu'(u) = 1_{[-1,1]}(u) = \begin{cases} 1, & u \in [-1,1] \\ 0, & \text{otherwise.} \end{cases}$$

Related to the previous exercise is the following question, which we adapt from [114, Exercise 20(e), p. 165].

Question 7.2.12. *Let $a_1, a_2, \ldots$ be a sequence of strictly positive numbers such that $\sum_{n=1}^{\infty} a_n < \infty$. Show that the infinite product*

$$F(t) = \prod_{n=1}^{\infty} \frac{\sin a_n t}{a_n t}$$

is convergent and defines a positive definite function. Find its representation (7.2.9).

Yet another example of interest is when $d\sigma$ is absolutely continuous with respect to Lebesgue measure, with density $\sigma'(u) = \frac{1}{\pi}\frac{1}{u^2+1}$. Then (see for instance [CAPB, p. 382]), one obtains the Green function of the one-dimensional Schrödinger operator (see (4.3.7))

$$K(t,s) = e^{-|t-s|}. \tag{7.2.11}$$

Question 7.2.13 (see [185, 186]). *Let $a > 0$ and let $K_a(t,s) = e^{-a|t-s|}$. The kernel*

$$M(t,s) = \int_{\mathbb{R}} K_a(t,u)K_a(u,s)du = \frac{1}{4a^3}\left(1+a|t-s|\right)\exp\left\{-a|t-s|\right\}$$

is positive definite on $\mathbb{R}$.

Equation (7.2.9) is of special interest when $d\sigma$ is absolutely continuous with respect to Lebesgue measure and when moreover the function $f(t)$ takes values in $[0,\infty)$ and is summable. It follows then from the formula (6.1.5) for the inverse Fourier transform that $f(t)$ is positive definite on the real line. As a first example we meet once more the function $e^{-|t|}$.

Question 7.2.14. *Show that the function $\frac{1}{1+t^2}$ is positive definite on $\mathbb{R}$.*

Hint: Compute the Fourier transform of $\frac{1}{1+t^2}$.

Another path to prove this result would be as follows:

Exercise 7.2.15. *Compute $\int_{[0,\infty)} e^{-u(1+t^2)}du$ and deduce that the function $\frac{1}{1+t^2}$ is positive definite on $\mathbb{R}$.*

Remarks 7.2.16.

(1) The method proposed in Exercise 7.2.15 illustrates a special case of a theorem of Schoenberg, which states that a function of the form $F(|t|)$ (where F is continuous on the real line) is positive definite if and only if it is of the form

$$F(r) = \int_0^\infty e^{-r^2 u}d\mu(u), \tag{7.2.12}$$

where $d\mu$ is a positive measure on the real line. See [280, 278], [69, Chapter 5], [118], and see [203] for extensions of this theorem. For every $p \in (0,2]$ and every positive finite measure μ on $\mathbb{R}^+$ the function

$$F(r) = \int_0^\infty e^{-r^p u}d\mu(u) \tag{7.2.13}$$

is positive definite on $\mathbb{R}_+$. For instance the choice $d\mu(u) = e^{-u}du$ gives that the kernels

$$\frac{1}{1+|t-s|} \quad \text{and} \quad \frac{1}{1+\sqrt{|t-s|}}$$

are positive definite on the real line (see [69, Exercise 2.12 (c), p. 79] for the first one). Functions of the form (7.2.13) have been characterized by Bretagnolle, Dacunha Castelle and Krivine; see [87, 88] and [69, p. 149]. See Remark 7.3.4 for the case of a jump measure with a single jump at $u = 1$.

(2) The positivity of the function $\frac{1}{1+t^2}$ has consequences in maybe unexpected domains. For instance, the fact that the kernel (7.2.10) is positive definite is proved in [224, p. 88] using the positivity of (7.2.15) and the infinite product represen- tation of $\cosh t$. This apparently more complicated method is conducive to prove the fact that more generally the inverse of entire functions of the form

$$\prod_{n=1}^{\infty}(1 + a_n t^2)$$

where $a_1, a_2, \ldots$ are positive numbers such that $\sum_{n=1}^{\infty} a_n < \infty$, are positive defi- nite.

It is much more difficult to prove that the functions

$$\left(\frac{1}{1 + t^2}\right)^a \qquad (7.2.14)$$

are positive definite[2] for (non-integer) $a > 1/2$. The result follows from the formula (see, e.g., [53, (4.6), p. 417])

$$\left(\frac{1}{1 + t^2}\right)^a = \int_{\mathbb{R}} e^{-itu} h_a(u) du$$

where the function $h_a(u)$ is defined by

$$h_a(u) = \frac{1}{2^{a-\frac{1}{2}}\sqrt{\pi}\Gamma(a)} K_{\frac{1}{2}-a}(|u|)|u|^{a-\frac{1}{2}},$$

where K denotes the modified Bessel function of the third kind, also called Mac- Donald's function. We refer to [215, pp. 107–111] for more information on these functions. To obtain that (7.2.14) is positive definite on $\mathbb{R}$ we need in particular the fact that K takes positive values for real argument.

We note that the multidimensional versions of the functions (7.2.14) are used in [184, 186] by H. Rabitz and his collaborators to solve numerically the bound- state Schrödinger equation.

The cases where in (7.2.9) $d\sigma$ is absolutely continuous with respect to Lebesgue's measure, and equal to

$$\sigma'(u) = \frac{1}{(u^2 + 1)^n} \quad \text{and} \quad \sigma'(u) = \frac{1}{(u^{2n} + 1)},$$

---

[2]The case of $a \in \mathbb{N}$ is trivial since a positive integer power of a positive definite function is positive definite; see also Question 7.2.17.

have been considered in Exercise 6.1.4. In the first case one has

$$f(t) = \frac{\pi e^{-|t|}}{2^{n-1}} \left(\sum_{k=0}^{n-1} |t|^{n-1-k} c_k \right),$$

where

$$c_k = \frac{\prod_{\ell=0}^{k-1}(n+\ell)}{k!(n-1-k)!2^k}, \quad k = 0, 1, \ldots, n-1.$$

It does not seem so obvious to check directly that this function is positive definite.

In the preceding examples the function $\sigma'(u)$ is rational without poles on the real line, and analytic at infinity. Such rational functions can be characterized in terms of realizations (see Section 1.7) and general formulas can be given for the associated Fourier transform, as we now explain. Recall first that a rational function analytic and vanishing at infinity can be written as

$$r(z) = C(zI_n - A)^{-1}B,$$

where $(A, B, C) \in \mathbb{C}^{1 \times n} \times \mathbb{C}^{n \times n} \times \mathbb{C}^{n \times 1}$. Of course it is not clear from the formula above when $r(z)$ is analytic and strictly positive on the real line. See Section 1.7 for a discussion.

The following question is based on [65, p. 284] (see also Exercise 6.1.5).

Question 7.2.17. *Let $\sigma(u)$ be absolutely continuous with respect to the Lebesgue measure, and assume that its derivative is rational, analytic on the real line and at infinity, and with minimal realization*

$$\sigma'(u) = C(uI_n - A)^{-1}B. \tag{7.2.15}$$

Let P denote the Riesz projection corresponding to the spectrum of A in $\mathbb{C}_+$, that is (see (1.5.2), [65, (0.2), p. 284])

$$P = \frac{1}{2\pi i} \int_{\mathbb{R}} (uI_n - A)^{-1}du.$$

Show that the kernel

$$K(t,s) = \begin{cases} iCe^{-i(t-s)A(I_n-P)}B, & t-s > 0, \\ -iCe^{-i(t-s)AP}B, & t-s < 0, \end{cases}$$

is positive definite on $\mathbb{R}$.

Yet another example, given by formula (2.1.30),

$$\int_{\mathbb{R}} e^{-\frac{u^2}{2}} e^{itu} du = \sqrt{2\pi} e^{-\frac{t^2}{2}}$$

has far-reaching analogs in the infinite-dimensional case. This is related to the Bochner–Minlos–Sazonov theorem. For instance, there exists a positive Borel measure P on the space $\mathscr{S}'_{\mathbb{R}}$ of real tempered distributions such that

$$e^{-\frac{\|s\|_2^2}{2}} = \int_{\mathscr{S}'_{\mathbb{R}}} e^{i\langle s's\rangle} dP(s'), \quad \forall s \in \mathscr{S}_{\mathbb{R}},$$

where $\mathscr{S}_{\mathbb{R}}$ denotes the space of real-valued Schwartz functions. See Definition 5.2.11.

In connection with this result, it is well to recall the following fact:

Question 7.2.18. *Let $\mathcal{H}$ be a separable infinite-dimensional real Hilbert space, with inner product $\langle \cdot, \cdot \rangle$ and norm $\| \cdot \|$. The function*

$$e^{-\frac{\|h\|^2}{2}}$$

is positive definite on $\mathcal{H}$, but there does not exist a positive Borel measure P on $\mathcal{H}$ such that

$$e^{-\frac{\|h\|^2}{2}} = \int_{\mathcal{H}} e^{i\langle h, u \rangle} dP(u). \tag{7.2.16}$$

Hint: Assume that such a measure exists. Show that it is a probability measure, and then set $h = e_n$ in (7.2.16) where $(e_n)_{n \in \mathbb{N}}$ is an orthonormal basis of $\mathcal{H}$, and apply the dominated convergence theorem to obtain a contradiction. See [CAPB, Exercise 15.5.5, p. 502].

In the following exercise, and as just above, $\mathscr{S}$ (resp. $\mathscr{S}_{\mathbb{R}}$) denotes the space of (resp. real-valued) Schwartz functions. See Definition 5.2.11. The result presented plays an important role in the study of Poisson processes. To elaborate on this point one needs more information on topological vector spaces, and in particular the notion of nuclear space and the Bochner–Minlos–Sazonov theorem. In the fourth item, $\mathscr{S}_{\mathbb{R}}$ is endowed with its Fréchet space structure; see Definition 5.2.11 for the latter.

Exercise 7.2.19.

(1) *Show that $\mathscr{S} \subset \mathbf{L}_1(\mathbb{R}, \mathcal{B}, dx)$.*
(2) *Let $s \in \mathscr{S}_{\mathbb{R}}$. Show that the integral*

$$\int_{\mathbb{R}} (e^{is(x)} - 1) dx$$

is absolutely convergent.
(3) *Show that the function*

$$F(s) = e^{\left(\int_{\mathbb{R}} (e^{is(x)} - 1) dx \right)} \tag{7.2.17}$$

is positive definite on $\mathscr{S}_{\mathbb{R}}$.
(4) *Show that F is continuous at the origin.*

Functions of the form (7.2.17) are a special case of characteristic functions of generalized stochastic processes with independent values at each "point". To describe these functions (as given in [148, Theorem 5, p. 282]), we recall two definitions. First, K denotes the space of test functions with compact support. Next, a function a belongs to the class Z if it is entire and if, for every $n \in \mathbb{N}$,

$$|z^n a(z)| \le C_{a,n} e^{k_a |z|}, \quad z \in \mathbb{C},$$

for some constants $C_{a,n}$ (depending on a and n) and k_a (depending on a). See [148, p. 22].

Remark 7.2.20 (see [148, Theorem 5, p. 282]). Characteristic functions of generalized stochastic processes with independent values at each "point" are functions of the form

$$L(s) = e^{\int_{\mathbb{R}} f(s(x)) dx}, \quad s \in K$$

where the function f is of the following form:

$$f(x) = \int_{\mathbb{R}} \left(e^{ixu} - a(u)(1 + iux) \right) d\sigma(u) + a_0 + ixa_1 - \frac{a_2}{2!} x^2,$$

where $a_0, a_1, a_2 \in \mathbb{R}$, $a_2 \ge 0$, $d\sigma$ is a positive measure such that

$$\int_{|u|>1} d\sigma(u) + \int_{|u| \le 1} u^2 d\sigma(u) < \infty,$$

and $a \in Z$ is such that $a - 1$ has a zero of order at least three at the origin.

Question 7.2.21. *Let $(X, \mathcal{A}, \mu)$ be a measured space, and let $\mathcal{A}_0$ denote the sets in $\mathcal{A}$ such that $\mu(A) < \infty$.*

(1) *Show that the function*

$$K(A, B) = \mu(A \cap B)$$

is positive definite on $\mathcal{A}_0$.

(2) *Show that the function*

$$K_0(A, B) = \begin{cases} \frac{\mu(A \cap B)}{\mu(A)\mu(B)}, & \text{if both } \mu(A) \text{ and } \mu(B) \text{ are not equal to } 0, \\ 0, & \text{otherwise}, \end{cases}$$

is positive definite on $\mathcal{A}_0$.

Recall that a probability space is a measured space $(\Omega, \mathcal{A}, P)$, where the measure P takes values in $[0, 1]$ and $P(\Omega) = 1$. The elements in the sigma-algebra $\mathcal{A}$ are called *measurable events*, and $P(A)$ is the probability that the event $A \in \mathcal{A}$ takes place. Note that

$$\mathbf{L}_2(\Omega, \mathcal{A}, P) \subset \mathbf{L}_1(\Omega, \mathcal{A}, P)$$

since $P(\Omega) < \infty$. We will use the notation

$$E(f) = \int_{\Omega} f(w)dP(w), \quad f \in \mathbf{L}_1(\Omega, \mathcal{A}, P).$$

Let now T denote some set. A second-order stochastic process indexed by T is a map $t \mapsto X_t$, where $X_t \in \mathbf{L}_2(\Omega, \mathcal{A}, P)$ for all $t \in T$. The covariance function of the second-order process $(X_t)_{t \in T}$ is the function

$$K(t, s) = E(X_t \overline{X_s}) = \langle X_t, X_s \rangle_P,$$

where we have denoted by $\langle \cdot, \cdot \rangle_P$ the inner product in $\mathbf{L}_2(\Omega, \mathcal{A}, P)$. By Exercise 7.1.7, the covariance function is positive definite on T. A key result of M. Loève (see [223, pp. 466–467]) is that, conversely, any positive definite function on a set T is the covariance function of a second-order stochastic process, and that this process can be chosen Gaussian.

Remark 7.2.22. Let now X_0 be a fixed real-valued random variable. The function $f(t) = E(e^{it X_0})$ is called the characteristic function of the random variable. It is continuous at the origin, and positive definite. The converse statement is true, and is one of the facets of Bochner's theorem.

We saw earlier in the section that the function $e^{-|t|}$ is positive definite (that is, the associated kernel $e^{-|t-s|}$ is positive definite on $\mathbb{R}$). See (7.2.11) and the related discussion. Another proof of this fact will be provided in the following section, using a result on conditionally negative functions (see Remark 7.3.4). We conclude this section with yet another way to see that $e^{-|t|}$ is positive definite. This is in fact a particular case of a theorem of Polya, and is based on Exercise 6.1.28.

Question 7.2.23.

(1) *Show that a function f satisfying the assumptions of Exercise 6.1.28 is positive definite on $\mathbb{R}$.*

(2) *Show that the function $e^{-|t|}$ is positive definite on $\mathbb{R}$.*

7.3 Conditionally negative functions

In this section we present exercises related to the notion of conditionally negative function (or kernel) (such functions are called *negative definite kernels* in the book [69], to which we refer for more information on the present topic). A (say scalar-valued, but more general cases are possible) function $M(z, w)$ is said to be conditionally negative on Ω if for every $n \in \mathbb{N}$, every choice of $w_1, \ldots, w_n \in \Omega$, and every choice of $c_1, \ldots, c_n \in \mathbb{C}$ such that

$$\sum_{j=1}^{n} c_j = 0,$$

it holds that

$$\sum_{j,k=1}^{n} \overline{c_j} M(w_j, w_k) c_k \leq 0.$$

Question 7.3.1. *In the notation of Question 7.2.21:*

(1) *Show that the function $\mu(A \cup B)$ is conditionally negative.*
(2) *Show that the function $\mu(A \Delta B)$ is conditionally negative, where Δ denotes the symmetric difference of sets.*

Hint: For (1) write

$$\mu(A \cup B) = \mu(A) + \mu(B) - \mu(A \cap B), \qquad (7.3.1)$$

and for (2)

$$\mu(A \Delta B) = \mu(A) + \mu(B) - 2\mu(A \cap B).$$

(See also (7.2.2) for the latter.)

Question 7.3.2.

(a) *Show that the functions*

$$M(t, s) = (t - s)^2 \quad and \quad |t - s|$$

are conditionally negative on $\Omega = \mathbb{R}$ (for the second function, see the hint after the exercise).

(b) *What can you say about the functions $e^{-xM(t,s)}$ for $x > 0$?*

For an application of the above, see Exercise 7.2.15.

The proof that the function $|t - s|$ is conditionally negative is a direct consequence of Exercise 7.3.6 below but seems otherwise quite illusive without a hint. The hint we now give is in fact a specialization of the proof of Question 7.3.6 to the present case.

Hint to prove that the function $|t - s|$ is conditionally negative on $\mathbb{R}$: Let $n \in \mathbb{N}$ and $t_1, \ldots, t_n \in \mathbb{N}$ and $c_1, \ldots, c_n \in \mathbb{C}$. Show that for every $x \geq 0$ we have

$$\varphi(x) \stackrel{\text{def.}}{=} \sum_{j,k=1}^{n} \overline{c_j} c_k e^{-x|t_j - t_k|} \geq 0, \quad (\text{see Question 7.2.23}).$$

Assume now that $\sum_{j=1}^{n} c_j = 0$. Show that $\varphi(0) = 0$. Compute $\varphi'(0)$ to conclude.

The preceding exercise corresponds to the values $p = 2$ and $p = 1$ in the much more difficult result presented now.

Exercise 7.3.3. *Let $p \in (0, 2]$. Show that the function $M(t, s) = |t - s|^p$ is conditionally negative on the real line.*

Hint: (see for instance [221, p. 154]). Recall (see [Exercise 5.3.5, p. 202, CAPB] or Exercise 7.2.9 above) that

$$\int_0^\infty \frac{1-\cos(tu)}{u^{p+1}}\,du = k_p |t|^p \tag{7.3.2}$$

where

$$k_p = \int_0^\infty \frac{1-\cos(u)}{u^{p+1}}\,du = \begin{cases} \dfrac{\cos(\pi p/2)\Gamma(1-p)}{(1-p)p}, & p \neq 1, \\ \dfrac{\pi}{2}, & p = 1. \end{cases}$$

Remark 7.3.4. It follows from the preceding exercise that the functions

$$e^{-|t|^p} \tag{7.3.3}$$

are positive definite in $\mathbb{R}$ for $p \in (0,2]$ (see also Remark 7.2.16 and formula (7.2.13)). It then follows from Remark 7.2.22 that there exists a random variable X such that

$$E(e^{itX}) = e^{-|t|^p}.$$

This is an example of the Lévy–Khintchine formula for random variables with infinitely divisible distribution. We will not elaborate on this point in the present work.

We present as questions two key results relating positive definite kernels and conditionally negative kernels. See [69, Lemma 2.1 and Theorem 2.2, p. 74] for more information.

Question 7.3.5 (see [69, Lemma 2.1, p. 74]). *Let $M(z,w)$ denote a Hermitian kernel on the set Ω, and let $w_0 \in \Omega$. Then, $M(z,w)$ is conditionally negative if and only if the function*

$$M(z,w_0) + M(w_0,w) - M(z,w) - M(w_0,w_0)$$

is positive definite. What can be said if $M(w_0,w_0) \geq 0$?

Question 7.3.6 (see [69, Theorem 2.2, p. 74]). *Let $M(z,w)$ denote a Hermitian kernel on the set Ω.*

(a) *Show that $M(z,w)$ is conditionally negative if and only if the kernel $e^{-xM(z,w)}$ is positive definite for every $x > 0$.*

(b) *Give another proof that the kernels (7.3.3) are conditionally negative.*

We can now present the solution to Menger's imbedding problem (Problem 3.9.29 above).

Theorem 7.3.7. *The metric space (E,d) can be isometrically imbedded inside ℓ_2 if and only if d^2 is conditionally negative on E.*

As another consequence of the previous exercises we have:

Question 7.3.8. *Show that for r continuous on the real line*

$$e^{-xr(t-s)} \geq 0 \quad \forall x > 0 \quad \Longleftrightarrow \quad r(t) + \overline{r(s)} - r(t-s) - r(0) \geq 0.$$

Exercise 7.3.9 (see [280, p. 536]).

(1) *Compute the integral*

$$\iint_{\mathbb{R}^2} \frac{e^{i(xu+yv)} \, du \, dv}{(1 + (u+v)^2)(1 + (u-v)^2)}, \quad x, y \in \mathbb{R}.$$

(2) *Show that the function*

$$M(x,y) = \max(|x|, |y|)$$

is conditionally negative on $\mathbb{R}$.

Hint: Following Schoenberg in [280, p. 536], make the change of variables

$$a = u + v \quad \text{and} \quad b = u - v.$$

From the proof of the preceding exercise one gets the following result:

Question 7.3.10. *Let* $f \in \mathbf{L}_1(\mathbb{R}, \mathcal{B}, dx)$ *be real-valued. Then the kernel (where* $\check{f}$ *denotes the inverse Fourier transform)*

$$K(x,y) = \check{f}(x-y)\check{f}(x+y)$$

is positive definite on $\mathbb{R}$.

Hint: It suffices to replace $\frac{1}{1+a^2}$ and $\frac{1}{1+b^2}$ by $f(a)$ and $f(b)$ in (7.8.8).

7.4   Vector-valued functions

In the definition of a positive definite function, one can assume $K(z,w)$ to be $\mathbb{C}^{n \times n}$-valued. Then, (7.1.1) is replaced by

$$\sum_{j,k=1}^{N} c_j^* K(w_j, w_k) c_k \geq 0 \tag{7.4.1}$$

where the c_j belong now to $\mathbb{C}^n$. More generally, one can assume that the values of $K(z,w)$ are continuous operators from a topological vector space $\mathcal{V}$ into its anti-dual $\mathcal{V}^*$ (that is, the space of anti-linear continuous functionals). We will use the notation $K(z,w) \in \mathbf{C}(\mathcal{V}, \mathcal{V}^*)$. The terms $\overline{c_j} K(w_j, w_k) c_k$ are then replaced by

$$\langle K(z,u)v, u \rangle_{\mathcal{V}^*, \mathcal{V}},$$

where $v, u \in \mathcal{V}$ and the brackets denote the duality between $\mathcal{V}$ and $\mathcal{V}^*$. The fact that $K(z, w)$ is Hermitian means that

$$\langle K(z,w)c, d \rangle_{\mathcal{V}^*, \mathcal{V}} = \overline{\langle K(w,z)d, c \rangle_{\mathcal{V}^*, \mathcal{V}}}, \quad c, d \in \mathcal{V}. \qquad (7.4.2)$$

This equality is made more explicit in the next exercise. We first recall the canonical injection τ from $\mathcal{V}$ into $\mathcal{V}^{**}$ is defined by:

$$\langle \tau(v), v_* \rangle_{\mathcal{V}^{**}, \mathcal{V}^*} = \overline{\langle v_*, v \rangle_{\mathcal{V}^*, \mathcal{V}}}, \quad \forall\, v \in \mathcal{V} \text{ and } v_* \in \mathcal{V}^*. \qquad (7.4.3)$$

It is also useful to recall the definition of the adjoint T^* of a continuous operator T from $\mathcal{V}$ into $\mathcal{V}^*$ (see, e.g., [79, II.49]):

$$\langle T^*(b_{**}), b \rangle_{\mathcal{V}^*, \mathcal{V}} = \langle b_{**}, T(b) \rangle_{\mathcal{V}^{**}, \mathcal{V}^*}, \quad b_{**} \in \mathcal{V}^{**} \text{ and } b \in \mathcal{V}. \qquad (7.4.4)$$

Exercise 7.4.1. *Let $K(z, w)$ be a $\mathbf{C}(\mathcal{V}, \mathcal{V}^*)$-valued function, positive definite on the set Ω. Compare $K(z, w)$ and $K(w, z)^*$.*

In the case of a Hilbert space, Riesz' representation theorem allows us to identify $\mathcal{V}$ and $\mathcal{V}^*$, and the brackets denote then the inner product.

Exercise 7.4.2. *Let $\mathcal{V}$ be a Hilbert space. Show that if $K(z, w)$ is positive definite on Ω, then $L(z, w) = K(z, w)^*$ need not be positive definite there.*

Hint: Take for instance $\Omega = \mathbb{C}^{n \times n}$ with $n > 1$ and the function $K(A, B) = A^* B$. This example appears in [44, Example 7.6, p. 454].

More generally, let $\mathcal{V}$ be a topological vector space, and let T_z be a $\mathbf{C}(\mathcal{V}, \mathcal{H})$-valued function. The $\mathbf{C}(\mathcal{V}, \mathcal{V}^*)$-valued function function

$$K(z, w) = T_w^* T_z \qquad (7.4.5)$$

is positive definite on Ω.

The product of two scalar positive definite functions is positive definite. The product of two positive matrices (of same size) need not be Hermitian, let alone positive, and this implies in particular that the product of two matrix-valued positive definite functions (of the same size) will not be in general positive definite. On the other hand the tensor product of two positive matrices is positive, as we recalled before Section 1.6 (see also [CABP, Exercise 14.3.3, p. 485]). This fact is the key to the following exercise.

Exercise 7.4.3. *Assume that K_1 and K_2 are positive definite kernels on the set Ω, and respectively $\mathbb{C}^{n_1 \times n_1}$ and $\mathbb{C}^{n_2 \times n_2}$-valued. Show that the $\mathbb{C}^{n_1 n_2 \times n_1 n_2}$-valued function*

$$K_1(z, w) \otimes K_2(z, w)$$

is positive definite in Ω.

Question 7.4.4. *In the notation of the previous exercise, let Γ_j be a $\mathbf{L}(\mathbb{C}^{n_j}, \mathcal{H}_j)$-valued function, where $\mathcal{H}_j$ is a Hilbert space, and let $K_j(z,w) = \Gamma_j(w)^*\Gamma_j(z)$, for $j = 1, 2$. Show that*

$$K_1(z,w) \otimes K_2(z,w) = (\Gamma_1(w) \otimes \Gamma_2(w))^*(\Gamma_1(w) \otimes \Gamma_2(w)).$$

In relation to the previous exercise, see M. Bożejko's paper [80].

One can ask the converse question: Given a $\mathbf{C}(\mathcal{V}, \mathcal{V}^*)$-valued function positive definite on a set Ω, can it be factorized via a Hilbert space as in (7.4.5). When $\mathcal{V} = \mathbb{C}$ the answer is yes, and the result is in fact the essence of the Aronszajn–Moore construction of a reproducing kernel Hilbert space associated with a positive definite kernel. More generally, the class of (locally convex) topological vector spaces for which a factorization of the form (7.4.5) always hold for positive definite functions has been characterized in [161]. These are the spaces with the factorization property. See Section 5.5. As already mentioned earlier, this family includes in particular Banach spaces and nuclear spaces.

7.5   Reproducing kernel Hilbert spaces

A Hilbert space $\mathcal{H}$ of functions defined on a set Ω is called a reproducing kernel Hilbert space if the point evaluations

$$f \mapsto f(w), \quad w \in \Omega$$

are bounded. By Riesz' representation theorem there exists a uniquely determined function $K(z,w)$ defined on $\Omega \times \Omega$, and with the following properties:

(1) For every $w \in \Omega$, the function

$$K_w : z \mapsto K(z,w)$$

belongs to $\mathcal{H}$, and

(2) For every $f \in \mathcal{H}$ and $w \in \Omega$,

$$\langle f, K_w \rangle_{\mathcal{H}} = f(w). \tag{7.5.1}$$

Question 7.5.1. *Show that the function $K(z,w)$ is positive definite in Ω.*

The function $K(z,w)$ is called the reproducing kernel of the space $\mathcal{H}$.

Conversely, the following fundamental result holds (see [52]):

Theorem 7.5.2. *Associated to a function $K(z,w)$ positive definite on a set Ω is a uniquely determined Hilbert space $\mathcal{H}(K)$, whose elements are functions on Ω, and with reproducing kernel $K(z,w)$.*

Remark 7.5.3. Quite often this space provides a useful framework within which one can proceed. A number of interesting and illustrative examples can be found in the book [71] of Berlinet and Thomas-Agnan. See in particular [71, pp. 298–326].

Question 7.5.4. *Using the existence of an associated reproducing kernel Hilbert space prove that every positive definite function can be written in the form* (7.1.3), *or, equivalently, in the form* (7.1.10).

Most of the reproducing kernel Hilbert spaces we will meet in this book are made of analytic functions. Specific properties of the analytic case are considered in Section 8.1. We begin with some examples of functions of a real variable.

Question 7.5.5. *The function*

$$K(t,s) = \begin{cases} 1, & \text{if } t = s, \\ 0, & \text{if } t \neq s, \end{cases}$$

is positive definite for $t, s \in [0,1]$. Show that the associated reproducing kernel Hilbert space consists of the functions of the form $\sum_{s \in [0,1]} K(t,s)c_s$ (where the c_s are complex numbers) which vanish everywhere but on a finite or countable set of points and such that $\sum_{s \in [0,1]} |c_s|^2 < \infty$. It is not separable.

Exercise 7.5.6. *Consider the Sobolev space $\mathcal{H}_{(m)}$ of all the functions with m continuous derivatives in the interval $[0,1]$ and first $m-1$ derivatives vanishing at 0, with the norm:*

$$[f,g]_{\mathcal{H}_{(m)}} = \int_0^1 f^{(m)}(x)g^{(m)}(x)dx,$$

and set:

$$G(t,x) = \begin{cases} \frac{(x-t)^{m-1}}{(m-1)!}, & t \leq x, \\ 0, & \text{else.} \end{cases}$$

Then $\mathcal{H}_{(m)}$ with the inner product $[,]_{\mathcal{H}_{(m)}}$ is a reproducing kernel Hilbert space with reproducing kernel:

$$K_m(x,y) = \int_0^1 G(u,x)G(u,y)du. \tag{7.5.2}$$

Exercise 7.5.7. *Show that, for $p = 1, 2, \ldots$ the spaces $\mathcal{K}_p$ defined in* (6.1.20):

$$\mathcal{K}_p = \left\{ f = \sum_{n=0}^{\infty} a_n \eta_n \; ; \; \sum_{n=0}^{\infty} (n+1)^{2p} |a_n|^2 < \infty \right\},$$

are reproducing kernel spaces of functions and find their respective reproducing kernels. What happens for $p = 0$?

Hint: It is useful to recall that the normalized Hermite functions are uniformly bounded: There exists a constant A such that

$$|\eta_n(u)| \le A, \quad \forall u \in \mathbb{R} \text{ and } \forall n \in \mathbb{N}_0. \tag{7.5.3}$$

See [177, (26), p. 435], [178, p. 90]. Note that the Hermite function is denoted by E_n in [177] while it is denoted by H_n in [178]. More precise bounds are given in these references, but will not be needed here.

When $K(z,w)$ is positive definite on a set Ω so is $tK(z,w)$ for every $t > 0$. As vector spaces the corresponding reproducing kernel Hilbert spaces coincide, but their norms are different when $t \ne 1$ as is shown in the next exercise.

Exercise 7.5.8. *Let $K(z,w)$ be positive definite on Ω and $t > 0$. Show that, as sets,*

$$\mathcal{H}(tK) = \mathcal{H}(K)$$

and that the corresponding norms of these spaces are related by

$$\|f\|_{\mathcal{H}(tK)}^2 = \frac{1}{t}\|f\|_{\mathcal{H}(K)}^2.$$

The following result is well known and allows us to gather under a common roof a wide family of reproducing kernel spaces, of which we mention here the Hardy and the Fock space. Still, for each specific case, the questions set in Remark 0.0.1 need to be addressed.

Question 7.5.9 (see for instance [9, Example 2.1.5]). *Let $L \subset \mathbb{N}_0$, and let $(\alpha_n)_{n \in L}$ be a sequence of strictly positive numbers such that the power series*

$$\sum_{n \in L} \frac{r^{2n}}{\alpha_n}$$

has a strictly positive radius of convergence. Then the set of functions of the form

$$F(z) = \sum_{n \in L} a_n z^n,$$

such that

$$\|F\| \overset{\text{def.}}{=} \sum_{n \in L} \alpha_n |a_n|^2$$

is a reproducing kernel Hilbert space, with reproducing kernel equal to

$$K(z,w) = \sum_{n \in L} \frac{z^n \overline{w}^n}{\alpha_n}.$$

The following two exercises illustrate this result. Other examples appear in the sequel.

Exercise 7.5.10 (see [37, Proposition 9.2]). *Let* $\alpha_n = (n!)^2$. *Show that the corresponding reproducing kernel Hilbert space is the space of entire functions F such that*

$$\frac{1}{\pi} \iint_{\mathbb{C}} |F(z)|^2 K_0(2|z|) dA(z) < \infty$$

where K_0 denotes the modified Bessel function of the second kind of order 0, also called MacDonald's function, or Bessel's function of the third kind.

Hint: Use the fact that the Mellin transform of $2K_0(2\sqrt{x})$ is the square of the function Gamma, meaning

$$\int_0^\infty x^{s-1} K_0(2\sqrt{x}) dx = (\Gamma(s))^2, \quad \text{where} \quad K_0(r) = \int_0^\infty e^{-r \cosh t} dt. \quad (7.5.4)$$

See [101, p. 50] (where a factor 2 is missing) and [215, Exercise 6, p. 14]. For the second formula in (7.5.4), see [215, (5.10.23), p. 119].

The first item in the following exercise is classical. See Section 8.2 for a more detailed study of the corresponding reproducing kernel Hilbert space. The second is much less classical and is taken from the work of Jorgensen and Pedersen. See [194].

Exercise 7.5.11.

(1) *Show that the function $\frac{1}{1-z\overline{w}}$ is positive definite in the open unit disk and characterize its associated reproducing kernel Hilbert space.*

(2) *Let $\Lambda \subset \mathbb{N}_0$ be defined by*

$$\Lambda = 4\Lambda \cup \{1 + 4\Lambda\}.$$

Compute

$$K_\Lambda(z, w) = \sum_{n \in \Lambda} z^n \overline{w}^n \quad (7.5.5)$$

in form of an infinite product.

Remark 7.5.12. Let us denote by $\mathbf{H}_2(\Lambda)$ the reproducing kernel Hilbert space with reproducing kernel (7.5.5). Jorgensen and Pedersen show in [194] that there exists a singular measure $d\mu$ on $[0, 1]$ such that

$$\mathbf{H}_2(\Lambda) = \mathbf{L}_2([0, 1], d\mu),$$

and that furthermore, $\{z^n, n \in \Lambda\}$ is an orthonormal basis of $\mathbf{L}_2([0, 1], d\mu)$. Finally, note that the elements of Λ are exactly the integers of the form

$$\sum b_i 4^i,$$

where the sum is finite and $b_i \in \{0, 1, 2, 3\}$.

Exercise 7.5.13. *Show that the function* (0.0.4)

$$K(t, s) = \min(t, s)$$

is positive definite on $[0, \infty)$ *and describe the associate reproducing kernel Hilbert space.*

There are Hilbert spaces of functions which are not reproducing kernel Hilbert spaces. The construction of such spaces involves the axiom of choice. Exercise 7.5.14 follows [40]. In the exercise, we consider the space $\mathbf{H}_2(\mathbb{D})$, and k_w denotes the function

$$k_w(z) = \frac{1}{1 - z\overline{w}}.$$

Exercise 7.5.14.

(1) *Build a dense set* $(z_n)_{n \in \mathbb{N}}$ *of pairwise distinct points in* $\mathbb{D}$, *and a family* $(w_{n,m})_{n \in \mathbb{N}}$ *of pairwise distinct points also in* $\mathbb{D}$, *all distinct from* $z_1, z_2, \ldots$, *and such that*

$$\lim_{m \to \infty} w_{n,m} = z_n.$$

(2) *Show that there is a Hamel basis in* $\mathbf{H}_2(\mathbb{D})$ *which contains the functions* $k_{w_{n,m}}$.

(3) *Define an operator* T *on* $\frac{1}{1 - z\overline{w}}$ *by*

$$T k_{w_{n,m}} = m k_{w_{n,m}},$$

and $Tf = f$ *if* f *is in the Hamel basis and* $f \neq k_{w_{n,m}}$. *Show that* T *is unbounded.*

(4) *Let* $\mathcal{H}_T = \operatorname{Ran} T$ *with the inner product*

$$\langle Tf, Tg \rangle_{\mathcal{H}_T} = \langle f, g \rangle_{\mathbf{H}_2(\mathbb{D})}.$$

Show that $\mathcal{H}_T = \operatorname{Ran} T$ *is a Hilbert space of functions, such the point evaluations* $Tf \mapsto (Tf)(z_n)$ *are not bounded.*

Remark 7.5.15. As vector spaces, $\mathcal{H}_T = \mathbf{H}_2(\mathbb{D})$. So in the previous exercise we endow the Hardy space $\mathbf{H}_2(\mathbb{D})$ with a Hilbert space structure in which point evaluations are unbounded for points in a dense subset of $\mathbb{D}$. Such results were proved first by Donoghue and Masani. See [117]. See also Remark 1.1.12 for a related discussion.

Question 7.5.16. *Characterize the reproducing kernel Hilbert spaces associated with the positive functions appearing in Exercise 7.2.21.*

Operator ranges and reproducing kernel spaces are closely related. The following exercise is a follow-up of Exercise 4.2.40.

Exercise 7.5.17. *In the notation of Exercise 4.2.40, if moreover $\mathcal{H}$ has reproducing kernel $k(z,w)$ then ran $\sqrt{\Gamma}$ has reproducing kernel the function K_w defined by*

$$K(z,w) = (\Gamma k_w)(z).$$

Question 7.5.18. *Let $K(z,w)$ be positive definite on the set Ω, with representation (7.1.3).*

(1) *(see also Exercise 7.1.17) Show that the function $e^{K(z,w)}$ admits a representation of the form (7.1.3), and express it in terms of the representation of $K(z,w)$.*

(2) *Characterize the associated reproducing kernel Hilbert space.*

Exercise 7.5.19.

(1) *The reproducing kernel Hilbert space associated with the positive kernel*

$$k(t-s) = \int_{\mathbb{R}} e^{iu(t-s)} d\sigma(u) \tag{7.5.6}$$

consists of the functions of the form

$$F(t) = \int_0^\infty e^{iut} f(u) d\sigma(u) \tag{7.5.7}$$

where f belongs to the closed linear span of the functions $u \mapsto e^{ius}$ in $\mathbf{L}_2(\mathbb{R}, \mathcal{B}, d\sigma)$ when s runs through $\mathbb{R}$, and norm

$$\|F\| = \|f\|_\sigma. \tag{7.5.8}$$

(2) *Explain the special case given in equation (7.2.11), that is $K(t,s) = e^{-|t-s|}$. In particular, show that the elements of $\mathcal{H}(K)$ are continuous and have a derivative in the sense of distributions.*

Hint: For (2), use Exercise 6.1.25.

In the previous exercise one could ask the characterization of the reproducing kernel Hilbert space associated with the restriction of F to an interval $[-T, T]$. Of course one can apply the theorem on restriction of positive functions. A more explicit description will involve the closed linear span Z_T of the functions $u \mapsto e^{iut}$ for $|t| \leq T$.

Theorem 7.5.20. *The reproducing kernel Hilbert space associated with the function k given by (7.5.6) and restricted to $[-T, T]$ is the set of functions of the form (7.5.7) with $f \in Z_T$ and norm (7.5.8).*

The spaces Z_T are Hilbert spaces of entire functions of the type introduced by de Branges. See the discussion after Exercise 6.1.22. There σ is a bit more general and the functions $u \mapsto \frac{e^{iut}-1}{u}$ are considered rather than the exponentials.

Question 7.5.21. *Let $K(z,w)$ be a function positive definite on a set Ω and let $\mathcal{H}(K)$ the associated reproducing kernel Hilbert space. Let*

$$K(z,w) = \sum_{a \in A} f_a(z)\overline{f_a(w)}$$

be a representation of the kernel in terms of an orthonormal basis of $\mathcal{H}(K)$. Let c_a be a sequence of numbers such that $0 < c_a < 1$. Then the functions

$$K_1(z,w) = \sum_{a \in A} c_a f_a(z)\overline{f_a(w)}$$

and

$$K_2(z,w) = \sum_{a \in A} (1 - c_a) f_a(z)\overline{f_a(w)}$$

are positive in Ω. Their sum is equal to K and we have:

$$\mathcal{H}(K_1) = \left\{ \sum_{a \in A} x_a f_a(z) \; ; \; \sum_{a \in A} \frac{|x_a|^2}{c_a} < \infty \right\}$$

and

$$\mathcal{H}(K_2) = \left\{ \sum_{a \in A} x_a f_a(z) \; ; \; \sum_{a \in A} \frac{|x_a|^2}{1 - c_a} < \infty \right\}.$$

Let $f(z) = \sum_{a \in A} x_a f_a(z)$ be in $\mathcal{H}(K)$. Then the decomposition $f = f_1 + f_2$ with

$$f_1(z) = \sum_{a \in A} x_a c_a f_a(z), \qquad f_2(z) = \sum_{a \in A} x_a (1 - c_a) f_a(z)$$

is such that $f_j \in \mathcal{H}(K_j)$ for $j = 1, 2$ and

$$\|f\|^2_{\mathcal{H}(K)} = \|f_1\|^2_{\mathcal{H}(K_1)} + \|f_2\|^2_{\mathcal{H}(K_2)}.$$

Discussion: By definition of the norms in $\mathcal{H}(K_1)$ and in $\mathcal{H}(K_2)$ we have:

$$\|f_1\|^2_{\mathcal{H}(K_1)} = \sum_{a \in A} \frac{|x_a|^2 c_a^2}{c_a} = \sum_{a \in A} |x_a|^2 c_a,$$

$$\|f_2\|^2_{\mathcal{H}(K_2)} = \sum_{a \in A} \frac{|x_a|^2 (1 - c_a)^2}{1 - c_a} = \sum_{a \in A} |x_a|^2 (1 - c_a).$$

When $c_a = c$ is independent of the index a, we get to the decomposition

$$K(z,w) = cK(z,w) + (1 - c)K(z,w).$$

The previous question is a special case of the following key result:

Theorem 7.5.22. *Let $K_1(z,w)$ and $K_2(z,w)$ be two $\mathbb{C}^{n\times n}$-valued functions positive definite on a given set Ω. The reproducing kernel Hilbert space $\mathcal{H}(K_1 + K_2)$ with reproducing kernel $K_1 + K_2$ is the sum of the reproducing kernel Hilbert spaces $\mathcal{H}(K_1)$ and $\mathcal{H}(K_2)$:*

$$\mathcal{H}(K_1 + K_2) = \mathcal{H}(K_1) + \mathcal{H}(K_2). \tag{7.5.9}$$

Furthermore

$$\|f\|^2_{\mathcal{H}(K_1+K_2)} = \min_{\substack{f=f_1+f_2 \\ f_i \in \mathcal{H}(K_i),\, i=1,2}} \|f_1\|^2_{\mathcal{H}(K_1)} + \|f_2\|^2_{\mathcal{H}(K_2)},$$

and the sum (7.5.9) is direct and orthogonal if and only if $\mathcal{H}(K_1) \cap \mathcal{H}(K_2) = \{0\}$.

This theorem will be used again and again in the sequel. Its proof goes along the line of the proof of Exercise 1.6.11, and can be divided into the following steps:

STEP 1: Let $K = K_1 + K_2$. Show that the linear relation spanned by the pairs

$$(K(z,w)c, (K_1(z,w)c, K_2(z,w)c)) \in \mathcal{H}(K) \times (\mathcal{H}(K_1) \times \mathcal{H}(K_2)), \quad c \in \mathbb{C}^n, \ w \in \Omega,$$

extends to the graph of an everywhere defined isometry, say T.

STEP 2: Compute T^* and show that T is unitary if and only if

$$\mathcal{H}(K_1) \cap \mathcal{H}(K_1) = \{0\}.$$

We now turn to the product of positive definite functions.

Exercise 7.5.23. *Let K_1 and K_2 be two complex-valued functions, positive definite on the sets Ω_1 and Ω_2 respectively. Show that the function*

$$K(z_1, z_2, w_1, w_2) = K_1(z_1, w_1)K_2(z_2, w_2)$$

is positive definite on $\Omega_1 \times \Omega_2$, and characterize the associated reproducing kernel Hilbert space.

Question 7.5.24. *Using the characterization given in the solution of the preceding exercise, give a description of the elements of the reproducing kernel Hilbert space with reproducing kernel given by*

$$K(z_1, z_2, w_1, w_2) = \frac{1}{(1 - z_1\overline{w_1})(1 - z_2\overline{w_2})}. \tag{7.5.10}$$

The space appearing in the previous question is the Hardy space of the bidisk.

The following question is an adaptation of results in [105], where real-valued functions are considered. It is motivated by applications of reproducing kernel spaces to statistics.

Question 7.5.25. Let K be a function definite positive in a set Ω, and let $w_1, \ldots,$ $w_N \in \Omega$. For $f \in \mathcal{H}(K)$ we set

$$X(f) = \left(\prod_{j=1}^{N} |f(w_j)| \right) e^{-\|f\|^2_{\mathcal{H}(K)}} \tag{7.5.11}$$

The aim of the question is to show that there exists $f_{\max} \in \mathcal{H}(K)$ such that

$$X(f_{\max}) = \max_{f \in \mathcal{H}(K)} X(f).$$

(1) Show that there exists a constant C (which depends on N and on the numbers $w_1, \ldots, w_N$) such that

$$X(f) \leq C, \quad \forall f \in \mathcal{H}(K).$$

(2) Show on an example (with $K \not\equiv 0$) that the expression (7.5.11) may be iden- tically equal to 0. Give a necessary and sufficient condition for the supremum

$$\sup_{f \in \mathcal{H}(K)} X(f) > 0.$$

(3) Assume that the condition in (2) holds. Show that there is a number $M > 0$ such that

$$\sup_{f \in \mathcal{H}(K)} X(f) = \sup_{\substack{f \in \mathcal{H}(K) \\ \|f\|_{\mathcal{H}(K)} \leq M}} X(f).$$

(4) Let $f_1, f_2 \ldots$ be a sequence of elements in $\mathcal{H}(K)$ such that

$$\lim_{n \to \infty} X(f_n) = \sup_{f \in \mathcal{H}(K)} X(f).$$

Show that we may assume this sequence to be weakly convergent.
(5) Show that the weak limit attains the maximum.

In the above-mentioned paper [105], stronger statements are made. First, it is assumed that a function with $X(f) > 0$ exists in a closed convex subspace of $\mathcal{H}(K)$ rather than looking for such a function in all of $\mathcal{H}(K)$. Since (see Exercise 4.1.21) a convex closed set is weakly closed, the existence of the maximum follows. Next, the maximum is shown to be unique using the second Fréchet derivative of the functional

$$\ln(X(f)^2).$$

The last exercise in this section is related to boundary values.

Exercise 7.5.26. Let $K(z, w)$ be a function positive definite in the open unit disk, with associated reproducing kernel Hilbert space $\mathcal{H}(K)$, let $\theta \in [0, 2\pi)$ and let $(r_n)_{n \in \mathbb{N}}$ be a sequence of numbers in $(0, 1)$ such that $\lim_{n \to \infty} r_n = 1$.

(1) *Assume that*

$$\sup_{n \in \mathbb{N}} K(r_n e^{i\theta}, r_n e^{i\theta}) < \infty \qquad (7.5.12)$$

Show that there is a subsequence $(n_k)_{k \in \mathbb{N}}$ *such that the map which to* $f \in \mathcal{H}(K)$ *associates*

$$\lim_{k \to \infty} f(r_{n_k} e^{i\theta}) \qquad (7.5.13)$$

is bounded.

(2) *Suppose that*

$$\lim_{n \to \infty} f(r_n e^{i\theta}) \qquad (7.5.14)$$

exists (as a complex number) for all $f \in \mathcal{H}(K)$. *Show that* (7.5.12) *holds.*

Hint: Use the uniform boundedness theorem.

7.6 Linear operators in reproducing kernel Hilbert spaces

As is well known not every bounded operator in the Lebesgue space $\mathbf{L}_2(\mathbb{R})$ is of the form

$$Tf(t) = \int_{\mathbb{R}} K(t,s)f(s)ds, \qquad (7.6.1)$$

for some function $K(t,s)$, that is, is defined by a kernel $K(t,s)$. We also mention that a linear operator from $\mathbf{L}_2(\mathbb{R})$ into itself is of Hilbert–Schmidt class (see Question 4.2.22) if and only if it can be written in the form (7.6.1) with a function $K(t,s)$ subject to

$$\iint_{\mathbb{R}^2} |K(t,s)|^2 dt ds < \infty.$$

See [257, Theorem VI.24, p. 211]. We take this opportunity to also mention that every linear continuous operator from the Schwartz space into its dual (the latter being endowed with the weak topology) is defined by a kernel, that is, by a distribution in two variables. This is the celebrated kernel theorem due to L. Schwartz. See for instance [167, Théorème 5, p. 73], and see Yger's book [324], and Zemanian's books [327, 326] for some of its applications to linear systems and signal theory.

In the case of a reproducing kernel Hilbert space we have:

Theorem 7.6.1. *Let* $K(x,y)$ *be a positive definite function on the set* Ω *with associated reproducing kernel Hilbert space* $\mathcal{H}(K)$ *and let* T *be a bounded operator from* $\mathcal{H}(K)$ *into itself. Then there exists a function* $A(x,y)$ *from* $\Omega \times \Omega$ *into* $\mathbb{C}$ *with the following properties:*

1. *For every* $y \in \Omega$ *the function* $x \mapsto A(x,y)$ *belongs to* $\mathcal{H}(K)$.
2. *It holds that*

$$Tf(x) = \langle f(\cdot), A(\cdot, x) \rangle_{\mathcal{H}(K)}. \qquad (7.6.2)$$

Not every function $A(x, y)$ from $\Omega \times \Omega$ into $\mathbb{C}$ with the first property in the theorem defines a bounded operator:

Theorem 7.6.2. *Let $K(x, y)$ be a positive definite function on the set Ω with associated reproducing kernel Hilbert space $\mathcal{H}(K)$. Let $A(x, y)$ be a function from $\Omega \times \Omega$ into $\mathbb{C}$ such that for every $y \in \Omega$ the function $x \mapsto A(x, y)$ belongs to $\mathcal{H}(K)$. Then the formula*

$$Tf(x) = \langle f(\cdot), A(\cdot, x) \rangle_{\mathcal{H}(K)} \tag{7.6.3}$$

defines a bounded operator in $\mathcal{H}(K)$ if and only if there exists a strictly positive number κ such that the function

$$K(x, y) - \kappa \langle A(\cdot, y), A(\cdot, x) \rangle_{\mathcal{H}(K)} \tag{7.6.4}$$

is positive definite on Ω.

Question 7.6.3. *What can be said when*

$$A(\cdot, x) = K(\cdot, \varphi(x)), \tag{7.6.5}$$

where φ is a function from Ω into itself?

Hint: See Exercise 7.6.6 below if needed.

In the case of a functional (that is, with range the complex numbers), equation (7.6.3) becomes $Tf = \langle \varphi, f \rangle$ where

$$\varphi(y) = \overline{T(K(\cdot, y))}.$$

This result is called the *representer theorem* and has important applications in the approximation of linear forms (this is a result of Golomb and Weinberger [159] and Boor and Lynch, [103]). See [226].

In terms of orthonormal basis one can go further:

Theorem 7.6.4. *In the notation of Theorem 7.6.2, let b_u (where the index u runs in a not necessarily finite or countable set U) be an orthonormal basis of $\mathcal{H}(K)$, and let*

$$A(x, y) = \sum_{u \in U} b_u(x) \overline{c_u(y)}.$$

Then (7.6.4) becomes

$$K(x, y) - \kappa \left(\sum_{u \in U} c_u(x) \overline{c_u(y)} \right). \tag{7.6.6}$$

When this function is positive definite on Ω the operator T defined by (7.6.3) is given by the formula

$$Tf(x) = \sum_{u \in U} c_u(x) f_u \quad \text{with} \quad f(x) = \sum_{u \in U} b_u(x) f_u.$$

Remarks 7.6.5. It is of interest to link the properties of T and of A. Furthermore, the fact that the kernel (7.6.6) is positive definite is equivalent to the fact that the reproducing kernel Hilbert space with reproducing kernel $\sum_{u \in U} c_u(x)\overline{c_u(y)}$ is continuously included in $\mathcal{H}(K)$.

A function s defined on Ω is called a multiplier if the kernel

$$(1 - s(z)\overline{s(w)})K(z,w)$$

is still positive definite in Ω.

Exercise 7.6.6. *Let $\mathcal{H}(K_1)$ and $\mathcal{H}(K_2)$ be two reproducing kernel Hilbert spaces of $\mathbb{C}^M$-valued and $\mathbb{C}^N$-valued functions, defined on the set Ω, and let φ be a function from Ω into itself. Let furthermore m denote a $\mathbb{C}^{M \times N}$-valued function defined on Ω.*

(1) *The operator*

$$(T_{m,\varphi}f)(z) = m(z)f(\varphi(z)) \tag{7.6.7}$$

is a contraction from $\mathcal{H}(K_1)$ into $\mathcal{H}(K_2)$ if and only if the kernel

$$K_2(z,w) - \frac{1}{c^2}m(z)K_1(\varphi(z),\varphi(w))m(w)^* \tag{7.6.8}$$

is positive definite in Ω for some $c > 0$. The smallest such c is equal to the norm of the operator $T_{m,\varphi}$.

(2) *What can be said when*

$$K_2(z,w) \equiv m(z)K_1(\varphi(z),\varphi(w))m(w)^*, \quad z,w \in \Omega. \tag{7.6.9}$$

(3) *Find the representation (7.6.2) of $T_{m,\varphi}$.*

Definition 7.6.7. Operators of the form $T_{m,\varphi}$ are called weighted composition operators.

Remark 7.6.8. We note the formula (7.8.17)

$$T^*_{m,\varphi}(K_2(\cdot,w)\xi) = K_1(\cdot,\varphi(w))m(w)^*\xi,$$

which is proved in the solution of the exercise.

Remark 7.6.9. Specializing (7.6.8) to $m(z) = I_N$ (resp. to $\varphi(z) = z$) one gets the characterization of composition operators (resp. of multiplication operators) in a reproducing kernel Hilbert space.

See also Remark 7.8.5 after the solution of the exercise.

An example of operator of the form (7.6.7) appears in Exercise 8.6.8, where it is shown that

$$Tf(z) = \frac{1}{z}f\left(-\frac{1}{z}\right)$$

defines a unitary transformation from the Hardy space $\mathbf{H}_2(\mathbb{C}_+)$ onto itself. For other examples of such an operator in this book, see (10.4.3) in Exercise 10.4.4 (for the Bergman space of polyanalytic functions in the disk), Exercise 8.7.3 (for the fractional Hardy space) and Exercise 11.1.9 (for the Fock space).

Question 7.6.10. *Consider the positive definite function (see Exercise 7.2.9),*

$$k_H(t, s) = |t|^{2H} + |s|^{2H} - |t - s|^{2H}, \quad t, s \in \mathbb{R},$$

where $H \in (0, 1)$ is preassigned, and let $a > 0$. Show that the map which to f associates the function

$$g(t) = a^{-H} f(at)$$

is unitary from $\mathcal{H}(k_H)$ onto itself.

Remark 7.6.11. As already mentioned, the function k_H is the correlation function of the fractional Brownian motion. The preceding question expresses that the fractional Brownian motion is a wide sense p-self-similar random process in the sense of Yazici and Kashyap; see [230], [323].

An important consequence of Exercise 7.6.6 is the following characterization of the elements of a reproducing kernel Hilbert space.

Exercise 7.6.12. *Let K be a $\mathbb{C}^{N \times N}$-valued function positive definite on the set Ω and let $\mathcal{H}(K)$ be the associated reproducing kernel Hilbert spaces of $\mathbb{C}^N$-valued functions defined on Ω. Show that the function $f : \Omega \longrightarrow \mathbb{C}^N$ belongs to $\mathcal{H}(K)$ if and only if there exists $M > 0$ such that the kernel*

$$K(z, w) - \frac{f(z)f(w)^*}{M^2} \tag{7.6.10}$$

is positive definite in Ω.

Hint: There are at least two ways to prove this result. One is to consider the decomposition

$$K(z, w) = \left(K(z, w) - \frac{f(z)f(w)^*}{M^2} \right) + \frac{f(z)f(w)^*}{M^2}$$

of $K(z, w)$ into a sum of two positive definite kernels. The second one is to apply the preceding exercise with $m(z) = f(z)$ and $K_1(z, w) = 1$.

As a direct consequence of Exercise 7.6.12 we have:

Exercise 7.6.13.

(1) *Let K be a positive definite on Ω and let $w_0 \in \Omega$ be such that $K(w_0, w_0) > 0$. Show that the kernel*

$$K(z, w) - \frac{K(z, w_0)K(w_0, w)}{K(w_0, w_0)}$$

is positive definite in Ω.

(2) *Show that the kernel*

$$e^{-|t-s|} - e^{-t-s} \tag{7.6.11}$$

is positive definite on $[0, \infty)$.

We note that the kernel (7.6.11) appears in particular in the paper [96].

More generally:

Exercise 7.6.14. *Let $K_1(z, w)$ and $K_2(z, w)$ be two positive definite kernels on the set Ω, and let φ denote a map from Ω into itself, and assume that the operator C_φ is bounded from $\mathcal{H}(K_1)$ into $\mathcal{H}(K_2)$. Let f be an entire function, with Taylor series $f(z) = \sum_{n=0}^\infty a_n z^n$ with $a_n \geq 0$ for all $n \in \mathbb{N}_0$. Show that there exists $M > 0$ such that C_φ is a contraction from $\mathcal{H}(f(K_1))$ into $\mathcal{H}(f(MK_2))$.*

To summarize:

Remark 7.6.15. There are a number of important problems which can be considered in a reproducing kernel Hilbert space, namely:

- Study of the multipliers.

- Study of the composition operators.

- More generally, study of the bounded operators of the form (7.6.7), that is:

$$(T_{m,\varphi} f)(z) = m(z) f(\varphi(z)),$$

where m is a function defined on Ω and φ sends Ω into itself.

- Interpolation in the space itself.

- Interpolation in the class of multipliers.

- Multiplicative decomposition of the kernel.

These problems have been studied for a long time, but new aspects and insights pop up regularly. We now discuss some of these. First, a simple question:

Question 7.6.16. *Let K be a positive definite kernel on a set Ω, and let $\mathcal{H}(K)$ be the associated reproducing kernel Hilbert space. Let $(w_1, s_1), \ldots, (w_N, s_N)$ be N pairs of elements in $\Omega \times \mathbb{C}$.*

(1) *Show that a necessary condition for a contractive multiplier s to exist such that*

$$s(w_k) = s_k, \quad k = 1, \ldots, N$$

is that the $N \times N$ Hermitian matrix with (j, k) entry equal to

$$(1 - s_j \overline{s_k}) K(w_j, w_k)$$

is non-negative.

(2) *Show by a counterexample that the condition in (1) is not sufficient in general.*

Remark 7.6.17. Kernels for which the above condition is also sufficient are called *complete Nevanlinna–Pick kernels.* They were characterized by J. Agler as those positive definite kernels k such that k^{-1} has one positive square in Ω, meaning that

$$\frac{1}{k(z,w)} = a(z)\overline{a(w)} - \langle b(z), b(w)\rangle_{\mathcal{H}},$$

where a is complex-valued, and b is a $\mathcal{H}$-valued function for some Hilbert space $\mathcal{H}$. For instance the kernels

$$k(z,w) = \frac{1}{1 - z\overline{w}}, \quad z, w, \in \mathbb{D}$$

and

$$k(h,k) = \frac{1}{1 - \langle h, k\rangle_{\mathcal{H}}},$$

where h, k varies in the open unit ball of some Hilbert space $\mathcal{H}$. See [254, 2]. Results of Kaluza and Lamperti (see Exercise 2.1.1 and the remark after that exercise) and of Baricz, Vesti, and Vuorinen (see in particular [62, Theorem 2.11, p. 11]) allow us to describe a wide class of complete kernels. See [287, p. 277], [59]. For instance the kernel

$$k(z,w) = \begin{cases} \frac{\tan(z\overline{w})}{z\overline{w}}, & z, w \in B\left(0, \sqrt{\frac{\pi}{2}}\right) \text{ such that } zw \neq 0, \\ 1 & z, w \in B\left(0, \sqrt{\frac{\pi}{2}}\right) \text{ such that } zw = 0. \end{cases}$$

is such a kernel. See [211, p. 91] and Question 2.1.4.

As is well known, an everywhere defined linear operator in a Hilbert space need not be continuous (we recalled a counterexample in Exercise 7.5.14 above). On the other hand we have:

Question 7.6.18. *Let Ω denote an open subset of $\mathbb{C}$ and let $\mathcal{H}$ be a reproducing kernel Hilbert space of functions analytic in Ω. Let $a \in \Omega$ and assume that the function $R_a f$ is defined by (1.7.3):*

$$(R_a f)(z) = \begin{cases} \dfrac{f(z) - f(a)}{z - a}, & z \neq a, \\ f'(a), & z = a \end{cases}$$

belongs to $\mathcal{H}$. Show that R_a is continuous.

Hint: Show that the operator is closed (that is, has a closed graph) and use the closed graph theorem (see Theorem 4.2.6).

Exercise 7.6.19. *Let $K(z,w)$ be a positive definite function on the set Ω, and let $\mathcal{H}(K)$ be the associated reproducing kernel Hilbert space, and assume that for every $w \in \Omega$ there exists $f_w \in \mathcal{H}(K)$ such that $f_w(w) \neq 0$. Let $(s_n)_{n\in\mathbb{N}}$ be a family of multipliers of $\mathcal{H}(K)$. Show that, via maybe a subsequence, $(s_n)_{n\in\mathbb{N}}$ converges pointwise to a multiplier. What can be said when the function $f(w) \equiv 1$ belongs to $\mathcal{H}(K)$?*

Hint: Use Exercise 4.2.26.

7.7 Finite-dimensional reproducing kernel spaces

The case of finite-dimensional reproducing kernel Hilbert spaces is of special interest, and hints at deep relationships with linear algebra. Let $\mathcal{H}$ denote a finite-dimensional space of $\mathbb{C}^N$-valued functions, defined on some set Ω, and let $f_1, \ldots, f_n$ denote a basis of the space and let P denote the $n \times n$ matrix with (ℓ, k) entry given by

$$P_{\ell,k} = \langle f_k, f_\ell \rangle_{\mathcal{H}}.$$

Then, P is strictly positive. Furthermore, let F denote the $\mathbb{C}^{N \times n}$-valued matrix-valued function

$$F(z) = \begin{pmatrix} f_1(z) & f_2(z) & \cdots & f_n(z) \end{pmatrix}.$$

The reproducing kernel of $\mathcal{H}$ is then given by the formula:

$$K(z,w) = F(z)P^{-1}F(w)^*. \tag{7.7.1}$$

Exercise 7.7.1. *Prove formula* (7.7.1).

When the basis is orthonormal, we have $P = I_n$ and the formula becomes

$$K(z,w) = \sum_{j=1}^{n} f_j(z)f_j(w)^*, \tag{7.7.2}$$

which is the finite-dimensional case of (7.1.10).

When $n = 1$, (7.7.1) becomes

$$K(z,w) = \frac{f(z)f(w)^*}{p},$$

where f is a basis of $\mathcal{H}$ and $p = \langle f, f \rangle_{\mathcal{H}}$. For instance, let $w \in \mathbb{C}_+$ and let

$$\mathscr{B}_w(z) = \frac{z-w}{z-\overline{w}}.$$

Then, the well-known formula (see for instance [CAPB, (1.1.52), p. 21]),

$$\frac{1 - \mathscr{B}_w(z)\overline{\mathscr{B}_w(v)}}{z - \overline{v}} = \frac{-2i\mathrm{Im}\, w}{(z - \overline{w})(\overline{v} - w)}, \quad z, v \in \mathbb{C}_+, \tag{7.7.3}$$

expresses the fact that the space spanned by $f(z) = \frac{1}{-2\pi i(z-\overline{w})}$ endowed with the inner product of the Hardy space of the open upper half-plane has reproducing kernel

$$\frac{f(z)\overline{f(v)}}{\|f\|^2_{\mathbf{H}_2(\mathbb{C}_+)}} = \frac{1 - \mathscr{B}_w(z)\overline{\mathscr{B}_w(v)}}{-2\pi i(z - \overline{v})},$$

with

$$\|f\|^2_{\mathbf{H}_2(\mathbb{C}_+)} = \frac{1}{-2\pi i(w - \overline{w})} = \frac{1}{4\pi(\mathrm{Im}\, w)}.$$

More generally, when the Blaschke factor $\mathscr{B}_w$ is replaced by a quotient of finite Blaschke products, formula (7.7.3) can be generalized as follows: Let $m \in \mathbb{N}$ and $w_1, \ldots, w_m \in \mathbb{C}$ be such that

$$w_j \neq \overline{w_k}, \quad \forall j, k \in \{1, \ldots, N\}, \tag{7.7.4}$$

and set

$$B(z) = \prod_{j=1}^{m} \frac{z - w_j}{z - \overline{w_j}}. \tag{7.7.5}$$

Let P denote the $m \times m$ matrix with (j, k) entry equal to

$$P_{jk} = \frac{1}{-i(w_k - \overline{w_j})}, \quad j, k = 1, \ldots, m. \tag{7.7.6}$$

Then it holds that (and here we denote by w rather than v the second variable)

$$\frac{1 - B(z)\overline{B(w)}}{-i(z - \overline{w})} = \left(\frac{1}{-i(z - \overline{w_1})} \quad \frac{1}{-i(z - \overline{w_2})} \quad \cdots \quad \frac{1}{-i(z - \overline{w_m})} \right) P^{-1} \begin{pmatrix} \frac{1}{i(\overline{w} - w_1)} \\ \frac{1}{i(\overline{w} - w_2)} \\ \vdots \\ \frac{1}{i(\overline{w} - w_m)} \end{pmatrix}. \tag{7.7.7}$$

Note that condition (7.7.4) insures the uniqueness of the solution of an underlying Lyapunov equation. It will automatically be in force when all the points are in the lower open half-plane or in the upper open half-plane.

To prove formula (7.7.7) one direct possibility is to compare the residues at both sides of this equation, and use formulas for matrices with entries of the form (7.7.6). We here choose another avenue and consider the finite-dimensional version of Exercise 4.2.46. We first make another remark:

Remark 7.7.2. There are analogs of Blaschke factors and of formula (7.7.7) when one moves from the plane (that is, the Riemann sphere) to a real compact Riemann surface. Two key players are then the prime form and Fay's identity. We send the reader to [129] for these and to [46] for the generalization of (7.7.7). We will elaborate on the underlying mathematics in the sequel to this book.

Exercise 7.7.3. Let $J \in \mathbb{C}^{n \times n}$ be invertible and such that $J = J^* = J^{-1}$ (that is, J is a signature matrix; see (1.5.4)). Let $(C, A) \in \mathbb{C}^{n \times m} \times \mathbb{C}^{m \times m}$ such that

$$\bigcap_{k=0}^{\infty} \ker C A^k = \{0\}. \tag{7.7.8}$$

(1) Show that (7.7.8) holds if and only if the following condition holds: Let $f \in \mathbb{C}^m$. Then,

$$C(I_m - zA)^{-1} f \equiv 0 \quad \Longrightarrow \quad f = 0. \tag{7.7.9}$$

7.7. Finite-dimensional reproducing kernel spaces

(2) *Let $P \in \mathbb{C}^{m \times m}$ be a Hermitian invertible matrix, and define*

$$\Theta(z) = I_n + izC(I_m - zA)^{-1}P^{-1}C^*J. \tag{7.7.10}$$

Show that it holds that

$$C(I_m - zA)^{-1}P^{-1}(I_m - wA)^{-*}C^* = \frac{J - \Theta(z)J\Theta(w)^*}{-i(z - \overline{w})}, \quad z, w \in \mathbb{C}, \tag{7.7.11}$$

if and only if P is a solution of the Lyapunov equation

$$PA - A^*P = iC^*JC. \tag{7.7.12}$$

(3) *Let*

$$\widetilde{\Theta}(z) = I_n - iC(zI_m - A)^{-1}P^{-1}C^*J.$$

Show that

$$C(zI_m - A)^{-1}P^{-1}(wI_m - A)^{-*}C^* = \frac{J - \widetilde{\Theta}(z)J\widetilde{\Theta}(w)^*}{-i(z - \overline{w})} \tag{7.7.13}$$

if and only if P is a solution of the Lyapunov equation

$$A^*P - PA = iC^*JC. \tag{7.7.14}$$

Theorem 1.5.11 allows us to characterize when (7.7.12) and (7.7.14) have a unique solution. The next question is a simple illustration of the case where non-uniqueness arises.

Question 7.7.4. *Let*

$$J = \begin{pmatrix} 1 & 0 \\ 0 & -1 \end{pmatrix}, \quad C = \begin{pmatrix} 1 & 1 \\ \alpha & \beta \end{pmatrix} \quad \text{and} \quad A = \operatorname{diag}(w, \overline{w}),$$

where $\alpha, \beta \in \mathbb{C}$ and $w \in \mathbb{C}_+$.

(1) *Show that (7.7.14) has a solution if and only if $\alpha\overline{\beta} = 1$.*
(2) *The matrix P is a strictly positive solution of (7.7.14) if and only if it is of the form*

$$P = \begin{pmatrix} \frac{1-|\alpha|^2}{-i(w-\overline{w})} & z \\ \overline{z} & \frac{1-|\beta|^2}{-i(\overline{w}-w)} \end{pmatrix},$$

where $|\alpha| < 1$ and $z \in \mathbb{C}$ is such that

$$|z|^2 < \frac{(1 - |\alpha|^2)(|\beta|^2 - 1)}{|w - \overline{w}|^2}.$$

A pair of matrices $(C, A) \in \mathbb{C}^{n \times m} \times \mathbb{C}^{m \times m}$ which satisfies condition (7.7.8) is said to be *observable*. Observability is an important concept in the theory of linear systems.

Question 7.7.5. *Let $(C, A) \in \mathbb{C}^{n \times m} \times \mathbb{C}^{m \times m}$. Show that*

$$\bigcap_{k=0}^{\infty} \ker CA^k = \bigcap_{k=0}^{m-1} \ker CA^k.$$

Hint: Use the Cayley–Hamilton theorem (see also the proof of Exercise 1.5.1 for a similar argument).

We note that the function $\Theta(1/z)$,

$$\Theta(1/z) = I_n + iC(zI_m - A)^{-1}P^{-1}C^*J,$$

satisfies

$$\Theta(z)J\Theta(z)^* = J, \quad z \in \mathbb{R} \cap \rho(A), \tag{7.7.15}$$

where $\rho(A)$ denotes the resolvent set of A, that is the set of points z where $(zI_m - A)$ is invertible. Property (7.7.15) is called J-unitarity on the real line and Θ is called a rational function J-unitary on the real line.

Remark 7.7.6. When $n = 1$, it is easy to check that a rational function is unitary on the real line if and only if it is a finite product of terms of the form

$$\mathscr{B}_{w_0}(z) = \frac{z - w_0}{z - \overline{w_0}}, \tag{7.7.16}$$

where $w_0 \neq \overline{w_0}$, times a unitary constant. When all the points w_0 belong to the open upper half-plane, an element (7.7.16) is called a Blaschke factor (see [CAPB, (1.1.44), p. 19 and (1.1.52), p. 21]) and the product is called a finite Blaschke product (in both cases, of the upper open half-plane). Thus a rational function is unitary on the real line if and only if it is a quotient of two finite Blaschke products. This is a very special case, set in the setting of the open upper half-plane, of a result of Kreĭn and Langer on functions s analytic in some open neighborhood of the open unit disk and such that the kernel

$$k_s(z, w) = \frac{1 - s(z)\overline{s(w)}}{1 - z\overline{w}} \tag{7.7.17}$$

has a finite number of negative squares. See [208], and Definition 7.1.3 for the notion of negative squares.

Question 7.7.7. *Assume $m = 1$ and $P > 0$. Using Exercise 7.7.3 find all corresponding rational J-unitary functions.*

Hint: Three cases occur, depending on whether Θ has a pole in the lower open half-plane, in the upper open half-plane, or on the real axis. The first case will occur only when J has at least one positive eigenvalue and the second case will occur only when J has at least one negative eigenvalue. The last case may happen only when J is indefinite.

Remarks 7.7.8.

(1) The elements corresponding to the above cases are called Blaschke–Potapov factors of the first, second and third kind respectively. Up to a normalizing factor, they are of the form (see also (1.7.17))

$$\Theta(z) = I_n + (\mathscr{B}_{w_0}(z) - 1)\frac{uu^*J}{u^*Ju}, \tag{7.7.18}$$

with $u \in \mathbb{C}^n$ such that $u^*Ju > 0$ and $w_0 \in \mathbb{C}_+$ for Blaschke–Potapov factors of the first kind, and $u^*Ju < 0$ and $\overline{w_0} \in \mathbb{C}_+$ for Blaschke–Potapov factors of the second kind.

Blaschke–Potapov factors of the third kind are of the form

$$\Theta(z) = I_n + \frac{ik}{z - x_0}uu^*, \tag{7.7.19}$$

where now $u^*Ju = 0$, $x_0 \in \mathbb{R}$ and $k > 0$.

Equation (7.7.13) reduces respectively to

$$\frac{J - \Theta(z)\Theta(w)^*}{-i(z - \overline{w})} = \begin{cases} \dfrac{uu^*}{u^*Ju} \cdot \dfrac{-2i(\operatorname{Im} w_0)}{(z - \overline{w_0})(\overline{w} - w_0)}, \\[3mm] \dfrac{kuu^*}{(z - x_0)(\overline{w} - x_0)}. \end{cases}$$

(2) Besides (7.7.15) Blaschke–Potapov factors satisfy

$$\Theta(z)J\Theta(z)^* \leq J, \quad z \in \mathbb{C}_+ \cap \rho(A), \tag{7.7.20}$$

and in particular are J-contractive (see Definition 4.2.47).

Definition 7.7.9. A function satisfying (7.7.15) and which satisfies also (7.7.20) is called J-inner on the real line and Θ is called a J-inner rational function.

It is an important result that any rational inner function is a finite product of terms of the kind discussed in Question 7.7.7. See V. Potapov's paper [253], where such results (and much more) are first proved (in the setting of the open unit disk, rather than the open half-plane). When P is Hermitian, but not positive, the situation is much more involved. For indefinite J, there are J-unitary rational functions of any degree without minimal J-unitary factorizations (see Section 1.7 for the notion of minimal factorization). See for instance [29] and Question 7.7.10 below. Still for such P but for $J = I_n$, the function Θ can be written as a quotient of two Blaschke–Potapov products. This is an illustration of the general result of Kreĭn and Langer mentioned above. See [208] for the latter.

Another case of interest is when $n = 2$ and $J = \begin{pmatrix} 1 & 0 \\ 0 & -1 \end{pmatrix}$ and Θ polynomial. Then a complete classification of the elementary factors exists, and is related to the generalized Schur algorithm. This classification was first given by C. Chamfy [97]. See also the works of Delsarte, Genin, Y. and Kamp, [107, 108] and the works of Azizov, Dijksma, Langer, Wanjala and the author [12, 17, 18].

Question 7.7.10. *Show that the function*

$$\Theta(z) = I_n + \frac{ik}{(z-x_0)^n} uu^*, \tag{7.7.21}$$

(where as above $u^* Ju = 0$, $x_0 \in \mathbb{R}$ and $k > 0$) is J-unitary on the real line for any integer n, but does not admit minimal J-unitary factorizations.

Formula (7.7.13) (and its open unit disk counterpart (7.7.30)) allows us to prove as special cases various well-known formulas for orthogonal polynomials, such as the Christofell–Darboux formula. As an example, see Exercise 7.7.14 below (see also the remark 7.8.6 after the proof).

Exercise 7.7.11.

(1) *Prove formula (7.7.7) using the previous exercise.*
(2) *Show that the (j,k) entry of the inverse of the matrix P defined by (7.7.6) is given by*

$$(P^{-1})_{jk} = \frac{\left((\overline{w_j} - w_j) \prod_{t \neq j} \frac{\overline{w_j} - w_t}{\overline{w_j} - \overline{w_t}}\right)\left((\overline{w_k} - w_k) \prod_{t \neq k} \frac{\overline{w_k} - w_t}{\overline{w_k} - \overline{w_t}}\right)}{-i(w_k - \overline{w_j})}.$$

$$\tag{7.7.22}$$

Hint: Set $A = \text{diag}(\overline{w_1}, \ldots, \overline{w_N})$. Find C and J such that (7.7.14) holds.

Remark 7.7.12. Let us summarize the strategy in the previous exercise:

1. One starts from a Hermitian matrix P satisfying a matrix equation.
2. One associates to P a backward shift invariant subspace $\mathcal{M}$, made of rational functions and with (possibly indefinite) metric defined by P.
3. One then computes the reproducing kernel of $\mathcal{M}$ in two different ways. The first way uses the formula for the reproducing kernel of a finite-dimensional reproducing kernel space (this is independent of the underlying structure), and the second way uses the matrix equation satisfied by P.

The last item in the remark above can also be replaced by an argument which does not (explicitly) uses the equation satisfied by P. Such an argument, outlined in the next question, is of special interest in the setting of real compact Riemann surfaces. See [46, Corollary 3.5, p. 305] for the latter.

Question 7.7.13. *Let $w_1, \ldots, w_m$ be distinct points in the complex plane such that*

$$w_j \neq \overline{w_k}, \quad j, k = 1, \ldots, m,$$

and let $\mathcal{M}$ be the span of the functions $f_1, \ldots, f_m$ with

$$f_k(z) = \frac{1}{-i(z - \overline{w_k})}, \quad k = 1, \ldots, m.$$

Show that there is a unique (possibly indefinite) inner product on $\mathcal{M}$ such that the reproducing kernel of $\mathcal{M}$ is of the form $k_B(z, w) = \frac{1 - B(z)\overline{B(w)}}{1 - z\overline{w}}$ for some scalar function B.

Hint: From the reproducing kernel formula one sees that B can be chosen equal to (7.7.5), that is $B(z) = \prod_{j=1}^{m} \frac{z - w_j}{z - \overline{w_j}}$. One then has

$$f_k(z) = k_B(z, w_k), \quad k = 1, \ldots, m$$

and the claim follows from

$$[k_B(\cdot, w), k_B(\cdot, v)] = k_B(v, w)$$

applied to $v, w = w_1, \ldots, w_m$.

Another case of importance is when P is a Hankel matrix.

Exercise 7.7.14. Let $h_0, \ldots, h_{2N}$ be real numbers, and let $P = H_N$ denote the $(N + 1) \times (N + 1)$ matrix with (j, k) entry equal to h_{j+k}, with $j, k \in \{0, \ldots, N\}$. Let $A \in \mathbb{R}^{(N+1) \times (N+1)}$ and $J \in \mathbb{C}^{2 \times 2}$ be defined by

$$A = \begin{pmatrix} 0 & 1 & 0 & \cdots & & 0 \\ 0 & 0 & 1 & 0 & & 0 \\ & & & \ddots & & \vdots \\ 0 & 0 & \cdots & & 0 & 1 \\ 0 & 0 & \cdots & & 0 & 0 \end{pmatrix} \quad \text{and} \quad J = \begin{pmatrix} 0 & -i \\ i & 0 \end{pmatrix}.$$

(1) Find C such that (7.7.12) holds.
(2) Consider the space of polynomials of degree less than or equal to N:

$$\mathcal{P}_N = \text{l.s.} \left\{ 1, \ldots, z^N \right\},$$

endowed with the inner product

$$\langle z^k, z^j \rangle_{\mathcal{P}_N} = h_{k+j}, \quad j, k = 0, \ldots, N. \tag{7.7.23}$$

Show that $\mathcal{P}_N$ is a reproducing kernel Hilbert space and that there exist polynomial functions A_N and B_N such that the reproducing kernel of $\mathcal{P}_N$ is of the form

$$\frac{B_N(z)\overline{A_N(w)} - A_N(z)\overline{B_N(w)}}{z - \overline{w}}. \tag{7.7.24}$$

Hint: For the second item, multiply formula (7.7.11) by $(1 \quad 0)$ on the left and by its transpose on the right, and use formula (7.7.10) for Θ. One finds in particular

$$A_N(z) = 1 - z \left(1 \quad z \quad \cdots \quad z^N\right) H_N^{-1} \begin{pmatrix} 0 \\ h_0 \\ \vdots \\ h_{N-1} \end{pmatrix} \qquad (7.7.25)$$

$$B_N(z) = z \left(1 \quad z \quad \cdots \quad z^N\right) H_N^{-1} \begin{pmatrix} 1 \\ 0 \\ \vdots \\ 0 \end{pmatrix}. \qquad (7.7.26)$$

We note that reproducing kernel spaces with a reproducing kernel of the form (7.7.24) (with A_N and B_N not necessarily polynomials) play an important role in analysis. L. de Branges and J. Rovnyak studied and characterized such spaces. See [83, 85, 86, 84].

Let us now add a real number h_{2N+1} such that the corresponding Hankel matrix H_{N+1} is strictly positive, and consider the space $\mathscr{P}_{N+1}$ with the corresponding inner product. We have:

Exercise 7.7.15. *Show that up to a multiplicative constant, $A_N(z)$ in (7.7.25) is equal to $P_{N+1}(z)$ with*

$$P_{N+1}(z) = \det \begin{pmatrix} 1 & z & \cdots & z^{N+1} \\ h_0 & h_1 & \cdots & h_{N+1} \\ & & & \\ h_N & h_{N+1} & \cdots & h_{2N+1} \end{pmatrix}. \qquad (7.7.27)$$

The following question exhibits an example of kernel of the form (7.7.24) appearing in the setting of Hermite polynomials.

Question 7.7.16. *Prove the reproducing kernel formula*

$$\sum_{u=0}^{n} \frac{H_u(z)\overline{H_u(w)}}{2^u u!} = \frac{H_{n+1}(z)\overline{H_n(w)} - H_n(z)\overline{H_{n+1}(w)}}{2^n n!(z - \overline{w})} \qquad (7.7.28)$$

for the Hermite polynomials.

Hint: A first approach is to follow [177, p. 440], using Mehler's formula (6.1.21) and the recursion (2.1.23). Another approach is to compute the inner product of z^n and z^m in the inner product associated with the weight e^{-x^2}, and apply Exercise 7.7.14. In both approaches the formula

$$\int_{\mathbb{R}} H_n(x)H_m(x)e^{-x^2} dx = \delta_{n,m}\sqrt{\pi}2^n m!$$

(see [177, (31), p. 436]) will be useful.

We now give the counterpart of Exercise 7.7.3 in the open unit disk case. In the statement we denote by Ω_A the set of points such that $\det(I_m - zA) \neq 0$.

Exercise 7.7.17. *Let $J \in \mathbb{C}^{n \times n}$ be a signature matrix, and let $(C, A) \in \mathbb{C}^{n \times m} \times \mathbb{C}^{m \times m}$ be an observable pair of matrices (see (7.7.8) for the latter). Let $P \in \mathbb{C}^{m \times m}$ be a Hermitian invertible matrix, let $\omega_0 \in \mathbb{T}$ be such that $(I_m - \omega_0 A)$ is invertible, and define*

$$\Theta(z) = I_n - (1 - z\overline{\omega_0})C(I_m - zA)^{-1}P^{-1}(I_m - \omega_0 A)^{-*}C^*J. \qquad (7.7.29)$$

Then, it holds that

$$C(I_m - zA)^{-1}P^{-1}(I_m - wA)^{-*}C^* = \frac{J - \Theta(z)J\Theta(w)^*}{1 - z\overline{w}}, \quad z, w \in \Omega_A \qquad (7.7.30)$$

if and only if P is a solution of the Stein equation

$$P - A^*PA = C^*JC. \qquad (7.7.31)$$

We note that Exercise 7.7.17 is used in particular in Exercise 9.2.5 in the solution of the Nevanlinna–Pick interpolation problem using reproducing kernel Hilbert spaces methods.

As in Exercise 7.7.3 we make the following remark: The function Θ satisfies

$$\Theta(z)J\Theta(z)^* = J, \quad z^{-1} \in \mathbb{T} \cap \rho(A) \qquad (7.7.32)$$

where $\rho(A)$ denotes the resolvent set of A. Property (7.7.15) is called J-unitarity on the unit circle and Θ is called a rational function J-unitary on the unit circle.

For an illustration of formula (7.7.30), see (8.8.10). More generally:

Question 7.7.18. *Illustrate Exercise 7.7.17 on functions of the form (1.7.17):*

Exercise 7.7.19.

(1) *Show by an example that (7.7.31) may have more than one solution.*
(2) *Assume that the spectral radius (see Definition 4.2.12) of A is strictly less than 1:*

$$\limsup_{m \to \infty} \|A^m\|^{1/m} < 1.$$

Show that (7.7.31) has a unique solution, given by

$$P = \sum_{m=0}^{\infty} A^{*m}C^*JCA^m. \qquad (7.7.33)$$

We conclude with the following question, which has connections with the Cuntz relations. See [34].

Question 7.7.20. *Let Θ be of the form (7.7.29) and J-unitary on the unit circle, and let $N \in \mathbb{N}$. Find a realization of the function $\Theta(z^N)$.*

7.8   Solutions of the exercises

Solution of Exercise 7.1.6: Let $z, w \in \Omega$, and

$$A = \begin{pmatrix} K(z,z) & K(z,w) \\ K(w,z) & K(w,w) \end{pmatrix}.$$

Since A is positive it is in particular Hermitian (see for instance [CAPB, p. 484]), $A = A^*$, and we get $K(z,w) = \overline{K(w,z)}$. The second claim follows from the fact that the sum of two positive matrices (of the same size) is still positive. The third claim follows from the fact that the Hadamard product (that is, coordinatewise product) of two positive matrices (of the same size) is still positive. See [CABP, Exercise 14.3.3, p. 485] if need be. As for the fourth claim, taking the conjugate of (7.1.1) we have

$$\sum_{j,k=1}^{N} c_j \overline{K(w_j, w_k) \overline{c_k}} \geq 0.$$

Replacing the numbers c_j by their conjugates we obtain the result.   □

Solution of Exercise 7.1.7:

(1) Let $N \in \mathbb{N}$ and $w_1, \ldots, w_N \in \Omega$ and $c_1, \ldots, c_N \in \mathbb{C}$. Then,

$$\sum_{j,k=1}^{N} \overline{c_j} K(w_j, w_k) c_k = \sum_{j,k=1}^{N} \overline{c_j} \left(\langle h_{w_k}, h_{w_j} \rangle_{\mathscr{H}} \right) c_k$$

$$= \langle f, f \rangle_{\mathscr{H}} \geq 0,$$

with $f = \sum_{j=1}^{N} h_{w_j} c_j$.

(2) Let

$$K(z,w) = \langle h_w, h_z \rangle_{\mathscr{H}} = \langle g_w, g_z \rangle_{\mathscr{G}} \tag{7.8.1}$$

be two minimal representations of $K(z,w)$, where $\mathscr{G}$ is also a Hilbert space. The idea is to define a map

$$U h_z = g_z \tag{7.8.2}$$

and show that it extends to a unitary map from $\mathscr{H}$ onto $\mathscr{G}$. The problem is that the vectors h_z are not assumed linearly independent in $\mathscr{H}$. To remedy this, one uses the notion of linear relation (see Definition 4.2.49). Consider the linear subspace R of $\mathscr{H} \times \mathscr{G}$ spanned by the pairs (h_z, g_z) when z runs through Ω. By definition, R is a relation. Its domain

$$\text{Dom}\, R = \{ h \in \mathscr{H} \, ; \, \exists g \in \mathscr{G} \text{ such that } (h, g) \in R \}$$

is dense. Furthermore, let $(f, g) \in R$. Thus

$$f = \sum_{j=1}^{N} c_j h_{z_j} \quad \text{and} \quad g = \sum_{j=1}^{N} c_j g_{z_j}$$

7.8. Solutions of the exercises

for some $N \in \mathbb{N}$ and $z_1, \ldots, z_N \in \Omega$ and $c_1, \ldots, c_N \in \mathbb{C}$. It follows from (7.8.1) that

$$\|f\|_{\mathscr{H}}^2 = \|g\|_{\mathscr{G}}^2 = \sum_{j,k=1}^{N} \overline{c_j} K(w_j, w_k) c_k,$$

that is, R is isometric. Recall, see Exercise 4.2.50, that a densely defined isometry relation defined on a pair of Hilbert spaces extends uniquely to the graph of an isometric operator. Therefore the map (7.8.2) indeed is well defined and extends to an isometry from $\mathscr{H}$ into $\mathscr{G}$. To show that U is unitary, one can for instance define in a similar way an isometry V from $\mathscr{G}$ into $\mathscr{H}$ by $V g_z = h_z$. Then, on dense domains and hence everywhere, $UV = I_{\mathscr{G}}$ and $VU = I_{\mathscr{H}}$, and so U is unitary, with inverse V.

(3) The minimality hypothesis implies that

$$F(z) \equiv 0 \qquad \Longleftrightarrow \qquad f = 0.$$

Thus f determines uniquely F in (7.1.4), and (7.1.5) indeed defines a norm, sometimes called the lifted norm. Furthermore, (7.1.3) shows that K_w is of the form (7.1.4) with $f = h_w$. The norm (7.1.5) is defined by the inner product of $\mathscr{H}$, and we have

$$\langle F, K_w \rangle_{\mathcal{H}(K)} = \langle f, h_w \rangle_{\mathscr{H}} = F(w),$$

that is, (7.1.6) holds. □

(4) We note that (7.1.7) is a mere rewriting of (4.2.16).

Solution of Exercise 7.1.8: In the first case it suffices to take $\mathscr{H} = \mathbb{C}^N$ and

$$h_z = \begin{pmatrix} \overline{z_1} \\ \overline{z_2} \\ \vdots \\ \overline{z_N} \end{pmatrix}.$$

The corresponding space $\mathcal{H}(K)$ consists of the linear functions of N variables

$$F(z) = \sum_{n=1}^{N} a_n z_n, \qquad a_1, \ldots, a_N \in \mathbb{C},$$

with norm $\|F\|_{\mathcal{H}(K)} = \sqrt{\sum_{n=1}^{N} |a_n|^2}$.

The second case is similar with $\ell_2(\mathbb{N})$ replacing $\mathbb{C}^N$. □

Solution of Exercise 7.1.9: It suffices to write

$$K(t, s) = \langle h_s, h_t \rangle_2,$$

with $h_t(u) = \sqrt{\alpha(u)} 1_{[0,t]}(u)$. □

Solution of Exercise 7.1.12: We first note that

$$\int_0^\infty \varphi'(t) 1_{[0,t]}(u) dt = \int_u^\infty \varphi'(t) dt = -\varphi(u). \tag{7.8.3}$$

Using Fubini's theorem and (7.8.3) we have:

$$\iint_{[0,\infty)^2} \varphi'(t)\psi'(s)K(t,s)dtds$$

$$= \iint_{[0,\infty)^2} \varphi'(t)\psi'(s) \left(\int_0^\infty 1_{[0,t]}(u)1_{[0,s]}(u)\alpha(u)du \right) dtds$$

$$= \int_0^\infty \alpha(u) \left(\iint_{[0,\infty)^2} \varphi'(t)\psi'(s)1_{[0,t]}(u)1_{[0,s]}(u)dtds \right) du$$

$$= \int_0^\infty \alpha(u) \left(\int_0^\infty \varphi'(t)1_{[0,t]}(u)dt \right) \left(\int_0^\infty \psi'(s)1_{[0,s]}(u)ds \right) du$$

$$= \int_0^\infty \alpha(u) \left(\int_u^\infty \varphi'(t)dt \right) \left(\int_u^\infty \psi'(s)ds \right) du$$

$$= \int_0^\infty \alpha(u)\varphi(u)\psi(u)du. \qquad \square$$

Solution of Exercise 7.1.14: In Exercise 2.1.33 we proved that

$$\sum_{n=0}^\infty f_n(z)\overline{f_n(w)} = \frac{\Gamma(1+\nu)}{(z+\overline{w})^{1+\nu}}, \quad z,w \in \mathbb{C}_r.$$

The left-hand side of this equation is positive definite, as a converging sum of positive definite functions, and this ends the proof. More precisely, the functions

$$K_M(z,w) = \sum_{n=0}^M f_n(z)\overline{f_n(w)}, \quad M = 0,1,2\ldots$$

are all positive definite in view of item (2) of Exercise 7.1.6. Thus for every choice of $N \in \mathbb{N}$ and $w_1,\ldots,w_N \in \mathbb{C}_r$ and every $c \in \mathbb{C}^N$ and with

$$A_{M,N} = \left(K_M(w_j, w_k) \right)_{j,k=1,\ldots,N}$$

we have $c^* A_{M,N} c \geq 0$. The result follows by letting $M \to \infty$. $\qquad \square$

Solution of Exercise 7.1.17:

(1) A finite product of summable families is a summable family, and its sum is independent of the way the elements are summed. Regrouping the products of N

elements of the sequence $(a_n)_{n\in\mathbb{N}}$ for which a_{k_j} appears k_j times $(j = 1, 2, \dots)$ we obtain (since there are $\frac{N!}{k!}$ such terms)

$$\left(\sum_{n=1}^{\infty} a_n\right)^N = \sum_{(i_1,\dots,i_N)\in\mathbb{N}^N} a_{i_1}\cdots a_{i_N} = \sum_{\substack{k\in\ell \\ |k|=N}} a^k \frac{N!}{k!}.$$

(2) Write the kernel in the form (7.1.10):

$$K(z,w) = \sum_{j\in J} f_j(z)\overline{f_j(w)}, \quad z, w \in \Omega.$$

The index need not be countable, but for a given $z \in \Omega$, there is at most a countable subset of J for which $f_j(z) \neq 0$, and as in (1) we can write

$$K(z,w)^N = \sum_{\substack{k\in\ell_J \\ |k|=N}} (f(z)\overline{f(w)})^k \frac{N!}{k!}$$

where now ℓ_J denotes the set of sequences of elements of $\mathbb{N}_0$ indexed by J, and for which at most a finite number of elements are not equal to 0. Thus

$$e^{K(z,w)} = \sum_{N=0}^{\infty}\left(\sum_{\substack{k\in\ell_J \\ |k|=N}} \frac{(f(z)\overline{f(w)})^k}{k!}\right),$$

which can be written as

$$e^{K(z,w)} = \sum_{k\in\ell_J} \frac{(f(z)\overline{f(w)})^k}{k!}.$$

(3) We now have $J = \mathbb{N}$ and $f_n(z) = z_n$, and

$$e^{\langle z,w\rangle_{\ell_2}} = \sum_{k\in\ell} \frac{z^k \overline{w}^k}{k!},$$

which can be rewritten as $\langle e_w, e_z\rangle_{\ell_2(\ell)}$, with

$$(e_z)_k = \frac{\overline{z}^k}{\sqrt{k!}}, \quad k \in \ell.$$

Solution of Exercise 7.1.18: We have for $v, w \in \mathbb{C}$

$$\begin{aligned}
\langle h_w, h_v\rangle_{\mathscr{H}} &= K(v,w) \\
&= k(v - w) \\
&= k((v + z) - (w + z)) \\
&= K(v + z, w + z) \\
&= \langle h_{w+z}, h_{v+z}\rangle_{\mathscr{H}}.
\end{aligned}$$

By item (2) of Exercise (7.1.7) there exists a unitary operator U_z such that (7.1.15) holds. The formulas

$$U_{z_1+z_2}(h_v) = h_{z_1+z_2+v} = U_{u_1}(h_{z_2+v}) = U_{z_1}(U_{z_2}(h_v)), \quad z_1, z_2, v \in \mathbb{C},$$

show that $(U_z)_{z \in \mathbb{C}}$ is a group of unitary operators.   □

Solution of Exercise 7.2.1: Write $q = e^r$ with $r < 0$. Using the hint we write:

$$K(A, B) = e^{r \sharp(A \triangle B)}$$

$$= e^r \{ \int_\Omega 1_A(x) d\mu(x) \} e^{-r} \{ \int_{A \cap B} d\mu(x) \} e^r \{ \int_\Omega 1_B f(x) d\mu(x) \}$$

The function

$$M(A, B) = \int_{A \cap B} d\mu(x) = \langle 1_A, 1_B \rangle_{d\mu}$$

is an inner product and hence is positive definite on Ω:

$$\sum_{j,k=1}^N \overline{c_j} c_k M(A_j, A_k) = \int_\Omega \left| \sum_{j=1}^N c_j 1_{A_j}(x) \right|^2 d\mu(x) \geq 0.$$

So $e^{-r} \{ \int_{A \cap B} d\mu(x) \}$ is also positive definite since $r < 0$. Write now:

$$\sharp(A \triangle B) = \int_\Omega 1_{A \triangle B}(x) d\mu(x)$$

$$= \int_\Omega (1_A(x) + 1_B(x) - 1_{A \cap B}(x),) d\mu(x),$$

and so:

$$q^{\sharp(A \delta B)} = q^{\sharp(A)} q^{\sharp(B)} q^{-\int_\Omega 1_{A \cap B}(x) d\mu(x)}$$

$$= q^{\sharp(A)} q^{\sharp(B)} e^{(-r(\int_\Omega 1_{A \cap B}(x) d\mu(x))},$$

which shows that the function $q^{\sharp(A \triangle B)}$ is positive definite on $\mathcal{P}(\Omega)$.   □

Solution of Exercise 7.2.4: Define

$$f_n = \left(\binom{n}{0}, \binom{n}{1}, \ldots, \binom{n}{n}, 0, 0, \ldots \right).$$

Formula (1.6.3) then leads to

$$K(m, n) = \langle f_m, f_n \rangle_{\ell_2}$$

and it follows from Exercise 7.1.7 that $K(m, n)$ is positive definite on $\mathbb{N}_0$.   □

7.8. *Solutions of the exercises*

Solution of Exercise 7.2.5: We write

$$F_2(z,w) = e^{z\overline{w}}(2 - |z - w|^2)$$

$$= 2\sum_{n=0}^{\infty} \frac{z^n\overline{w}^n}{n!} + \underbrace{\sum_{n=1}^{\infty} \frac{z^n\overline{w}^n}{(n-1)!}}_{\text{equal to } z\overline{w}e^{z\overline{w}}}$$

$$+ \underbrace{\overline{z}w + |z|^2|w|^2 \sum_{n=0}^{\infty} \frac{z^n\overline{w}^n}{(n+1)!}}_{\text{equal to } e^{z\overline{w}}\overline{z}w}$$

$$- (|z|^2 + |w|^2) \sum_{n=0}^{\infty} \frac{z^n\overline{w}^n}{n!}$$

$$= 2 + \overline{z}w + |z|^2|w|^2 - (|z|^2 + |w|^2)$$

$$+ \sum_{n=1}^{\infty} \frac{z^n\overline{w}^n}{(n+1)!} \left\{ 2(n+1) + n(n+1) + |z|^2|w|^2 - (n+1)(|z|^2 + |w|^2) \right\}$$

$$= \sum_{n=0}^{\infty} a_n(z,w) \frac{z^n\overline{w}^n}{(n+1)!},$$

where

$$a_n(z,w) = \begin{cases} 2 + \overline{z}w + |z|^2|w|^2 - (|z|^2 + |w|^2) \text{ for } n = 0, \\ 2(n+1) + n(n+1) + |z|^2|w|^2 - (n+1)(|z|^2 + |w|^2) \\ \qquad\qquad\qquad\qquad\qquad\qquad \text{for } n = 1, 2, \ldots. \end{cases}$$

To conclude we remark that,

$$a_0(z,w) = 1 + \overline{z}w + (|z|^2 - 1)(|w|^2 - 1),$$

and that for $n = 1, 2, \ldots,$

$$a_n(z,w) = n + 1 + (|z|^2 - (n+1))(|w|^2 - (n+1)).$$

This expresses $F_2(z,w)$ as a sum of "squares" as in (7.1.10). $\square$

Solution of Exercise 7.2.6: The three claims follow directly from Exercise 7.1.7, with $\mathcal{H} = \mathbf{L}_2(d\sigma)$ and $\mathcal{H} = \mathbb{R}$ (cases (1) and (2)) and $\mathcal{H} = \mathbb{C}_+$ (case (3)), and

$$h_t(u) = \begin{cases} e^{-iut} & \text{(first case)}, \\ \begin{cases} \frac{e^{-iut}-1}{u}, & u \neq 0 \\ -it, & u = 0 \end{cases} & \text{(second case)}, \\ e^{-iu\overline{z}} & \text{(third case)}. \end{cases}$$

In all three cases (and using (6.3.2) in the third case) it is clear that $h_t \in \mathbf{L}_2(d\sigma)$. $\square$

Solution of Exercise 7.2.8: Rewriting (for $u \neq 0$)

$$\left(e^{itu} - 1 - \frac{itu}{u^2 + 1}\right)\frac{1}{u^2} = \frac{(e^{itu} - 1)(u^2 + 1) - itu}{(u^2 + 1)u^2}$$

$$= \frac{e^{itu} - 1 - itu}{u^2} + \frac{itu}{u^2 + 1},$$

we see that the function inside the integral in (7.2.8) makes sense at the origin. We now compute (7.2.7):

$$r(t) + \overline{r(s)} - r(t - s) = \int_{\mathbb{R}} \left(-e^{itu} + 1 + \frac{itu}{u^2 + 1} - e^{-isu} + 1\right.$$

$$\left. - \frac{isu}{u^2 + 1} + e^{i(t-s)u} - 1 - \frac{i(t - s)u}{u^2 + 1}\right) d\sigma(u)$$

$$= \int_{\mathbb{R}} \left(\frac{e^{i(t-s)u} - e^{itu} - e^{-isu} + 1}{u^2}\right) d\sigma(u)$$

$$= \int_{\mathbb{R}} \frac{e^{iut} - 1}{u} \frac{e^{-ius} - 1}{u} d\sigma(u),$$

which is (7.2.6).

$\square$

Solution of Exercise 7.2.9: The function $\sigma'(u)$ is even and so the formula for r reduces to

$$r(t) = \int_{\mathbb{R}} (1 - \cos(tu))\frac{\sigma'(u)}{u^2} du = 2\int_0^\infty \frac{1 - \cos(tu)}{u^{1+2H}} du.$$

Setting $tu = v$ and assuming first $t > 0$ we have

$$r(t) = |t|^{2H} \cdot \left(2\int_0^\infty \frac{1 - \cos(v)}{v^{1+2H}} dv\right).$$

This last integral was computed in [Exercise 5.3.5, p. 202, CABP] and is equal to

$$\frac{\cos(\pi H)\Gamma(2 - 2H)}{(1 - 2H)H}$$

when $H \neq \frac{1}{2}$.

$\square$

Solution of Exercise 7.2.10: Since $\left|\frac{\sinh u}{\cosh u}\right| < 1$ for $u \in \mathbb{R}$ the positivity of the given kernel follows from

$$\frac{1}{\cosh(t - s)} = \frac{1}{(\cosh t)(1 - \tanh t \tanh s)(\cosh s)}$$

$$= \frac{1}{(\cosh t)(\cosh s)} \frac{1}{1 - \tanh t \tanh s}$$

$$= \frac{1}{(\cosh t)(\cosh s)} \left(\sum_{n=0}^\infty \tanh^n t \tanh^n s\right).$$

7.8. *Solutions of the exercises*

The positivity is also a direct consequence of Exercise 6.1.9. Indeed, formula (6.1.5) for the inverse Fourier transform applied to (6.4.7) gives

$$\frac{1}{\cosh t} = \int_{\mathbb{R}} e^{iut} d\mu(u)$$

□

with $d\mu(u) = \frac{1}{\cosh\left(\frac{\pi u}{2}\right)} du$.

Solution of Exercise 7.2.11: Take σ to be absolutely continuous with respect to Lebesgue measure, and with density equal to

$$\sigma'(u) = \begin{cases} \frac{1}{2}, & u \in [-1, 1], \\ 0, & |u| > 1. \end{cases}$$

Then

$$\int_{\mathbb{R}} e^{itu} d\sigma(u) = \frac{1}{2} \int_{-1}^{1} e^{itu} du = \begin{cases} \frac{\sin(t)}{t}, & t \neq 0, \\ 1, & t = 0, \end{cases}$$

□

and this ends the proof.

Solution of Exercise 7.2.15: We have

$$\int_{[0,\infty)} e^{-u(1+t^2)} du = \frac{1}{1+t^2}, \quad t \in \mathbb{R}.$$

Hence

$$\frac{1}{1+(t-s)^2} = \int_{[0,\infty)} e^{-u(1+(t-s)^2)} du$$

$$= \int_{[0,\infty)} e^{-u} e^{-ut^2} e^{-us^2} e^{2uts} du$$

$$= \int_{[0,\infty)} e^{-u} e^{-ut^2} e^{-us^2} \left(\sum_{a=0}^{\infty} \frac{(2uts)^a}{a!} \right) du$$

$$= \sum_{a=0}^{\infty} \frac{2^a t^a s^a}{a!} \int_{[0,\infty)} e^{-u} u^a e^{-ut^2} e^{-us^2} du,$$

□

which exhibits $\frac{1}{1+(t-s)^2}$ as a sum of positive definite functions.

Remark 7.8.1. In fact in the previous exercise we have expressed the function $\frac{1}{1+t^2}$ as an integral of exponentials of conditionally negative functions. See Question 7.3.2. Such an approach is conducive to more examples. See for instance [69, Exercise 2.1.2 (c)], and more generally Chapter 3 in that book.

Solution of Exercise 7.2.19:

(1) Let $s \in \mathscr{S}$. Since $K \overset{\text{def.}}{=} \max_{x \in \mathbb{R}} |(x^2 + 1)s(x)| < \infty$ we have

$$|s(x)| \le \frac{K}{x^2 + 1}, \quad x \in \mathbb{R},$$

and so $\mathscr{S} \subset \mathbf{L}_1(\mathbb{R}, \mathcal{B}, dx)$.

(2) Recall that, for real t,

$$|e^{it} - 1| = \left| i \int_0^t e^{iu} du \right| \le |t|, \tag{7.8.4}$$

and so, for a real-valued Schwartz function s,

$$|e^{is(x)} - 1| \le |s(x)|, \quad \forall x \in \mathbb{R}. \tag{7.8.5}$$

This proves the claim in view of item (1).

(3) Let $s_1, s_2 \in \mathscr{S}_{\mathbb{R}}$. We have:

$$e^{\left(\int_{\mathbb{R}} (e^{i(s_1(x) - s_2(x))} - 1) dx \right)} = e^{\left(\int_{\mathbb{R}} ((e^{is_1(x)} - 1)(e^{-is_2(x)} - 1) + e^{is_1(x)} - 1 + e^{-is_2(x)} - 1) dx \right)}$$

$$= e^{\int_{\mathbb{R}} (e^{is_1(x)} - 1) dx}$$

$$\times e^{\left(\int_{\mathbb{R}} (e^{is_1(x)} - 1)(e^{-is_2(x)} - 1) dx \right)}$$

$$\times e^{\int_{\mathbb{R}} (e^{-is_2(x)} - 1) dx}.$$

Now, the kernel

$$\int_{\mathbb{R}} (e^{is_1(x)} - 1)(e^{-is_2(x)} - 1) dx$$

is an inner product, and so is positive definite on $\mathscr{S}_{\mathbb{R}}$, and so is its exponential

$$e^{\left(\int_{\mathbb{R}} (e^{is_1(x)} - 1)(e^{-is_2(x)} - 1) dx \right)}.$$

The claim then follows since the multiplication by $e^{\int_{\mathbb{R}} (e^{is_1(x)} - 1) dx} e^{\int_{\mathbb{R}} (e^{-is_2(x)} - 1) dx}$ does not affect the definite positivity property.

(4) Recall first the well-known and elementary estimate

$$|e^z - 1| \le |z|(e - 1) \tag{7.8.6}$$

for $|z| \le 1$, obtained from the power series expansion

$$e^z - 1 = z \left(\sum_{n=1}^{\infty} \frac{z^{n-1}}{n!} \right).$$

Let $s \in \mathscr{S}_{\mathbb{R}}$ be such that

$$|(x^2 + 1)s(x)| \le \frac{1}{\pi}, \quad x \in \mathbb{R}. \tag{7.8.7}$$

Then, with $z = \int_{\mathbb{R}}(e^{is(x)} - 1)dx$, and using (7.8.5), we have $|z| \leq 1$. Using (7.8.6) and (7.8.4) for such functions s, we have:

$$|e^{\int_{\mathbb{R}}(e^{is(x)}-1)dx} - 1| \leq |\int_{\mathbb{R}}(e^{is(x)} - 1)dx| \cdot (e - 1)$$

$$\leq \left(\int_{\mathbb{R}} |e^{is(x)} - 1|dx\right)(e - 1)$$

$$\leq \left(\int_{\mathbb{R}} |s(x)|dx\right)(e - 1)$$

$$\leq \left(\int_{\mathbb{R}} \frac{dx}{x^2 + 1}\right)\left(\max_{x \in \mathbb{R}}(x^2 + 1)|s(x)|\right)(e - 1).$$

The set

$$\left\{s \in \mathscr{S}_{\mathbb{R}} \; ; \; \max_{x \in \mathbb{R}} x^2|s(x)| < \epsilon/2 \text{ and } \max_{x \in \mathbb{R}} |s(x)| < \epsilon/2\right\}$$

is open in $\mathscr{S}_{\mathbb{R}}$, and its functions satisfy (7.8.7) for ϵ small enough. For such ϵ an element s in this set satisfies

$$|e^{\int_{\mathbb{R}}(e^{is(x)}-1)dx} - 1| \leq \epsilon \left(\int_{\mathbb{R}} \frac{dx}{x^2 + 1}\right)(e - 1),$$

and this proves the asserted continuity. $\qquad\square$

Solution of Exercise 7.3.3: Following an argument in the book [221, p. 154] of Li and Queffélec, and using (7.3.2) we write for $N \in \mathbb{N}$ and $t_1, \ldots, t_N \in \mathbb{R}$ and $c_1, \ldots, c_N \in \mathbb{C}$ such that $\sum_{j=1}^{N} c_j = 0$:

$$\sum_{j,k=1}^{N} c_j\overline{c_k}|t_j - t_k|^p = k_p \int_0^\infty \frac{\sum_{j,k=1}^{N} c_j\overline{c_k} - c_j\overline{c_k}\cos((t_j - t_k)u)}{u^{p+1}}du$$

$$= -k_p \int_0^\infty \frac{\sum_{j,k=1}^{N} c_j\overline{c_k}\cos((t_j - t_k)u)}{u^{p+1}}du$$

$$= -k_p \int_0^\infty \frac{\sum_{j,k=1}^{N} c_j\overline{c_k}(\cos(t_j u)\cos(t_k u) + \sin(t_j u)\sin(t_k u))}{u^{p+1}}du$$

$$= -k_p \int_0^\infty \frac{|\sum_{j=1}^{N} c_j \cos(t_j u)|^2 + |\sum_{j=1}^{N} c_j \sin(t_j u)|^2}{u^{p+1}}du. \qquad\square$$

Solution of Exercise 7.3.9: Following the hint we make the change of variables

$$a = u + v,$$
$$b = u - v.$$

The given integral then splits and is equal to

$$\iint_{\mathbb{R}^2} \frac{e^{i(xu+yv)}\,dudv}{(1+(u+v)^2)(1+(u-v)^2)} = \frac{1}{2}\left(\int_{\mathbb{R}} \frac{e^{i\frac{a(x+y)}{2}}}{1+a^2}\,da\right)\left(\int_{\mathbb{R}} \frac{e^{i\frac{b(x-y)}{2}}}{1+b^2}\,db\right).$$

$$(7.8.8)$$

Using (7.2.11) we have

$$\iint_{\mathbb{R}^2} \frac{e^{i(xu+yv)}\,dudv}{(1+(u+v)^2)(1+(u-v)^2)} = \frac{\pi^2}{2}e^{-\frac{|x+y|+|x-y|}{2}},$$

and hence the result since

$$\frac{|x+y|+|x-y|}{2} = \max(|x|,|y|).$$

This proves (1). Item (2) follows from Exercise 7.3.2.   □

Solution of Exercise 7.4.1: We start from (7.4.2):

$$\langle K(z,w)c,d\rangle_{\mathcal{V}^*,\mathcal{V}} = \overline{\langle K(w,z)d,c\rangle_{\mathcal{V}^*,\mathcal{V}}}, \quad c,d \in \mathcal{V}.$$

Let τ be the canonical injection from $\mathcal{V}$ into $\mathcal{V}^{**}$. Then (see (7.4.3))

$$\langle K(w,z)d,c\rangle_{\mathcal{V}^*,\mathcal{V}} = \overline{\langle K(z,w)c,d\rangle_{\mathcal{V}^*,\mathcal{V}}}$$
$$= \langle \tau(d), K(z,w)c\rangle_{\mathcal{V}^{**},\mathcal{V}^*}$$

(and, by definition of the adjoint $K(z,w)^* \in \mathbf{C}(\mathcal{V}^{**},\mathcal{V}^*)$; see (7.4.4))

$$= \langle K(z,w)^*(\tau(d)),c\rangle_{\mathcal{V}^*,\mathcal{V}},$$

and hence we have:

$$K(z,w)^*\tau = K(w,z), \quad z,w \in \Omega.$$ □

Solution of Exercise 7.4.2: We take $\Omega = \mathbb{C}^{n\times n}$ with $n > 1$ and

$$K(A,B) = A^*B.$$

Then for $N \in \mathbb{N}$ and $A_1,\ldots,A_N \in \mathbb{C}^{n\times n}$ and $c_1,\ldots,c_N \in \mathbb{C}^n$ we have

$$\sum_{j,k=1}^{N} \langle K(A_k,A_j)c_j,c_k\rangle_{\mathbb{C}^n} = \sum_{j,k=1}^{N} \langle A_k^*A_jc_j,c_k\rangle_{\mathbb{C}^n}$$
$$= \sum_{j,k=1}^{N} \langle A_jc_j,A_kc_k\rangle_{\mathbb{C}^n}$$
$$= \left\langle \sum_{j=1}^{N} A_jc_j,\sum_{k=1}^{N} A_kc_k\right\rangle_{\mathbb{C}^n} \geq 0,$$

7.8. Solutions of the exercises

so the function $K(A, B)$ is positive definite. On the other hand, let

$$L(A, B) = K(A, B)^* = B^* A.$$

Let $N = 2$ and A_1, A_2, c_1, c_2 be such that

$$A_1 c_1 = A_2 c_2 = 0, \quad A_1 c_2 \neq 0, \quad A_2 c_1 \neq 0,$$

and

$$\operatorname{Re} \langle A_1 c_2, A_2 c_1 \rangle_{\mathbb{C}^n} < 0.$$

For instance, $n = 2$ and

$$A_1 = \begin{pmatrix} 1 & 0 \\ 0 & 0 \end{pmatrix}, \quad A_2 = \begin{pmatrix} 0 & 1 \\ 0 & 0 \end{pmatrix}, \quad c_1 = \begin{pmatrix} 0 \\ 1 \end{pmatrix}, \quad c_2 = -\begin{pmatrix} 1 \\ 0 \end{pmatrix}.$$

Then

$$\operatorname{Re} \langle A_1 c_2, A_2 c_1 \rangle_{\mathbb{C}^2} = -2.$$

Such a choice will show that $L(A, B)$ is not positive definite. Indeed,

$$
\begin{aligned}
\sum_{j,k=1}^{2} \langle L(A_k, A_j) c_j, c_k \rangle_{\mathbb{C}^2} &= \langle L(A_1, A_1) c_1, c_1 \rangle_{\mathbb{C}^2} + \langle L(A_2, A_2) c_2, c_2 \rangle_{\mathbb{C}^2} \\
&\quad + \langle L(A_2, A_1) c_1, c_2 \rangle_{\mathbb{C}^2} + \langle L(A_1, A_2) c_2, c_1 \rangle_{\mathbb{C}^2} \\
&= c_1^* A_1^* A_1 c_1 + c_2^* A_2^* A_2 c_2 + \\
&\quad + c_2^* A_1^* A_2 c_1 + c_1^* A_2^* A_1 c_2 \\
&= c_2^* A_1^* A_2 c_1 + c_1^* A_2^* A_1 c_2 \\
&< 0.
\end{aligned}
$$

$\square$

Solution of Exercise 7.4.3: Recall first that the tensor product of $A \in \mathbb{C}^{p \times p}$ and $B \in \mathbb{C}^{q \times q}$ is the matrix $A \otimes B \in \mathbb{C}^{pq \times pq}$ defined by

$$
A \otimes B = \begin{pmatrix}
a_{11} B & a_{12} B & \cdots & a_{1p} B \\
a_{21} B & a_{22} B & \cdots & a_{2p} B \\
\vdots & & & \vdots \\
a_{p1} B & a_{p2} B & \cdots & a_{pp} B
\end{pmatrix}.
$$

The tensor product of two positive matrices is still positive (here too we refer to [CABP, Exercise 14.3.3, p. 485]), and the claim follows using

$$A = (K_1(w_i, w_j))_{i,j=1,\ldots,N} \quad \text{and} \quad B = (K_2(w_i, w_j))_{i,j=1,\ldots,N},$$

where $N \in \mathbb{N}$ and $w_1, \ldots, w_N \in \Omega$.

$\square$

Solution of Exercise 7.5.6: Let $a \in [0,1]$. Taylor's formula with remainder (see for instance [48, p. 279] if need be) gives, for $f \in \mathcal{H}_{(m)}$,

$$f(a) = f(0) + af'(0) + a^2 \frac{f^{(2)}(0)}{2!} + a^{m-1} \frac{f^{(m-1)}(0)}{(m-1)!} + \cdots$$

$$\cdots + \int_0^a \frac{(a-t)^{m-1}}{(m-1)!} f^{(m)}(t)dt$$

$$= \int_0^1 G(t,a) f^{(m)}(t)dt,$$

where

$$G(t,a) = \begin{cases} \frac{(a-t)^{m-1}}{(m-1)!}, & t \leq a, \\ 0, & \text{else.} \end{cases}$$

Thus, for $f \in \mathcal{H}_{(m)}$ it stands that:

$$f(a) = \int_0^1 G(t,a) f^{(m)}(t)dt$$

from which the claim follows, since the function

$$K(x,y) = \int_0^1 G(t,x)G(t,y)dt = \int_0^{x \wedge y} \frac{(x-t)^{m-1}(y-t)^{m-1}}{(m-1)!(m-1)!}dt$$

belongs to $\mathcal{H}_{(m)}$. To verify this last claim note that, for $x < y$,

$$\frac{\partial K(x,y)}{\partial x} = \int_0^x \frac{(x-t)^{m-2}(y-t)^{m-1}}{(m-2)!(m-1)!}dt$$

$$\frac{\partial^2 K(x,y)}{\partial x^2} = \int_0^x \frac{(x-t)^{m-3}(y-t)^{m-1}}{(m-3)!(m-1)!}dt$$

$$\vdots$$

$$\frac{\partial^{m-1} K(x,y)}{\partial x^{m-1}} = \int_0^x \frac{(y-t)^{m-1}}{(m-1)!}dt$$

$$\frac{\partial^m K(x,y)}{\partial x^m} = \frac{(y-x)^{m-1}}{(m-1)!}.$$

$\square$

Solution of Exercise 7.5.7: Consider a series of the form

$$\sum_{n=0}^{\infty} a_n \eta_n \tag{7.8.9}$$

with $(a_n)_{n \in \mathbb{N}_0}$ satisfying

$$\sum_{n=0}^{\infty} (n+1)^{2p} |a_n|^2 < \infty,$$

where $p \geq 1$. Then the bound (7.5.3) and the Cauchy–Schwarz inequality imply that

$$\left| \sum_{n=0}^{\infty} a_n \eta_n \right| \leq A \left(\sum_{n=0}^{\infty} (|a_n|(n+1)^p) \frac{1}{(n+1)^p} \right)$$

$$\leq \left(\sum_{n=0}^{\infty} (n+1)^{2p} |a_n|^2 \right)^{\frac{1}{2}} \left(\sum_{n=0}^{\infty} \frac{1}{(n+1)^{2p}} \right)^{\frac{1}{2}} < \infty,$$

and so (7.8.9) converges uniformly and absolutely on $\mathbb{R}$. The space $\mathcal{K}_p$ is isomorphic to a weighted ℓ_2 space and hence is a Hilbert space. Its reproducing kernel is equal to the sum

$$\sum_{n=0}^{\infty} \frac{\eta_n(t)\eta_n(s)}{(n+1)^{2p}}. \tag{7.8.10}$$

When $p = 0$ the sum (7.8.10) converges in the sense of distributions to the kernel $\delta(t-s)$:

$$\sum_{n=0}^{\infty} \eta_n(t)\eta_n(s) = \delta(t-s). \tag{7.8.11}$$

See the remark below about the term *kernel*. $\qquad\square$

Remarks 7.8.2.

(1) In (7.8.11) we use the term *kernel* in the sense of Schwartz' kernel theorem that is, a distribution of two variables (or of two sets of variables in the case of several variables). See for instance [148, 324] for more discussions on the kernel theorem. See also the discussion at the beginning of Section 7.6.

(2) There seems to be no closed formula for the sum (7.8.10). If in the definition of $\mathcal{K}_p$ one replaces the weights $(n^2 + 1)$ by 2^n, $n = 0, 1, \ldots$, one can then use Mehler's formula to characterize the corresponding space. See [41].

Solution of Exercise 7.5.8: Let, with $N \in \mathbb{N}$, $w_1, \ldots, w_N \in \Omega$ and $c_1, \ldots, c_N \in \mathbb{C}$,

$$f(z) = \sum_{j=1}^{N} tK(z, w_j)c_j = \sum_{j=1}^{N} K(z, w_j)c_j t \in \mathcal{H}(tK).$$

Then,

$$\|f\|_{\mathcal{H}(tK)}^2 = \sum_{i,j=1}^{N} \overline{c_i} t K(z_i, z_j) c_j = \frac{1}{t} \sum_{i,j=1}^{N} \overline{c_i} t K(z_i, z_j) c_j t = \frac{1}{t} \|f\|_{\mathcal{H}(K)}^2.$$

This concludes the proof since such functions f are dense in both spaces. $\qquad\square$

Remark 7.8.3. Exercise 7.5.8 and its proof are given for scalar functions, but the proof is readily adapted to the vector case.

Solution of Exercise 7.5.10: We look for a planar measure of the form $d\mu(z) = r a(r) dr d\theta$ such that

$$\iint_{\mathbb{C}} z^n \bar{z}^m d\mu(z) = \int_0^{2\pi} e^{i(n-m)\theta} d\theta \int_0^\infty z^n \bar{z}^m a(r) r dr = \delta_{n,m} (n!)^2,$$

that is

$$2\pi \int_0^\infty r^{2n+1} a(r) dr = (n!)^2, \quad n \in \mathbb{N}_0.$$

The change of variable $\rho = r^2$ gives

$$\pi \int_0^\infty \rho^n a(\sqrt{\rho}) d\rho = (\Gamma(n+1))^2, \quad n \in \mathbb{N}_0.$$

In view of (7.5.4) with $s = n+1$, $n \in \mathbb{N}_0$, we can take

$$a(\sqrt{\rho}) = \frac{2}{\pi} K_0(2\sqrt{\rho}),$$

and hence the result.

$\square$

Remark 7.8.4. In the preceding argument we did not address the question of uniqueness of the underlying measure.

Solution of Exercise 7.5.11:

(2) Let $f(z) = \sum_{n \in \Lambda} z^n$. Since

$$4\Lambda \cap \{1 + 4\Lambda\} = \emptyset,$$

we have

$$f(z) = f(z^4) + z f(z^4),$$

so that

$$f(z) = \prod_{p=0}^\infty (1 + z^{4^p}).$$

$\square$

Solution of Exercise 7.5.13: The formula

$$\min(t, s) = \langle 1_{[0,t]}, 1_{[0,s]} \rangle_{\mathbf{L}_2(\mathbb{R}_+, dx)}$$

expresses the given function as an inner product, and hence as a positive definite function by item (1) of Exercise 7.1.7. One can also remark that $\min(t, s)$ is the restriction to $[0, \infty)$ of the function (6.1.28) to express it as an inner product. It follows from item (3) of Exercise 7.1.7 and from Exercise 6.3.3 (or, more precisely, from Remark 6.3.4) that

$$\mathcal{H}(K) = \left\{ F(t) = \int_0^t f(u) du \, ; \, f \in \mathbf{L}_2(\mathbb{R}_+, dx) \right\},$$

7.8. *Solutions of the exercises*

with the norm
$$\|F\| = \|f\|_2.$$

Such functions are by definition restrictions to $[0, \infty)$ of functions in the Sobolev space of functions in $\mathbf{L}_2(\mathbb{R}, \mathcal{B}, dx)$ with distributional derivative of the Fourier transform also in $\mathbf{L}_2(\mathbb{R}, \mathcal{B}, dx)$. □

Solution of Exercise 7.5.14:

(1) Take the z_n to be the points in $\mathbb{D}$ with rational coordinates and
$$w_{n,m} = t_m z_n,$$

where $(t_m)_{m \in \mathbb{N}}$ is a sequence of pointwise different irrational numbers in $\mathbb{D} \setminus \{0\}$ and tending to 1 as m goes to ∞. for instance
$$t_m = \frac{m + 1 + \pi}{1 + (m+1)\pi}, \quad m = 1, 2, \ldots.$$

(2) The functions $k_{w_{n,m}}$ are linearly independent and are included, by the axiom of choice, in a Hamel basis.

(3) Assume that the point evaluation at z_n is bounded. By Riesz' theorem there exists $g_n \in \mathbf{H}_2(\mathbb{D})$ such that, for all $f \in \mathbf{H}_2(\mathbb{D})$:
$$(Tf)(z_n) = \langle Tf, Tg_n \rangle_{\mathcal{H}_T}$$
$$= \langle f, g_n \rangle_{\mathbf{H}_2(\mathbb{D})}.$$

The choice $f = k_{w_{n,m}}$ leads to
$$\frac{m}{1 - z_n \overline{w_{n,m}}} = \langle k_{w_{n,m}}, g_n \rangle_{\mathbf{H}_2(\mathbb{D})},$$

and so
$$g_n(w_{n,m}) = \frac{m}{1 - w_{n,m}\overline{z_n}}.$$

It follows that g_n is not continuous at the point z_n, and this contradicts the fact that $g_n \in \mathbf{H}_2(\mathbb{D})$. □

Solution of Exercise 7.5.17: The function $K(\cdot, w)$ belongs to the range of Γ and hence to the range of $\sqrt{\Gamma}$. Furthermore, by definition of the inner product we have for $f \in \mathcal{H}$:
$$\langle \sqrt{\Gamma} f, K_w \rangle_{\operatorname{ran} \sqrt{\Gamma}} = \langle f, (I - \pi)\sqrt{\Gamma} k_w \rangle_{\mathcal{H}}$$
$$= \langle \sqrt{\Gamma} f, k_w \rangle_{\mathcal{H}}$$
$$= (\sqrt{\Gamma} f)(w).$$
□

Solution of Exercise 7.5.19:

(1) This is a direct consequence of Exercise 7.1.7.

(2) It follows from the first item that $\mathcal{H}(K)$ consists of functions of the form

$$F(t) = \int_{\mathbb{R}} \frac{e^{itu} f(u)}{1 + u^2} du,$$

where $f \in \mathbf{L}_2(\mathbb{R}, \mathcal{B}, \frac{du}{u^2+1})$. We check directly that f uniquely determines F. Indeed, $\mathbf{L}_2(\mathbb{R}, \mathcal{B}, \frac{du}{u^2+1}) \subset \mathbf{L}_1(\mathbb{R}, \mathcal{B}, \frac{du}{u^2+1})$ and we have

$$\int_{\mathbb{R}} \frac{|f(u)|}{1 + u^2} du < \infty.$$

It follows that F is continuous (see for instance [Exercise 15.6.3, p. 503, CABP]). Since the function g:

$$g: \quad u \mapsto \frac{u f(u)}{1 + u^2} \in \mathbf{L}_2(\mathbb{R}, \mathcal{B}, du)$$

it follows from item (2) of Exercise 6.1.25 that F has a distributional derivative, which is $\widehat{g}$. Since

$$\|F\|^2_{\mathbf{L}_2} + \|\widehat{g}\|^2_{\mathbf{L}_2} = \int_{\mathbb{R}} \frac{|f(u)|^2 + u^2|f(u)|^2}{(1 + u^2)^2} du = \int_{\mathbb{R}} \frac{|f(u)|^2}{1 + u^2} du = \|F\|^2_{\mathcal{H}(K)}$$

we see that $\mathcal{H}(K)$ is equal to the Sobolev space of functions $F \in \mathbf{L}_2(\mathbb{R}, \mathcal{B}, dx)$ with distributional derivative also in $\mathbf{L}_2(\mathbb{R}, \mathcal{B}, dx)$.   □

Solution of Exercise 7.5.23: Let $\mathcal{H}(K_i)$ be the reproducing kernel Hilbert spaces associated with K_i, $i = 1, 2$. Let $(f_t)_{t \in T}$ and $(g_s)_{s \in S}$ be orthonormal basis of $\mathcal{H}(K_1)$ and $\mathcal{H}(K_2)$ respectively. Then the formula

$$K(z_1, z_2, w_1, w_2) = \sum_{(t,s) \in T \times S} f_t(z_1) g_s(z_2) \overline{f_t(w_1) g_s(w_2)}$$

exhibits the function K as a sum of positive definite functions, and one obtains from this representation the characterization of elements in $\mathcal{H}(K)$ as follows: The function $f(z_1, z_2)$ belongs to the space $\mathcal{H}(K)$ if and only if it can be written in one of the three equivalent ways:

(1)
$$f(z_1, z_2) = \sum_{(t,s) \in T \times S} c_{t,s} f_t(z_1) g_s(z_2),$$

where $c_{st} \in \mathbb{C}$ are such that

$$\sum_{(t,s) \in T \times S} |c_{st}|^2 < \infty. \qquad (7.8.12)$$

7.8. *Solutions of the exercises*

(2)
$$f(z_1, z_2) = \sum_{t \in T} f_t(z_1) c_t(z_2),$$

where the functions $c_t \in \mathcal{H}(K_2)$ and are such that

$$\sum_{t \in T} \|c_t\|^2_{\mathcal{H}(K_2)} < \infty. \tag{7.8.13}$$

(3)
$$f(z_1, z_2) = \sum_{s \in S} g_s(z_1) d_s(z_2),$$

where the functions $d_s \in \mathcal{H}(K_1)$ and are such that

$$\sum_{s \in S} \|d_s\|^2_{\mathcal{H}(K_1)} < \infty. \tag{7.8.14}$$

Then the norm of f is given by either of the equal quantities (7.8.12)–(7.8.14). □

Solution of Exercise 7.5.26:

(1) We assume that $\sup_{n \in \mathbb{N}} K(r_n e^{i\theta}, r_n e^{i\theta}) < \infty$. Thus the family of functions $K(\cdot, r_n e^{i\theta})$ is uniformly bounded in norm in $\mathcal{H}(K)$. Since closed balls in $\mathcal{H}(K)$ are weakly compact and metrizable, there is a subsequence $(n_k)_{k \in \mathbb{N}}$ such that $K(\cdot, r_n e^{i\theta})$ tends weakly to some element, say g, in $\mathcal{H}(K)$. Thus for every $f \in \mathcal{H}(K)$

$$\lim_{k \to \infty} \langle f, K(\cdot, r_{n_k} e^{i\theta}) \rangle = \langle f, g \rangle$$

that is

$$\lim_{k \to \infty} f(r_{n_k} e^{i\theta}) = \langle f, g \rangle,$$

and in particular (7.5.13) defines a bounded operator.

(2) Suppose that (7.5.14) exists for all $f \in \mathcal{H}(K)$. We define an operator T_n by

$$T_n(f) = f(r_n e^{i\theta}) = \langle f, K(\cdot, r_n e^{i\theta}) \rangle.$$

By hypothesis, for every $f \in \mathcal{H}(K)$, we have $\sup_{n \in \mathbb{N}} |T_n(f)| < \infty$. The uniform boundedness principle implies that $\sup_{n \in \mathbb{N}} \|T_n\| < \infty$, that is (7.5.12) holds (see Theorem 4.1.19 for the norm of the map T_n if need be). □

Solution of Exercise 7.6.6:

(1) We first assume that the operator is bounded. Then there is a $c > 0$ such that

$$c^2 I_{\mathcal{H}(K_2)} - T_{m,\varphi} T^*_{m,\varphi} \geq 0, \tag{7.8.15}$$

and the smallest such c is equal to $\|T_{m,\varphi}\|$ (see Exercise 4.2.20). By continuity of the operator, (7.8.15) is equivalent to

$$\langle f,\, (c^2 I_{\mathcal{H}(K_2)} - T_{m,\varphi}T^*_{m,\varphi})f\rangle_{\mathcal{H}(K_2)} \geq 0$$

for every function f of the form

$$f(z) = \sum_{u=1}^{U} K_2(z, w_u)\xi_u, \tag{7.8.16}$$

where $U \in \mathbb{N}$ and $w_1,\dots,w_U \in \Omega$ and $\xi_1,\dots,\xi_U \in \mathbb{C}^N$. To conclude we compute $T^*_{m,\varphi}$ on the kernels. For $w, z \in \Omega$ and $\xi \in \mathbb{C}^N$, $\eta \in \mathbb{C}^M$ we have:

$$
\begin{aligned}
\langle T^*_{m,\varphi}(K_2(\cdot, w)\xi), K_1(\cdot, z)\eta\rangle_{\mathcal{H}(K_1)} &= \langle K_2(\cdot, w)\xi,\, T_{m,\varphi}(K_1(\cdot, z)\eta)\rangle_{\mathcal{H}(K_2)} \\
&= \langle K_2(\cdot, w)\xi,\, m(\cdot)(K_1(\varphi(\cdot), z)\eta)\rangle_{\mathcal{H}(K_2)} \\
&= \overline{\langle m(\cdot)(K_1(\varphi(\cdot), z)\eta),\, K_2(\cdot, w)\xi\rangle_{\mathcal{H}(K_2)}} \\
&= \overline{\xi^* m(w) K_1(\varphi(w), z)\eta} \\
&= \eta^*\left(K_1(z, \varphi(w))m(w)^*\xi\right),
\end{aligned}
$$

so that

$$T^*_{m,\varphi}(K_2(\cdot, w)\xi) = K_1(\cdot, \varphi(w))m(w)^*\xi. \tag{7.8.17}$$

Thus, with f of the form (7.8.16),

$$
\begin{aligned}
&\langle f,\, (c^2 I_{\mathcal{H}(K_2)} - T_{m,\varphi}T^*_{m,\varphi})f\rangle_{\mathcal{H}(K_2)} \\
&= c^2\left(\sum_{u,v=1}^{U} \xi_u^* K_2(w_u, w_v)\xi_v\right) - \sum_{u,v=1}^{U} \xi_u^* m(w_u)K_1(\varphi(w_u), \varphi(w_v))m(w_v)^*\xi_v \\
&= \sum_{u,v=1}^{U} \xi_u^* K(w_u, w_v)\xi_v,
\end{aligned}
$$

with

$$K(z, w) = K_2(z, w) - m(z)K_1(\varphi(z), \varphi(w))m(w)^*,$$

and hence the kernel (7.6.8) is positive definite in Ω.

Conversely, assume that K is positive definite. The linear relation $R \subset \mathcal{H}(K_2) \times \mathcal{H}(K_1)$ spanned by the pairs of the form

$$(K_2(\cdot, w)\xi, K_1(\cdot, \varphi(w))m(w)^*\xi), \quad w \in \Omega,\ \xi \in \mathbb{C}^N,$$

is densely defined (see Definition 4.2.49). It is contractive since the kernel K is positive definite. Indeed, let $(f, g) \in R$ with f as in (7.8.16) and

$$g(\cdot) = \sum_{u=1}^{U} K_1(\cdot, \varphi(w_u))m(w_u)^*\xi_u.$$

We have

$$\langle f, f \rangle_{\mathcal{H}(K_2)} - \langle g, g \rangle_{\mathcal{H}(K_1)} = \sum_{u,v=1}^{U} \xi_u^* K(w_u, w_v) \xi_v \geq 0.$$

Hence, R extends to the graph of an everywhere defined contraction X (see Exercise 4.2.50). For $f \in \mathcal{H}(K_1)$ and $w \in \Omega$ and $\xi \in \mathbb{C}^N$ we have

$$\begin{aligned}
\xi^*(X^*f)(w) &= \langle X^*f, K_2(\cdot, w)\xi \rangle_{\mathcal{H}(K_2)} \\
&= \langle f, X(K_2(\cdot, w)\xi) \rangle_{\mathcal{H}(K_1)} \\
&= \langle f, K_1(\cdot, \varphi(w))m(w)^*\xi \rangle_{\mathcal{H}(K_1)} \\
&= \xi^* m(w) f(\varphi(w)),
\end{aligned}$$

and so $X^* = T_{m,\varphi}$.

(2) The map $T_{m,\varphi}$ is then unitary from $\mathcal{H}(K_1)$ onto $\mathcal{H}(K_2)$.

(3) As in (7.6.2) we consider the scalar case, $M = N = 1$. From

$$m(\nu)f(\varphi(\nu)) = \langle f(\cdot), K_1(\cdot, \varphi(\nu))\overline{m(\nu)} \rangle_{\mathcal{H}(K_1)}, \quad \nu \in \Omega,$$

we see that $A(\cdot, \nu) = K_1(\cdot, \varphi(\nu))\overline{m(\nu)}$. $\qquad\square$

Remark 7.8.5. With K_1 and Ω as above, positive definite kernels of the form

$$K(z, w) = K_1(z, w) - \sum_{u=1}^{U} m_u(z) k_2(\varphi(z), \varphi(w)) m_u(w)^*, \quad z, w \in \Omega,$$

where the functions $m_1, \ldots, m_N$ are $\mathbb{C}^{M \times N}$-valued and $k_2(z, w)$ is positive definite in Ω and $\mathbb{C}^{N \times N}$-valued, are of course a special case of (7.6.8) with

$$m(z) = \begin{pmatrix} m_1(z) & \cdots & m_U(z) \end{pmatrix} \quad \text{and} \quad K_2(z, w) = \mathrm{diag}\,(k_2(z, w), \ldots, k_2(z, w)).$$

This case is of special importance when $K(z, w) \equiv 0$. Then a decomposition of the form

$$\mathcal{H}(K_1) = \sum_{u=1}^{U} m_u \mathcal{H}(k_{2,\varphi}), \quad \text{where} \quad k_{2,\varphi}(z, w) = k_2(\varphi(z), \varphi(w)),$$

holds, and every element $f \in \mathcal{H}(K_1)$ can be written (in a possibly non unique way) as

$$f(z) = \sum_{u=1}^{U} m_u(z) g_u(\varphi(z)),$$

where $g_1, \ldots, g_U \in \mathcal{H}(k_2)$. Such decompositions are used to solve *linear combination interpolation problems* of the type given in Problem 8.5.3. See [35, 36] for more on these problems.

Solution of Exercise 7.6.12: By Exercise 7.6.6, the function f belongs to $\mathcal{H}(K)$ if and only if the operator M_f is bounded from $\mathbb{C}$ into $\mathcal{H}(K)$, that is if and only the kernel (7.6.10) is positive definite for some $M > 0$. That the smallest such M is the norm of f follows also from Exercise 7.6.6. $\square$

Solution of Exercise 7.6.13:

(1) It suffices to apply the preceding exercise with $f(z) = K(z, w_0)$. Then, $\|f\|^2_{\mathcal{H}(K)} = K(w_0, w_0)$.

(2) The kernel $e^{-|t-s|}$ is positive in $\mathbb{R}$ and so also on $[0, \infty)$. Applying item (1) to $w_0 = t = 0$ we obtain the required result. $\square$

Solution of Exercise 7.6.14: By Exercise 7.6.6 there exists $M > 0$ such that the kernel

$$K_1(z, w) - MK_2(\varphi(z), \varphi(w)) \tag{7.8.18}$$

is positive definite in Ω. For $n \in \mathbb{N}$ the function

$$\sum_{k=0}^{n-1} M^{n-1-k}(K_1(z, w))^k (K_2(\varphi(z), \varphi(w)))^{n-1-k}$$

is positive definite in Ω since it is a sum of products of positive definite functions. Thus the function

$$(K_1(z, w))^n - (MK_2(\varphi(z), \varphi(w)))^n = (K_1(z, w) - MK_2(\varphi(z), \varphi(w)))$$
$$\times \left(\sum_{k=0}^{n-1} M^{n-1-k}(K_1(z, w))^k (K_2(\varphi(z), \varphi(w)))^{n-1-k}\right)$$

is positive definite in Ω. Hence,

$$f(K_1(z, w) - f(MK_2(\varphi(z), \varphi(w)))$$
$$= \sum_{n=1}^{\infty} a_n \left((K_1(z, w))^n - (MK_2(\varphi(z), \varphi(w)))^n\right)$$

is positive definite, and the claim follows from Exercise 7.6.6. $\square$

Solution of Exercise 7.6.19: By Exercise 4.2.26 there is a subsequence of the family of operators $(M_{s_n})_{n \in \mathbb{N}}$ converging weakly to some contraction operator, say T. Thus, for every $f \in \mathcal{H}(K)$ and $w \in \Omega$, and denoting the subsequence by $(n_k)_{k \in \mathbb{N}}$,

$$\lim_{k \to \infty} \langle s_{n_k}(\cdot)f(\cdot), K(\cdot, w)\rangle = (Tf)(w).$$

Thus, for every $w \in \Omega$,

$$\lim_{k \to \infty} s_{n_k}(w)f(w) = (Tf)(w).$$

By hypothesis, for every $w \in \mathcal{H}(K)$ there exists $f_w \in \mathcal{H}(K)$ such that $f_w(w) \neq 0$.
So

$$\lim_{k \to \infty} s_{n_k}(w) = \frac{(T f_w)(w)}{f_w(w)}, \quad \forall w \in \Omega.$$

Denoting by $s(w)$ the above limit we have $T = M_s$.

When one can choose $f(z) \equiv 1$ we have $s(w) = (T1)(w)$. □

Solution of Exercise 7.7.1: The function $z \mapsto K(z,w)c$ belongs to the linear span of the functions f_j and therefore belongs to $\mathcal{H}$. Furthermore, let $f(z) = F(z)d$ be a function in $\mathcal{H}$. Since

$$K(z,w)c = F(z)P^{-1}F(w)^*c$$

we have

$$[f(\cdot), K(\cdot,w)c]_{\mathcal{H}} = [F(\cdot)d, F(\cdot)P^{-1}F(w)^*c]_{\mathcal{H}}$$
$$= (P^{-1}F(w)^*c)^*Pd = c^*F(w)d = c^*f(w). \quad \square$$

The proof still works, *mutatis mutandis*, when the matrix P is Hermitian and invertible, but not necessarily positive. The function $K(z,w)$ is not positive anymore, but has a finite number of positive squares, i.e., there is a fixed number κ (the number of negative eigenvalues of P) such that all the Hermitian matrices of the form $(K(w_i, w_j))_{i,j=1,\dots,r}$ have at most κ strictly negative eigenvalues, and exactly κ such eigenvalues for some choice of $r, w_1, \dots, w_r$. See Definition 7.1.3. This notion is a natural extension of the notion of positive definite function and was introduced by M.G. Kreĭn. The metric is nonpositive, and the space is a reproducing kernel Pontryagin space; see [55], [293], for more on these notions.

In the above result, Ω is an arbitrary set. An important case is the situation where $\Omega \subset \mathbb{C}$ and where the functions are rational. It is then of interest to determine the form of the reproducing kernel $K(z,w)$ under some supplementary conditions. This has been extensively studied by H. Dym to clarify the distinction between general reproducing kernel Hilbert and Kreĭn spaces and the special class of spaces considered by L. de Branges; see, e.g., [123], [124], and the references therein.

Finally, we mention that the Gram matrix also makes sense in the setting of spaces in duality; then it need not be Hermitian anymore. See [89, pp. 17–20] for a discussion.

Solution of Exercise 7.7.3:

(1) For z in a neighborhood of the origin we have

$$C(I_m - zA)^{-1}f = \sum_{k=0}^{\infty} CA^k f.$$

Thus

$$C(I_m - zA)^{-1}f \equiv 0 \quad \Longleftrightarrow \quad f \in \cap_{k=0}^{\infty} \ker CA^k,$$

which ends the proof.

(2) Let z, w be either equal to 0 or such that their inverse belongs to the resolvent set of A (that is, such that the matrices $(I_m - zA)$ and $(I_m - wA)$ are invertible). Then,

$$
\begin{aligned}
J &- \Theta(z)J\Theta(w)^* = J \\
&- (I_n + izC(I_m - zA)^{-1}P^{-1}C^*J)J(I_n + iwC(I_m - wA)^{-1}P^{-1}C^*J)^* \\
&= C(I_m - zA)^{-1}P^{-1}\{-iz(I_m - \overline{w}A^*)P + iP\overline{w}(I_m - zA) \\
&\quad + (iz)(i\overline{w})C^*JC\}\, P^{-1}(I_m - wA)^{-*}C \\
&= C(I_m - zA)^{-1}P^{-1}\{-i(z - \overline{w})P \\
&\quad + iz\overline{w}(A^*P - PA + iC^*JC)\}\, P^{-1}(I_m - wA)^{-*}C,
\end{aligned}
$$

which can be rewritten as

$$
\begin{aligned}
\frac{J - \Theta(z)J\Theta(w)^*}{-i(z - \overline{w})} &= C(I_m - zA)^{-1}P^{-1}(I_m - wA)^{-*}C \\
&+ \frac{iz\overline{w}}{-i(z - \overline{w})}C(I_m - zA)^{-1}P^{-1}(A^*P - PA + iC^*JC)P^{-1}(I_m - wA)^{-*}C.
\end{aligned}
$$

$$(7.8.19)$$

Thus equation (7.7.12) implies (7.7.11), even without the observability condition (7.7.8). Suppose now that (7.7.11) holds. Then we have from (7.8.19)

$$\frac{iz\overline{w}}{-i(z - \overline{w})}C(I_m - zA)^{-1}P^{-1}(A^*P - PA + iC^*JC)P^{-1}(I_m - wA)^{-*}C \equiv 0,$$

i.e.,

$$C(I_m - zA)^{-1}P^{-1}(A^*P - PA + iC^*JC)P^{-1}(I_m - wA)^{-*}C \equiv 0,$$

where z, w are such that the corresponding inverses exist. Taking into account (7.7.9) twice, first for the variable z and then for the variable w, we see first that

$$(A^*P - PA + iC^*JC)P^{-1}(I_m - wA)^{-*}C \equiv 0,$$

and then, $A^*P - PA + iC^*JC = 0$, which is (7.7.12).

(3) It suffices to replace z and w by $1/z$ and $1/w$ respectively and J by $-J$ in (7.7.10) and (7.7.11).

$\square$

Solution of Exercise 7.7.11:

(1) With A as in the hint, we have

$$A^*P - PA = iC^*C, \quad \text{with} \quad C = \begin{pmatrix} 1 & 1 & \cdots & 1 \end{pmatrix}.$$

7.8. *Solutions of the exercises*

Thus (7.7.14) holds with this choice of A and C and $J = 1$. Furthermore

$$C(zI_m - A)^{-1} = (-i) \times \left(\frac{1}{-i(z-\overline{w_1})} \quad \frac{1}{-i(z-\overline{w_2})} \quad \cdots \quad \frac{1}{-i(z-\overline{w_m})} \right).$$

Still with A and C as above, let

$$B(z) = 1 + iC(zI_m - A)^{-1}P^{-1}C^*.$$

The function B is rational, analytic at infinity, with value at infinity $B(\infty) = 1$, and with poles at the points $\overline{w_1}, \ldots, \overline{w_m}$. Furthermore, (7.7.13) leads to

$$C(zI_m - A)^{-1}P^{-1}(wI_m - A)^{-*}C^* = \frac{1 - B(z)\overline{B(w)}}{-i(z - \overline{w})},$$

and so B has modulus one on the real line. It follows that B is equal to B_0 given by (7.7.5).

(2) We rewrite (7.7.7) as

$$\sum_{j,k=1}^{m} \frac{(P^{-1})_{jk}}{(z - \overline{w_j})(\overline{w} - w_k)} = \frac{1 - B(z)\overline{B(w)}}{-i(z - \overline{w})}.$$

Hence

$$(P^{-1})_{jk} = \lim_{\substack{z \to \overline{w_j} \\ w \to \overline{w_k}}} (z - \overline{w_j})(\overline{w} - w_k)\frac{1 - B(z)\overline{B(w)}}{-i(z - \overline{w})}, \quad j, k = 1, \ldots, m.$$

The result follows since

$$\lim_{z \to \overline{w_j}} (z - \overline{w_j})B(z) = (\overline{w_j} - w_j) \prod_{t \neq j} \frac{\overline{w_j} - w_t}{\overline{w_j} - \overline{w_t}}. \qquad \square$$

Remark 7.8.6. Let $H_j = (\overline{w_j} - w_j)\prod_{t \neq j} \frac{\overline{w_j} - w_t}{\overline{w_j} - \overline{w_t}}$. Formula (7.7.22) becomes

$$(P^{-1})_{jk} = \frac{H_j \overline{H_k}}{-i(w_k - \overline{w_j})}.$$

To prove directly that these formulas indeed define the inverse we need to show that

$$\sum_{v=1}^{m} \frac{H_v \overline{H_k}}{(w_j - \overline{w_v})(\overline{w_v} - w_k)} = \delta_{j,k}, \quad j, k = 1, \ldots, m. \qquad (7.8.20)$$

We now prove this formula for $j \neq k$. Then (7.8.20) reads

$$\left[\left(\sum_{v=1}^{m} \frac{H_v}{w_j - \overline{w_v}} \right) + \left(\sum_{v=1}^{m} \frac{H_k}{\overline{w_v} - w_k} \right) \right] \frac{\overline{H_t}}{w_j - \overline{w_t}} = 0.$$

But the exactity relation (see [CAPB, Exercise 7.3.6, p. 326]) applied to the functions $\frac{B(z)}{z-w_t}$ (with $t = 1,\ldots,m$) gives

$$\sum_{v=1}^{m} \frac{H_t}{\overline{w}_v - w_t} = 0$$

and hence the required result for $j \neq k$.

Solution of Exercise 7.7.14:

(1) With P and A as in the exercise we have

$$PA = \begin{pmatrix} 0 & h_0 & h_1 & \cdots & h_{N-1} \\ 0 & h_1 & h_2 & \cdots & h_N \\ & & & & \\ & & & & \\ 0 & h_N & h_{N+1} & \cdots & h_{2N-1} \end{pmatrix}.$$

Thus

$$PA - A^*P = \begin{pmatrix} 0 & h_0 & h_1 & \cdots & h_{N-1} \\ -h_0 & 0 & 0 & \cdots & 0 \\ -h_1 & 0 & & & \\ & & & & \\ -h_{N-1} & 0 & 0 & \cdots & 0 \end{pmatrix} = iC^*JC$$

with

$$iJ = \begin{pmatrix} 0 & 1 \\ -1 & 0 \end{pmatrix} \quad \text{and} \quad C = \begin{pmatrix} 1 & 0 & 0 & \cdots & 0 \\ 0 & h_0 & h_1 & \cdots & h_{N-1} \end{pmatrix}.$$

(2) We denote by Θ_N the function given by (7.7.10); it is a polynomial since $A^{N+1} = 0_{(N+1)\times(N+1)}$. Furthermore

$$(I_{N+1} - zA)^{-1} = \begin{pmatrix} 1 & z & z^2 & \cdots & z^{N-1} & z^N \\ 0 & 1 & z & \cdots & z^{N-2} & z^{N-1} \\ 0 & 0 & 1 & & & \\ & & & \ddots & & \\ & & & & 1 & z \\ 0 & 0 & \cdots & & 0 & 1 \end{pmatrix},$$

and so

$$(1 \quad 0)\, C(I_{N+1} - zA)^{-1} = (1 \quad z \quad z^2 \quad \cdots \quad z^{N-1} \quad z^N).$$

Formula (7.7.11) leads to

$$(1 \quad z \quad z^2 \quad \cdots \quad z^{N-1} \quad z^N)\, H_N^{-1} (1 \quad w \quad w^2 \quad \cdots \quad w^{N-1} \quad w^N)^*$$

$$= \frac{-(1 \quad 0)\,\Theta_N(z) \begin{pmatrix} 0 & 1 \\ -1 & 0 \end{pmatrix} \Theta_N(w)^* \begin{pmatrix} 1 \\ 0 \end{pmatrix}}{z - \overline{w}}$$

$$= \frac{B_N(z)\overline{A_N(w)} - A_N(z)\overline{B_N(w)}}{z - \overline{w}},$$

$(7.8.21)$

with

$$\begin{pmatrix} A_N(z) & B_N(z) \end{pmatrix} = \begin{pmatrix} 1 & 0 \end{pmatrix} \Theta_N(z),$$

that is, A_N and B_N are given by formulas (7.7.25) and (7.7.26).

On the other hand the space $\mathscr{P}_N$ endowed with the inner product (7.7.23) is a Hilbert space since $H_N > 0$. It is a reproducing kernel Hilbert space since its elements are functions and it is finite dimensional. Its reproducing kernel is given by formula (7.7.1), that is, by the left-hand side of (7.8.21). $\square$

Remark 7.8.7. The right-hand side of (7.8.21) expresses the structured form of the kernel, coming from the fact that the Gram matrix is Hankel. This is one example of a recurring theme, which appears various times in this book. *The structure of the Gram matrix reflects on the structure of the reproducing kernel.* Note that the solution of the exercise uses Exercise 7.7.3, where one can find a general example of such a relation between the Gram matrix and the reproducing kernel.

Solution of Exercise 7.7.15: By definition of the inner product in $\mathscr{P}_{N+1}$ we have

$$\langle P_{N+1}, z^j \rangle_{\mathscr{P}_{N+1}} = \det \begin{pmatrix} h_j & h_{j+1} & \cdots & h_{j+N+1} \\ h_0 & h_1 & \cdots & h_{N+1} \\ & & & \\ h_j & h_{j+1} & \cdots & h_{j+N+1} \\ & & & \\ h_N & h_{N+1} & \cdots & h_{2N+1} \end{pmatrix} = 0, \quad j = 0, \ldots, N.$$

Thus P_{N+1} is orthogonal to $\mathscr{P}_N$. We now show that the same holds for $A_N(z)$. We have for $j \in \{0, \ldots, N\}$

$$\langle A_N, z^j \rangle_{\mathscr{P}_{N+1}} = h_j - \begin{pmatrix} h_{j+1} & h_{j+2} & \cdots & h_{j+N+1} \end{pmatrix} H_N^{-1} \begin{pmatrix} 0 \\ h_0 \\ \vdots \\ h_N \end{pmatrix}$$

$$= h_j - \underbrace{\begin{pmatrix} 0 & 0 & \cdots & 0 & 1 & 0 & \cdots & 0 \end{pmatrix}}_{\text{the 1 is at the } j+1\text{th entry}} \begin{pmatrix} 0 \\ h_0 \\ \vdots \\ h_N \end{pmatrix}$$

$$= 0.$$

But, up to a multiplicative constant, there is only one polynomial of degree $N+1$ orthogonal to $\mathscr{P}_N$, and this ends the proof. $\square$

Remark 7.8.8. Let

$$G_N(z) = \begin{pmatrix} 1 & z & \cdots & z^N \end{pmatrix} H_N^{-1} \begin{pmatrix} 1 \\ 0 \\ \vdots \\ 0 \end{pmatrix}.$$

We have for $j = 0, \ldots, N$:

$$
\langle G_N, z^j \rangle_{\mathcal{P}_N} = \begin{cases} 1, & \text{if } j = 0, \\ 0, & \text{otherwise.} \end{cases}
$$

Solution of Exercise 7.7.17: With $F(z) = C(I_m - zA)^{-1}$ we have

$$
\begin{aligned}
J - \Theta(z)J\Theta(w)^* &= (1 - z\overline{w_0})F(z)P^{-1}F(\omega_0)^* + (1 - \omega_0\overline{w})F(\omega_0)P^{-1}F(w)^* \\
&\quad - (1 - z\overline{w_0})(1 - \omega_0\overline{w})F(z)P^{-1}F(\omega_0)^* J F(\omega_0)P^{-1}F(w)^* \\
&= F(z)P^{-1}(I_m - \omega_0 A)^{-*} \\
&\quad \times \big\{ (1 - z\overline{w_0})(I_m - wA)^* P(I_m - \omega_0 A) \\
&\quad\quad + (1 - \omega_0\overline{w})(I_m - \omega_0 A)^* P(I_m - zA) \\
&\quad\quad - (1 - z\overline{w_0})(1 - \omega_0\overline{w})C^* J C \big\} \\
&\quad \times (I_m - \omega_0 A)^{-1}P^{-1}F(w)^*.
\end{aligned}
$$

In view of the observability of the pair (C, A), the reproducing kernel formula

$$
\frac{J - \Theta(z)J\Theta(w)^*}{(1 - z\overline{w})} = F(z)P^{-1}F(w)^*
$$

will hold if and only if we have

$$
\begin{aligned}
(1 - z\overline{w})(I_m - \omega_0 A)^* P(I_m - \omega_0 A) &= (1 - z\overline{w_0})(I_m - wA)^* P(I_m - \omega_0 A) \\
&\quad + (1 - \omega_0\overline{w})(I_m - \omega_0 A)^* P(I_m - zA) \\
&\quad - (1 - z\overline{w_0})(1 - \omega_0\overline{w})C^* J C,
\end{aligned}
$$

or equivalently

$$
\begin{aligned}
(1 - z\overline{w})(I_m - \omega_0 A)^* P(I_m - \omega_0 A) &= (1 - z\overline{w_0})(I_m - wA)^* P(I_m - \omega_0 A) \\
&\quad + (1 - \omega_0\overline{w})(I_m - \omega_0 A)^* P(I_m - zA) \\
&\quad + (1 - z\overline{w_0})(1 - \omega_0\overline{w})(P + A^* P A) \\
&\quad + (1 - z\overline{w_0})(1 - \omega_0\overline{w})(C^* J C - P - A^* P A).
\end{aligned}
$$

But an easy, albeit unpleasant, computation shows that

$$
\begin{aligned}
(1 - z\overline{w})(I_m - \omega_0 A)^* P(I_m - \omega_0 A) &= (1 - z\overline{w_0})(I_m - wA)^* P(I_m - \omega_0 A) \\
&\quad + (1 - \omega_0\overline{w})(I_m - \omega_0 A)^* P(I_m - zA) \\
&\quad + (1 - z\overline{w_0})(1 - \omega_0\overline{w})(P + A^* P A).
\end{aligned}
$$

Thus we get the formula

$$
\begin{aligned}
\frac{J - \Theta(z)J\Theta(w)^*}{(1 - z\overline{w})} &= F(z)P^{-1}F(w)^* \\
&\quad + \frac{(1 - z\overline{w_0})(1 - \omega_0\overline{w})}{1 - z\overline{w}} F(z)P^{-1}(I_m - \omega_0 A)^{-*}\Delta(I_m - \omega_0 A)^{-1}P^{-1}F(w)^*,
\end{aligned}
$$

where we have set $\Delta = C^* J C - P - A^* P A$. Thus formula (7.7.30) is in force if and only if

$$F(z) P^{-1} (I_m - w_0 A)^{-*} \Delta (I_m - w_0 A)^{-1} P^{-1} F(w)^* \equiv 0.$$

Since the pair (C, A) is observable, this will be the case if and only if $\Delta = 0$, that is, if and only if the Stein equation (7.7.31) holds. $\square$

Solution of Exercise 7.7.19:

(1) Take

$$J = \begin{pmatrix} 1 & 0 \\ 0 & -1 \end{pmatrix}, \quad C = \begin{pmatrix} 1 \\ -1 \end{pmatrix}, \quad A = 1.$$

Then equation (7.7.31) becomes $P - P = 0$, and any real (or even complex number) solves it.

(2) Let $r \in (0, 1)$. Let $r \in (r(T), 1)$. By definition of the lim sup, there exists n_0 such that:

$$n \geq n_0 \implies \sup_{m \geq n} \|A^m\|^{1/m} \leq r.$$

In particular $\|A^m\| \leq r^m$ for $m \geq n_0$. It follows that

$$\|A^m C^* J C A^m\| \leq r^{2m} \|C^* J C\|,$$

and so the series (7.7.33) converges in norm to a matrix P which satisfies the Stein equation. This is the only solution. Indeed, let P_1 and P_2 be two solutions of (7.7.31). Then the difference $Q = P_1 - P_2$ satisfies

$$Q = A^* Q A = A^2 Q A^{*2} = \cdots = A^m Q A^{*m} = \cdots,$$

and letting $m \to \infty$ gives $Q = 0$. $\square$

Chapter 8

Hardy Spaces

In this chapter we present exercises related to Hardy spaces. These spaces played a key role in operator theory in particular in view of Beurling's theorem, and more generally, of its vector-valued version, namely the Beurling–Lax theorem. They also play a key role in the theory of linear systems. The fractional Hardy spaces play an important role in the theory of self-similar systems.

8.1 Reproducing kernel Hilbert spaces of analytic functions

We first specify some of the results discussed in Sections 7.1 and 7.5 to the case of analytic kernels.

Exercise 8.1.1. *Let $\Omega \subset \mathbb{C}$ be an open set and let $K(z, w)$ be a function analytic in z and $\overline{w}$ in Ω and positive definite there.*

(1) *Show that the elements of the associated reproducing kernel Hilbert space are analytic in Ω.*

(2) *Show that for every $n \in \mathbb{N}$ and $w_0 \in \Omega$, the function*

$$z \mapsto \frac{\partial^n K}{\partial \overline{w}^n}(z, w_0)$$

belongs to the associated reproducing kernel Hilbert space $\mathcal{H}(K)$ and that

$$\langle f(\cdot), \frac{\partial^n K}{\partial \overline{w}^n}(\cdot, w_0)\rangle = f^{(n)}(w_0), \quad \forall f \in \mathcal{H}(K) \quad and \quad \forall w_0 \in \Omega. \tag{8.1.1}$$

We will rewrite in a shorter way (8.1.1) as

$$\left\langle f, \frac{\partial^n K}{\partial \overline{w}^n} \right\rangle = f^{(n)}(w).$$

Recall that we have denoted by $\partial \Omega$ the boundary of a set Ω; see (3.1.7).

Exercise 8.1.2. *In the notation of the previous exercise, assume that $\partial\Omega \neq \emptyset$ and let $w \in \partial\Omega$. Let $(w_n)_{n \in \mathbb{N}}$ be a sequence of numbers in Ω with limit w and such that*

$$\sup_{n \in \mathbb{N}} K(w_n, w_n) < \infty. \tag{8.1.2}$$

(1) Show that there exists a subsequence $(w_{n_k})_{k \in \mathbb{N}}$ such that the limit

$$\lim_{k \to \infty} f(w_{n_k})$$

exists for all $f \in \mathcal{H}(K)$, and defines a bounded map from $\mathcal{H}(K)$ into $\mathbb{C}$.
(2) Express it in terms of a limit of kernels.

8.2 The Hardy space of the open unit disk

The Hardy space $\mathbf{H}_2(\mathbb{D})$ (see Definition 8.2.1 below) provides a convenient setting to describe shift-invariant subspaces of $\ell_2(\mathbb{N}_0)$ (see Section 8.3), and this was the main motivation for introducing this space. It now has applications to numerous other problems in analysis and digital signal processing. Indeed, a sequence in $\ell_2(\mathbb{N}_0)$ represents a finite energy discrete signal, and its associated power series belongs to $\mathbf{H}_2(\mathbb{D})$. This allows us to translate various problems in signal processing into problems in the setting of function theory in the open unit disk.

We recall (see Question 2.1.20) that, given a function f analytic in the open unit disk, the function (2.1.24)

$$M_2(r) = \frac{1}{2\pi} \int_0^{2\pi} |f(re^{it})|^2 dt, \quad r \in (0,1),$$

is increasing.

Definition 8.2.1. The space $\mathbf{H}_2(\mathbb{D})$ is the set of functions analytic in $\mathbb{D}$ and such that

$$\sup_{r \in (0,1)} \frac{1}{2\pi} \int_0^{2\pi} |f(re^{it})|^2 dt < \infty, \tag{8.2.1}$$

or equivalently, such that

$$\sum_{n=0}^{\infty} |a_n|^2 < \infty, \tag{8.2.2}$$

holds, where $f(z) = \sum_{n=0}^{\infty} a_n z^n$ is the power series of f at the origin.

We also refer to [CABP] for the following exercise; see Exercise 5.6.11, p. 213 there. A proof is given here.

Exercise 8.2.2. *Show that the two definitions of $\mathbf{H}_2(\mathbb{D})$ are equivalent.*

The second definition brings us back to item (1) of Exercise 7.5.11, and we have here an important example of a recurring theme (see Remark 0.0.1): We are given a Hilbert space of functions which:

(1) has a reproducing kernel, namely

$$k_w(z) = \frac{1}{1 - z\overline{w}}, \tag{8.2.3}$$

see Exercise 7.5.11.
(2) It has an analytic description in terms of power series; see (8.2.2).
(3) It has a geometric description; here, (8.2.1).
(4) Last, but not least, has a transform interpretation (the z-transform)

$$(a_n)_{n \in \mathbb{N}_0} \mapsto \left(z \mapsto \sum_{n=0}^{\infty} a_n z^n \right),$$

from $\ell_2(\mathbb{N}_0)$ onto $\mathbf{H}_2(\mathbb{D})$.

Each of these four equivalent characterizations thus reveals a different facet of the space. A similar situation holds in other cases, such as the Bergman space (see Section 10.1), the Bargmann–Segal–Fock space (see Exercise 11.1), and others such as the Dirichlet space, to name a few.

The reproducing kernel description leads to (use Exercise 7.6.12):

Exercise 8.2.3. *Let f be defined in the open unit disk. Then f belongs to the Hardy space and has norm less than or equal to 1 if and only if the kernel*

$$\frac{1}{1 - z\overline{w}} - f(z)\overline{f(w)} \tag{8.2.4}$$

is positive definite in $\mathbb{D}$.

The previous result has an important consequence, namely a representation result for elements of the unit ball of $\mathbf{H}_2(\mathbb{D})$. Indeed, the kernel (8.2.4) can be rewritten as

$$\frac{a(z)a(w)^* - b(z)\overline{b(w)}}{1 - z\overline{w}},$$

with

$$a(z) = \begin{pmatrix} 1 & zf(z) \end{pmatrix},$$
$$b(z) = f(z).$$

It is tempting to say that one can "factorize via a contraction", that is write

$$b(z) = a(z)s(z),$$

where s is $\mathbb{C}^2$-valued, analytic and contractive in the open unit disk. This is indeed the case and s can be chosen analytic in $\mathbb{D}$. This is a consequence of a result of Leech. See for instance [261].

More generally, replacing $\mathbb{D}$ by a subset E of the open unit disk, we have:

Question 8.2.4. *Let f be defined in $E \subset \mathbb{D}$. Then f is the restriction of a function of the Hardy space of norm less than or equal to 1 if and only if the kernel (8.2.4) is positive definite in E.*

We now turn to an exercise related to the boundary behaviour of functions in the Hardy space. This exercise is used in Exercise 9.3.1, which discusses Carathéodory's theorem.

Exercise 8.2.5.

(1) *Let $\theta_0 \in [0, 2\pi)$. Assume that $h \in \mathbf{H}_2(\mathbb{D})$ is such that the function*

$$z \mapsto \frac{h(z)}{z - e^{i\theta_0}} \tag{8.2.5}$$

is also in $\mathbf{H}_2(\mathbb{D})$. Show that

$$\lim_{\substack{r \to 1 \\ r \in (0,1)}} h(re^{i\theta_0}) = 0.$$

(2) *More generally, if $\theta_0, \ldots, \theta_n \in [0, 2\pi)$ are pairwise distinct and such that the function*

$$z \mapsto \frac{h(z)}{\prod_{u=0}^{n}(z - e^{i\theta_u})} \tag{8.2.6}$$

is also in $\mathbf{H}_2(\mathbb{D})$. Show that

$$\lim_{\substack{r \to 1 \\ r \in (0,1)}} h(re^{i\theta_u}) = 0, \quad u = 0, \ldots, n. \tag{8.2.7}$$

In a similar vein one has:

Question 8.2.6. *Let H be a $\mathbb{C}^{n \times m}$-valued function, with entries in $\mathbf{H}_2(\mathbb{D})$. Assume that for $\theta_0 \in [0, 2\pi)$, $c \in \mathbb{C}^m$ and $d \in \mathbb{C}^n$ the function*

$$z \mapsto \frac{H(z)c - d}{z - e^{i\theta_0}} \in (\mathbf{H}_2(\mathbb{D}))^n.$$

Then,

$$\lim_{\substack{r \to 1 \\ r \in (0,1)}} \|H(re^{i\theta_0})c - d\| = 0,$$

where $\| \cdot \|$ denotes any norm in $\mathbb{C}^n$.

Let $n \in \mathbb{N}_0$. The function

$$k_n(z, w) = \frac{1}{(1 - z\overline{w})^{n+1}} \tag{8.2.8}$$

is a product of positive definite functions and hence is positive definite in the open unit disk. The associated reproducing kernel Hilbert space is called the weighted Bergman space. More generally:

Question 8.2.7. *Let* $\nu > -1$. *Show that the function*

$$k_\nu(z, w) = \frac{1}{(1 - z\overline{w})^{1+\nu}}$$

is positive definite in the open unit disk. The associated reproducing kernel Hilbert space is called the fractional Hardy space of the open unit disk.

8.3 Some operator theory in $\mathbf{H}_2(\mathbb{D})$

First a remark. The map I defined by

$$(I(a))(z) = \sum_{n=0}^{\infty} a_n z^n \tag{8.3.1}$$

is unitary and sends $\ell_2(\mathbb{N}_0)$ onto $\mathbf{H}_2(\mathbb{D})$, and the shift operator

$$Z(a_0, a_1, \dots) = (0, a_0, a_1, \dots)$$

is sent to the operator M_z of multiplication by z in $\mathbf{H}_2(\mathbb{D})$:

$$IZI^* = M_z.$$

Exercise 8.3.1.

(1) *Show that the operator* M_z *of multiplication by* z *is an isometry from* $\mathbf{H}_2(\mathbb{D})$ *into itself, and compute its adjoint.*

(2) *Same question for the Blaschke factor* $b_a(z) = \frac{z-a}{1-z\overline{a}}$, *where* $a \in \mathbb{D}$.

(3) *Same question for the finite Blaschke product* $b(z) = \prod_{n=1}^{N} \frac{z-w_n}{1-z\overline{w_n}}$. *In particular, when the zeros of the Blaschke product are all different from* 0 *and simple, prove that there exist complex numbers* $c_0, \dots c_N$ *such that*

$$(M_b^* f)(z) = \overline{c_0} f(z) + \sum_{n=1}^{N} \overline{c_n} \frac{z f(z) - w_n f(w_n)}{z - w_n}, \tag{8.3.2}$$

for $f \in \mathbf{H}_2(\mathbb{D})$.

Remark 8.3.2. See Question 1.7.2 and formula (1.7.5) in connection with (8.3.2). Furthermore, note that each $n = 1, \dots, N$, the function

$$z \mapsto \frac{z f(z) - w_n f(w_n)}{z - w_n}$$

has a removable singularity at $z = w_n$, with value at that point equal to

$$(z f(z))'|_{z=w_n} = f(w_n) + w_n f'(w_n).$$

Exercise 8.3.3. *Compute the spectra of M_z and of M_z^* in $\mathbf{H}_2(\mathbb{D})$.*

Exercise 8.3.4.

(1) *Show that the resolvent set of $\operatorname{Re} M_z$ (in $\mathbf{H}_2(\mathbb{D})$) is $\mathbb{C} \setminus [-1, 1]$.*
(2) *Show that the operator $\operatorname{Re} M_z$ has no point spectrum.*
(3) *Show that it has no residual spectrum.*

Remark 8.3.5. We note that $\operatorname{Re} M_z$ has no point spectrum in the Fock space (where it is an unbounded operator). See Exercise 11.1.8.

The following exercise can be proved using Exercise 7.7.17. The key point is to identify the function Θ defined there by (7.7.29) with the finite Blaschke product b defined in the statement of the exercise. In item (3), note that equality holds in (8.3.3) when all the $w_n \neq 0$.

Exercise 8.3.6. *Let $w_1, \ldots, w_N$ be distinct points in $\mathbb{D}$, and let $\mathcal{M}$ denote the linear span of the functions $k_{w_1}, \ldots, k_{w_N}$, endowed with the $\mathbf{H}_2(\mathbb{D})$ inner product.*

(1) *Show that $\mathcal{M}$ is the finite-dimensional reproducing kernel Hilbert space with reproducing kernel*

$$k_b(z, w) = \frac{1 - b(z)\overline{b(w)}}{1 - z\overline{w}},$$

with $b(z) = \prod_{n=1}^{N} \frac{z - w_n}{1 - z\overline{w_n}}$.

(2) *Show that*

$$\mathcal{M} = \mathbf{H}_2(\mathbb{D}) \ominus b\mathbf{H}_2(\mathbb{D}).$$

(3) *Show that for every $a \in \mathbb{D}$,*

$$R_a \mathcal{M} \subset \mathcal{M}, \tag{8.3.3}$$

where R_a is defined by (1.7.3), that is:

$$(R_a f)(z) = \begin{cases} \dfrac{f(z) - f(a)}{z - a}, & z \neq a, \\ f'(a), & z = a. \end{cases}$$

Hint: Apply Exercise 7.7.17 with

$$J = 1, \quad C = \underbrace{(1 \;\; 1 \;\; \cdots \;\; 1)}_{N \text{ times}} \quad \text{and} \quad A = \operatorname{diag}(\overline{w_1}, \overline{w_2}, \ldots, \overline{w_N}).$$

Exercise 8.3.7. *Let $B(z) = \prod_{n=1}^{N} b_{a_n}(z)$, where $a_1, \ldots$ are not necessarily different points in $\mathbb{D}$, and set $B_0(z) \equiv 1$ and $B_k(z) = \prod_{n=1}^{k} b_{a_n}(z)$, $k = 1, 2, \ldots, N$.*

(1) *Show that*

$$\mathbf{H}_2(\mathbb{D}) \ominus B\mathbf{H}_2(\mathbb{D}) = \oplus_{k=1}^{N} B_{k-1}(\mathbf{H}_2(\mathbb{D}) \ominus b_{a_k}\mathbf{H}_2(\mathbb{D})). \tag{8.3.4}$$

(2) *Using (8.3.4), find an orthogonal basis for $\mathbf{H}_2(\mathbb{D}) \ominus B\mathbf{H}_2(\mathbb{D})$.*

Exercise 8.3.6 suggests the following question.

Problem 8.3.8. *Find the structure of the closed R_0 invariant subspaces of $\mathbf{H_2}(\mathbb{D})$.*

The famous Beurling theorem characterizes closed M_z-invariant subspaces of $\mathbf{H_2}(\mathbb{D})$. We first present a finite-dimensional version of this theorem.

Exercise 8.3.9. *Let $\mathcal{N}$ be a closed M_z-invariant subspace of $\mathbf{H_2}(\mathbb{D})$, and assume that $\mathcal{N}$ has finite co-dimension, that is, $\dim \mathbf{H_2}(\mathbb{D}) \ominus \mathcal{N} < \infty$. Show that there is a finite Blaschke product b (with possibly multiple zeros) such that $\mathcal{N} = b\mathbf{H_2}(\mathbb{D})$.*

Hint: Consider $\mathcal{N}^{\perp}$ and show first that a space of functions analytic at the origin is R_0-invariant if and only if it is spanned by the entry of a row function of the form $C(I_N - zA)^{-1}$, where the pair $(C, A) \in \mathbb{C}^{1 \times N} \times \mathbb{C}^{N \times N}$ is observable, that is

$$\cap_{m=0}^{\infty} \ker CA^m = \{0\} \, . \tag{8.3.5}$$

Then show that the spectrum of A is inside the open unit disk.

Recall that b_a was defined by (1.2.9), that is:

$$b_a(z) = \frac{z - a}{1 - \overline{a}z} \, .$$

Question 8.3.10. *In the preceding question write*

$$b(z) = c \prod_{k=1}^{K} (b_{a_k}(z))^{n_k},$$

where $c \in \mathbb{T}$ and where the points $a_1, \ldots$ are pairwise different. Characterize $\mathcal{N}$ in terms of interpolation conditions.

The case of spaces $\mathcal{N}$ of possibly infinite co-dimension is much more involved, and is the content of Beurling's theorem.

Exercise 8.3.11. *Let $\mathcal{M} \subsetneq \mathbf{H_2}(\mathbb{D})$ be closed and invariant under the operator M_z of multiplication by z, and let $\mathcal{N} = \mathcal{M} \ominus z\mathcal{M}$.*

(1) *Show that $\dim \mathcal{N} = 1$.*
(2) *Show that*

$$\mathcal{M} = \oplus_{n=0}^{\infty} z^n \mathcal{N}.$$

(3) *Conclude that there is a function j analytic in $\mathbb{D}$ such that $\mathcal{M} = j\mathbf{H_2}(\mathbb{D})$.*

Remark 8.3.12. The function j in the preceding exercise is inner, meaning that its boundary values, which exist almost everywhere, are almost everywhere of modulus one. We will not consider these important problems here.

Multipliers of the Hardy space are characterized in the following question (see the questions mentioned in Remark 7.6.15).

Question 8.3.13.

(1) *Let s be defined in a subset Ω of the open unit disk. Show that s is the restriction to Ω of a function analytic and contractive in $\mathbb{D}$ if and only if the kernel (7.7.17):*

$$k_s(z,w) = \frac{1 - s(z)\overline{s(w)}}{1 - z\overline{w}}$$

is positive definite in Ω.

(2) *Interpret the previous result when $\Omega = \mathbb{D}$.*

(3) *Show that the operator of multiplication by s is a contraction in $\mathbf{H}_2(\mathbb{D})$ if and only if the kernel k_s is positive definite.*

We conclude with an important result on multipliers. It has key implications in the theory of linear systems. Recall that M_z denotes the operator of multiplication by z.

Exercise 8.3.14. *Let T be a linear bounded operator from $\mathbf{H}_2(\mathbb{D})$ into itself, which commutes with M_z, and let $\psi = T1 \in \mathbf{H}_2(\mathbb{D})$.*

(1) *Show that for any polynomial p it holds that*

$$Tp = \psi p. \tag{8.3.6}$$

(2) *Show that polynomials are dense in $\mathbf{H}_2(\mathbb{D})$.*

(3) *Let $f \in \mathbf{H}_2(\mathbb{D})$ and let $(p_n)_{n\in\mathbb{N}}$ be a sequence of polynomials tending to f in the $\mathbf{H}_2(\mathbb{D})$-norm. Show that, for every $z \in \mathbb{D}$,*

$$\lim_{n\to\infty} p_n(z) = f(z),$$
$$\lim_{n\to\infty} (Tp_n)(z) = (Tf)(z). \tag{8.3.7}$$

(4) *Conclude that (8.3.6) holds for all $f \in \mathbf{H}_2(\mathbb{D})$.*

(5) *Using the previous exercise, or computing directly T^* on the reproducing kernel, show that ψ is bounded in modulus in the open unit disk.*

8.4 Composition operators

The first exercise is a classic, and can be found for instance in [86, p. 50]. See Remark 2.3.8 for a related discussion. In the proof of the first two exercises use is made of the reproducing kernel $\mathcal{H}(s)$ with reproducing kernel (7.7.17). These spaces were introduced and studied by de Branges and Rovnyak in [86]. See Chapter 9. In the proofs we use the fact that the kernel (7.7.17) is positive definite in $\mathbb{D}$ for s analytic and contractive there. See Question 8.3.13.

Exercise 8.4.1. *Let s be a function analytic and contractive in the open unit disk, and suppose that $s(0) = 0$. Prove that the composition operator $f \mapsto f \circ s$ is a contraction from the Hardy space $\mathbf{H}_2(\mathbb{D})$ into itself.*

Hint: Write $s(z) = z\sigma(z)$ where σ is analytic and contractive in the open unit disk, and apply Exercise 7.6.6.

The condition $s(0) = 0$ in the previous exercise can be relaxed. This is the topic of the next exercise

Exercise 8.4.2. *Let s be analytic and strictly contractive[1] in $\mathbb{D}$, let $\omega_0 \in \mathbb{D}$, and let f_0 denote the function:*

$$z \mapsto \frac{1 - s(z)\overline{s(\omega_0)}}{1 - z\overline{\omega_0}}.$$

(1) *Show that there exists $k_0 > 0$ such that the kernel*

$$\frac{1 - \dfrac{1}{k_0 f_0(z)\overline{f_0(w)}}}{1 - z\overline{w}} \tag{8.4.1}$$

is positive definite in $\mathbb{D}$.

(2) *Show that there exists $k_1 > 0$ such that the kernel*

$$\frac{k_1}{f_0(z)\overline{f_0(w)}} \frac{1}{1 - z\overline{w}} - \frac{1}{1 - s(z)\overline{s(w)}} \tag{8.4.2}$$

is positive definite in $\mathbb{D}$.

(3) *Show that the composition operator C_s is a bounded operator from the Hardy space $\mathbf{H}_2(\mathbb{D})$ into itself.*

(4) *Find its representation* (7.6.2).

Question 8.4.3. *Prove formula* (2.3.16).

8.5 Cuntz relations

Let $N \in \mathbb{N}$, let $\mathcal{H}$ be a Hilbert space and let $S_1, \ldots, S_N$ be N isometries of $\mathcal{H}$:

$$S_j^* S_j = I_{\mathcal{H}}, \quad j = 1, \ldots, N.$$

If furthermore it holds that

$$\sum_{j=1}^{N} S_j S_j^* = I_{\mathcal{H}},$$

the isometries are said to satisfy the Cuntz relations. When $N = 1$ we merely have a unitary map from $\mathcal{H}$ into itself, and the case of interest is when $N \geq 2$.

Let $v \in \mathbb{D}$ and $b_v(z) = \frac{z-v}{1-z\overline{v}}$. We recall that the function $e_v(z) = \frac{\sqrt{1-|v|^2}}{1-z\overline{v}}$ form an orthonormal basis of the space $\mathbf{H}_2(\mathbb{D}) \ominus b_v \mathbf{H}_2(\mathbb{D})$ and that the weighted composition operator

$$(T_v f)(z) = e_v(z) f(b_v(z)) \tag{8.5.1}$$

---

[1] (that is, in view of the maximum modulus principle, different from a unitary constant)

is a unitary map from $\mathbf{H}_2(\mathbb{D}) \ominus b_v \mathbf{H}_2(\mathbb{D})$ onto itself. In the following question and exercise, taken from [35], we explore what happens if one replaces b_v by a finite Blaschke product b and the function e_v by an orthonormal basis of $\mathbf{H}_2(\mathbb{D}) \ominus b\mathbf{H}_2(\mathbb{D})$. In the following, we assume for simplicity that the Blaschke product has pairwise disjoint zeros. This hypothesis can easily be removed, at the price of some extra notation.

Question 8.5.1. *Let b be the finite Blaschke product*

$$b(z) = \prod_{k=1}^{N} b_{v_k},$$

where the points $v_1, \dots, v_N$ are assumed pairwise disjoint, and let $e_1, \dots, e_N$ be an orthonormal basis of $\mathbf{H}_2(\mathbb{D}) \ominus b\mathbf{H}_2(\mathbb{D})$. Show that the operators

$$S_j f = e_j f(b), \quad j = 1, \dots, N,$$

satisfy the Cuntz relations in the Hardy space.

Exercise 8.5.2. *Let b be the finite Blaschke product*

$$b(z) = \prod_{k=1}^{N} b_{v_k},$$

where the points $v_1, \dots, v_N$ are assumed pairwise disjoint.
(1) *Show that the kernel*

$$\frac{1}{1 - b(z)\overline{b(w)}}$$

is positive definite in $\mathbb{D}$ and describe the associated reproducing kernel Hilbert space.
(2) *Let $e_1, \dots, e_N$ denote an orthonormal basis of $\mathbf{H}_2(\mathbb{D}) \ominus b\mathbf{H}_2(\mathbb{D})$. Writing*

$$\frac{1}{1 - z\overline{w}} = \frac{1}{1 - b(z)\overline{b(w)}} \times \frac{1 - b(z)\overline{b(w)}}{1 - z\overline{w}}, \tag{8.5.2}$$

show that any element $f \in \mathbf{H}_2(\mathbb{D})$ can be written in a unique way as

$$f(z) = \sum_{n=1}^{N} e_n(z) f_n(b(z)), \quad f_n \in \mathbf{H}_2(\mathbb{D}), \tag{8.5.3}$$

and

$$\|f\|_{\mathbf{H}_2(\mathbb{D})}^2 = \sum_{n=1}^{N} \|f_n\|_{\mathbf{H}_2(\mathbb{D})}^2.$$

Using (8.5.3) one can solve the following multipoint interpolation problem (see [35]):

Problem 8.5.3. *Given N pairwise points $w_1, \ldots, w_N$, and numbers $a_1, \ldots, a_N, c$, describe the set of all functions $f \in \mathbf{H}_2(\mathbb{D})$ such that*

$$\sum_{u=1}^{N} a_u f(w_u) = c. \tag{8.5.4}$$

Question 8.5.4. *Using (8.5.3) reduce Problem 8.5.3 to a one point tangential interpolation problem in $(\mathbf{H}_2(\mathbb{D}))^N$.*

Remark 8.5.5. When $N = 2$ one can use Question 1.2.8 to solve (8.5.4). See the work by Bolotnikov, Rodman and the author [13].

Related to Problem 8.5.6 see also the following section.

Problem 8.5.6. *Let $M \in \mathbb{N}$. What can be said when one replaces (8.5.2) by its M-th power:*

$$\frac{1}{(1 - z\overline{w})^M} = \frac{1}{(1 - b(z)\overline{b(w)})^M} \times \left(\frac{1 - b(z)\overline{b(w)}}{1 - z\overline{w}} \right)^M. \tag{8.5.5}$$

8.6 The Hardy space of the open upper half-plane

We already met the Hardy space of the unit disk in Section 8.2. Here we focus on the open upper half-plane case. Recall that the Lebesgue space $\mathbf{L}_2(\mathbb{R}, \mathcal{B}, dx)$ is not a space of functions, let alone a space of analytic functions. Still, it admits a key decomposition into an orthogonal sum of two spaces, one of (equivalence classes) of functions with, in each equivalence class, one representative whose values are almost everywhere the (non-tangential) boundary values of a function analytic in the open lower half-plane $\mathbb{C}_-$, and the other of (equivalent classes) of functions with one representative whose values are almost everywhere the (non-tangential) boundary values of a function analytic in the open upper half-plane $\mathbb{C}_+$. Another natural orthogonal decomposition is obtained when writing an element in $\mathbf{L}_2(\mathbb{R}, \mathcal{B}, dx)$ as a sum of a function with support in $(-\infty, 0]$ and of a function with support in $[0, \infty)$:

$$f(u) = 1_{(-\infty, 0]}(u) f(u) + 1_{[0, \infty)}(u) f(u). \tag{8.6.1}$$

These two decompositions are related by the Fourier transform. These various connections form the main topic of the exercises of the present section.

Let

$$f_n(t) = \frac{1}{\sqrt{\pi}} \frac{1}{t + i} \left(\frac{t - i}{t + i} \right)^n, \quad n \in \mathbb{Z}. \tag{8.6.2}$$

The difficulty in the following exercise is not to show that the functions f_n form an orthonormal set (the mixed inner products are easily computed using the residue theorem), but lies in proving the completeness of this set.

Exercise 8.6.1.

(1) *Show that the family* $(f_n)_{n \in \mathbb{Z}}$ *is an orthonormal basis of* $\mathbf{L}_2(\mathbb{R}, \mathcal{B}, dx)$.

(2) *Show that*

$$\frac{1}{-2\pi i (z - \overline{w})} = \sum_{n=0}^{\infty} f_n(z) \overline{f_n(w)}, \quad z, w \in \mathbb{C}_+. \tag{8.6.3}$$

(3) *Compute*

$$\sum_{n=1}^{\infty} f_{-n}(z) \overline{f_{-n}(w)}, \quad z, w \in \mathbb{C}_-,$$

where $\mathbb{C}_-$ *denotes the open lower half-plane.*

Hint: To prove completeness, prove that any f orthogonal to all the f_n satisfies (6.2.10):

$$\int_{\mathbb{R}} \frac{f(t)}{t - z} du = 0, \quad \forall z \in \mathbb{C} \setminus \mathbb{R},$$

and apply Exercise 6.2.9.

Another way to prove completeness would be to use approximation by continuous functions with compact support and then by rational functions. Yet another way could be by using Vitali's theorem (see [270, Theorem 26, p. 25] and Exercise 6.3.3 above).

We note that the inverse Fourier transform of the $f_n, n = 0, 1, 2, \dots$ are called Laguerre functions (see also [126, Exercise 2, p. 35], and the reference there to Y.W. Lee's book [217]).

We denote by $\mathbf{H}_2(\mathbb{C}_+)$ (or $\mathbf{H}_2$ for short) the closed linear span of the $f_n, n = 0, 1, 2, \dots$, and by $\mathbf{H}_2(\mathbb{C}_-)$ (or $\mathbf{H}_2^-$ for short) the closed linear span of the $f_n, n = -1, -2, -3, \dots$. The space $\mathbf{H}_2(\mathbb{C}_+)$ is the Hardy space of the open right half-plane. In view of (8.6.3) and of the formula (7.1.13) for the reproducing kernel we have:

Theorem 8.6.2. *The space* $\mathbf{H}_2(\mathbb{C}_+)$ *is the reproducing kernel Hilbert space with reproducing kernel* $\frac{1}{-2\pi i (z - \overline{w})}$.

Exercise 8.6.3. *The map*

$$f \mapsto \frac{1}{2\pi} \widehat{f} \tag{8.6.4}$$

is unitary from $\mathbf{L}_2(\mathbb{R}_-, \mathcal{B}, du)$ *onto* $\mathbf{H}_2^+$, *and from* $\mathbf{L}_2(\mathbb{R}_+, \mathcal{B}, du)$ *onto* $\mathbf{H}_2^-$.

Hint: We have for $z, w \in \mathbb{C}_+$

$$\frac{1}{-2\pi i(z - \overline{w})} = \frac{1}{2\pi} \int_{(-\infty, 0]} e^{-i(z-\overline{w})t} dt,$$

and so for $N \in \mathbb{N}$, $c_1, \ldots, c_N \in \mathbb{C}$ and $w_1, \ldots, w_N \in \mathbb{C}_+$ we have

$$F(z) = \sum_{n=1}^N \frac{c_n}{-2\pi i(z - \overline{w_n})} = \frac{1}{2\pi} \int_{(-\infty, 0]} e^{-itz} \underbrace{\left(\sum_{n=1}^N c_n e^{i\overline{w_n}t} \right)}_{f(t)} dt.$$

Use then (6.1.4) to show that

$$\|F\|_{\mathbf{H}_2^+}^2 = \sum_{n,m=1}^N \frac{c_n \overline{c_m}}{-2\pi i(w_m - \overline{w_n})} = 2\pi \|f\|_{\mathbf{L}_2(\mathbb{R}_-)}^2 = \|\widehat{f}\|_{\mathbf{L}_2(\mathbb{R})}^2.$$

Conclude by a density argument.

By Exercise 8.6.3, and after a change of variable, one sees that there exist functions $f_0, f_1, \ldots \in \mathbf{L}_2(\mathbb{R}_+, dx)$ such that

$$\frac{1}{\sqrt{\pi}} \left(\frac{t - i}{t + t} \right)^n = \frac{1}{2\pi} \int_{[0,\infty)} e^{itu} f_n(u) du, \quad n = 0, 1, \ldots \tag{8.6.5}$$

and

$$\langle f_n, f_m \rangle_{\mathbf{L}_2(\mathbb{R}_+, dx)} = \delta_{n,m}, \quad n.m \in \mathbb{N}_0. \tag{8.6.6}$$

Question 8.6.4. *Show that the functions f_n are of the form*

$$f_n(u) = e^{-u} p_n(u), \quad n = 0, 1, \ldots$$

where p_n is a polynomial of degree n.

Hint: Following [217, p. 474], use (2.1.30).

We have in particular

$$\int_{[0,\infty)} e^{-2u} p_n(u) p_m(u) du = \delta_{n,m}, \quad n.m \in \mathbb{N}_0,$$

and the functions f_n and p_n are (up to a change of variable) the Laguerre functions and Laguerre polynomials.

Question 8.6.5. *Let $f_+ \in \mathbf{L}_2(\mathbb{R}_+, dx)$. The expression*

$$F_+(z) = \frac{1}{2\pi} \int_0^\infty e^{izu} f_+(u) du \tag{8.6.7}$$

defines a function analytic in the open upper half-plane $\mathbb{C}_+$. *The formula*

$$\langle F_+, G_+ \rangle = \langle f, g \rangle_{\mathbf{L}_2(\mathbb{R}_+, dx)}$$

defines an inner product on the set of functions of the form (8.6.7), *and this set is a Hilbert space with respect to the associated norm. Show that this space is a Hardy space.*

Question 8.6.6. *Show that a function f analytic in $\mathbb{C}_+$ belongs to $\mathbf{H}_2$ if and only if*

$$\sup_{y>0} \int_{\mathbb{R}} |f(x+iy)|^2 dx < \infty.$$

Question 8.6.7. *Recall that the functions γ_t $(t \in \mathbb{R})$ were defined in* (6.1.26):

$$\gamma_t(u) = \begin{cases} \frac{e^{itu}-1}{iu}, & u \neq 0, \\ t & u = 0. \end{cases}$$

What is the closed linear span of $\{\gamma_t, t \in I\}$ in $\mathbf{L}_2(\mathbb{R}, \mathcal{B}, dx)$ for

$$I = [0, \infty), \quad I = (-\infty, 0] \quad and \quad I = [0, T], \ (T > 0).$$

Exercise 8.6.8. *Show that the weighted composition operator which to f associates the function*

$$Tf(z) = \frac{1}{z}f\left(-\frac{1}{z}\right) \tag{8.6.8}$$

is unitary from $\mathbf{L}_2(\mathbb{R}, \mathcal{B}, dx)$ into itself and from $\mathbf{H}_2(\mathbb{C}_+)$ into itself.

Hint: Compute Tf_n, where f_n is given by (8.6.2), and conclude that the operator (8.6.8), a priori only densely defined on $\mathbf{L}_2(\mathbb{R}, \mathcal{B}, dx)$ admits a unitary extension to that space. One can also use Exercise 7.6.6 in the case of $\mathbf{H}_2(\mathbb{C}_+)$.

We note that finding appropriate extensions of the Hardy space and the associated Schur functions is a fruitful strategy in Schur analysis.

Question 8.6.9. *Let R_a be defined by* (1.7.3). *Show that for $f, g \in \mathbf{H}_2(\mathbb{C}_+)$ and $a, b \in \mathbb{C}_+$ it holds that*

$$\langle R_a f, g \rangle - \langle f, R_b g \rangle - (a - \overline{b})\langle R_a, R_b \rangle = 2\pi i \overline{g(b)} f(a). \tag{8.6.9}$$

Hint: Compute (8.6.9) on an appropriate dense set.

Remark 8.6.10. In some sense, the identity (8.6.9) is a weakening of the inner product in the Hardy space $\mathbf{H}_2(\mathbb{C}_+)$. It allows us to extend part of the theory of this space to wider settings. The identity was introduced by L. de Branges (see [83]) and is called de Branges' identity (for the half-plane; a similar identity holds for the disk case).

Question 8.6.11. *Let $\mathcal{H}$ be a reproducing kernel Hilbert space of functions analytic in $\Omega \subset \mathbb{C}_+$, and suppose that (8.6.9) holds on a dense set. Show that the operators R_a $(a \in \Omega)$ have continuous extensions to $\mathcal{H}$.*

Hint: Show that (8.6.9) implies an inequality of the form

$$\|R_a f\|^2 \leq K_a \|R_a f\| \cdot \|f\| + C_a \|f\|^2$$

for some strictly positive constants K_a and C_a, for f on the given dense set.

Define an inner function to be a function j analytic in the open unit disk and such that the operator M_j of multiplication by j is an isometry from the space $\mathbf{H}_2(\mathbb{C}_+)$ into itself.

Question 8.6.12. *Let j be an inner function of the open upper half-plane. Show that the space $\mathbf{H}_2(\mathbb{C}_+) \ominus j\mathbf{H}_2(\mathbb{C}_+)$ is the reproducing kernel Hilbert space with reproducing kernel*

$$\frac{1 - j(z)\overline{j(w)}}{-2\pi i(z - \overline{w})}.$$

We now recall that the functions f_n have been defined by (8.6.2):

$$f_n(z) = \frac{1}{\sqrt{\pi}} \frac{1}{z+i} \left(\frac{z-i}{z+i}\right)^n, \quad n \in \mathbb{Z}.$$

Question 8.6.13. *Let $n \in \mathbb{N}$ and define:*

$$B_N(z) = \left(\frac{z-i}{z+i}\right)^N.$$

Show that $\mathbf{H}_2(\mathbb{C}_+) \ominus B_N \mathbf{H}_2(\mathbb{C}_+)$ is spanned by the functions $f_0, \ldots, f_{N-1}$, and write formula (7.7.1) in this setting.

Question 8.6.14. *Show that the operators R_a (defined by (1.7.3)) are bounded in $\mathbf{H}_2(\mathbb{C}_+)$ for $a \in \mathbb{C}_+$, and find the structure of the finite-dimensional R_a-invariant subspaces of $\mathbf{H}_2(\mathbb{C}_+)$.*

We conclude this section with a question inspired from the definition of the Schwartz functions in terms of their Hermite coefficients (see Remark 6.1.18) and which may have connections with infinite-dimensional analysis and the theory of distributions.

Problem 8.6.15. *Characterize those functions $f = \sum_{n=0}^{\infty} a_n f_n \in \mathbf{H}_2(\mathbb{C}_+)$ such that*

$$\sum_{n=0}^{\infty} |a_n|^2 (n+1)^{2p} < \infty, \quad p = 0, 1, \ldots, \tag{8.6.10}$$

and such that

$$\sum_{n=0}^{\infty} |a_n|^2 2^{2p} < \infty, \quad p = 0, 1, \ldots, \tag{8.6.11}$$

and their respective duals.

8.7   The fractional Hardy space $\mathcal{H}_2^\nu$ of the open upper half-plane

The following exercise introduces the fractional Hardy space of the open upper half-plane. A first characterization of the space in terms of a transform can be obtained using formula (8.7.2) given in the hint. We recall the various aspects of the space at the end of the section. Some of the analysis relies on the paper [230] by Mamadou Mboup.

Exercise 8.7.1 (see [84, p. 216]). *Let $\nu > -1$. Show that the function*

$$\frac{\Gamma(1+\nu)}{(-i(z-\overline{w}))^{1+\nu}} \tag{8.7.1}$$

is positive definite in $\mathbb{C}_+$, and describe the associated reproducing kernel Hilbert space.

Hint: First prove the formula

$$\int_0^\infty t^\nu e^{-zt} dt = \frac{\Gamma(1+\nu)}{z^{1+\nu}}, \quad \operatorname{Re} z > 0. \tag{8.7.2}$$

Remark 8.7.2. (8.7.2) is an example of a Laplace transform. More generally the Laplace transform is defined by

$$F(p) = \int_0^\infty e^{-pt} f(t) dt$$

for $f \in \mathbf{L}_1([0, \infty), dx)$.

Exercise 8.7.3. *Show that the Hankel matrix (1.6.6) is strictly positive.*

The reproducing kernel Hilbert space associated with the kernel (8.7.1) will be denoted by $\mathcal{H}_2^\nu$. It is called the fractional Hardy space of the open upper half-plane.

The following exercise is [230, Proposition 3.4, p. 279]. As pointed out in that paper, the invariance under the operator T_ν (defined in (8.7.3)) allows us to express self-similarity of a system in the frequency domain in a way similar to the time domain. We refer to [230] for more details. We also note that in [84, §50], L. de Branges studies spaces of entire functions invariant under the map $a^{\frac{\nu+1}{2}} T_\nu$, and for which this map is unitary. Paley–Wiener spaces are an example of such spaces.

Exercise 8.7.4. *Let $\nu > -1$. Show that for every $a \in (0, 1)$ the function*

$$T_\nu(F)(z) = a^{\frac{1+\nu}{2}} F(az) \tag{8.7.3}$$

belongs to $\mathcal{H}_2^\nu$ and has the same norm as F.

As mentioned in the introduction, a reproducing kernel Hilbert space can often be characterized in a number of equivalent ways, each one more appropriate for a specific application. The following theorem illustrates this fact for the space $\mathcal{H}_2^\nu$. In the statement, the functions $f_n^{(\nu)}$ have been defined in (2.1.41).

Theorem 8.7.5. *Let F be analytic in $\mathbb{C}_+$. The following are equivalent:*

(1) *F belongs to $\mathcal{H}_2^\nu$.*
(2) *F can be written as*

$$F(z) = \int_{[0,\infty)} e^{izt} t^{\frac{\nu}{2}} f(t) dt$$

for some $f \in \mathbf{L}_2(\mathbb{R}_+, dx)$, and

$$\|F\|_{\mathcal{H}_2^\nu}^2 = \|f\|^2.$$

(3) *F can be written as*

$$F(z) = \sum_{n=0}^\infty a_n f_n^{(\nu)}(iz),$$

with

$$\|F\|_{\mathcal{H}_2^\nu}^2 = \sum_{n=0}^\infty |a_n|^2.$$

8.8 Solutions of the exercises

Solution of Exercise 8.1.1: We only consider (1), and leave the second item to the reader. Let $w \in \Omega$ and let $(h_n)_{n\in\mathbb{N}}$ be a sequence of numbers going to 0 and such that $w + h_n \in \Omega$ for $n = 1, 2, \ldots$. Since the function $K(z, w)$ is analytic in z and $\overline{w}$ we can write, for numbers $a, b, \ldots, e$ which represent the partial derivatives at (w, w),

$$K(w + h_n, w + h_n) = K(w, w) + ah_n + b\overline{h_n} + h_n^2 c + |h_n|^2 d + e\overline{h_n}^2 + o(|h_n|^2),$$
$$K(w + h_n, w) = K(w, w) + ah_n + h_n^2 c + |h_n|^2 d + o(|h_n|^2),$$
$$K(w, w + h_n) = K(w, w) + b\overline{h_n} + e\overline{h_n}^2 + o(|h_n|^2).$$

Thus

$$\frac{K(w + h_n, w + h_h) + K(w, w) - K(w, w + h_n) - K(w + h_n, w)}{|h_n|^2}$$

$$= \frac{\partial^2 K}{\partial w \partial \overline{w}}(w, w) + \frac{o(|h_n|^2)}{|h_n|^2},$$

since $e = \frac{\partial^2 K}{\partial w \partial \overline{w}}(w, w)$.

Let now

$$f_n(z) = \frac{K(z, w + h_n) - K(z, w)}{h_n}, \quad n = 1, 2, \ldots$$

Then, $f_n \in \mathcal{H}(K)$ and

$$\|f_n\|^2 = \frac{K(w + h_n, w + h_h) + K(w, w) - K(w, w + h_n) - K(w + h_n, w)}{|h_n|^2}.$$

Thus the family $(f_n)_{n \in \mathbb{N}}$ is bounded in norm in $\mathcal{H}(K)$ and admits a weak limit (via a subsequence, which we will still denote by $(f_n)_{n \in \mathbb{N}}$) to some $g \in \mathcal{H}(K)$. We have for $z \in \Omega$

$$g(z) = \lim_{n \to \infty} \langle f_n(\cdot), K(\cdot, z) \rangle$$

$$= \lim_{n \to \infty} \frac{K(z, w + h_n) - K(z, w)}{\overline{h_n}}$$

$$= \frac{\partial K}{\partial \overline{w}}(z, w).$$

Hence the function $z \mapsto \frac{\partial K}{\partial \overline{w}}(z, w)$ belongs to $\mathcal{H}(K)$ and for $f \in \mathcal{H}(K)$

$$\langle f, g \rangle = \lim_{n \to \infty} \langle f, f_n \rangle = f'(w). \qquad \square$$

Solution of Exercise 8.1.2:

(1) The family of elements $K_{w_1}, K_{w_2}, \ldots$ of $\mathcal{H}(K)$ is uniformly bounded and so admits a weakly converging subsequence since a closed ball of a Hilbert space is weakly compact and metrizable. Let $(w_{n_k})_{k \in \mathbb{N}}$ be a subsequence of the original sequence and let $g \in \mathcal{H}(K)$ such that the weak limit of the functions $K_{w_{n_k}}$ is g. Thus

$$\langle f, g \rangle = \lim_{k \to \infty} \langle f, K_{w_{n_k}} \rangle, \quad \forall f \in \mathcal{H}(K).$$

This ends the proof since $\langle f, K_{w_k} \rangle = f(w_{n_k})$.

(2) Let $z \in \Omega$. We have

$$g(z) = \langle g, K_z \rangle$$

$$= \lim_{k \to \infty} \langle K_{w_{n_k}}, K_z \rangle$$

$$= \lim_{k \to \infty} K(z, w_{n_k}). \qquad \square$$

Solution of Exercise 8.2.2: We use the dominated convergence and the monotone convergence theorem (see [CABP]). One can of course use more direct, but a bit

8.8. *Solutions of the exercises*

longer, arguments. A shorter argument uses Parseval's equality. Let f be analytic in the open unit disk, with power series centered at the origin equal to

$$f(z) = \sum_{n=0}^{\infty} a_n z^n, \quad z \in \mathbb{D}.$$

By the Cauchy–Schwarz inequality, we have for $r \in (0,1)$, $t \in [0, 2\pi]$ and $N \in \mathbb{N}$:

$$\left| \sum_{n=0}^{N} a_n r^n e^{int} \right|^2 \le \left(\sum_{n=0}^{N} |a_n|^2 \right) \left(\sum_{n=0}^{N} r^{2n} \right) \le \frac{\left(\sum_{n=0}^{\infty} |a_n|^2 \right)}{1 - r^2}.$$

Furthermore, for every $t \in [0, 2\pi]$,

$$\lim_{n \to \infty} \left| \sum_{n=0}^{N} a_n r^n e^{int} \right|^2 = |f(re^{it})|^2.$$

The dominated convergence theorem insures that

$$\frac{1}{2\pi} \int_0^{2\pi} |f(re^{it})|^2 dt = \lim_{N \to \infty} \frac{1}{2\pi} \int_0^{2\pi} \left| \sum_{n=0}^{N} a_n r^n e^{int} \right|^2 dt$$

$$= \lim_{N \to \infty} \sum_{n=0}^{N} r^{2n} |a_n|^2$$

$$= \sum_{n=0}^{\infty} r^{2n} |a_n|^2.$$

The monotone convergence theorem implies then that

$$\lim_{r \uparrow 1} \int_0^{2\pi} |f(re^{it})|^2 dt = \sum_{n=0}^{\infty} |a_n|^2$$

where both sides will be simultaneously finite or infinite. We leave to the reader to check that

$$\lim_{r \uparrow 1} \int_0^{2\pi} |f(re^{it})|^2 dt = \lim_{r \to 1} \int_0^{2\pi} |f(re^{it})|^2 dt$$

since the first integral exists. $\square$

Solution of Exercise 8.2.3: This is an application of Exercise 7.6.12 with $\Omega = E$ and $K(z,w) = \frac{1}{1 - z\overline{w}}$. $\square$

Solution of Exercise 8.2.5:

(1) Using the reproducing kernel property at the point $re^{i\theta_0}$ for the function g defined by (8.2.5)

$$g(z) = \frac{h(z)}{z - e^{i\theta_0}},$$

we have

$$g(re^{i\theta_0}) = \frac{h(re^{i\theta_0})}{re^{i\theta_0} - e^{i\theta_0}} = \langle g, k_{re^{i\theta_0}} \rangle,$$

and so

$$\frac{|h(e^{i\theta_0})|}{|r-1|} \leq \|g\| \cdot \sqrt{\frac{1}{1-r^2}}.$$

It follows that

$$|h(re^{i\theta_0})| \leq \|g\| \cdot \sqrt{\frac{1}{1-r^2}} \cdot (1-r) = \sqrt{1-r} \cdot \frac{\|g\|}{\sqrt{1+r}},$$

and hence the result.

(2) We prove (8.2.7) for $u = 0$ for simplicity. The argument is similar to the one in (1). We now set

$$g(z) = \frac{h(z)}{\prod_{v=0}^{n}(z - e^{i\theta_v})},$$

and have, for $r \in [0,1)$,

$$g(re^{i\theta_0}) = \frac{h(re^{i\theta_0})}{(re^{i\theta_0} - e^{i\theta_0})\prod_{v=1}^{n}(re^{i\theta_0} - e^{i\theta_v})} = \langle g, k_{re^{i\theta_0}} \rangle,$$

and so

$$\frac{|h(re^{i\theta_0})|}{|r-1|} \leq \|g\| \cdot \sqrt{\frac{1}{1-r^2}} \cdot 2^n.$$

The result follows as in (1). $\qquad\square$

Solution of Exercise 8.3.1:

(1) Let $f \in \mathbf{H}_2(\mathbb{D})$, with power series expansion $f(z) = \sum_{n=0}^{\infty} a_n z^n$. We have $(M_z f)(z) = \sum_{n=0}^{\infty} a_n z^{n+1}$ and hence

$$\|M_z f\|^2 = \sum_{n=0}^{\infty} |a_n|^2 = \|f\|^2.$$

Formula (7.8.17) gives (with $k_w(z) = \frac{1}{1-z\overline{w}}$ and R_0 defined by (1.7.3))

$$(M_z^* k_w)(z) = \overline{w} \cdot \frac{1}{1-z\overline{w}} = (R_0 k_w)(z),$$

and so $M_z^* = R_0$. This can also be checked by direct computation.

(2) More generally, let $a \in \mathbb{D}$. We have for $z \in \mathbb{D}$:

$$\frac{z-a}{1-z\overline{a}} = -a + \left(\frac{z-a}{1-z\overline{a}} + a\right)$$

$$= -a + \frac{z(1-|a|^2)}{1-z\overline{a}}$$

$$= -a + (1-|a|^2)\left(\sum_{n=0}^{\infty} z^{n+1}\overline{a}^n\right).$$

So

$$\langle b_a, b_a \rangle = |a|^2 + (1 - |a|^2)^2 \left(\sum_{n=0}^{\infty} |a|^{2n} \right)$$

$$= 1 + (1 - |a|^2)^2 \frac{1}{1 - |a|^2}$$

$$= 1,$$

and

$$\langle M_z b_a, b_a \rangle = -\overline{a} \cdot 0 + (1 - |a|^2) \cdot (-\overline{a}) + (1 - |a|^2)^2 \left(\sum_{n=0}^{\infty} \overline{a} |a|^{2n} \right)$$

$$= (1 - |a|^2) \left(-\overline{a} + (1 - |a|^2) \frac{\overline{a}}{1 - |a|^2} \right)$$

$$= 0.$$

Since M_z is an isometry it follows that, for $n, m \in \mathbb{N}_0$,

$$\langle M_z^n b_a, M_z^m b_a \rangle = \delta_{n,m}$$

where $\delta_{n,m}$ denotes Kronecker's symbol. It follows that for any polynomial p we have

$$\langle b_a p, b_a p \rangle = \langle p, p \rangle.$$

Since the polynomials are dense in $\mathbf{H}_2(\mathbb{D})$ the operator M_{b_a} extends to a uniquely defined isometry from $\mathbf{H}_2(\mathbb{D})$ into itself. Its adjoint can be computed on the kernels k_w using formula (7.8.17):

$$(M_{b_a}^* k_w)(z) = \frac{\overline{b_a(w)}}{1 - z\overline{w}}.$$

(3) One can proceed as above and view M_b as the composition $M_{b_1} M_{b_2} \cdots M_{b_N}$. One can also proceed as follows; consider the partial fraction expansion of b:

$$b(z) = c_0 + \sum_{n=1}^{N} \frac{c_n}{1 - z\overline{w}_n} \tag{8.8.1}$$

where, for simplicity, we assumed the zeros of b simple and different from 0 and obtain (8.3.2)

$$(M_b^* f)(z) = \overline{c}_0 f(z) + \sum_{n=1}^{N} \overline{c}_n \frac{z f(z) - w_n f(w_n)}{z - w_n},$$

for $f \in \mathbf{H}_2(\mathbb{D})$. This formula can easily be checked on monomials $f(z)$ as follows, and then is valid for all $f \in \mathbf{H}_2(\mathbb{D})$ by continuity. Let thus $w \in \mathbb{D}$ and $M \in \mathbb{N}$. We have (with some abuse of notation)

$$\langle M_b^* z^M, k_w \rangle = \langle z^M, b k_w \rangle$$

$$= \left\langle z^M, \left(c_0 + \sum_{n=1}^{N} \frac{c_n}{1 - z\overline{w_n}} \right) \frac{1}{1 - z\overline{w}} \right\rangle$$

$$= \overline{c_0} w^M + \sum_{n=1}^{N} \overline{c_n} \left\langle z^M, \frac{1}{z(\overline{w_n} - \overline{w})} \left(\frac{1}{1 - z\overline{w_n}} - \frac{1}{1 - z\overline{w}} \right) \right\rangle$$

$$= \overline{c_0} w^M + \sum_{n=1}^{N} \overline{c_n} \left\langle z^{M+1}, \frac{1}{(\overline{w_n} - \overline{w})} \left(\frac{1}{1 - z\overline{w_n}} - \frac{1}{1 - z\overline{w}} \right) \right\rangle$$

(where we have used the fact that M_z is an isometry in $\mathbf{H}_2(\mathbb{D})$)

$$= \overline{c_0} w^M + \sum_{n=1}^{N} \overline{c_n} \frac{w_n^{M+1} - w^{M+1}}{w_n - w},$$

which is exactly (8.3.2) for $f(z) = z^M$. $\qquad\square$

Remarks 8.8.1.

(1) Formula (8.3.2) is in fact valid for any function b of the form (8.8.1), not necessarily a Blaschke product.

(2) One can also check formula (8.3.2) for $b(z) = \frac{1}{1 - z\overline{w_n}}$ and $f(z) = \frac{1}{1 - z\overline{u}}$ (of course, this b is not a Blaschke product, but part of the expansion (8.8.1)). One then obtains

$$(M_b^* f)(z) = \frac{1}{1 - \overline{u}w_n} \cdot \frac{1}{1 - z\overline{u}}$$

on the one hand and

$$\frac{zf(z) - w_n f(w_n)}{z - w_n} = \frac{\dfrac{z}{1 - z\overline{u}} - \dfrac{w_n}{1 - w_n\overline{u}}}{z - w_n} = \frac{\dfrac{z - zw_n\overline{u} - w_n + z\overline{u}w_n}{(1 - z\overline{u})(1 - w_n\overline{u})}}{z - w_n}$$

$$= \frac{1}{1 - \overline{u}w_n} \cdot \frac{1}{1 - z\overline{u}}$$

on the other hand.

Solution of Exercise 8.3.3: The spectra of M_z and M_z^* are inside the closed unit disk. We show that in fact they consist of the whole closed unit disk in both cases.

(1) The operator M_z: The forward shift operator has no point spectrum since, for any $\lambda \in \mathbb{C}$, the equation

$$zf(z) = \lambda f(z)$$

has only solution $f \equiv 0$ in the Hardy space.

8.8. *Solutions of the exercises*

Let $\lambda \in \mathbb{D}$ and
$$k_\lambda(z) = \frac{1}{1 - z\overline{\lambda}}. \tag{8.8.2}$$

By the reproducing kernel property
$$\langle (M_z - \lambda)f, k_\lambda \rangle_{\mathbf{H}_2(\mathbb{D})} = 0,$$

and so $\mathbb{D}$ is made of points in the residual spectrum. We now show that the unit circle $\mathbb{T}$ consists of points in the essential spectrum. Let $\lambda \in \mathbb{T}$, and let $g \in \mathbf{H}_2(\mathbb{D})$ be such that
$$\langle (M_z - \lambda)f, g \rangle_{\mathbf{H}_2(\mathbb{D})} = 0, \quad \forall f \in \mathbf{H}_2(\mathbb{D}).$$

The choice $f(z) = z^n$ leads to
$$\lambda g_n = g_{n+1}, \quad n = 0, 1, \dots \text{ where } g(z) = \sum_{n=0}^\infty g_n z^n.$$

Thus $g(z) = \frac{g_0}{1 - z\overline{\lambda}}$, which will not belong to the Hardy space, unless $g_0 = 0$. Since
$$\sigma(M_z) = \sigma_p(M_z) \cup \sigma_e(M_z) \cup \sigma_r(M_z),$$

we have:
$$\sigma_p(M_z) = \emptyset, \quad \sigma_e(M_z) = \mathbb{T}, \quad \text{and} \quad \sigma_r(M_z) = \mathbb{D}. \tag{8.8.3}$$

(2) The operator R_0: With k_λ as in (8.8.2) one readily sees that
$$R_0 k_\lambda = \overline{\lambda} k_\lambda, \tag{8.8.4}$$

and so the open unit disk consists of eigenvalues. That the unit disk cannot include part of the point spectrum is clear since the solution of the equation $R_\lambda f = \lambda f$ are multiples of k_λ. Since the spectrum is closed we have that $\mathbb{T}$ belongs to either the residual or the essential spectrum of R_0. We show that for every $\lambda \in \mathbb{T}$, the range ran$(R_0 - \lambda I)$ is dense. This will show that $\mathbb{T}$ consists of points in the essential spectrum (we do not need to show that the above range is not closed since we already know that λ is a spectrum point). Let g be orthogonal to ran$(R_0 - \lambda I)$. In particular
$$\langle (R_0 - \lambda I)k_w, g \rangle = 0, \quad \forall w \in \mathbb{D}.$$

In view of (8.8.4) we have $(w - \lambda)g(w) = 0$ and so $g \equiv 0$.

Thus
$$\sigma_p(R_0) = \mathbb{D}, \quad \sigma_e(R_0) = \mathbb{T}, \quad \text{and} \quad \sigma_r(R_0) = \emptyset. \tag{8.8.5}$$
$\square$

Remark 8.8.2. It is interesting to compare (8.8.3) and (8.8.5), taking into account that
$$\sigma_r(T) \subset \sigma_p(T^*),$$
$$\sigma_p(T) \subset \sigma_p(T^*) \cup \sigma_r(T^*)$$

for every bounded operator acting in a Hilbert space. See [257, Proposition, p. 194] for the latter.

Solution of Exercise 8.3.4:

(1) The operator norm of M_z is equal to 1 and so $\|\operatorname{Re} M_z\| \leq 1$ (see Question 4.2.21) and the spectrum (which is real since $\operatorname{Re} M_z$ is self-adjoint) is inside $[-1, 1]$. We now show that all points in $[-1, 1]$ are in the spectrum of $\operatorname{Re} M_z$. Let $\lambda \in [-1, 1]$ and $u \in \mathbf{H}_2(\mathbb{D})$, and consider the equation

$$\left(\frac{M_z + R_0}{2} - \lambda I \right) f = u,$$

where the unknown $f \in \mathbf{H}_2(\mathbb{D})$. Then,

$$f(z) = \frac{2zu(z) + f(0)}{z^2 - 2\lambda z + 1}.$$

Set $\lambda = \cos\theta$ for some real number θ. Then

$$f(z) = \frac{2zu(z) + f(0)}{(z - e^{i\theta})(z - e^{-i\theta})}, \quad z \in \mathbb{D}.$$

By Exercise 8.2.5 applied to $h(z) = 2zu(z) + f(0)$ and $\theta_0 = -\theta_1 = \theta$ we have

$$2e^{i\theta} \cdot \lim_{r \to 1} u(re^{i\theta}) = 2e^{-i\theta} \cdot \lim_{r \to 1} u(re^{-i\theta}) = -f(0).$$

But it is easy to find functions for which this condition will not hold. For instance $u(z) \equiv 1$ and $f(0) = -2e^{i\theta}$ leads to

$$f(z) = \frac{2}{z - e^{i\theta}}, \quad z \in \mathbb{D},$$

which does not belong to $\mathbf{H}_2(\mathbb{D})$.

(2) Let λ be an eigenvalue of $\operatorname{Re} M_z$, with corresponding eigenvector f. Then,

$$\frac{f(z) - f(0) + z^2 f(z)}{z} = 2\lambda f(z),$$

where the left-hand side is equal to $f'(0)$ when $z = 0$. Thus,

$$f(z) = \frac{f(0)}{z^2 - 2z\lambda + 1}, \quad z \in \mathbb{D}.$$

In view of (1), we have $\lambda \in [-1, 1]$. Set as above $\lambda = \cos\theta$ for some real number θ. Then

$$f(z) = \frac{f(0)}{(z - e^{i\theta})(z - e^{-i\theta})}, \quad z \in \mathbb{D},$$

which does not belong to $\mathbf{H}_2(\mathbb{D})$ unless $f(0) = 0$ since the function

$$z \mapsto \frac{1}{(z - e^{i\theta})(z - e^{-i\theta})}$$

does not belong to $\mathbf{H}_2(\mathbb{D})$. To verify this last point write

$$\frac{1}{(z - e^{i\theta})(z - e^{-i\theta})} = \frac{1}{(1-z)^2} = \sum_{n=0}^{\infty}(n+1)z^n, \quad \text{if } e^{i\theta} = 1, \qquad (8.8.6)$$

and

$$\frac{1}{(z - e^{i\theta})(z - e^{-i\theta})} = \frac{1}{2i\sin\theta}\left(\frac{1}{e^{-i\theta} - z} - \frac{1}{e^{i\theta} - z}\right)$$

$$= \frac{1}{2i\sin\theta}\left(\sum_{n=0}^{\infty}e^{i(n+1)\theta}z^n - \sum_{n=0}^{\infty}e^{-i(n+1)\theta}z^n\right)$$

$$= \sum_{n=0}^{\infty}\frac{\sin(n+1)\theta}{\sin\theta}z^n$$

otherwise. In either case, the sequence of Taylor coefficients does not belong to ℓ_2.
(3) We now show that $\operatorname{Re} M_z$ has no residual spectrum. Let $\lambda \in [-1,1]$ and let
$u \in \mathbf{H}_2(\mathbb{D})$ be orthogonal to the range of $\operatorname{Re} M_z - \lambda I$. Then, for all $f \in \mathbf{H}_2(\mathbb{D})$ we
have

$$\langle u, (\operatorname{Re} M_z - \lambda I)f\rangle_{\mathbf{H}_2(\mathbb{D})} = 0.$$

The choice $f(z) = z^n$ leads to (with $u(z) = \sum_{n=0}^{\infty} u_n z^n$ and $u_{-1} = 0$)

$$u_{n+1} + u_{n-1} = \lambda u_n, \quad n = 0, 1, \ldots$$

It follows that

$$u(z) = \frac{u_0 + zu_1 - \lambda zu_0}{z^2 - \lambda z + 1},$$

but such a function does not belong to $\mathbf{H}_2(\mathbb{D})$ (unless $u_0 = u_1 = 0$) since the
product of the zeros of the polynomial $z^2 - \lambda z + 1$ is equal to 1. $\square$

Solution of Exercise 8.3.6:
(1) We first note that

$$\begin{pmatrix} k_{w_1}(z) & k_{w_2}(z) & \cdots & k_{w_N}(z) \end{pmatrix} = C(I_N - zA)^{-1},$$

with C and A as in the hint, that is,

$$C = \underbrace{\begin{pmatrix} 1 & 1 & \cdots & 1 \end{pmatrix}}_{N \text{ times}} \quad \text{and} \quad A = \operatorname{diag}\left(\overline{w_1}, \overline{w_2}, \ldots, \overline{w_N}\right).$$

Furthermore, the Gram matrix of the space $\mathcal{M}$ with respect to the $\mathbf{H}_2(\mathbb{D})$ inner
product is equal to

$$P_{\ell j} = \langle k_{w_j}, k_{w_\ell}\rangle = \frac{1}{1 - w_\ell\overline{w_j}}, \quad \ell, j = 1, \ldots, N.$$

Since for $m \in \mathbb{N}_0$,

$$A^{*m}C = \begin{pmatrix} w_1^m \\ w_2^m \\ \vdots \\ w_N^m \end{pmatrix},$$

we have that P is given by

$$P = \sum_{m=0}^{\infty} A^{*m}C^*CA^m,$$

(see Exercise 7.7.19), and so is the (unique) solution of (7.7.31). Formula (7.7.30) with Θ given by (7.7.29) gives then the reproducing kernel of $\mathcal{M}$. In the present case, Θ has no poles in the closed unit disk. Since it is rational and unitary on the unit circle, it is a finite Blaschke product, see [CAPB, Exercise 11.5.3, p. 438]. Since the points w_n are assumed different, there is at most one n for which $w_n = 0$. The finite poles of Θ are at the points $1/\overline{w_n}$ for those n for which $w_n \neq 0$. Hence Θ is equal, up to a multiplicative unitary constant, to the Blaschke product given in the statement of the exercise.

(2) Let $f \in \mathbf{H}_2(\mathbb{D})$. We have

$$f \in \mathcal{M}^\perp \iff \langle f, k_{w_n} \rangle = 0, \quad n = 1, \ldots, N$$
$$\iff f(w_n) = 0, \quad n = 1, \ldots, N$$
$$\iff f(z) = \left(\prod_{n=1}^{N} (z - w_n) \right) g(z),$$
$$\iff f(z) = \left(\prod_{n=1}^{N} \frac{z - w_n}{1 - z\overline{w_n}} \right) h(z)$$

where we use Exercise 2.1.18 and where h is analytic in $\mathbb{D}$,

$$\iff f(z) = b(z)h(z)$$

where h is analytic in $\mathbb{D}$.

To conclude we show that h is in $\mathbf{H}_2(\mathbb{D})$. Since the function f/b has only removable singularities inside the open unit disk, it is enough to show that

$$\sup_{\substack{r \in (0,1) \\ r > \max_{n=1,\ldots,N} |w_n|}} \int_0^{2\pi} \left| \frac{f(re^{it})}{b(re^{it})} \right|^2 dt < \infty.$$

This follows from the bound

$$\frac{1}{|b(z)|} \leq \prod_{n=1}^{N} \frac{1 + |w_n|}{r_0 - |w_n|},$$

valid for $|z| \geq r_0 > \max_{n=1,\ldots,N} |w_n|$.

8.8. *Solutions of the exercises*

(3) This follows from

$$(R_0 k_{w_n})(z) = \overline{w_n} k_{w_n}(z), \quad n = 1, 2, \ldots, N,$$

and from the resolvent identity (1.7.4) (see Question 4.2.10). $\qquad\square$

Remark 8.8.3. In fact it holds that

$$\sup_{\substack{r \in (0,1) \\ r > \max_{n=1,\ldots,N} |w_n|}} \int_0^{2\pi} \left| \frac{f(re^{it})}{b(re^{it})} \right|^2 dt = \sup_{\substack{r \in (0,1) \\ r > \max_{n=1,\ldots,N} |w_n|}} \int_0^{2\pi} |f(re^{it})|^2 dt.$$

This equality will follow once we know that $h \in \mathbf{H}_2(\mathbb{D})$. It could also be proved directly using for instance the dominated convergence theorem if we assume known the existence of non-tangential limits for f.

Solution of Exercise 8.3.7:

(1) Using Theorem 7.5.22, we see that the decomposition

$$\frac{1 - B(z)\overline{B(w)}}{1 - z\overline{w}} = \sum_{n=1}^N B_{n-1}(z) \frac{1 - b_{a_n}(z)\overline{b_{a_n}(w)}}{1 - z\overline{w}} \overline{B_{n-1}(w)}$$

of the positive definite function $\frac{1 - B(z)\overline{B(w)}}{1 - z\overline{w}}$ into a sum of N positive definite functions leads to the vector space decomposition

$$\mathbf{H}_2(\mathbb{D}) \ominus B\mathbf{H}_2(\mathbb{D}) = \sum_{n=1}^N B_{n-1}(\mathbf{H}_2(\mathbb{D}) \ominus b_{a_n}\mathbf{H}_2(\mathbb{D})). \tag{8.8.7}$$

To show that this sum is orthogonal, let $k, \ell \in \{1, \ldots, N\}$ and assume $k < \ell$. Recall that $\mathbf{H}_2(\mathbb{D}) \ominus b_{a_k}\mathbf{H}_2(\mathbb{D})$ is spanned by the function $\frac{1}{1-z\overline{a_k}}$ and similarly for $\mathbf{H}_2(\mathbb{D}) \ominus b_{a_\ell}\mathbf{H}_2(\mathbb{D})$. Let $F_k(z) = \frac{B_{k-1}(z)}{1-z\overline{a_k}}$ and $F_\ell(z) = \frac{B_{\ell-1}(z)}{1-z\overline{a_\ell}}$. Then (with the convention that $b_{a_1} \cdots b_{a_{k-1}} = 1$ if $k = 1$):

$$\langle F_k, F_\ell \rangle = \left\langle \frac{1}{1-z\overline{a_k}}, \frac{b_{a_k}(z) \cdots b_{a_{\ell-1}}(z)}{1-z\overline{a_\ell}} \right\rangle = 0,$$

where we have used that M_{b_a} is an isometry from $\mathbf{H}_2(\mathbb{D})$ into itself and Cauchy's formula (or, equivalently, the reproducing kernel property).

(2) Since $\|F_n\|^2 = \frac{1}{1-|a_n|^2}$ the functions $\frac{1}{\sqrt{1-|a_n|^2}} F_n$, $n = 1, \ldots, N$ form an orthonormal basis of $\mathbf{H}_2(\mathbb{D}) \ominus B\mathbf{H}_2(\mathbb{D})$. $\qquad\square$

Solution of Exercise 8.3.9: We follow the hint given after the exercise, and divide the proof in a number of steps. We set $\mathcal{M} = \mathcal{N}^\perp$.

STEP 1: *The space $\mathcal{M}$ is R_0-invariant.*

This is because R_0 is the adjoint of M_z in the Hardy space $\mathbf{H}_2(\mathbb{D})$.

STEP 2: *There exists an observable pair of matrices* $(C, A) \in \mathbb{C}^{1 \times N} \times \mathbb{C}^{N \times N}$ *such that the entries of the row function* $F(z) = C(I_N - zA)^{-1}$ *form a basis* $f_1, \ldots, f_N$ *of* $\mathcal{M}$.

Indeed, let $f_1, \ldots, f_N$ be a basis of $\mathcal{M}$. Since $\mathcal{M}$ is R_0-invariant, for every $j \in \{1, \ldots, n\}$ there exists a vector $a_j \in \mathbb{C}^N$ such that $R_0 f_j = F a_j$ and so $R_0 F = FA$ with $A = \begin{pmatrix} a_1 & \cdots & a_N \end{pmatrix}$. Writing

$$\frac{F(z) - F(0)}{z} = F(z)A$$

for $z \neq 0$ we get the required form for F with $C = F(0)$.

STEP 3: *The spectrum of* A *is inside the open unit disk.*

Indeed, we have $R_0 F = FA$. Let now η be an eigenvector of A, corresponding to the eigenvalue λ. Then,

$$R_0 F \eta = F A \eta = \lambda F \eta,$$

and so $F\eta$ is an eigenvector of R_0. From $R_0 F \eta = \lambda F \eta$ follows that

$$F(z)\eta = \frac{F(0)\eta}{1 - \lambda z}. \tag{8.8.8}$$

We note that $F(0)\eta \neq 0$. Otherwise (8.8.8) implies $F(z)\eta \equiv 0$, and so $\eta = 0$. But (8.8.8) belongs to $\mathbf{H}_2(\mathbb{D})$ if and only if $\lambda \in \mathbb{D}$.

STEP 4: *Let* P *be the Gram matrix of the basis* $f_1, \ldots, f_N$, *and let*

$$b(z) = 1 - (1 - z)C(I_N - zA)^{-1}P^{-1}(I_N - A)^{-*}C^*.$$

We show that P is the unique solution of the equation

$$P - A^*PA = C^*C, \tag{8.8.9}$$

and that

$$F(z)P^{-1}F(w)^* = \frac{1 - b(z)\overline{b(w)}}{1 - z\overline{w}}, \quad z, w \in \mathbb{D}. \tag{8.8.10}$$

An element of $\mathcal{M}$ can be written as $F\eta$ for a unique $\eta \in \mathbb{C}^N$. Let now ξ be another element of $\mathbb{C}^N$. Since

$$F(z)\eta = \sum_{n=0}^{\infty}(CA^n\eta)z^n \quad \text{and} \quad F(z)\xi = \sum_{n=0}^{\infty}(CA^n\xi)z^n,$$

we have by definition of the inner product in $\mathbf{H}_2(\mathbb{D})$ (see (8.2.2))

$$\langle F(\cdot)\eta, F(\cdot)\xi \rangle_{\mathbf{H}_2(\mathbb{D})} = \sum_{n=0}^{\infty}(CA^n\xi)^*(CA^n\eta)$$

$$= \xi^*\left(\sum_{n=0}^{\infty}A^{*n}C^*CA^n\right)\eta$$

8.8. Solutions of the exercises

and so the Gram matrix is equal to

$$\sum_{n=0}^{\infty} A^{*n} C^* C A^n, \tag{8.8.11}$$

and equation (8.8.9) has a unique solution, given by this series, since $\sigma(A) \subset \mathbb{D}$.

Finally, we note that (8.8.10) is a special case of formula (7.7.30). See Exercise 7.7.17. (It is also another example of the formula (7.7.1) for a finite-dimensional reproducing kernel Hilbert space.)

STEP 5: *Show that* $\mathcal{M}$ *is the span of the functions* $z \mapsto \frac{1-b(z)\overline{b(w)}}{1-z\overline{w}}$ *when* w *runs through* $\mathbb{D}$.

This comes from (8.8.10) since the span of the vectors $P^{-1} F(w)^* \eta$ when η runs through $\mathbb{C}^N$ and w runs through $\mathbb{D}$ is all of $\mathbb{C}^N$. To check this last point take $\xi \in \mathbb{C}^N$ such that

$$\xi^* P^{-1} F(w)^* \eta = 0, \quad \forall \eta \in \mathbb{C}^N \text{ and } w \in \mathbb{D}.$$

Then,

$$F(w) P^{-1} \xi \equiv 0$$

and this implies $\xi = 0$ since the entries of F form a basis of $\mathcal{M}$.

STEP 6: *It holds that*

$$\mathbf{H}_2(\mathbb{D}) = (M_b \mathbf{H}_2(\mathbb{D})) \oplus ((I - M_b M_b^*) \mathbf{H}_2(\mathbb{D})).$$

Since multiplication by b is an isometry from $\mathbf{H}_2(\mathbb{D})$ into itself we have that the sum is indeed orthogonal. To see that it is equal to all of $\mathbf{H}_2(\mathbb{D})$, note that ran $M_b M_b^* \subset$ ran M_b (and that in fact there is equality).

STEP 7: *Show that* $\mathcal{N} = b\mathbf{H}_2(\mathbb{D})$.

This is because

$$\left((I - M_b M_b^*) \left(\frac{1}{1 - \zeta \overline{w}} \right) \right) (z) = \frac{1 - b(z)\overline{b(w)}}{1 - z\overline{w}}. \qquad \square$$

Solution of Exercise 8.3.11:

(1) Let f_1 and f_2 be in $\mathcal{N}$, both not identically equal to 0. We assume that $f_1(0) = f_2(0)$. Assume that there is $n_0 > 0$ and $a_{n_0} \neq 0$ such that

$$f_1(z) - f_2(z) = a_{n_0} z^{n_0} + \cdots.$$

For any $g \in \mathcal{M}$,

$$\langle f_1 - f_2, M_{z^{n_0}} g \rangle = 0.$$

By definition of the inner product, this gives $a_{n_0} + a_{n_0+1} z + \cdots \equiv 0$, and so $a_{n_0} = 0$, contradicting our assumption. Thus $f_1 = f_2$.

(2) From (1) we have

$$\mathcal{M} = \left(\oplus_{k=1}^{N} z^k \mathcal{N}\right) \oplus z^{N+1} \mathcal{M}, \quad \forall N \in \mathbb{N}.$$

It follows that an element of $\mathcal{M} \ominus \left(\oplus_{k=1}^{N} z^k \mathcal{N}\right)$ will be in $\cap_{N=1}^{\infty} z^{N+1} \mathcal{M}$, and hence equal to 0.

(3) Let j be a function spanning $\mathcal{N}$. It follows from (2) that any $m \in \mathcal{N}$ can be written as

$$m(z) = \left(\sum_{n=0}^{\infty} a_n z^n\right) j(z),$$

and that

$$\|m\|^2 = \left(\sum_{n=0}^{\infty} |a_n|^2\right) \|j\|^2. \qquad \square$$

Solution of Exercise 8.3.14:

(1) From $T M_z = M_z T$ we get that $T M_{z^m} = M_{z^m} T$ for all $m \in \mathbb{N}$. Applying to $f = 1$ we get that, for every $w \in \mathbb{D}$,

$$(Tp)(w) = p(w)\left((T1)(w)\right),$$

for $p(z) = z^m$. The result follows by linearity.

(2) Let $f \in \mathbf{H}_2(\mathbb{D})$, with power series expansion $f(z) = \sum_{m=0}^{\infty} a_m z^m$, and let $p_n(z) = \sum_{m=0}^{n} a_m z^m$. Then

$$\|f - p_n\|_{\mathbf{H}_2(\mathbb{D})}^2 = \sum_{m=n+1}^{\infty} |a_m|^2 \longrightarrow 0, \quad \text{as } n \to \infty.$$

(3) Since T is continuous we have $\lim_{n\to\infty} \|T p_n - T f\|_{\mathbf{H}_2(\mathbb{D})} = 0$. But in a reproducing kernel Hilbert space, convergence in norm implies pointwise convergence, and this implies (8.3.7).

(4) From (8.3.7) we have

$$(Tf)(z) = \lim_{n\to\infty} (Tp_n)(z) = \lim_{n\to\infty} \psi(z) p_n(z) = \psi(z) f(z). \qquad (8.8.12)$$

(5) Since T is bounded by hypothesis, equation (8.8.12) implies that ψ is a multiplier, and so is bounded. To check this directly, recall that (with $k_w(z) = \frac{1}{1 - z\overline{w}}$)

$$(T^* k_w) = \overline{\psi(w)} k_w, \quad w \in \mathbb{D}.$$

Since there exists a $M < \infty$ such that $\|T^* k_w\|_{\mathbf{H}_2(\mathbb{D})} \leq M \|k_w\|_{\mathbf{H}_2(\mathbb{D})}$ we get

$$\frac{|\psi(w)|}{\sqrt{1 - |w|^2}} \leq \frac{M}{\sqrt{1 - |w|^2}}, \quad w \in \mathbb{D},$$

and so ψ is bounded in absolute value by M.

$\square$

Solution of Exercise 8.4.1: By Schwarz' lemma we can write $s(z) = z\sigma(z)$ where σ is analytic and bounded by one in modulus in the open unit disk. We write

$$\frac{1}{1-z\overline{w}} - \frac{1}{1-s(z)\overline{s(w)}} = \frac{1}{1-z\overline{w}} - \frac{1}{1-z\sigma(z)\overline{w}\overline{\sigma(w)}}$$

$$= \frac{z\overline{w}(1-\sigma(z)\overline{\sigma(w)})}{(1-z\overline{w})(1-z\sigma(z)\overline{w}\overline{\sigma(w)})}$$

$$= \frac{z\overline{w}}{1-z\sigma(z)\overline{w}\overline{\sigma(w)}} \frac{1-\sigma(z)\overline{\sigma(w)}}{(1-z\overline{w})}.$$

This formula expresses the kernel

$$\frac{1}{1-z\overline{w}} - \frac{1}{1-s(z)\overline{s(w)}}$$

as a product of two positive definite kernels in the open unit disk, and the claim follows from Exercise 7.6.6. □

Solution of Exercise 8.4.2:

(1) The function

$$(f_0(z))^{-1} = \frac{1-z\overline{w_0}}{1-s(z)\overline{s(w_0)}}$$

is analytic and bounded in the open unit disk. More precisely

$$|f_0(z)|^{-1} \le \frac{1+|w_0|}{1-|s(w_0)|}, \quad \forall z \in \mathbb{D}.$$

By Exercise 7.6.12 we can take $k_0 = \|f_0^{-1}\|_\infty^2$.

(2) The function f_0 belongs to the reproducing kernel Hilbert space $\mathcal{H}(s)$, with norm

$$\|f_0\|_{\mathcal{H}(s)} = \sqrt{\frac{1-|s(w_0)|^2}{1-|w_0|^2}} \overset{\text{def.}}{=} \sqrt{k_1}.$$

By Exercise 7.6.6 (or by Exercise 7.6.12) the kernel

$$\frac{1-s(z)\overline{s(w)}}{1-z\overline{w}} - \frac{f_0(z)\overline{f_0(w)}}{k_1} \tag{8.8.13}$$

is positive definite in $\mathbb{D}$. Since the kernel

$$\frac{k_1}{f_0(z)\overline{f_0(w)}} \frac{1}{1-s(z)\overline{s(w)}} \tag{8.8.14}$$

is positive definite in $\mathbb{D}$ the result follows by multiplying the kernels (8.8.13) and (8.8.14).

(3) We will use Exercise 7.6.6 and compute:

$$\frac{1}{1-z\overline{w}} - \frac{\frac{1}{k_0 k_1}}{1-s(z)\overline{s(w)}} = \frac{1-\frac{1}{k_0 f_0(z)\overline{f_0(w)}}}{1-z\overline{w}}$$

$$+ \frac{\frac{1}{k_0 f_0(z)\overline{f_0(w)}}}{1-z\overline{w}} - \frac{\frac{1}{k_0 k_1}}{1-s(z)\overline{s(w)}}$$

$$= \frac{1-\frac{1}{k_0 f_0(z)\overline{f_0(w)}}}{1-z\overline{w}}$$

$$+ \frac{1}{k_0 k_1}\left\{\frac{k_1}{f_0(z)\overline{f_0(w)}}\frac{1}{1-z\overline{w}} - \frac{1}{1-s(z)\overline{s(w)}}\right\},$$

which is positive definite in $\mathbb{D}$ since the kernels (8.4.1) and (8.4.2) are positive definite there.

(4) By formula (7.6.5) (or by direct inspection), we have

$$A(z,w) = \frac{1}{1-z\overline{s(w)}}. \qquad \square$$

Remark 8.8.4. We note that one gets the following bound on the operator norm of C_s:

$$\|C_s\| \le \sqrt{k_0 k_1},$$

and so

$$\|C_s\|^2 \le \inf_{\omega_0 \in \mathbb{D}}\left\{\sup_{z \in \mathbb{D}}\left|\frac{1-s(z)\overline{s(\omega_0)}}{1-z\overline{\omega_0}}\right|^2 \cdot \frac{1-|s(\omega_0)|^2}{1-|\omega_0|^2}\right\}.$$

Solution of Exercise 8.5.2:

(1) It suffices to write

$$\frac{1}{1-b(z)\overline{b(w)}} = \sum_{n=0}^{\infty}(b(z))^n(\overline{b(w)})^n, \quad z,w \in \mathbb{D}.$$

For $f \in \mathbf{H}_2(\mathbb{D})$ (or in fact for f analytic in $\mathbb{D}$) we have:

$$f(b(z)) \equiv 0 \implies f \equiv 0.$$

It follows that the space $\mathbf{H}_2(b)$ of functions of the form $F = f(b)$ with $f \in \mathbf{H}_2(\mathbb{D})$ and norm

$$\|F\|_{\mathbf{H}_2(b)} = \|f\|_{\mathbf{H}_2(\mathbb{D})}$$

is a Hilbert space. The function $k_{b,w}$

$$k_{b,w}(z) = \frac{1}{1-b(z)\overline{b(w)}}$$

belongs to $\mathbf{H}_2(b)$ for every $w \in \mathbb{D}$ and we have (with $k_w(z) = \frac{1}{1-z\overline{w}}$ and $\langle \cdot, \cdot \rangle_2$ denoting the inner product in $\mathbf{H}_2(\mathbb{D})$),

$$\langle F, k_{b,w} \rangle_{\mathbf{H}_2(b)} = \langle f, k_{b(w)} \rangle_2 = f(b(w)) = F(w).$$

(2) Let $e_1, \ldots, e_N$ be an orthonormal basis of the reproducing kernel Hilbert space with reproducing kernel $\frac{1-b(z)\overline{b(w)}}{1-z\overline{w}}$ (that is, of $\mathbf{H}_2(\mathbb{D}) \ominus b\mathbf{H}_2(\mathbb{D})$). We have (see Exercise 7.7.1 and formula (7.7.2) if need be)

$$\frac{1-b(z)\overline{b(w)}}{1-z\overline{w}} = \sum_{n=1}^{N} e_n(z)\overline{e(w)},$$

and thus

$$\frac{1}{1-z\overline{w}} = \sum_{n=1}^{N} \frac{e_n(z)\overline{e(w)}}{1-b(z)\overline{b(w)}}.$$

It follows from Theorem 7.5.22 that

$$\mathbf{H}_2(\mathbb{D}) = \sum_{n=1}^{N} e_n \mathbf{H}_2(b).$$

To show that the sum is orthogonal, it is enough to verify that

$$\langle e_j b^n, e_k b^m \rangle_2 = \delta_{n,m} \delta_{j,k}.$$

This is done as follows. The result is clear when $n = m$ and $j = k$. Assume now $n > m$. By properties of the inner product we have

$$\langle e_j b^n, e_k b^m \rangle_2 = \langle e_j b^{n-m}, e_k \rangle_2 = 0,$$

since $e_j b^{n-m} \in b\mathbf{H}_2(\mathbb{D})$ and $e_k \in \mathbf{H}_2(\mathbb{D}) \ominus b\mathbf{H}_2(\mathbb{D})$. The other cases are treated in much the same way. $\square$

Solution of Exercise 8.6.1:

(1) We note that

$$\left|\frac{u-i}{u+i}\right| = 1, \quad u \in \mathbb{R},$$

so that

$$\langle f_n, f_m \rangle_2 = \frac{1}{\pi} \int_{\mathbb{R}} \frac{1}{(u+i)(u-i)} \left(\frac{u-i}{u+i}\right)^{n-m} du,$$

and in particular

$$\langle f_n, f_n \rangle_2 = \frac{1}{\pi} \int_{\mathbb{R}} \frac{du}{u^2+1} = 1.$$

We now assume $n > m$. Then,

$$\langle f_n, f_m \rangle_2 = \frac{1}{\pi} \int_{\mathbb{R}} \frac{(u-i)^{n-m-1}}{(u+i)^{n-m+1}} du.$$

We compute this integral using Cauchy's theorem. For $R > 0$, consider the contour γ_R made of the concatenation of the interval $[-R, R]$ and of the part C_R^+ of the circle C_R (centered at the origin and with radius R) in the open upper half-plane. By Cauchy's theorem,

$$\int_{\gamma_R} \frac{(z-i)^{n-m-1}}{(z+i)^{n-m+1}} dz = 0.$$

For $R > 1$ and $z \in \mathbb{C}_+$ we have:

$$\left| \frac{(z-i)^{n-m-1}}{(z+i)^{n-m+1}} \right| \leq \frac{(R+1)^{n-m-1}}{(R-1)^{n-m+1}},$$

and so we have

$$\left| \int_{C_R^+} \frac{(z-i)^{n-m-1}}{(z+i)^{n-m+1}} dz \right| \leq \pi R \frac{(R+1)^{n-m-1}}{(R-1)^{n-m+1}} \to 0$$

as R goes to infinity. It follows that

$$\lim_{R \to \infty} \int_{[-R,R]} \frac{(z-i)^{n-m-1}}{(z+i)^{n-m+1}} dz = 0,$$

and hence the result. The case $n < m$ follows since $\langle f_n, f_m \rangle_2 = \overline{\langle f_m, f_n \rangle_2}$. We now turn to the proof that the f_n form an orthonormal basis of $\mathbf{L}_2(\mathbb{R}, \mathcal{B}, dx)$. Let $f \in \mathbf{L}_2(\mathbb{R}, \mathcal{B}, dx)$ be such that $\langle f, f_n \rangle_2 = 0$ for all $n \in \mathbb{Z}$, and let $\epsilon \in \mathbb{D}$. The dominated convergence theorem allows us to interchange integral and sum in the following equation:

$$\sum_{n=0}^{\infty} \epsilon^n \int_{\mathbb{R}} \frac{f(u)}{u-i} \left(\frac{u+i}{u-i} \right)^n du = \int_{\mathbb{R}} \frac{f(u)}{u-i} \left(\sum_{n=0}^{\infty} \epsilon^n \left(\frac{u+i}{u-i} \right)^n \right) du$$

$$= \int_{\mathbb{R}} \frac{f(u)}{u-i} \frac{1}{1 - \epsilon \frac{u+i}{u-i}} du$$

$$= \int_{\mathbb{R}} \frac{f(u)}{u-i - \epsilon(u+i)} du$$

$$= \frac{1}{1-\epsilon} \int_{\mathbb{R}} \frac{f(u)}{u-\omega} du,$$

with

$$\omega = i \frac{1+\epsilon}{1-\epsilon}.$$

Note that Im $\omega = \frac{1-|\epsilon|^2}{|1-\epsilon|^2}$, and so the orthogonality of f to $f_0, f_1, f_2, \ldots$ implies (6.2.10) for ω in the open upper half-plane. In the same way, the fact that $\langle f, f_n \rangle_2 = 0$ for $n = -1, -2, \ldots$ implies that (6.2.10) for ω in the open lower half-plane. Therefore, for any $\omega \in \mathbb{C}_+$,

$$\int_{\mathbb{R}} \frac{f(t)}{t-\omega} dt - \int_{\mathbb{R}} \frac{f(t)}{t-\overline{\omega}} dt = 0,$$

so that

$$\int_{\mathbb{R}} \frac{f(t)}{|t-\omega|^2} dt = 0, \quad \forall \omega \in \mathbb{C} \setminus \mathbb{R}.$$

Applying the Stieltjes–Perron inversion formula (see Proposition 6.2.7) we conclude that $f(t) = 0$ almost everywhere.

(2) For $z, w \in \mathbb{C}_+$ we can write:

$$\sum_{n=0}^{\infty} f_n(z)\overline{f_n(w)} = \frac{1}{\pi(z+i)(\overline{w}-i)} \frac{1}{1 - \dfrac{z-i}{z+i}\dfrac{\overline{w}+i}{\overline{w}-i}}$$

$$= \frac{1}{\pi((\overline{w}-i)(z+i) - (\overline{w}+i)(z-i))}$$

$$= \frac{1}{-2\pi i(z-\overline{w})},$$

which is the reproducing kernel of the space $\mathbf{H}_2(\mathbb{C}_+)$.

(3) A similar computation gives for $z, w \in \mathbb{C}_-$,

$$\sum_{n=1}^{\infty} f_{-n}(z)\overline{f_{-n}(w)} = \frac{1}{2\pi i(z-\overline{w})},$$

which is the of the reproducing kernel of the space $\mathbf{H}_2(\mathbb{C}_-)$. $\qquad\square$

Solution of Exercise 8.6.3: Following the hint we have for $z, w \in \mathbb{C}_+$,

$$\frac{1}{-2\pi i(z-\overline{w})} = \frac{1}{2\pi} \int_{(-\infty,0]} e^{-i(z-\overline{w})t} dt,$$

and so for $N \in \mathbb{N}$, $c_1, \ldots, c_N \in \mathbb{C}$ and $w_1, \ldots, w_N \in \mathbb{C}_+$ we have

$$F(z) = \sum_{n=1}^{N} \frac{c_n}{-2\pi i(z-\overline{w_n})} = \frac{1}{2\pi} \int_{(-\infty,0]} e^{-itz} \underbrace{\left(\sum_{n=1}^{N} c_n e^{i\overline{w_n}t}\right)}_{f(t)} dt.$$

We now use (6.1.4) and get

$$\|F\|_{\mathbf{H}_2^+}^2 = \sum_{n,m=1}^{N} \frac{c_n\overline{c_m}}{-2\pi i(w_m - \overline{w_n})} = 2\pi\|f\|_{\mathbf{L}_2(\mathbb{R}_-)}^2 = \|\widehat{f}\|_{\mathbf{L}_2(\mathbb{R})}^2.$$

We conclude by a density argument. Assume that there is $f \in \mathbf{L}_2(\mathbb{R}_+, dx)$ such that

$$\int_{(-\infty,0]} e^{-izt} f(t) dt = 0, \quad \forall z \in \mathbb{C}_+. \tag{8.8.15}$$

Let $z = s + ih$ for some fixed $h > 0$. Then, (8.8.15) becomes

$$\int_{(-\infty,0]} e^{-ist} e^{ht} f(t) dt = 0, \quad \forall s \in \mathbb{R}.$$

Taking conjugate we get

$$\int_{[0,\infty)} e^{-ist} e^{-ht} \overline{f(-t)} dt = 0, \quad \forall s \in \mathbb{R}.$$

Adding these two equalities we get that $f(t) \equiv 0$. $\qquad\square$

Solution of Exercise 8.6.8: Let f_n be defined by (8.6.2),

$$f_n(z) = \frac{1}{\sqrt{\pi}} \frac{1}{z+i} \left(\frac{z-i}{z+i} \right)^n, \quad n \in \mathbb{Z},$$

and denote by T the operator $Tf(z) = \frac{1}{z} f(-1/z)$. We have

$$
\begin{aligned}
Tf_n(z) &= \frac{1}{\sqrt{\pi}} \frac{1}{z} \frac{1}{(-\frac{1}{z}+i)} \left(\frac{-\frac{1}{z}-i}{-\frac{1}{z}+i} \right)^n \\
&= \frac{1}{\sqrt{\pi}} \frac{1}{i(z+i)} \left(\frac{iz+1}{iz-1} \right)^n (-1)^n \\
&= -i(-1)^n f_n(z).
\end{aligned}
$$

Thus, we have that $Tf \in \mathbf{L}_2(\mathbb{R}, \mathcal{B}, dx)$ and has the same norm as f when f is a finite linear combination of f_n. This ends the proof (in both cases at hand) since a densely defined unitary map between Hilbert spaces has a (unique) everywhere unitary extension. In the case of $\mathbf{H}_2(\mathbb{C}_+)$ one can also use Exercise 7.6.6 with

$$m(z) = \frac{1}{z} \quad \text{and} \quad \varphi(z) = -\frac{1}{z},$$

and note that

$$\frac{1}{-2\pi i(z-\overline{w})} - \frac{1}{z} \left(\frac{1}{-2\pi i(-\frac{1}{z}+\frac{1}{\overline{w}})} \right) \frac{1}{\overline{w}} = 0, \quad z, w \in \mathbb{C}_+. \qquad\square$$

Solution of Exercise 8.7.1: To prove (8.7.2) we first recall the formula for the Gamma function

$$\Gamma(s) = \int_0^\infty t^{s-1} e^{-t} dt, \quad \text{Re } s > 0.$$

8.8. *Solutions of the exercises*

Thus for real $\nu > -1$,

$$\int_0^\infty t^\nu e^{-t} dt = \Gamma(\nu + 1). \tag{8.8.16}$$

For $\alpha > 0$ the change of variable $t \mapsto \alpha t$ leads to

$$\int_0^\infty t^\nu e^{-\alpha t} dt = \frac{\Gamma(\nu + 1)}{\alpha^{\nu+1}},$$

which is formula (8.7.2) restricted to $(0, \infty)$. In order to extend this formula by analytic continuation to the open right half-plane we need to show that both sides of (8.7.2) define functions which are analytic there. This is clear for the right-hand side. The case of the left-hand side has been treated in Exercise 2.1.31.

Let now $z, w \in \mathbb{C}_+$. Then $-i(z - \overline{w}) \in \mathbb{C}_r$ and formula (8.7.2) gives

$$\frac{\Gamma(1 + \nu)}{(-i(z - \overline{w}))^{1+\nu}} = \int_0^\infty t^\nu e^{i(z-\overline{w})t} dt = \langle e^{izt}, e^{iwt} \rangle_{\mathbf{L}^2([0,\infty),t^\nu dt)} \tag{8.8.17}$$

and this expresses the function (8.7.1) as an inner product, and so (8.7.1) is positive definite in $\mathbb{C}_+$.

From Exercise 2.1.33, with $f_n^{(\nu)}$ as in (2.1.41) and with a change of variable $z \mapsto iz$, we have that

$$\frac{\Gamma(1 + \nu)}{(-i(z - \overline{w}))^{1+\nu}} = \sum_{n=0}^\infty f_n^{(\nu)}(iz) \overline{f_n^{(\nu)}(iw)}.$$

It follows that the functions $f_0^{(\nu)}(iz), f_2^{(\nu)}(iz), \ldots$ form an orthonormal basis of the associated reproducing kernel Hilbert space $\mathcal{H}_2^\nu$. $\square$

Remark 8.8.5. The function (8.8.17) is the reproducing kernel of the fractional Hardy space of the open upper half-plane. For $\nu > 0$ this space can be characterized as the set of functions f analytic in $\mathbb{C}_+$ and such that the integral

$$\int_0^\infty \int_{\mathbb{R}} |f(x + iy)|^2 y^{\nu-1} dx dy < \infty.$$

The integral is then equal to $2\pi 2^{-\nu} \|f\|_{\mathcal{H}_2^\nu}^2$. See for instance [84, Problème 265, p. 216].

Solution of Exercise 8.7.3: It suffices to set $x = i(j + \frac{1}{2})$ and $y = i(k + \frac{1}{2})$ ($k, j = 0, \ldots, N$) in (8.8.17). $\square$

Solution of Exercise 8.7.4: We use Exercise 7.6.7 with $m(z) \equiv a^{\frac{\nu+1}{2}}$ and $\varphi(z) = az$. Then,

$$K(z, w) - m(z) K(\varphi(z), \varphi(w)) \overline{m(w)}$$
$$= \frac{\Gamma(1 + \nu)}{(-i(z - \overline{w}))^{1+\nu}} - a^{\nu+1} \frac{\Gamma(1 + \nu)}{(-i(az - \overline{aw}))^{1+\nu}} \equiv 0. \qquad \square$$

Chapter 9

de Branges–Rovnyak Spaces

The de Branges–Rovnyak spaces are reproducing kernel spaces which appear in various fields, from operator theory to interpolation theory, stochastic processes and system theory. In this chapter we merely outline some of their elementary properties.

9.1 de Branges–Rovnyak spaces

The theory of de Branges–Rovnyak spaces originated with the work [86] of 1966, where de Branges and Rovnyak introduced reproducing kernel Hilbert spaces of functions analytic in the open unit disk and with reproducing kernel (7.7.17), that is,

$$\frac{1 - s(z)\overline{s(w)}}{1 - z\overline{w}}.$$

In other words, s is a Schur multiplier of the Hardy space of the open unit disk, or equivalently, s is analytic and contractive in the open unit disk. See [86]. Later, in joint and independent work, they introduced a variety of other kernels, of various forms, but which can be encompassed in a single formula as

$$K(z, w) = \frac{X(z)JX(w)^*}{\rho_w(z)}. \tag{9.1.1}$$

In this expression, the denominator is of the form

$$\rho_w(z) = a(z)\overline{a(w)} - b(z)\overline{b(w)},$$

where a and b are analytic in a connected set U such that the sets

$$U_+ = \{z \in U, \ |a(z)| > |b(z)|\}$$
$$U_- = \{z \in U, \ |a(z)| < |b(z)|\}$$
$$U_0 = \{z \in U, \ |a(z)| = |b(z)|\}$$

are all non-empty, the function X is matrix-valued and meromorphic in Ω_+ and J is a signature matrix, that is, is both Hermitian and self-adjoint (in fact if both U_+ and U_- are non-empty then U_0 is also non-empty; see [CAPB, Exercise 4.1.12, p. 148]). Such a formulation and general setting is developed in particular in [24], [26]. Typical examples of kernels of the form (9.1.1) are

$$\frac{1 - s(z)\overline{s(w)}}{1 - z\overline{w}}, \quad \frac{\varphi(z) + \overline{\varphi(w)}}{1 - z\overline{w}} \quad \text{and} \quad \frac{J - \Theta(z)J\Theta(w)^*}{1 - z\overline{w}}.$$

See [84], [85], [86] for the original papers, which consider the cases $\rho_w(z) = 1 - z\overline{w}$ and $\rho_w(z) = -i(z - \overline{w})$ and [25], [26] for the theory in the general setting of denominators of the form

$$a(z)\overline{a(w)} - b(z)\overline{b(w)}.$$

The book [9] explores relationships of these spaces with the theory of linear systems and the book [122] studies interpolation theory using these spaces. de Branges spaces of entire functions play an important role in prediction theory and in inverse problems: see [126], [125] and [22, 23].

To give a specific (but by no mean unique) motivation to study these spaces we start from the Hardy space of the open unit disk $\mathbf{H}_2(\mathbb{D})$. Beurling's theorem (see Theorem 8.3.11) can be rewritten as:

Theorem 9.1.1. *A closed subspace* $\mathcal{M} \subset \mathbf{H}_2(\mathbb{D})$ *is backward shift invariant if and only if it is of the form*

$$\mathbf{H}_2(\mathbb{D}) \ominus j\mathbf{H}_2(\mathbb{D}) \tag{9.1.2}$$

where j is an inner function.

The space (9.1.2) is the reproducing kernel Hilbert space with reproducing kernel

$$\frac{1 - j(z)\overline{j(w)}}{1 - z\overline{w}}, \quad z, w \in \mathbb{D}, \tag{9.1.3}$$

and so, equivalently:

Theorem 9.1.2. *There is a one-to-one correspondence between closed backward shift invariant subspaces of $\mathbf{H}_2(\mathbb{D})$ and reproducing kernel Hilbert spaces with a reproducing kernel of the form (9.1.3) and j inner.*

This raises a number of questions and possible generalizations:

1. What happens if we assume $\mathcal{M}$ contractively included in $\mathbf{H}_2(\mathbb{D})$ rather than isometrically?

2. Recall that the closed linear span of the polynomials in $\mathbf{L}_2(\mathbb{T}, d\lambda)$ is $\mathbf{H}_2(\mathbb{D})$. What happens if we replace $\mathbf{L}_2(\mathbb{T}, d\lambda)$ by $\mathbf{L}_2(\mathbb{T}, d\mu)$ where μ is a positive Borel measure on $\mathbb{T}$. Let $\mathbf{H}_2(\mathbb{D}, d\mu)$ denote this closure. What is the structure of R_0-invariant closed subspaces of $\mathbf{H}_2(\mathbb{D}, d\mu)$?

3. As a particular case of the preceding question, what is the structure of the space of polynomials of degree less than or equal to N, when endowed with the inner product of $\mathbf{L}_2(\mathbb{T}, d\mu)$. This problem has important connections with the theory of second-order stationary processes and signal processing.

4. What is the structure of R_0-invariant Hilbert spaces of analytic functions contractively (or maybe only continuously) included in $\mathbf{H}_2(\mathbb{D}, d\mu)$?

5. What happens if j is assumed contractive rather than inner. Then, we know that the kernel k_s is still positive definite in $\mathbb{D}$.

6. What happens if rather than R_0-invariance we consider the following weakened invariance property

$$f \in \mathcal{M} \text{ and } f(0) = 0 \implies \frac{f(z)}{z} \in \mathcal{M}.$$

These questions have been answered by L. de Branges and L. de Branges and J. Rovnyak in various papers. Another avenue of research is to consider these questions when the Hardy space is replaced by the Bergman space, or the Dirichlet space.

Recall that we have denoted by R_0 the backward shift operator; see (1.7.3). We denote by $\|\cdot\|_2$ the norm in $\mathbf{H}_2(\mathbb{D})$.

Exercise 9.1.3. *Show that the space $\mathbf{H}_2(\mathbb{D})$ is R_0-invariant and that the equality*

$$\|R_0 f\|_2^2 = \|f\|_2^2 - |f(0)|^2 \tag{9.1.4}$$

holds for all $f \in \mathbf{H}_2(\mathbb{D})$.

An interesting question is what can be said if this equality is relaxed, for instance to an inequality, or more generally if $\mathcal{H}$ is a reproducing kernel Hilbert space of functions analytic in a neighborhood of the origin, R_0-invariant, and for which

$$\|R_0 f\|_{\mathcal{H}}^2 \le \|f\|_{\mathcal{H}}^2 - |f(0)|^2 \tag{9.1.5}$$

holds for all $f \in \mathcal{H}$.

Exercise 9.1.4. *Let a_1 and a_2 denote two different points in the open unit disk $\mathbb{D}$, and let*

$$s(z) = \frac{1}{\sqrt{2}} \begin{pmatrix} b_{a_1}(z) & b_{a_2}(z) \end{pmatrix}.$$

(1) *Show that the function $\frac{1 - s(z)s(w)^*}{1 - z\overline{w}}$ is positive definite in $\mathbb{D}$, and characterize the associated reproducing kernel Hilbert space (which we will denote by $\mathcal{H}(s)$).*

(2) *Show that $\mathcal{H}(s)$ is R_0-invariant and that (9.1.5) holds in $\mathcal{H}(s)$. Does equality hold in (9.1.5)?*

Exercise 9.1.5. *Let $\mathcal{H}$ be an R_0-invariant Hilbert space of functions analytic in a neighborhood of the origin, and in which inequality (9.1.5) holds. Then the elements of $\mathcal{H}$ can be analytically extended to $\mathbb{D}$.*

We conclude by mentioning two important results on $\mathcal{H}(s)$ spaces.

Exercise 9.1.6. *Let $\mathcal{H}$ be an R_0-invariant Hilbert space of functions analytic in $\mathbb{D}$, and in which inequality (9.1.5) holds. Then $\mathcal{H}$ is a reproducing kernel Hilbert space, and its reproducing kernel is of the form*

$$K_{\mathcal{H}}(z, w) = \frac{1 - s(z)s(w)^*}{1 - z\overline{w}}$$

where s is a (possibly) ℓ_2-valued function analytic in $\mathbb{D}$. In particular, $\mathcal{H}$ is contractively included in $\mathbf{H}_2(\mathbb{D})$.

Hint: (see [19]) Set $Cf = f(0)$, and rewrite (9.1.5) as

$$R_0^* R_0 + C^* C \leq I_{\mathcal{H}},$$

and deduce that the operator

$$\begin{pmatrix} R_0 \\ C \end{pmatrix} : \mathcal{H} \longrightarrow \mathcal{H} \oplus \mathbb{C} \tag{9.1.6}$$

is a contraction. Finally extend the operator (9.1.6) to a unitary operator

$$\begin{pmatrix} R_0 & B \\ C & D \end{pmatrix} : \mathcal{H} \oplus \mathcal{G} \longrightarrow \mathcal{H} \oplus \mathbb{C}$$

where $\mathcal{G}$ is a Hilbert space, and set $s(z) = D + zC(I - zR_0)^{-1}B$.

We conclude with:

Question 9.1.7. *A function s defined in the open unit disk is analytic and contractive there if and only if it can be written as*

$$s(z) = D + zC(I_{\mathcal{H}} - zA)^{-1}B, \tag{9.1.7}$$

where $\mathcal{H}$ is a Hilbert space and where the operator

$$\begin{pmatrix} A & B \\ C & D \end{pmatrix} : \begin{pmatrix} \mathcal{H} \\ \mathbb{C} \end{pmatrix} \Longrightarrow \begin{pmatrix} \mathcal{H} \\ \mathbb{C} \end{pmatrix}$$

is coisometric.

Hints: One direction is clear. If the function s can be written as (9.1.7) then an easy computation leads to (with the notation $M^{-*} = (M^{-1})^*$)

$$k_s(z, w) = \frac{1 - s(z)\overline{s(w)}}{1 - z\overline{w}} = C(I_{\mathcal{H}} - zA)^{-1}(I_{\mathcal{H}} - wA)^{-*}C^*.$$

Conversely, consider the space $\mathcal{H}(s)$ with reproducing kernel k_s. The linear relation (see Definition 4.2.49) in $(\mathcal{H}(s) \oplus \mathbb{C}) \times (\mathcal{H}(s) \oplus \mathbb{C})$ spanned by the functions

$$\left(\begin{pmatrix} k_s(\cdot, w)c \\ d \end{pmatrix}, \begin{pmatrix} \frac{k_s(\cdot, w) - k_s(\cdot, 0)}{\overline{w}} c + k_s(\cdot, 0)d \\ \frac{s(w) - s(0)}{\overline{w}} c + \overline{s(0)}d \end{pmatrix} \right), \quad c, d \in \mathbb{C}, \ w \in \mathbb{D} \setminus \{0\},$$

is densely defined and isometric. It extends (see Exercise 4.2.50) to the graph of an everywhere defined isometry, whose adjoint gives the required realization. See [19, §2.2] for more details (in the setting of indefinite inner product spaces).

The results in Question 9.1.7 originate with the work of de Branges and Rovnyak, and is of fundamental importance for the following reason: The kernel k_s and the realization formula (9.1.7) have counterparts in a wide range of situations, most of them when one leaves the realm of one complex variable. We now list some of them (note that the references given are only indicative, and is in no way exhaustive):

1. In one complex variable, when the kernel k_s is assumed to have a finite number of negative squares. See the work of Kreĭn and Langer [208].
2. In the setting of compact Riemann surfaces, and models of pairs of commuting non-self-adjoint operators. See [222, 316]
3. In the theory of time-varying systems (and then the complex numbers are replaced by diagonal operators). See [110, 16, 109]
4. In the theory of multiscale linear systems, when the integers are replaced by the nodes of a homogeneous tree. See [21]
5. In several complex variables, in particular in the setting of the polydisk and the ball. In particular, in the polydisk setting, the counterpart of Schur functions is the class of Schur–Agler functions, which are functions s analytic in the polydisk $\mathbb{D}^N$, which admit a representation of the form

$$1 - s(z)\overline{s(w)} = \sum_{u=1}^{N} (1 - z_u \overline{w_u}) k_u(z, w), \quad (\text{with } z = (z_1, \dots, z_N)), \quad (9.1.8)$$

where $k_1, \dots, k_u$ are positive definite in $\mathbb{D}^N$. See [3].
6. In the theory of functions of several noncommuting variables. See for instance the work of Ball and Bolotnikov and Ball and Vinnikov, and Kalyuzhnyi-Verbovetskiĭ and Vinnikov [199].
7. In hypercomplex analysis, in the setting of Fueter series. See the work of Shapiro, Volok and the author [45], in the setting of slice-hyperholomorphic functions (see the work [14] with Sabadini and Colombo) and in the setting of bicomplex numbers, see the work [27] with Luna-Elizarrarás, Shapiro and Struppa.

and other cases too, such as white noise space analysis.

In each of these cases, a version of the equivalence in Question 9.1.7 holds.

9.2 Interpolation

There are numerous approaches to interpolation problems for Schur functions. We here use the tools presented earlier in the book to use the theory of de Branges spaces to solve the Nevanlinna–Pick interpolation problem in its simplest form. This is a very simple instance of a general matrix-valued two-sided interpolation problem.

Problem 9.2.1. *Let* $(w_1, s_1), \ldots, (w_N, s_N)$ *be* N *pairs of elements in* $\mathbb{D} \times \mathbb{D}$.

(1) *Find a necessary and sufficient condition for a Schur function* s *to exist such that*

$$s(w_k) = s_k, \quad k = 1, \ldots, N.$$

(2) *Describe the set of all solutions when the above condition is in force.*

In preparation of the various exercises and questions which present the solution to Problem 9.2.1 we introduce some notation.

We denote by P the $N \times N$ Hermitian matrix with (j, k) entry

$$P_{jk} = \frac{1 - s_j \overline{s_k}}{1 - w_j \overline{w_k}}.$$

Furthermore, we set

$$C = \begin{pmatrix} 1 & 1 & \cdots & 1 \\ s_1 & s_2 & \cdots & s_N \end{pmatrix}, \quad A = \mathrm{diag}\,(\overline{w_1}, \ldots, \overline{w_N}), \quad \text{and} \quad J = \begin{pmatrix} 1 & 0 \\ 0 & -1 \end{pmatrix}.$$

Finally we denote by $\mathcal{M}$ the space spanned by the columns of the matrix function

$$F(z) = C(I_N - zA)^{-1},$$

and endow $\mathcal{M}$ with the inner product defined by P, meaning that

$$\langle Fc, Fd \rangle_{\mathcal{M}} = d^* Pc, \quad c, d \in \mathbb{C}^N.$$

Exercise 9.2.2. $P \geq 0$ *is a necessary condition for Problem 9.2.1 to have a solution.*

Exercise 9.2.3. *It holds that*

$$P - A^* PA = C^* JC.$$

Exercise 9.2.4. *Assume that Problem 9.2.1 has a solution* s. *Show that the map*

$$f \mapsto \begin{pmatrix} 1 & -s \end{pmatrix} f \tag{9.2.1}$$

is an isometry from $\mathcal{M}$ *into* $\mathcal{H}(s)$.

In the following four questions we assume $P > 0$. Question 9.2.5 is a particular case of Exercise 7.7.17.

Question 9.2.5. *Assume $P > 0$ and let*

$$\Theta(z) = I_2 - (1 - z)C(I_N - zA)^{-1}P^{-1}(I_N - A^*)^{-1}C^*J.$$

Show that

$$F(z)P^{-1}F(w)^* = \frac{J - \Theta(z)J\Theta(w)^*}{1 - z\overline{w}}.$$

The argument in the following exercise seems to first have appeared in the work of de Branges and Rovnyak; see [85, Theorem 13, p. 305]. It is the key to the reproducing kernel approach to interpolation problem.

Exercise 9.2.6. *Let $\Theta = \begin{pmatrix} a & b \\ c & d \end{pmatrix}$, and assume that s is a solution to the interpolation problem. Show that there is a Schur function ε such that*

$$s = \frac{a\varepsilon + b}{c\varepsilon + d}. \tag{9.2.2}$$

Hint: Use Theorem 7.6.6. In the proof it is useful to check that d^{-1} is invertible in $\mathbb{D}$ and that both d^{-1} and $d^{-1}c$ are Schur functions.

Exercise 9.2.7. *Show that*

$$\begin{pmatrix} 1 & -s_j \end{pmatrix} \Theta(w_j) = \begin{pmatrix} 0 & 0 \end{pmatrix}, \quad j = 1, \ldots, N.$$

Hint: Check that

$$\begin{pmatrix} 1 & -s_j \end{pmatrix} C(I_N - w_j A)^{-1}$$

is the jth line of the matrix P.

Exercise 9.2.8. *Show that any function of the form (9.2.2) is a solution to the interpolation problem.*

Question 9.2.9. *Assume now that P is singular. Show that there is a unique solution, and that this solution is a finite Blaschke product.*

9.3 Carathéodory's theorem

In this section we follow Sarason's analysis from [274, p. 48] and discuss Carathéodory's theorem on the angular derivative using reproducing kernel spaces. To simplify the exposition we consider radial limits rather than angular limits.

Exercise 9.3.1. *Let s be a Schur function and let $e^{it_0} \in \mathbb{T}$ be such that*

$$\sup_{r \in (0,1)} \frac{1 - |s(re^{it_0})|^2}{1 - r^2} < \infty. \tag{9.3.1}$$

Show that there is $c \in \mathbb{T}$ such that the function

$$z \mapsto \frac{1 - s(z)\overline{c}}{1 - ze^{-it_0}}$$

belongs to $\mathcal{H}(s)$.

Exercise 9.3.2. *In the notation of the previous exercise, show that the limit*

$$\lim_{\substack{r \to 1 \\ r \in (0,1)}} \frac{1 - s(re^{it_0})\overline{c}}{1 - r} \tag{9.3.2}$$

exists and is positive.

Question 9.3.3. *What happens when the limit (9.3.2) is equal to 0?*

9.4   A few words on Schur analysis

Functions analytic and contractive in the open unit disk are part of classical mathematics, as is illustrated by the works of Schur [281, 282], Takagi [303, 304] and
Bloch [75], to name a few. They bear various names and we will call them Schur
functions here. We will denote by $\mathcal{S}$ the family of Schur functions. They play an
important role in numerous areas of mathematics, and can be characterized in a
number of way. Via an iterative procedure now called the Schur algorithm, Schur
associated in 1917 to a function $s \in \mathcal{S}$ a (possibly finite) sequence of numbers in
the open unit disk (with additionally a number of modulus one if the sequence
is finite), which uniquely characterizes s. An important, but not so well known,
consequence of the Schur algorithm is in topological analysis, with a proof of the
power expansion for an analytic function without using integration. See [252] and
see [320, 102] for more information on topological analysis. The study of various questions associated with the Schur algorithm is called Schur analysis. In
the present book, Schur analysis does not really come into play, but this section
reviews some of its aspects, and is based in part on the survey [15].

An element $s \in \mathcal{S}$ can be characterized in various ways. Recall (see for
instance [180, pp. 67–68]) that a function f analytic and bounded in the open
unit disk admits a multiplicative representation $f(z) = i(z)o(z)$ into an inner
function $i(z)$ and an outer function $o(z)$, the inner function being itself a product
of a constant c of modulus 1, a Blaschke product $b(z)$ and of a singular inner
function $j(z)$: thus,

$$f(z) = cb(z)j(z)o(z),$$

with

$$b(z) = z^p \prod_{n \in J} \frac{\overline{z_n}}{z_n} \frac{z_n - z}{1 - \overline{z_n}z},$$

$$j(z) = e^{-\frac{1}{2\pi} \int_{[0,2\pi]} \frac{e^{it}+z}{e^{it}-z} d\mu(t)},$$

$$o(z) = e^{\frac{1}{2\pi} \int_{[0,2\pi]} \frac{e^{it}+z}{e^{it}-z} \ln|f(e^{it})|dt}.$$

In these expressions, $p \in \mathbb{N}_0$, $J \subset \mathbb{N}$ and the points $z_n \in \mathbb{D} \setminus \{0\}$. Furthermore, $d\mu$
is a finite singular measure. When the function at hand is a Schur function, the
outer part is also a Schur function.

The above representation is fundamental in function theory and in operator theory, and admits generalizations to the matrix-valued and operator-valued cases. See [253]. On the other hand, it does not seem to be the best tool to solve classical interpolation problems such as the Nevanlinna–Pick interpolation problem (see Section 9.2, where the theory of de Branges–Rovnyak spaces is used rather than the Schur algorithm), or Carathéodory–Fejér interpolation problem:

Problem 9.4.1 (the Carathéodory–Fejér interpolation problem). *Given* $N \in \mathbb{N}_0$ *and* $a_0, \ldots, a_N \in \mathbb{C}$, *find necessary and sufficient conditions for a function* $s \in \mathcal{S}$ *to exist such that*

$$\frac{s^{(n)}(0)}{n!} = a_n, \quad n = 0, \ldots, N,$$

and describe the set of all solutions when these conditions are in force.

This problem arose at the beginning of the previous century as a step to solve the trigonometric moment problem, as we explain below.

In his 1917 paper [281] Schur associates to $s \in \mathcal{S}$ a series of functions $s_1, s_2, \ldots$ of $\mathcal{S}$ via the recursion

$$s_0(z) = s(z)$$

$$s_{n+1}(z) = \begin{cases} \dfrac{s_n(z) - s_n(0)}{z(1 - \overline{s_n(0)}s_n(z))}, & z \in \mathbb{D} \setminus \{0\}, \\ s_n'(0), & z = 0, \end{cases} \quad n = 0, 1, \ldots \tag{9.4.1}$$

called the Schur algorithm. The recursion stops at rank n if $|s_n(0)| = 1$. This algorithm, originally developed to solve classical interpolation problem, has been shown to have numerous applications in signal processing. The Schur algorithm ends after a finite number of iterations if and only if f is a finite Blaschke product.

As just mentioned above, the motivation of Schur was the trigonometric moment problem which is equivalent to an interpolation problem in the class of functions analytic in the open unit disk, and with a positive real part there. We will denote by $\mathcal{C}$ this family of functions. These functions are called Carathéodory functions.

Exercise 9.4.2. *Let* $s \not\equiv -1 \in \mathcal{S}$. *Show that the function*

$$\varphi(z) = \frac{1 - s(z)}{1 + s(z)} \tag{9.4.2}$$

belongs to $\mathcal{C}$, *and compare the power expansions at the origin of* s *and* φ.

The previous exercise is simple, but very important. To see this, recall first the definition of the trigonometric moment problem:

Problem 9.4.3. *Given* $t_0, \ldots, t_N \in \mathbb{C}$, *find a necessary and sufficient condition for a positive Borel measure on* $[0, 2\pi]$ *to exist such that*

$$\int_0^{2\pi} e^{int} d\mu(t) = t_n \quad n = 0, \ldots, N,$$

and describe the set of all solutions when this condition is in force.

Question 9.4.4. *Using the Herglotz representation formula for Carathéodory functions (see* [CAPB, (5.5.9), p. 207]*), show that the trigonometric moment problem is equivalent to the Carathéodory-Fejér problem for Schur functions.*

The following question is adapted from [32, pp. 28–29].

Question 9.4.5. *Let* $(a, b, c) \in \mathbb{C}^{N \times N} \times \mathbb{C}^{N \times 1} \times \mathbb{C}^{1 \times N}$, *and assume that the spectral radius of* a *is strictly less than one.*

(1) *Show that the equations*

$$\Delta - a^* \Delta a = b b^* \quad and \quad \Omega - a^* \Omega a = c^* c$$

admit unique solutions Δ *and* Ω, *and that these are Hermitian matrices.*

(2) *For* $n \in \mathbb{N}_0$ *such that*

$$\det \left(I_N - \Delta a^{*(n+1)} \Omega a^{n+1} \right) \neq 0, \tag{9.4.3}$$

we define

$$\rho_n = -c a^n \left(I_N - \Delta a^{*(n+1)} \Omega a^{n+1} \right)^{-1} b, \quad n = 0, 1, 2, \ldots$$

and

$$R_n(z) = c \left((I_N - \Delta a^{*(n+1)} \Omega a^{n+1}) - z(I_N - \Delta a^{*n} \Omega a^n) a \right)^{-1} b.$$

(a) *Show that there is at most a finite number of integers such that* (9.4.3) *fails to hold.*

(b) *Let* n_0 *be such that* (9.4.3) *holds for all* $n \geq n_0$. *Show that, in a neighborhood* V *of the origin,*

$$R_{n+1}(z) = \begin{cases} \frac{R_n(z) - \rho_n}{z(1 - \overline{\rho_n} R_n(z))}, & z \in V \setminus \{0\}, \\ \rho_{n+1}, & z = 0. \end{cases}$$

As an example we have (see [32]):

Question 9.4.6. *Let* $s(z) = \frac{5}{33 - 16z}$. *Show that*

$$\rho_n = \frac{15 \cdot 2^n}{(5 \cdot 2^{n+1} - 1)(5 \cdot 2^{n+1} + 1)}, \quad n = 0, 1, \ldots$$

The following is Exercise 2.3.8, p. 68 of [CAPB], to which we refer for a proof.[1] The result appears in [227]; see also [192, Theorem 7, p. 67]. It is an example where the Schur algorithm *is not* the right tool to use.

Exercise 9.4.7. *Show that the non-trivial Moebius map* $\varphi(z) = \dfrac{az+b}{cz+d}$ *maps the open unit disk into itself if and only if*

$$|\bar{a}c - \bar{b}d| + |ad - bc| \leq |d|^2 - |c|^2. \tag{9.4.4}$$

Another question of importance where the Schur algorithm does not seem to be the appropriate tool to use is the following result of Landau. See Landau's paper [212] and also the paper [98] of Chen and Choi for recent advances on the subject.

Theorem 9.4.8. *Let s be a Schur function (not equal to a unitary constant), with power series expansion at the origin* $s(z) = \sum_{n=0}^{\infty} s_n z^n$ *Then it holds that*

$$\left| \sum_{n=0}^{N} s_n \right| \leq G_N, \quad N = 0, 1, \dots \quad \text{where} \quad G_N = \sum_{n=0}^{N} \frac{1}{16^n} \binom{2n}{n}, \quad N = 0, 1, \dots$$

are the Landau constants.

9.5 Solutions of the exercises

Solution of Exercise 9.1.3: Let $f \in \mathbf{H}_2(\mathbb{D})$, with power series

$$f(z) = \sum_{n=0}^{\infty} a_n z^n, \quad \text{and} \quad \|f\|_2^2 = \sum_{n=0}^{\infty} |a_n|^2.$$

Then,

$$(R_0 f)(z) = \sum_{n=1}^{\infty} a_n z^n, \quad \text{and} \quad \|R_0 f\|_2^2 = \sum_{n=1}^{\infty} |a_n|^2.$$

The equality (9.1.4) follows since $f(0) = a_0$.

$\square$

Solution of Exercise 9.1.4:
(1) Recall the notation $k_s(z, w) = \dfrac{1 - s(z)s(w)^*}{1 - z\overline{w}}$, and similarly for $k_{b_{a_1}}$ and $k_{b_{a_2}}$ (see (7.7.17)). From the formula

$$\frac{1 - b_{a_1}(z)\overline{b_{a_1}}(w)}{1 - z\overline{w}} = \frac{1 - |a_1|^2}{(1 - z\overline{a_1})(1 - \overline{w}a_1)}$$

---

[1] We mention a misprint in the proof in [CAPB]; adjoints are missing in the completing of the square argument, and the center of the image of the unit circle is $\frac{b\overline{d} - \overline{a}c}{|d|^2 - |c|^2}$, and not as written. This does not change the arguments.

(see for instance [CABP, (1.1.49), p. 21]) and

$$\frac{1 - s(z)s(w)^*}{1 - z\overline{w}} = \frac{1}{2}\left\{\frac{1 - b_{a_1}(z)\overline{b_{a_1}}(w)}{1 - z\overline{w}} + \frac{1 - b_{a_2}(z)\overline{b_{a_2}}(w)}{1 - z\overline{w}}\right\},$$

we see that the function $k_s(z, w)$ is positive definite in $\mathbb{D}$. The reproducing kernel Hilbert space associated with the functions $\frac{1}{2}k_{b_{a_1}}(z, w)$ and $\frac{1}{2}k_{b_{a_2}}(z, w)$ are spanned by the function $\frac{1}{1 - z\overline{a_1}}$ and $\frac{1}{1 - z\overline{a_2}}$ respectively, and so have a zero intersection. By Theorem 7.5.22 the space $\mathcal{H}(k_s)$ is the orthogonal sum of the spaces $\mathcal{H}(\frac{1}{2}k_{b_{a_1}})$ and $\mathcal{H}(\frac{1}{2}k_{b_{a_2}})$. More precisely,

$$f \in \mathcal{H}(s) \iff f(z) = \lambda_1\frac{1}{1 - z\overline{a_1}} + \lambda_2\frac{1}{1 - z\overline{a_2}},$$

where $\lambda_1, \lambda_2 \in \mathbb{C}$, and with norm

$$\|f\|^2 = 2\left(\frac{|\lambda_1|^2}{1 - |a_1|^2} + \frac{|\lambda_2|^2}{1 - |a_2|^2}\right).$$

See Exercise 7.5.8 for the latter.

(2) We have

$$R_0 f(z) = \lambda_1\frac{\overline{a_1}}{1 - z\overline{a_1}} + \lambda_2\frac{\overline{a_2}}{1 - z\overline{a_2}},$$

and so

$$\begin{aligned}\|R_0 f\|^2 &= 2\left(\frac{|a_1\lambda_1|^2}{1 - |a_1|^2} + \frac{|a_2\lambda_2|^2}{1 - |a_2|^2}\right)\\&= \|f\|^2 - 2(|\lambda_1|^2 + |\lambda_2|^2)\\&\leq \|f\|^2 - |f(0)|^2,\end{aligned}$$

since

$$|f(0)|^2 = |\lambda_1 + \lambda_2|^2 \leq 2(|\lambda_1|^2 + |\lambda_2|^2). \tag{9.5.1}$$

Furthermore,

$$2(|\lambda_1|^2 + |\lambda_2|^2) - |\lambda_1 + \lambda_2|^2 = |\lambda_1 - \lambda_2|^2,$$

and so inequality (9.5.1) is strict for $\lambda_1 \neq \lambda_2$. So equality will not hold in (9.1.5) in general. $\qquad\square$

Solution of Exercise 9.1.6: We follow the hint. The operator (9.1.6) is a contraction and so is its adjoint. Thus

$$\begin{pmatrix}R_0\\C\end{pmatrix}\begin{pmatrix}R_0\\C\end{pmatrix}^* \leq I_{\mathcal{H}\oplus\mathbb{C}}.$$

9.5. *Solutions of the exercises*

A contractive operator has a positive squareroot (see Exercise 4.2.31). We have

$$I_{\mathcal{H} \oplus \mathbb{C}} = \begin{pmatrix} R_0 \\ C \end{pmatrix} \begin{pmatrix} R_0 \\ C \end{pmatrix}^* + \begin{pmatrix} B \\ D \end{pmatrix} \begin{pmatrix} B \\ D \end{pmatrix}^*,$$

where we decompose the positive squareroot though a Hilbert space $\mathcal{G}$ which is isomorphic to

$$\mathrm{ran} \sqrt{I_{\mathcal{H} \oplus \mathbb{C}} - \begin{pmatrix} R_0 \\ C \end{pmatrix} \begin{pmatrix} R_0 \\ C \end{pmatrix}^*}.$$

Then the function $s(z) = D + zC(I - zR_0)^{-1}B$ satisfies

$$C(I - zR_0)^{-1}(I - wR_0)^{-*}C^* = \frac{1 - s(z)s(w)^*}{1 - z\overline{w}}.$$

This ends the proof since the reproducing kernel of $\mathcal{H}$ is given by

$$C(I - zR_0)^{-1}(I - wR_0)^{-*}C^*. \qquad \square$$

Solution of Exercise 9.2.2: This is a special case of Question 7.6.16 with $K(z,w) = \frac{1}{1-z\overline{w}}$. $\qquad \square$

Solution of Exercise 9.2.3: The (j,k) entry of the matrix $P - A^*PA$ is

$$\begin{aligned} P_{jk} - w_j P_{jk}\overline{w_k} &= \frac{1 - s_j\overline{s_k}}{1 - w_j\overline{w_k}} - w_j \frac{1 - s_j\overline{s_k}}{1 - w_j\overline{w_k}}\overline{w_k} \\ &= 1 - s_j\overline{s_k} \\ &= \begin{pmatrix} 1 \\ s_j \end{pmatrix}^* \begin{pmatrix} 1 & 0 \\ 0 & -1 \end{pmatrix} \begin{pmatrix} 1 \\ s_k \end{pmatrix}, \end{aligned}$$

which is the (j,k) entry of the matrix $C^* JC$. $\qquad \square$

Solution of Exercise 9.2.4: The space $\mathcal{M}$ is spanned by the functions $f_1, \ldots, f_N$, where

$$f_j(z) = \frac{\begin{pmatrix} 1 \\ s_j \end{pmatrix}}{1 - z\overline{w_j}}, \quad j = 1, \ldots, N.$$

Assuming s to be a solution to the interpolation problem (note that at this stage we do not know yet that solutions exist at all) we have

$$f_j(z) = \frac{\begin{pmatrix} 1 \\ s(w_j) \end{pmatrix}}{1 - z\overline{w_j}}, \quad j = 1, \ldots, N.$$

Thus

$$\begin{pmatrix}1 & -s(z)\end{pmatrix} f_j(z) = \frac{1 - s(z)\overline{s(w_j)}}{1 - z\overline{w_j}} = k_s(z, w_j) \quad j = 1, \dots, N, \tag{9.5.2}$$

and the map (9.2.1) sends $\mathcal{M}$ into $\mathcal{H}(s)$. To check that (9.2.1) is an isometry, and taking into account (9.5.2), we write

$$\begin{aligned}
\left\langle \begin{pmatrix}1 & -s\end{pmatrix} f_k, \begin{pmatrix}1 & -s\end{pmatrix} f_j \right\rangle_{\mathcal{H}(s)} &= \langle k_s(\cdot, w_k), k_s(\cdot, w_j) \rangle_{\mathcal{H}(s)} \\
&= k_s(w_j, w_k) \\
&= \frac{1 - s(w_j)\overline{s(w_k)}}{1 - w_j \overline{w_k}} \\
&= \frac{1 - s_j \overline{s_k}}{1 - w_j \overline{w_k}} \\
&= P_{jk} \\
&= \langle f_k, f_j \rangle_{\mathcal{M}}. \qquad \qquad \square
\end{aligned}$$

Solution of Exercise 9.2.6: From Exercise 7.6.6 with $\varphi(z) = z$ and

$$K_1(z, w) = \frac{J - \Theta(z) J \Theta(w)^*}{1 - z\overline{w}}, \quad K_2(z, w) = k_s(z, w)$$

and

$$m(z) = \begin{pmatrix}1 & -s(z)\end{pmatrix}$$

we have that the kernel

$$k_s(z, w) - \begin{pmatrix}1 & -s(z)\end{pmatrix} \frac{J - \Theta(z) J \Theta(w)^*}{1 - z\overline{w}} \begin{pmatrix}1 & -s(w)\end{pmatrix}^*$$

is positive definite in $\mathbb{D}$. Since

$$\frac{\begin{pmatrix}1 & -s(z)\end{pmatrix} J \begin{pmatrix}1 & -s(w)\end{pmatrix}^*}{1 - z\overline{w}} = k_s(z, w)$$

we see that the kernel

$$\begin{pmatrix}1 & -s(z)\end{pmatrix} \frac{\Theta(z) J \Theta(w)^*}{1 - z\overline{w}} \begin{pmatrix}1 & -s(w)\end{pmatrix}^* \geq 0$$

in $\mathbb{D}$, that is,

$$\frac{(a(z) - s(z)c(z))\overline{(a(w) - s(w)c(w))} - (b(z) - s(z)d(z))\overline{(b(w) - s(w)d(w))}}{1 - z\overline{w}} \geq 0$$

in $\mathbb{D}$. Since $\Theta(1) = I_2$ we have that $a(z) - s(z)c(z) \not\equiv 0$ and so, with

$$\varepsilon(z) = -\frac{b(z) - s(z)d(z)}{a(z) - s(z)c(z)} \tag{9.5.3}$$

we have that $k_\varepsilon(z,w)$ is positive definite on $\mathbb{D} \setminus Z(a - sc)$, where we denote by $Z(a - sc)$ the set of zeros of $a - sc$. The function ε is in particular bounded in modulus in $\mathbb{D} \setminus Z(a - sc)$, and Riemann's removable singularity theorem insures that the singularities of ε are removable. Thus ε is a Schur function.

Since Θ is J-contractive we have

$$\begin{pmatrix} 0 & 1 \end{pmatrix} \Theta(z) J \begin{pmatrix} 0 & 1 \end{pmatrix}^* \le \begin{pmatrix} 0 & 1 \end{pmatrix} J \begin{pmatrix} 0 & 1 \end{pmatrix}^*, \quad z \in \mathbb{D},$$

that is

$$|c(z)|^2 - |d(z)|^2 \le -1, \quad z \in \mathbb{D},$$

or

$$1 + |c(z)|^2 \le |d(z)|^2, \quad z \in \mathbb{D}.$$

It follows that d^{-1} is invertible in $\mathbb{D}$ and that $d^{-1}c$ is a Schur function, strictly contractive in $\mathbb{D}$ and so $c(z)\varepsilon(z) + d(z)$ is invertible in $\mathbb{D}$. Then (9.2.2) follows from (9.5.3). $\qquad \square$

Solution of Exercise 9.2.7: This is a direct computation. We follow the hint and write

$$\begin{pmatrix} 1 & -s_j \end{pmatrix} C(I_N - w_j A)^{-1} = \begin{pmatrix} 1 - s_j \overline{s_1} & 1 - s_j \overline{s_2} & \cdots & 1 - s_j \overline{s_N} \end{pmatrix}$$

$$\times \operatorname{diag}\left(\frac{1}{1 - w_j \overline{w_1}}, \frac{1}{1 - w_j \overline{w_2}}, \ldots, \frac{1}{1 - w_j \overline{w_N}} \right)$$

$$= \begin{pmatrix} \frac{1 - s_j \overline{s_1}}{1 - w_j \overline{w_1}}, & \frac{1 - s_j \overline{s_2}}{1 - w_j \overline{w_2}}, & \cdots & \frac{1 - s_j \overline{s_N}}{1 - w_j \overline{w_N}} \end{pmatrix}$$

$$= \begin{pmatrix} P_{j1} & P_{j2} & \cdots & P_{jN} \end{pmatrix}.$$

Thus

$$\begin{pmatrix} 1 & -s_j \end{pmatrix} C(I_N - w_j A)^{-1} P^{-1} = \underbrace{\begin{pmatrix} 0 & 0 & \cdots & 1 & 0 & \cdots \end{pmatrix}}_{j\text{th entry equal to 1 and the other ones equal to 0}},$$

and

$$\begin{pmatrix} 1 & -s_j \end{pmatrix} \Theta(w_j)$$

$$= \begin{pmatrix} 1 & -s_j \end{pmatrix} \left(I_2 - (1 - w_j) C(I_N - w_j A)^{-1} P^{-1} (I_N - A^*)^{-1} C^* J \right)$$

$$= \begin{pmatrix} 1 & -s_j \end{pmatrix} - (1 - w_j) \begin{pmatrix} 0 & 0 & \cdots & 1 & 0 & \cdots \end{pmatrix}$$

$$\times \operatorname{diag}\left(\frac{1}{1 - w_1}, \frac{1}{1 - w_2}, \ldots, \frac{1}{1 - w_N} \right) C^* J$$

$$= \begin{pmatrix} 0 & 0 \end{pmatrix}$$

since the jth row of $C^* J$ is equal to $\begin{pmatrix} 1 & -s_j \end{pmatrix}$. $\qquad \square$

Solution of Exercise 9.2.8: Let ε be a Schur function. We have:

$$\left(\begin{pmatrix} 1 & -s(z) \end{pmatrix} \Theta(z) \begin{pmatrix} 1 \\ \varepsilon(z) \end{pmatrix} \right)_{z=w_j} = 0.$$

Thus

$$(a(z)\varepsilon(z) + b(z)) - s(z)(c(z)\varepsilon(z) + d(z)) = 0 \quad \text{for} \quad z = w_1, \ldots, w_N.$$

The result follows since $c\varepsilon + d$ is invertible in the open unit disk (see Exercise 9.2.6) and since the image of the open unit disk under the linear fractional transformation defined by a J-contractive matrix is inside the open unit disk. □

Solution of Exercise 9.3.1: For every $r \in (0,1)$ we have that

$$k_s(\cdot, re^{it_0}) \in \mathcal{H}(s) \quad \text{and} \quad \|k_s(:, re^{it_0})\|_{\mathcal{H}(s)} = \frac{1 - |s(re^{it_0})|^2}{1 - r^2}.$$

Condition (9.3.1) expresses that the family $(k_s(\cdot, re^{it_0}))_{r \in (0,1)}$ is uniformly bounded in norm in $\mathcal{H}(s)$ and therefore has a weakly convergent subsequence (say, indexed by the sequences $r_1, r_2, \ldots$ of points in $(0,1)$), whose limit we denote by g. Since weak convergence implies pointwise convergence we have

$$g(z) = \lim_{n \to \infty} \langle k_s(\cdot, r_n e^{it_0}), k_s(\cdot, z) \rangle_{\mathcal{H}(s)}$$

$$= \lim_{n \to \infty} \frac{1 - s(z)\overline{s(r_n e^{it_0})}}{1 - z r_n e^{-it_0}}, \quad \forall z \in \mathbb{D}.$$

Since (in view of (9.3.1)) $s \not\equiv 0$ we have that $c = \lim_{n \to \infty} s(r_n e^{it_0})$ exists, and is of modulus 1 since $\mathbb{T}$ is compact. Thus the function

$$g(z) = \frac{1 - s(z)\overline{c}}{1 - z e^{-it_0}}$$

belongs to $\mathcal{H}(s)$.

To conclude we show that the limit does not depend on the given subsequence. Assume that both

$$\frac{1 - s(z)\overline{c_1}}{1 - z e^{-it_0}} \quad \text{and} \quad \frac{1 - s(z)\overline{c_2}}{1 - z e^{-it_0}}$$

belong to the space $\mathcal{H}(s)$ for two different points c_1 and c_2 on the unit circle. Then, the function

$$z \mapsto \frac{s(z)}{e^{it_0} - z}$$

belongs to $\mathcal{H}(s)$, and in particular in the Hardy space $\mathbf{H}_2(\mathbb{D})$. But this forces

$$\lim_{\substack{r \to 1 \\ r \in (0,1)}} s(re^{it_0}) = 0.$$

See Exercise 8.2.5 for the latter. □

9.5. *Solutions of the exercises*

Solution of Exercise 9.3.2: By definition of the weak limit we have

$$
\begin{aligned}
0 &\leq \langle g(\cdot), g(\cdot) \rangle_{\mathcal{H}(s)} \\
&= \lim_{n \to \infty} \langle g(\cdot), k_s(\cdot, r_n e^{it_0}) \rangle_{\mathcal{H}(s)} \\
&= \lim_{n \to \infty} g(r_n e^{it_0}) \\
&= \lim_{n \to \infty} \frac{1 - s(r_n e^{it_0}) \overline{c}}{1 - r_n}.
\end{aligned}
$$

As in the previous exercise, the limit does not depend in fact on the given sequence. $\square$

Solution of Exercise 9.4.2: Since $s(z) \not\equiv -1$, the maximum modulus principle insures that $s(z) \neq -1$ for all $z \in \mathbb{D}$ (note that, of course, s could be a constant of modulus 1, but not the function $s(z) \equiv -1$). Therefore φ is defined for all points in $\mathbb{D}$ and we have

$$
\operatorname{Re} \varphi(z) = \frac{1 - |s(z)|^2}{|1 + s(z)|^2} \geq 0, \quad z \in \mathbb{D}.
$$

From (9.4.2) we have

$$
(1 + \varphi(z))(1 + s(z)) = 2
$$

from which we obtain the relation between the power series developments of φ and s at the origin. If $\varphi(z) = \sum_{n=0}^{\infty} \varphi_n z^n$ and $s(z) = \sum_{n=0}^{\infty} s_n z^n$ we have

$$
\begin{aligned}
(1 + \varphi_0)(1 + s_0) &= 2, \\
(1 + \varphi_0)s_1 + \varphi_1(1 + s_0) &= 0, \\
(1 + \varphi_0)s_2 + \varphi_1 s_1 + \varphi_2(1 + s_0) &= 0,
\end{aligned}
$$

$$
\vdots
$$

$\square$

Chapter 10

Bergman Spaces

As already said, Bergman spaces seem to form the first example of reproducing kernel Hilbert spaces of analytic functions. In this chapter we discuss the case of the disk and the polyanalytic cases. More involved cases, such as the annulus and the ellipse, are only mentioned *en passant*. We urge the reader to look at these cases.

10.1 The Bergman space of analytic functions analytic in $\mathbb{D}$

In this section we focus on the case $\Omega = \mathbb{D}$ and $p = 2$ in Exercise 4.1.17. As already mentioned the cases where Ω is an ellipse or an annulus are treated in Sections 10.2 and 10.3 respectively.

Exercise 10.1.1. *Let f be analytic in the open unit disk, with power expansion $f(z) = \sum_{n=0}^{\infty} f_n z^n$. Show that $\sum_{n=0}^{\infty} \frac{|f_n|^2}{n+1} < \infty$ if and only if $\iint_{\mathbb{D}} |f(z)|^2 dxdy < \infty$, and that, when one of these expressions is finite, it holds that:*

$$\sum_{n=0}^{\infty} \frac{|f_n|^2}{n+1} = \frac{1}{\pi} \iint_{\mathbb{D}} |f(z)|^2 dxdy. \tag{10.1.1}$$

Exercise 10.1.2. *In the notation of Exercise 4.1.17, assume that $p = 2$. Let $w \in \Omega$. Show that there exists $K_w \in \mathcal{B}_2(\Omega)$ such that*

$$f(w) = \langle f, K_w \rangle_{\mathcal{B}_2(\Omega)}, \quad \forall f \in \mathcal{B}_2(\Omega). \tag{10.1.2}$$

Exercise 10.1.3. *Show that the backward shift operator R_0 (defined by (1.7.3) with $a = 0$) is bounded from the Bergman space $\mathcal{B}_2(\mathbb{D})$ into itself.*

Exercise 10.1.4. Let s be analytic and strictly contractive in $\mathbb{D}$.

(1) Show that C_s is a contraction from the Bergman space into itself, and show the bound

$$\|C_s\|_{\mathbf{B}_2(\mathbb{D})} \leq \|C_s\|_{\mathbf{H}_2(\mathbb{D})}^2. \tag{10.1.3}$$

(2) Prove directly the claims in (1) for $s(z) = b_a(z)$, where b_a is the Blaschke factor based at $a \in \mathbb{D}$, that is, $b_a(z) = \frac{z-a}{1-z\overline{a}}$ (see (1.2.9)).

Question 10.1.5. Let s be analytic and strictly contractive in $\mathbb{D}$. Is there a converse inequality to (10.1.3)?

In view of the following question, recall that the weighted Bergman space was defined as the reproducing kernel Hilbert space with reproducing kernel (8.2.8).

Question 10.1.6. Let $N \in \{2, 3, \ldots\}$.

(1) Show that the positive plane Borel measure μ_N absolutely continuous with respect to Lebesgue measure and with density

$$\mu_N'(x, y) = \frac{(N-1)}{\pi}(1 - |z|^2)^{N-2}$$

is such that the reproducing kernel Hilbert space $\mathcal{B}_N$ with reproducing kernel $\frac{1}{(1-z\overline{w})^N}$ is isometrically included in $\mathbf{L}_2(\mathbb{D}, \mathcal{B}, \mu_N)$.

(2) Characterize the elements of $\mathcal{B}_N$ in terms of their power series at the origin.

In the previous exercise, μ_N was given. On the other hand, a detailed hint is provided in order to solve the following problem.

Exercise 10.1.7. Let $N \in \mathbb{N}$. Show that there is a measure $d\mu_N$ on $\mathbb{D}$, of the form

$$d\mu_N(z) = f_N(r)dr d\theta$$

such that the reproducing kernel Hilbert space with reproducing kernel $\frac{1}{(1-z\overline{w})^N}$ is isometrically included in $\mathbf{L}_2(\mathbb{D}, \mathcal{B}, \mu_N)$.

Hint: We first recall the following result on the moment problem on an interval. See [209, Theorem 2.5, p. 64].

Theorem 10.1.8. Given $a, b \in \mathbb{R}$ with $a < b$ and given a sequence $(c_n)_{n \in \mathbb{N}_0}$ of real numbers, there exists a positive Borel measure on $[a, b]$ such that

$$\int_{[a,b]} u^n d\mu(u) = c_n, \quad n = 0, 1, 2, \ldots$$

if and only if the functions

$$K(\ell, k) = c_{\ell+k} \tag{10.1.4}$$
$$K_{a,b}(\ell, k) = (a+b)c_{\ell+k+1} - ac_{\ell+k} - c_{\ell+k+2} \tag{10.1.5}$$

are positive definite on $\mathbb{N}_0$.

With this result at hand, we suggest the following strategy:

(1) Show that there exist strictly positive numbers $d_{n,N}$ such that

$$\frac{1}{(1 - z\overline{w})^N} = \sum_{n=0}^{\infty} \frac{z^n \overline{w}^n}{d_{n,N}}.$$

(2) Let $\mathcal{H}_N$ denote the reproducing kernel Hilbert space of functions analytic in the open unit disk with reproducing kernel $\frac{1}{(1-z\overline{w})^N}$. Show that

$$\langle z^\ell, z^k \rangle_{\mathcal{H}_N} = \delta_{\ell,k} d_{\ell,N}. \tag{10.1.6}$$

(3) Taking into account the special form of μ_N, translate (10.1.6) into a moment problem on the interval $[0, 1]$ and apply the previous theorem.

10.2 The Bergman space of analytic functions analytic in an ellipse

Recall that the Bergman spaces associated with an open set Ω have been defined in Exercise 4.1.17. We now discuss the density of the polynomials inside the space $\mathcal{B}_2(\Omega)$, when Ω is a bounded open set. This is important in view of the formula for the kernel function associated with Ω. In [244, p. 254] and using Riemann's mapping theorem, Zeev Nehari gives such a criteria. Nehari's result is:

Theorem 10.2.1 ([244, p. 254]). *Let Ω be a bounded domain and assume that $\mathbb{C} \setminus \Omega$ is the closure of a domain. Then the polynomials are dense in $\mathcal{B}_2(\Omega, dxdy)$.*

Note that an annulus will not meet the hypothesis of the theorem.

Exercise 10.2.2. *Show that the polynomials are not dense in the space $\mathcal{B}_2(A_R)$, where $R > 1$ and (as in (2.1.14)) A_R denotes the annulus*

$$A_R = \{z \in \mathbb{C} \,; 1 < |z| < R\}.$$

We present as an exercise another result of this type, based on Mergelyan's theorem (see for instance [266, Théorème 20.5]).

Theorem 10.2.3 (Mergelyan). *Let K be a compact subset of the complex plane such that $\mathbb{C} \setminus K$ is connected. Then any function analytic in the interior of K and continuous on K can be uniformly approximated by polynomials.*

Exercise 10.2.4. *Let K be a compact subset of the complex plane such that $\mathbb{C} \setminus K$ is connected. Show that the polynomials are dense in $\mathcal{B}_2(K, dxdy)$.*

Hint: Approximate K by an increasing sequence $K_1, K_2, \ldots$ of compact sets with connected complements, and included in $\overset{\circ}{K}$. On each K_n apply Mergelyan's theorem to f restricted to K_n.

Question 10.2.5 (see [244, p. 259]). *Let $a, b > 0$ be such that $r = a + b > 1$, and let $\mathscr{E}$ denote the ellipse*

$$\frac{x^2}{a^2} + \frac{y^2}{b^2} < 1.$$

The Bergman kernel associated with $\mathscr{E}$ is given by the formula

$$K(z, w) = \frac{4}{\pi} \sum_{n=0}^{\infty} \frac{(n+1)U_n(z)\overline{U_n(w)}}{r^{n+1} - r^{-(n+1)}}, \tag{10.2.1}$$

where $U_0, U_1, \ldots$ denote the Tchebycheff polynomials of the second kind.

10.3 The Bergman space of the annulus

The Bergman space of the annulus is studied in details in particular in [70, pp. 2 and 10]; where $R > 1$. There, and with the annulus $r < |z| < 1$, S. Bergman shows that the reproducing kernel of the associated Bergman space is given by the formula (5.6.5):

$$\frac{1}{\pi z \overline{w}} \left\{ \wp(\ln(z\overline{w})) + \frac{\eta_1}{\pi i} - \frac{1}{2 \ln r} \right\}.$$

In this expression, η_1 is a constant whose definition we will not recall here and $\wp$ denotes a Weierstrass function (with appropriate periods). The expression $\wp(\ln(z\overline{w}))$ is well defined in view of the periodicity of $\wp$.

Recall that the annulus A_R has been defined in (2.1.14) as the set of complex numbers such that $1 < |z| < R$,

Example 10.3.1. *Let $R > 1$, and let f be analytic in A_R.*

(1) *Compute*

$$\iint_{A_R} z^n \overline{z}^m \, dx dy, \quad m, n \in \mathbb{Z}. \tag{10.3.1}$$

(2) *Express in terms of the coefficient of the Laurent series of f the fact that*

$$\iint_{A_R} |f(z)|^2 \, dx dy < \infty.$$

10.4 The Bergman spaces of polyanalytic functions

Let now $\Omega = \mathbb{D}$ and $d\mu = dxdy$, the area measure in $\mathbb{C}$. There is an interesting family of Hilbert spaces between $\mathcal{B}_2(\mathbb{D})$ and $\mathbf{L}_2(\mathbb{D}, dxdy)$, namely the Bergman spaces of polyanalytic functions. See Balk's book [57, p. 169].

Definition 10.4.1. Let $N \in \mathbb{N}$. We denote by $\mathcal{B}_2^N(\mathbb{D})$ the closure in $\mathbf{L}_2(\mathbb{D}, dxdy)$ of the functions of the form

$$f(z) = \sum_{j=0}^{N} \bar{z}^j p_j(z)$$

where $p_0, \ldots, p_N$ are polynomials in z. When $p = 0$, we have $\mathcal{B}_2^0(\mathbb{D}) = \mathcal{B}_2(\mathbb{D})$.

Exercise 10.4.2. *Show that the range of the operator $M_{\bar{z}}$ of multiplication by $\bar{z}$ from $\mathcal{B}_2(\mathbb{D})$ into $\mathbf{L}_2(\mathbb{D}, dxdy)$ is closed.*

Exercise 10.4.3. *Show that the spaces $\bar{z}^j \mathcal{B}_2(\mathbb{D})$ are closed in $\mathbf{L}_2(\mathbb{D}, dxdy)$, but that their sum is not closed.*

Hint: Use Theorem 4.1.25.

It follows in particular from the preceding exercise that $\mathcal{B}_2^N(\mathbb{D})$ is *not* the space of functions of the form

$$f(z) = \sum_{j=0}^{N} \bar{z}^j h_j(z) \tag{10.4.1}$$

where the h_j belong to the Bergman space $\mathcal{B}_2(\mathbb{D})$.

It is not clear at this stage that the spaces $\mathcal{B}_2^N(\mathbb{D})$ are spaces of functions, let alone reproducing kernel Hilbert spaces. We consider these questions in the following exercises. We denote by $\mathcal{L}(\mathcal{B}_2)$ the space of the functions of the form (10.4.1), with $h_1, h_2, \ldots \in \mathcal{B}_2(\mathbb{D})$.

Exercise 10.4.4.

(1) *Find the function $k_0 \in \mathcal{B}_2^N(\mathbb{D})$ such that*

$$f(0) = \langle f, k_0 \rangle_{\mathcal{B}_2^N(\mathbb{D})}, \tag{10.4.2}$$

for all $f \in \mathcal{B}_2^N(\mathbb{D})$ of the form (10.4.1) with $h_1, h_2, \ldots$ polynomials.

(2) *Let $f \in \mathcal{B}_2^N(\mathbb{D})$ and $w \in \mathbb{D}$, and let b_w be a Blaschke factor (see (1.2.9)):*

$$b_w(z) = \frac{z - w}{1 - z\bar{w}}.$$

Show that the function

$$g_w(z) = f(b_{-w}(z)) \cdot \frac{(1 + \bar{z}w)^N}{(1 + z\bar{w})^{N+2}} \tag{10.4.3}$$

also belongs to $\mathcal{B}_2^N(\mathbb{D})$.

(3) *Show that for every $w \in \mathbb{D}$ there exists $k_w \in \mathcal{B}_2^N(\mathbb{D})$ such that*

$$f(w) = \langle f, k_w \rangle_{\mathcal{B}_2^N(\mathbb{D})}, \tag{10.4.4}$$

for all $f \in \mathcal{B}_2^N(\mathbb{D})$ of the form (10.4.1) with $h_1, h_2, \ldots$ polynomials.

Hints: (see Košelev's paper [207]) Remark that the functions $z^{2j} = \bar{z}^j z^j$ ($j = 0, 1, \ldots, N$) belong to $\mathcal{B}_2^N$ and look for h of the form

$$h(z) = \sum_{i=0}^{N} a_i |z|^{2i} \tag{10.4.5}$$

for appropriate choice of real constants $a_0, \ldots, a_N$.

In the proof of (2) it is enough to prove that the composition map $f \mapsto f(b_w)$ maps $\mathcal{B}_2(\mathbb{D})$ into itself. This can be done directly computing the norm of $f(b_w)$ in $\mathcal{B}_2(\mathbb{D})$.

Remark 10.4.5. The operator defined by (10.4.3) is of the form (7.6.7). It is in fact continuous thanks to Exercise 7.6.18.

Exercise 10.4.6. *Show that $\mathcal{B}_2^N(\mathbb{D})$ is a reproducing kernel Hilbert space.*

Remark 10.4.7. We note that the functions $e_{k,j}$ defined by (2.1.11) form an orthonormal basis of $\mathcal{B}_2^N(\mathbb{D})$ for $k = 0, \ldots, N$ and $j \in \mathbb{N}_0$. Furthermore an explicit form of the reproducing kernel is given by (0.0.6). See [207] and [57, (6.41), p. 169].

10.5 Solutions of the exercises

Solution of Exercise 10.1.1: Let $\rho \in (0, 1)$. For z such that $|z| = r < \rho$ we have:

$$|f(z)| \leq \sum_{n=0}^{\infty} |f_n| \rho^n = M_\rho < \infty.$$

Furthermore let, with $z = re^{it}$,

$$F_N(z) = \sum_{\substack{n,m \in \mathbb{N}_0 \\ n+m \leq N}} f_n \overline{f_m} r^{n+m} e^{it(n-m)}. \tag{10.5.1}$$

We have

$$\lim_{N \to \infty} F_N(z) = |f(z)|^2,$$

and

$$|F_N(z)| \leq M_\rho^2.$$

The dominated convergence theorem allows us to conclude that

$$\lim_{N \to \infty} \iint_{|z| \leq \rho} F_N(z) dx dy = \iint_{|z| \leq \rho} \lim_{N \to \infty} F_N(z) dx dy$$

$$= \iint_{|z| \leq \rho} |f(z)|^2 (z) dx dy.$$

On the other hand,

$$\iint_{|z|\leq\rho} F_N(z)\,dxdy = \sum_{\substack{n,m\in\mathbb{N}_0 \\ n+m\leq N}} \int_{t=0}^{2\pi}\int_{r=0}^{\rho} f_n\overline{f_m}r^{n+m}e^{it(n-m)}\,rdrdt,$$

and

$$\int_{t=0}^{2\pi}\int_{r=0}^{\rho} r^{n+m}e^{it(n-m)}\,rdrdt, = \begin{cases} 0, & \text{if } n\neq m, \\ 2\pi\frac{\rho^{2n+2}}{2n+2}, & \text{if } n=m. \end{cases}$$

Hence for $\rho\in(0,1)$ we have:

$$\sum_{n=0}^{\infty}\frac{|f_n|^2\rho^{2n+2}}{n+1} = \frac{1}{\pi}\iint_{|z|\leq\rho}|f(z)|^2\,dxdy.$$

One concludes by using the dominated convergence theorem to let $\rho\to1$. $\square$

Solution of Exercise 10.1.2: The existence of K_w follows from Riesz' theorem. The function K_w can also be computed directly from the power series characterization of the Bergman space:

$$K_w(z) = \sum_{n=0}^{\infty}(n+1)z^n\overline{w}^n. \tag{10.5.2}$$

$\square$

Remark 10.5.1. It follows from (10.5.2) that

$$\langle z^n, z^m\rangle_{\mathcal{B}_2(\mathbb{D})} = \frac{1}{n+1}\delta_{m,n}.$$

This formula can also be obtained from a direct computation of the inner product in terms of a double integral. We see in particular that the functions

$$\sqrt{n+1}z^n, \quad n = 0,1,2,\dots$$

form an orthonormal basis of the Bergman space $\mathcal{B}_2(\mathbb{D})$.

Solution of Exercise 10.1.3: Let $f\in\mathcal{B}_2(\mathbb{D})$ with Taylor expansion $f(z) = \sum_{n=0}^{\infty} f_n z^n$. Then, $(R_0f)(z) = \sum_{n=0}^{\infty} f_{n+1}z^n$ and we have

$$\begin{aligned} \|R_0f\|_{\mathcal{B}_2(\mathbb{D})}^2 &= \sum_{n=0}^{\infty}\frac{|f_{n+1}|^2}{n+1} \\ &= \sum_{n=0}^{\infty}\frac{|f_{n+1}|^2}{n+2}\frac{n+2}{n+1} \leq 2\sum_{n=0}^{\infty}\frac{|f_{n+1}|^2}{n+2} \leq 2\|f\|_{\mathcal{B}_2(\mathbb{D})}^2. \end{aligned}$$

$\square$

Solution of Exercise 10.1.4:

(1) We have for $z, w \in \mathbb{D}$:

$$
\frac{1}{(1-z\overline{w})^2} - \frac{1}{(1-s(z)\overline{s(w)})^2} = \left(\frac{1}{1-z\overline{w}} - \frac{1}{1-s(z)\overline{s(w)}} \right)
$$
$$
\times \left(\frac{1}{1-z\overline{w}} + \frac{1}{1-s(z)\overline{s(w)}} \right). \tag{10.5.3}
$$

By hypothesis, and using Exercise 7.6.6, the kernel

$$
\frac{1}{1-z\overline{w}} - \frac{1}{1-s(z)\overline{s(w)}}
$$

is positive definite in $\mathbb{D}$. Thus (10.5.3) expresses the kernel

$$
\frac{1}{(1-z\overline{w})^2} - \frac{1}{(1-s(z)\overline{s(w)})^2}
$$

as a product of two positive definite kernels, and the claim follows from Exercise 7.6.6.

We now prove (10.1.3). Let $M > 0$. We have

$$
\frac{1}{(1-z\overline{w})^2} - \frac{M}{(1-s(z)\overline{s(w)})^2} = \left(\frac{1}{1-z\overline{w}} + \frac{\sqrt{M}}{1-s(z)\overline{s(w)}} \right)
$$
$$
\times \left(\frac{1}{1-z\overline{w}} - \frac{\sqrt{M}}{1-s(z)\overline{s(w)}} \right). \tag{10.5.4}
$$

The kernel $\left(\frac{1}{1-z\overline{w}} + \frac{\sqrt{M}}{1-s(z)\overline{s(w)}} \right)$ is positive definite in $\mathbb{D}$ for any $M \geq 0$, while the kernel

$$
\frac{1}{1-z\overline{w}} - \frac{\sqrt{M}}{1-s(z)\overline{s(w)}} \tag{10.5.5}
$$

is positive definite for $M \leq \frac{1}{\|C_s\|^4_{\mathbf{H}_2(\mathbb{D})}}$. For such M the kernel (10.5.4) is positive definite in $\mathbb{D}$. To show the bound (10.1.3) we note that any M for which (10.5.4) is positive definite in $\mathbb{D}$ satisfies

$$
M\|C_s\|^2_{\mathbf{B}_2(\mathbb{D})} \leq 1.
$$

Since the kernel (10.5.5) is positive definite in $\mathbb{D}$ for

$$
M = \frac{1}{\|C_s\|^4_{\mathbf{H}_2(\mathbb{D})}}
$$

we obtain the desired inequality.

10.5. *Solutions of the exercises*

(2) We recall that (see [CAPB, Exercise 1.1.20, p. 21])

$$\frac{1 - b_a(z)\overline{b_a(w)}}{1 - z\overline{w}} = \frac{1 - |a|^2}{(1 - z\overline{a})(1 - a\overline{w})}.$$

Let

$$C = \frac{1 - |a|^2}{(1 + |a|)^2} = \frac{1 - |a|}{1 + |a|}.$$

We have

$$\frac{1}{(1 - z\overline{w})^2} - \frac{C^2}{(1 - b_a(z)\overline{b_a(w)})^2} = \frac{1}{(1 - z\overline{w})^2}\left(1 - \frac{C^2}{\left(\frac{1 - b_a(z)\overline{b_a(w)}}{1 - z\overline{w}}\right)^2}\right)$$

$$= \frac{1}{(1 - z\overline{w})^2}\left(1 - \frac{C^2}{\left(\frac{1 - |a|^2}{(1 - z\overline{a})(1 - a\overline{w})}\right)^2}\right)$$

$$= \frac{1}{1 - z\overline{w}}\left(1 + \frac{C(1 - z\overline{a})(1 - a\overline{w})}{1 - |a|^2}\right)\left(\frac{1 - \frac{\sqrt{C}(1 - z\overline{a})}{\sqrt{1 - |a|^2}}\frac{\sqrt{C}(1 - \overline{w}a)}{\sqrt{1 - |a|^2}}}{1 - z\overline{w}}\right)$$

$$= k_1(z, w)k_2(z, w)k_3(z, w),$$

with $k_1(z, w) = \frac{1}{1 - z\overline{w}}$, and (since $\sqrt{\frac{C}{1 - |a|^2}} = \frac{1}{1 + |a|}$),

$$k_2(z, w) = 1 + r(z)\overline{r(w)}, \qquad k_3(z, w) = \frac{1 - r(z)\overline{r(w)}}{1 - z\overline{w}},$$

where $r(z) = \frac{1 - z\overline{a}}{1 + |a|}$.

The kernels k_1 and k_2 are clearly positive definite in $\mathbb{D}$. So is the kernel k_3 since $\sup_{z \in \mathbb{D}} |r(z)| = 1$. It follows that the kernel

$$\frac{1}{(1 - z\overline{w})^2} - \frac{C^2}{(1 - b_a(z)\overline{b_a(w)})^2}$$

is positive definite in $\mathbb{D}$ and so $\|C_{b_a}\|_{\mathbf{B}_2(\mathbb{D})} \le \frac{1 + |a|}{1 - |a|}$. But the equality (with r as above)

$$\frac{1}{1 - z\overline{w}} - \frac{C}{1 - b_a(z)\overline{b_a(w)}} = \frac{1 - r(z)\overline{r(w)}}{1 - z\overline{w}}$$

shows that $\|C_{b_a}\|_{\mathbf{H}_2(\mathbb{D})} = \sqrt{\frac{1 + |a|}{1 - |a|}}$, and hence we obtain the required inequality. $\square$

Solution of Exercise 10.1.7:

(1) One has

$$\frac{1}{(1-z)^N} = \sum_{n=0}^{\infty} \frac{(N+n-1)!}{(N-1)!n!} z^n, \quad z \in \mathbb{D},$$

so that $d_{N,n} = \frac{(N-1)!n!}{(N+n-1)!}$.

(2) We thus have

$$2\pi \int_{[0,1]} r^{2n} d\mu_N(r) = \frac{(N-1)!n!}{(N+n-1)!} = \frac{(N-1)!}{(N+n-1)\cdots(n+1)}.$$

The change of variable $r^2 = u$ suggests that we look for a positive Borel measure on $[0,1]$ such that

$$\int_{[0,1]} u^n d\nu_N(u) = \frac{1}{(N+n-1)\cdots(n+1)}.$$

(3) We now apply Theorem 10.1.8 with $c_n = \frac{1}{(N+n-1)\cdots(n+1)}$. Since the function $\frac{1}{n+m+1}$ is positive definite on the real numbers we conclude that each of the functions

$$\frac{1}{N+n+m-1} = \frac{1}{(n+\frac{N-1}{2})+(m+\frac{N-1}{2})}$$

is positive definite on $\mathbb{R}$ and so is the function c_{n+m} defined by (10.1.4), since it is then a product of positive definite functions. We now consider the function (10.1.5). With $a = 0$ and $b = 1$ we have:

$$c_{n+m+1} - c_{n+m+2}$$
$$= \frac{1}{(N+n+m)\cdots(m+n+2)} - \frac{1}{(N+n+m+1)\cdots(m+n+3)}$$
$$= \frac{N-1}{(N+1+m+n)\cdots(m+n+2)}$$

which is positive definite, as is seen using the same argument as for the function (10.1.4). □

Solution of Exercise 10.2.4: There exists an increasing sequence $K_1, K_2, \ldots$ of compact sets, with connected complements, and such that $\overset{\circ}{K}_n \neq \emptyset$ and

$$K_1 \subset K_2 \subset \cdots \quad \text{and} \quad \cup_{n=1}^{\infty} K_n = K.$$

By Mergelyan's theorem there exist polynomials $p_1, p_2, \ldots$ such that

$$|f(z) - p_n(z)| \leq \frac{1}{n}, \quad z \in K_n, \ n = 1, 2, \ldots$$ □

10.5. *Solutions of the exercises*

Solution of Exercise 10.3.1:

(1) Using polar coordinates to compute the integral at hand we get

$$\iint_{A_R} z^n \bar{z}^m \, dx dy = \int_1^R \int_0^{2\pi} r^{n+m+1} e^{it(n-m)} \, dt dr$$

$$= \begin{cases} 0, & n \neq m, \\ \pi \frac{R^{2n+2}-1}{n+1}, & n = m \neq 1, \\ 2\pi \ln R, & n = m = -1. \end{cases}$$

(2) Let

$$f(z) = \sum_{n=0}^{\infty} a_n z^n + \sum_{n=1}^{\infty} \frac{b_n}{z^n}$$

denote the Laurent series of f in A_R. Furthermore, let for $N \in \mathbb{N}$,

$$g_N(z) = \left| \sum_{n=0}^{M} a_n z^n + \sum_{n=1}^{N} \frac{b_n}{z^n} \right|^2,$$

and let r_1 and r_2 be such that $1 < r_1 < r_2 < R$. The Laurent expansion of f converges absolutely and uniformly in the closed annulus

$$A_{r_1, r_2} = \{ z \in \mathbb{C} : r_1 \leq |z| \leq r_2 \},$$

and we have in particular:

$$|f(z)| \leq \sum_{n=0}^{\infty} |a_n| r_2^n + \sum_{n=1}^{\infty} \frac{|b_n|}{r_1^n}$$

and

$$\lim_{N \to \infty} g_N(z) = |f(z)|^2, \quad \forall z \in A_{r_1, r_2}.$$

Thus the dominated convergence theorem allows us to write:

$$\iint_{A_{r_1,r_2}} |f(z)|^2 dx dy = 2\pi \left(\int_{r_1}^{r_2} \sum_{n=0}^{\infty} |a_n|^2 r^{2n+1} dr \right.$$
$$\left. + |b_1|^2 (\ln r_2 - \ln r_1) + \sum_{n=2}^{\infty} \int_{r_1}^{r_2} |b_n|^2 r^{-2n+1} dr \right) \qquad (10.5.6)$$

Let now, as in the previous exercises, r_{1N} be a decreasing sequence of numbers such that $\lim_{N \to \infty} r_{1N} = r_1$ and r_{2N} be an increasing sequence of numbers such that $\lim_{N \to \infty} r_{2N} = r_2$ (and assume $r_{1N} < r_{2M}$ for every choice of $N, M \in \mathbb{N}$), and let

$$g_N(z) = |f(z)|^2 \mathbf{1}_{A_{r_{1N}, r_{2N}}}(z).$$

(10.5.6) leads to

$$\iint_{A_R} g_N(z)dxdy = 2\pi \left(\int_{r_1}^{r_2} \sum_{n=0}^{\infty} |a_n|^2 r^{2n+1} dr + \right.$$

$$\left. + |b_1|^2 (\ln r_2 - \ln r_1) + \sum_{n=2}^{\infty} \int_{r_1}^{r_2} |b_n|^2 r^{-2n+1} dr \right)$$

and the monotone convergence theorem allows us to conclude that

$$\pi \left(\left(\sum_{n=1}^{\infty} |a_n|^2 \frac{R^{2n+2}-1}{n+1} \right) + 2|b_1|^2 \ln R + \left(\sum_{n=2}^{\infty} |b_n|^2 \frac{R^{2-2n}-1}{1-n} \right) \right) < \infty$$

is a necessary and sufficient condition for the function f analytic in A_R to belong to $\mathbf{L}_2(A_R, dxdy)$. $\qquad\square$

We note that the functions

$$\frac{1}{\sqrt{2\pi \ln R}} \frac{1}{z} \quad \text{and} \quad \sqrt{\frac{n+1}{\pi(R^{2n+2}-1)}} z^n, \quad n \in \mathbb{Z} \setminus \{-1\}$$

form an orthonormal basis of the Bergman space of the annulus. The computation of the associated reproducing kernel

$$K(z,w) = \frac{1}{(2\pi \ln R)z\overline{w}} + \frac{1}{\pi} \sum_{\substack{n \in \mathbb{Z} \\ n \neq -1}} \frac{(n+1)z^n \overline{w}^n}{R^{2n+2}-1}$$

is done in [70, p. 2 and p. 10]. See (5.6.5) above.

Solution of Exercise 10.4.2: We have

$$\langle \overline{z}z^n, \overline{z}z^m \rangle_{\mathbf{L}_2(\mathbb{D}, dxdy)} = \langle z^{n+1}, z^{m+1} \rangle_{\mathbf{L}_2(\mathbb{D}, dxdy)} = \delta_{n,m} \frac{\pi}{n+2}.$$

Let now $h(z) = \sum_{n=0}^{\infty} h_n z^n \in \mathcal{B}_2(\mathbb{D})$. Then,

$$\|M_{\overline{z}}h\|^2 = \sum_{n=0}^{\infty} \frac{|h_n|^2}{n+2} \geq \sum_{n=0}^{\infty} \frac{|h_n|^2}{2n+2} = \frac{1}{2}\|h\|^2.$$

The inequality

$$\|M_{\overline{z}}h\|^2 \geq \frac{1}{2}\|h\|^2 \tag{10.5.7}$$

implies that the range is closed. Indeed, let $(M_{\overline{z}}h_n)_{n \in \mathbb{N}}$ be a Cauchy sequence in the range of $M_{\overline{z}}$. Then (10.5.7) implies that $(h_n)_{n \in \mathbb{N}}$ is a Cauchy sequence in the Bergman space and thus converges to some element, say h, of $\mathcal{B}_2(\mathbb{D})$. By continuity of $M_{\overline{z}}$ we then have $\lim_{n\to\infty} M_{\overline{z}}h_n = M_{\overline{z}}h$. $\qquad\square$

Solution of Exercise 10.4.3: We give the proof for $N = 1$ and will apply Theorem 4.1.25 with $\mathcal{H}_1 = \mathcal{B}_2(\mathbb{D})$ and $\mathcal{H}_2 = \overline{z}\mathcal{B}_2(\mathbb{D})$. From the previous exercise we know that $\mathcal{H}_2$ is closed in $\mathbf{L}_2(\mathbb{D}, \mathcal{B}, dxdy)$.

Assume that inequality (4.1.19) holds for some $c > 0$. Then (with some abuse of notation for the functions)

$$\|z^n - \overline{z}z^{n+1}\|^2_{\mathbf{L}_2(\mathbb{D},\mathcal{B},dxdy)} \geq c^2 \|z^n\|^2_{\mathbf{L}_2(\mathbb{D},\mathcal{B},dxdy)}, \quad n = 0, 1, \ldots$$

for some $c > 0$. Thus

$$\frac{1}{n+1} + \frac{1}{n+3} - \frac{2}{n+2} \geq c^2 \frac{1}{n+1}.$$

Multiplying both sides of this inequality by n and letting $n \to \infty$ we get $c = 0$, which is not possible by hypothesis. $\qquad\square$

Solution of Exercise 10.4.4:

(1) Let $f \in \mathcal{L}_N(\mathcal{B}_2)$ and write

$$f(z) = f_0(z) + \overline{z}f_1(z) + \cdots + \overline{z}^N f_N(z),$$

where the functions $f_0, \ldots, f_N \in \mathcal{B}_2(\mathbb{D})$. Set $f_i(z) = \sum_{n=0}^{\infty} f_{in}z^n$. We note that

$$\langle \overline{z}^j f_j, |z|^{2i}\rangle_{\mathcal{B}_2^N(\mathbb{D})} = \langle z^i f_j, z^{i+j}\rangle_{\mathcal{B}_2(\mathbb{D})} = f_{jj}\frac{\pi}{i+j+1}, \tag{10.5.8}$$

for $i, j = 0, \ldots, N$. Thus, with h of the form (10.4.5), we have

$$\langle f, h \rangle = \sum_{i,j=0}^{N} a_i f_{jj}\frac{\pi}{i+j+1} = \sum_{j=0}^{N} f_{jj}\left(\sum_{i=0}^{N} a_i \frac{\pi}{i+j+1}\right).$$

This last expression will be equal to $f(0) = f_{00}$ for all $f \in \mathcal{L}_N(\mathcal{B}_2(\mathbb{D}))$ if and only if $a_0, \ldots, a_N$ are chosen such that

$$\begin{pmatrix} \frac{1}{1} & \frac{1}{2} & \cdots & \frac{1}{1+N} \\ \frac{1}{2} & \frac{1}{3} & \cdots & \frac{1}{2+N} \\ & & & \\ \frac{1}{N+1} & \frac{1}{N+2} & \cdots & \frac{1}{2N+1} \end{pmatrix} \begin{pmatrix} a_0 \\ a_1 \\ a_2 \\ \vdots \\ a_N \end{pmatrix} = \begin{pmatrix} \frac{1}{\pi} \\ 0 \\ 0 \\ \vdots \\ 0 \end{pmatrix}.$$

This is always possible since the Hankel matrix (1.6.5)

$$\begin{pmatrix} \frac{1}{1} & \frac{1}{2} & \cdots & \frac{1}{1+N} \\ \frac{1}{2} & \frac{1}{3} & \cdots & \frac{1}{2+N} \\ & & & \\ \frac{1}{N+1} & \frac{1}{N+2} & \cdots & \frac{1}{2N+1} \end{pmatrix}$$

is invertible (and in fact strictly positive; see Exercise 4.2.34).

(2) With h as in (1) and g_w defined by (10.4.3) with $f \in \mathcal{B}_2^N(\mathbb{D})$, we have:

$$f(w) = g_w(0)$$

$$= \iint_{\mathbb{D}} g_w(z)\overline{h(z)}\,dxdy = \iint_{\mathbb{D}} g_w(b_w(\zeta))\overline{h(b_w(\zeta))}|b'_w(\zeta)|^2\,dadb$$

(where we have made the change of variable $z = b_w(\zeta)$ with Jacobian equal to $|b'_w(\zeta)|^2$, and we set $\zeta = a + ib$)

$$= \iint_{\mathbb{D}} f(\zeta)\frac{\dfrac{(1-|w|^2)^N}{(1-\overline{\zeta}w)^N}}{\dfrac{(1-|w|^2)^{N+2}}{(1-\zeta\overline{w})^{N+2}}}\,\overline{h(b_w(\zeta))}\frac{(1-|w|^2)^2}{|1-\zeta\overline{w}|^4}\,dadb$$

(since $b'_w(\zeta) = \frac{1-|w|^2}{(1-\zeta\overline{w})^2}$)

$$= \iint_{\mathbb{D}} f(\zeta)\overline{k_w(\zeta)}\,dadb \quad \text{with} \quad k_w(\zeta) = \frac{(1-\overline{\zeta}w)^N}{(1-\zeta\overline{w})^{N+2}}h(b_w(\zeta)).$$

To conclude, it follows from

$$k_w(\zeta) = \frac{(1-\overline{\zeta}w)^N}{(1-\zeta\overline{w})^{N+2}}\left(\sum_{i=0}^{N} a_i(b_w(\zeta))^i\overline{(b_w(\zeta))}^i\right)$$

$$= \sum_{i=0}^{N}\underbrace{\left((1-\zeta\overline{w})^{N-i}(\overline{\zeta}-\overline{w})^i\right)}_{\text{polynomial of degree } N \text{ in } \overline{\zeta}}\underbrace{\left(\frac{a_i(\zeta-w)^i}{(1-\zeta\overline{w})^{N+2+i}}\right)}_{\in\,\mathcal{B}_2}$$

that the function $k_w \in \mathcal{B}_2^N(\mathbb{D})$.
$\square$

Remark 10.5.2. The properties of the inner product of $\mathbf{L}_2(\mathbb{D}, dxdy)$ really used are

$$\langle zf(z), g(z)\rangle = \langle f(z), \overline{z}g(z)\rangle, \tag{10.5.9}$$

$$\langle z^i, z^j\rangle = 0, \quad \text{for } i \neq j, \tag{10.5.10}$$

and that, with $\alpha_n = \langle z^n, z^n\rangle$, the Hankel matrix

$$H = (\alpha_{i+j})_{i,j=0,\ldots,N}$$

is invertible. We note that this last property will not hold in the case of the Lebesgue space $\mathbf{L}_2(\mathbb{T}, dx)$, which corresponds to the kernel $\frac{1}{1-z\overline{w}}$, that is to the Hardy space.

Solution of Exercise 10.4.6: Formula (10.4.4) implies that Cauchy sequences of functions of the form (10.4.1) with $h_1, h_2, \ldots$ polynomials converge also pointwise.
$\square$

Chapter 11

Fock Spaces

Bargmann–Fock–Segal spaces of analytic functions play an important role in quantum mechanics; we refer for instance to the book of Yurii Neretin [246]. In this chapter we collect some exercises related to these spaces and to their polyanalytic extensions. Important related tools, such as the Bargmann transform (see Exercise 2.1.27), were introduced in the earlier chapters.

11.1   The Bargmann–Fock–Segal spaces of analytic functions

As we saw earlier in the book, the operators

$$Pf(z) = zf(z) \quad \text{and} \quad Qf(z) = f'(z)$$

satisfy the equation $QP - PQ = I$, and are continuous in the Schwartz space and in the space of functions analytic in an open set Ω. On the other hand they are never simultaneously continuous in a normed space (say, of functions analytic in some open set). Motivated by the work of Fock [134] on quantum field theory, Bargmann [60, 61] introduced a Hilbert space of entire functions where these operators are adjoint of each other. They are then both unbounded.

Before giving the definition of the Bargmann space (also called Fock, or Bargmann–Fock–Segal space) we present one preliminary exercise. In [CAPB, Exercise 5.6.9 (2), p. 213], the following question is posed: *Show that there are no entire functions (different from the function identically equal to 0) such that*

$$\iint_{\mathbb{C}} |f(z)|^2 dx dy < \infty.$$

Similarly we ask now:

Exercise 11.1.1. *Show that there are no nonzero entire function f such that*

$$\iint_{\mathbb{C}} |f(z)|^2 e^{x^2-y^2}\,dxdy < \infty. \tag{11.1.1}$$

Hint: Use [CAPB, Exercise 5.6.9 (2), p. 213].

Remark 11.1.2. As for the Hardy space, the Bargmann space defined below will admit four characterizations:

(1) The first is geometric. This is the second item in Exercise 11.1.3.
(2) The second is in terms of reproducing kernel. This is Exercise 11.1.4.
(3) The third is analytic. This is the first item in Exercise 11.1.3.
(4) The fourth is in terms of a transform (the Bargmann transform). This is Exercise 2.1.27.

Exercise 11.1.3. *Let f be an entire function, with power series expansion $f(z) = \sum_{n=0}^{\infty} f_n z^n$. Show that*

$$\sum_{n=0}^{\infty} n!|f_n|^2 < \infty \tag{11.1.2}$$

if and only if $\iint_{\mathbb{C}} |f(z)|^2 e^{-|z|^2}\,dxdy < \infty$, and that, when one of these expressions is finite, it holds that:

$$\sum_{n=0}^{\infty} n!|f_n|^2 = \frac{1}{\pi} \iint_{\mathbb{C}} |f(z)|^2 e^{-|z|^2}\,dxdy. \tag{11.1.3}$$

The space of entire functions f for which (11.1.3) is finite is a Hilbert space called the Bargmann space, Fock space or Bargmann–Fock–Segal space. We will denote it by $\mathcal{F}$. More precisely, it is the symmetric Fock space associated to $\mathbb{C}$ (the full Fock space associated with $\mathbb{C}$ being the space $\ell_2(\mathbb{N}_0)$). The Fock space was given in the previous exercise a geometric and analytic interpretation. To complete the picture we now give:

Exercise 11.1.4.

(1) The Fock space is the reproducing kernel Hilbert space with reproducing kernel $e^{z\overline{w}}$.
(2) The polynomials are dense in the Fock space.

Exercise 11.1.5. *Using formula (8.1.1), or by direct computations, compute (with $u = x + iy$)*

$$\frac{1}{\pi} \iint_{\mathbb{C}} \overline{u}^m u^n e^{z\overline{u}} e^{-|u|^2}\,dxdy. \tag{11.1.4}$$

Hint: Apply (8.1.1) to $f(u) = u^m$ and $K(z,w) = e^{z\overline{w}}$.

Analytic description is given in terms of the Bargmann transform. See Exercise 2.1.27. Summarizing these three results we have:

Theorem 11.1.6. *Let f be an entire function, with power series expansion $f(z) = \sum_{n=0}^{\infty} a_n z^n$. The following are equivalent:*

(1) *f satisfies*

$$\iint_{\mathbb{C}} |f(z)|^2 e^{-|z|^2}\, dx dy < \infty.$$

(2) *f satisfies*

$$\sum_{n=0}^{\infty} n! |f_n|^2 < \infty.$$

(3) *There exists $g \in \mathbf{L}_2(\mathbb{R}, dx)$ such that*

$$f(z) = \int_{\mathbb{R}} g(u)\overline{h_z(u)}\,du,$$

where

$$h_z(u) = \frac{1}{\pi^{\frac{1}{4}}} e^{\{-\frac{1}{2}(\bar{z}^2 + u^2) + \sqrt{2}\bar{z}u\}}.$$

Exercise 11.1.7.

(1) *Show directly (that is, without resorting to Exercise 4.3.5) that the operators (5.6.1) (denoted here by M_z and ∂_z respectively)*

$$(M_z f)(z) = z f(z) \quad \text{and} \quad (\partial_z f)(z) = f'(z) \tag{11.1.5}$$

are not bounded in the Fock space.

(2) *Show that $M_z^* = \partial_z$ and $\partial_z^* = M_z$.*

(3) *What are the eigenvectors of the operators $M_z \partial_z$ and $\partial_z M_z$?*

Exercise 11.1.8. *Show that the operator $\operatorname{Re} M_z$ has no point spectrum in the Fock space.*

Exercise 11.1.9. *Let $w_0 \in \mathbb{C}$. Show that the weighted composition operator*

$$(T_{w_0} f)(z) = e^{-\frac{|w_0|^2}{2}} e^{-z\overline{w_0}} f(z + w_0) \tag{11.1.6}$$

is unitary from the Fock space onto itself, and compute its inverse.

Hint: Use Exercise 7.6.6 to prove that T_{w_0} is coisometric and compute $T_{-w_0} T_{w_0}$.

Remark 11.1.10. The analysis and results in this section extend without much difficulties to the case of several complex variables, the corresponding positive definite function being $e^{\sum_{n=1}^{N} z_n \overline{w_n}}$. The situation is much more involved in the case of an infinite number of variables, that is, when the positive definite function is (7.1.14):

$$K(z, w) = e^{\langle z, w\rangle_{\ell_2}}, \quad z, w \in \ell_2.$$

One enters then the realm of infinite-dimensional analysis.

The counterpart of Exercise 11.1.9 in the reproducing kernel Hilbert space with reproducing kernel (7.1.14) (that is, in the symmetric Fock space associated with ℓ_2) is:

Question 11.1.11. *Let $\mathcal{F}(\ell_2)$ be the reproducing kernel Hilbert space with reproducing kernel (7.1.14), and let $u \in \ell_2$. Show that the map which to $f \in \mathcal{F}(\ell_2)$ associates the function*

$$a \mapsto e^{-\frac{\|u\|_2^2}{2} - \langle a, u\rangle_2} f(a + u)$$

is a unitary map from $\mathcal{F}(\ell_2)$ into itself.

11.2   The Bargmann–Fock–Segal spaces of polyanalytic functions

We defined the Bergman spaces of polyanalytic functions in Section 10.4. See Exercise 10.4.4. The Fock space of polyanalytic functions is discussed in [57, pp. 169–170], and is the topic of the exercise below. We denoted by $\mathcal{L}_N(\mathcal{F})$ the set of functions of the form (10.4.1), that is,

$$f(z) = \sum_{j=0}^{N} \bar{z}^j h_j(z)$$

with $h_0, h_1, \ldots, h_N \in \mathcal{F}$.

Exercise 11.2.1. *Let $N \in \mathbb{N}_0$. We denote by $\mathcal{F}^N$ the closure in $\mathbf{L}_2(\mathbb{R}^2, e^{-|z|^2} dx dy)$ of the functions*

$$z^n \bar{z}^j, \quad n \in \mathbb{N}_0 \quad and \quad j = 0, \ldots, N.$$

(1) Find the function $k_0 \in \mathcal{F}^N$ such that

$$f(0) = \langle f, k_0 \rangle_{\mathcal{F}^N}, \tag{11.2.1}$$

for all $f \in \mathcal{L}_N(\mathcal{F})$.

(2) Let $f \in \mathcal{L}_N(\mathcal{F})$ and $w \in \mathbb{C}$. Show that the function

$$g_w(z) = e^{z\bar{w}} f(z + w) \tag{11.2.2}$$

is of the same form.

(3) Show that for every $w \in \mathbb{C}$ there exists $k_w \in \mathcal{F}^N$ such that

$$f(w) = \langle f, k_w \rangle_{\mathcal{F}^N}, \tag{11.2.3}$$

for all $f \in \mathcal{F}^N$ of the form (10.4.1) with $h_0, h_1, \ldots, h_N$ polynomials.

Hints: As in Exercise 10.4.4, remark that the functions $z^{2j} = \overline{z}^j z^j$ $(j = 0, 1, \ldots, N)$ belong to $\mathcal{F}^N$ and look for k_0 of the form (10.4.5) for an appropriate choice of real constants $a_0, \ldots, a_N$. In the proof the $(N+1) \times (N+1)$ matrix with (m, n) entry equal to $(m + n)!$ appears. Use Exercise 1.6.2 to show that this matrix is strictly positive.

Remark 11.2.2. In fact the function k_w sought for in item (3) in the previous exercise is given by (0.0.7).

The proof of the following question is as the proof of Exercise 10.4.6.

Question 11.2.3. *Show that $\mathcal{F}^N$ is a reproducing kernel Hilbert space.*

Exercises 10.4.4, 10.4.6, 11.2.1 and Question 11.2.3 suggest the following problems:

Question 11.2.4.

(1) *Given a positive measure on $\Omega \subset \mathbb{C}$ such that the closed linear span of the polynomials in $\mathbf{L}_2(\Omega, d\mu)$ is a reproducing kernel Hilbert space, define (if possible) the corresponding Hilbert space of polyanalytic functions and find its reproducing kernel. The case of an annulus should be of special interest.*

(2) *Consider the polyanalytic versions of the Dirichlet space and of other spaces of the kind considered in this book.*

(3) *More generally, if $\mathcal{H}$ is a reproducing kernel Hilbert space of functions analytic in a set $\Omega \subset \mathbb{C}$, characterize the closure (in some appropriate sense) of the functions of the form (10.4.1) with $h_1, \ldots, h_N \in \mathcal{H}$.*

11.3 Solutions of the exercises

Solution of Exercise 11.1.1: We note that

$$e^{x^2 - y^2} = |e^{z^2}|,$$

and so (11.1.1) can be rewritten as

$$\iint_{\mathbb{C}} |f(z)e^{z^2/2}|^2 \, dx \, dy < \infty.$$

By [CAPB, Exercise 5.6.9 (2), p. 213], $f(z)e^{z^2/2} \equiv 0$, and so $f(z) \equiv 0$. $\qquad\square$

Solution of Exercise 11.1.3: The proof is similar to that of Exercise 10.1.1. First fix $\rho \in (0, \infty)$, and define F_N as in (10.5.1). Then a change of variable and the dominated convergence theorem leads to:

$$\lim_{N \to \infty} \frac{1}{\pi} \iint_{|z| \leq \rho} |F_N(z)|^2 e^{-|z|^2} dx dy = \frac{1}{\pi} \iint_{|z| \leq \rho} |f(z)|^2 e^{-|z|^2} dx dy$$

$$= 2 \sum_{n=0}^{\infty} |f_n|^2 \int_0^{\rho} r^{2n+1} e^{-r^2} dr$$

$$= \sum_{n=0}^{\infty} |f_n|^2 \int_0^{\sqrt{\rho}} u^n e^{-u} du.$$

The monotone convergence theorem now allows to let $\rho \to \infty$ and to conclude since

$$\int_0^{\infty} u^n e^{-u} du = \Gamma(n+1) = n!, \qquad (11.3.1)$$

where Γ denotes the Gamma function.

Remark 11.3.1. We note that (11.3.1) corresponds to $\nu = n$ in (8.8.16) and to $t = 0$ in the integral (2.1.36). $\square$

Solution of Exercise 11.1.4:

(1) For $w \in \mathbb{C}$ the function

$$e^{z\overline{w}} = \sum_{n=0}^{\infty} z^n \frac{\overline{w}^n}{n!}$$

belongs to $\mathcal{F}$ since

$$\sum_{n=0}^{\infty} n! \left| \frac{\overline{w}^n}{n!} \right|^2 = \sum_{n=0}^{\infty} \frac{|w|^{2n}}{n!} < \infty, \quad \forall w \in \mathbb{C}.$$

Furthermore, by definition of the inner product we have for $f(z) = \sum_{n=0}^{\infty} a_n z^n \in \mathcal{F}$:

$$\langle f(z), e^{z\overline{w}} \rangle_{\mathcal{F}} = \sum_{n=0}^{\infty} n! a_n \frac{\overline{\overline{w}^n}}{n!} = \sum_{n=0}^{\infty} a_n w^n = f(w).$$

(2) It follows from the previous item that

$$\langle z^n, z^m \rangle_{\mathcal{F}} = n! \delta_{m,n}.$$

The formula

$$e^{z\overline{w}} = \sum_{n=0}^{\infty} \frac{z^n}{\sqrt{n!}} \frac{\overline{w}^n}{\sqrt{n!}}$$

implies then the asserted density. $\square$

Solution of Exercise 11.1.5: Applying (8.1.1) to $f(z) = u^n$ and $K(u,z) = e^{u\overline{z}}$ we have

$$\frac{1}{\pi} \iint_{\mathbb{C}} \overline{u}^m u^n e^{z\overline{u}} e^{-|u|^2} \, dx dy = \left\langle u^n, \frac{\partial^m}{\partial \overline{z}^m} e^{u\overline{z}} \right\rangle_{\mathcal{F}} = (z^n)^{(m)}$$

$$= \begin{cases} 0, & \text{if } m > n, \\ n(n-1)\cdots(n-m+1)z^{n-m}, & \text{if } m \geq n. \end{cases} \qquad \square$$

Solution of Exercise 11.1.7:

(1) A first way to see that M_z is unbounded is via the characterization of multipliers in a reproducing kernel Hilbert space (see Exercise 7.6.6). Since for any $M > 0$

$$\lim_{z \to \infty} e^{|z|^2}(M - |z|^2) = -\infty,$$

there does not exist a $M > 0$ such that the kernel $e^{z\overline{w}}(M - z\overline{w})$ is positive definite in $\mathbb{C}$, and so M_z is unbounded.

Another quicker way is to note that

$$\frac{\|M_z(z^n)\|_{\mathcal{F}}^2}{\|z^n\|_{\mathcal{F}}^2} = \frac{\|z^{n+1}\|_{\mathcal{F}}^2}{\|z^n\|_{\mathcal{F}}^2} = \frac{(n+1)!}{n!} = n+1 \to \infty \quad \text{as} \quad n \to \infty.$$

As for the differentiation operator ∂_z, it is also unbounded since

$$\frac{\|(z^n)'\|_{\mathcal{F}}^2}{\|z^n\|_{\mathcal{F}}^2} = n^2 \frac{\|z^{n-1}\|_{\mathcal{F}}^2}{\|z^n\|_{\mathcal{F}}^2} = n^2 \frac{(n-1)!}{n!} = n \to \infty \quad \text{as} \quad n \to \infty.$$

(2) Recall that the domain of the adjoint operator M_z^* is the space of elements $g \in \mathcal{F}$ such that the map

$$f \mapsto \langle M_z f, g \rangle_{\mathcal{F}} \qquad (11.3.2)$$

extends to a continuous linear functional on $\mathcal{F}$. Let $f \in \mathcal{F}$ with power series expansion $f(z) = \sum_{n=0}^{\infty} a_n z^n$. Then, $f \in \text{Dom } M_z$ if and only if

$$\sum_{n=1}^{\infty} |a_{n-1}|^2 n! < \infty.$$

For such an f and for $g \in \mathcal{F}$ with powers series expansion $g(z) = \sum_{n=0}^{\infty} b_n z^n$, we have

$$\langle M_z f, g \rangle_{\mathcal{F}} = \sum_{n=1}^{\infty} a_{n-1} \overline{b_n} n! = \sum_{n=0}^{\infty} a_n ((n+1)\overline{b_{n+1}}) n!.$$

Thus, (11.3.2) is bounded if and only if

$$\partial_z g(z) = \sum_{n=0}^{\infty} (n+1) b_{n+1} z^n \in \mathcal{F}.$$

This shows that $M_z^* = \partial_z$. The other claim is proved in the same way.

(3) We look for solutions in $\mathcal{F}$ of the equations

$$zf'(z) = \lambda f(z) \quad \text{and} \quad zf'(z) + f(z) = \lambda f(z)$$

for some complex number λ. Note that these equations are identical (up to replacing λ by $\lambda - 1$ in the first equation). To solve the first equation, it suffices to look for a solution of the form $f(z) = \sum_{n=0}^{\infty} a_n z^n$, where the coefficients a_n are subject to (11.1.2). Thus, for the first equation, we have:

$$\sum_{n=1}^{\infty} n a_n z^n = \lambda \sum_{n=0}^{\infty} a_n z^n.$$

Comparing coefficients, we get

$$0 = \lambda a_0 \quad \text{and} \quad n a_n = \lambda a_n, \quad n = 1, 2, \dots$$

and it follows that the eigenvalues of $M_z \partial_z$ are exactly the functions

$$f_p(z) = z^p, \quad p = 0, 1, \dots,$$

with corresponding eigenvalue p. Finally, the functions f_{p+1} are the eigenfunctions of the operator $\partial_z M_z$, with corresponding eigenvalues $p+1$ where $p = 0, 1, \dots$. $\quad\square$

Solution of Exercise 11.1.8: By Exercise 2.1.14, the solutions of the equation

$$\operatorname{Re} M_z u = \mu u$$

are of the form $u(z) = c e^{\mu z - \frac{z^2}{2}}$, and such functions are in not the Fock space for $c \neq 0$. Indeed, with $\mu = a + ib$,

$$|u(z)| = |c| e^{(\operatorname{Re}(\mu z)) + \frac{y^2 - x^2}{2}} = |c| e^{ax - by + \frac{y^2 - x^2}{2}}.$$

So

$$|u(z)|^2 e^{-|z|^2} = |c|^2 e^{2ax - 2x^2} e^{-2by},$$

which is not integrable with respect to the Lebesgue measure of the plane (since, for every real b, the function $y \mapsto e^{-2by}$ does not belong to $\mathbf{L}_2(\mathbb{R}, \mathcal{B}, dy)$). $\quad\square$

Solution of Exercise 11.1.9: We use Exercise 7.6.6 with

$$m(z) = e^{-\frac{|w_0|^2}{2}} e^{-z\overline{w_0}}, \quad \varphi(z) = z + w_0, \quad \text{and} \quad K_1(z, w) = K_2(z, w) = e^{z\overline{w}}.$$

Then,

$$e^{z\overline{w}} - e^{-\frac{|w_0|^2}{2}} e^{-z\overline{w_0}} e^{(z+w_0)(\overline{w}+\overline{w_0})} e^{-\overline{w}w_0} e^{-\frac{|w_0|^2}{2}} = 0.$$

Thus equation (7.6.9) holds:

$$K_2(z,w) = m(z)K_1(\varphi(z)), \varphi(w))\overline{m(w)}, \quad \forall\, z, w \in \mathbb{C},$$

and so, by Exercise 7.6.6, T_{w_0} is unitary from the Fock space onto itself. Furthermore, for $f \in \mathcal{F}$ it holds that

$$
\begin{aligned}
(T_{-w_0}T_{w_0}f)(z) &= e^{-\frac{|w_0|^2}{2}} e^{z\overline{w_0}} (T_{w_0}f)(z - w_0) \\
&= e^{-\frac{|w_0|^2}{2}} e^{z\overline{w_0}} e^{-\frac{|w_0|^2}{2}} e^{-(z-w_0)\overline{w_0}} f(z + w_0 - w_0) \\
&= f(z).
\end{aligned}
$$

Thus,

$$T_{-w_0}T_{w_0} = T_{w_0}T_{-w_0} = I_{\mathcal{F}}.$$

It follows that $T_{w_0}^{-1} = T_{-w_0}$.

Remark 11.3.2. More generally we note that

$$T_{w_0}T_{w_1} = e^{\frac{\overline{w_0}w_1 - \overline{w_1}w_0}{2}} T_{w_0+w_1}, \quad \forall\, w_0, w_1 \in \mathbb{C}.$$

Therefore we have

$$T_{w_0}T_{w_1} = e^{\overline{w_0}w_1 - \overline{w_1}w_0} T_{w_1}T_{w_0}, \quad \forall\, w_0, w_1 \in \mathbb{C}.$$

Solution of Exercise 11.2.1:

(1) In a way similar to the proof of Exercise 10.4.4 we have:

$$
\begin{pmatrix}
0! & 1! & \cdots & N! \\
1! & 2! & \cdots & (N+1)! \\
 & & & \\
N! & (N+1)! & \cdots & (2N)!
\end{pmatrix}
\begin{pmatrix}
a_0 \\
a_1 \\
a_2 \\
\vdots \\
a_N
\end{pmatrix}
=
\begin{pmatrix}
\frac{1}{\pi} \\
0 \\
0 \\
\vdots \\
0
\end{pmatrix}.
\tag{11.3.3}
$$

It is always possible to solve this system of equations since the Hankel matrix appearing in (11.3.3) is invertible. To check this last fact, set

$$
H_N =
\begin{pmatrix}
0! & 1! & \cdots & N! \\
1! & 2! & \cdots & (N+1)! \\
 & & & \\
N! & (N+1)! & \cdots & (2N)!
\end{pmatrix}
\tag{11.3.4}
$$

and let M_N be as in Exercise 1.6.2. Then

$$M_N = D_N H_N D_N$$

484

Chapter 11. Fock Spaces

where $D_N = \text{diag}\left(\frac{1}{0!}, \frac{1}{1!}, \frac{1}{2!}, \ldots, \frac{1}{N!}\right)$. So H_H is strictly positive since M_N is strictly positive (see Exercise 1.6.2 for the latter).

(2) First note that

$$\overline{(z+w)^k} = \sum_{u=0}^{k} \overline{z}^u \overline{w}^{k-u} \binom{k}{u}.$$

The claim follows by noting furthermore that for any polynomial p and any $c \in \mathbb{C}$ the function $z \mapsto e^{cz} p(z)$ belongs to $\mathcal{F}$.

(3) Let now $f(z) = \sum_{j=0}^{N} \overline{z}^j p_j(z)$, where $p_0, \ldots, p_N$ are polynomials in z, and let

$$F(z) = e^{-z\overline{w}} f(z+w).$$

In view of the preceding remarks, $F \in \mathcal{L}_N(\mathcal{F})$. By (1) we have

$$F(0) = \langle F, k_0 \rangle_{\mathcal{F}^N},$$

that is

$$f(w) = \frac{1}{\pi} \iint_{\mathbb{C}} F(z) \overline{k_0(z)} e^{-|z|^2} dx dy.$$

Hence (and with the abuse of notation as in (0.0.8)),

$$f(w) = \langle e^{-z\overline{w}} f(z+w), k_0(z) \rangle_{\mathcal{F}^N}$$
$$= \frac{1}{\pi} \iint_{\mathbb{C}} e^{-z\overline{w}} f(z+w) \overline{k_0(z)} e^{-|z|^2} dx dy$$
$$= \frac{1}{\pi} \iint_{\mathbb{C}} f(z) e^{-(z-w)\overline{w}} \overline{k_0(z-w)} e^{-|z-w|^2} dx dy$$
$$= \frac{1}{\pi} \iint_{\mathbb{C}} f(z) \overline{(k_0(z-w) e^{z\overline{w}})} e^{-|z|^2} dx dy$$

which gives

$$k_w(z) = k_0(z-w) e^{z\overline{w}}$$

since

$$e^{-(z-w)\overline{w}} e^{-|z-w|^2} = e^{\overline{z}w - |z|^2)}.$$

□

Remark 11.3.3. The reproducing kernel of $\mathcal{F}^N$ is given in [57, p. 170] to be equal to

$$F_N(z, w) = e^{z\overline{w}} \left(\sum_{k=0}^{N} (-1)^k \binom{N+1}{k+1} \frac{1}{k!} |z-w|^{2k} \right).$$

This formula is obtained also from the above analysis since (see Exercise 1.6.4) the first column of M_N^{-1} (where M_N is the Hankel matrix defined by (11.3.4)) is the $(N+1) \times 1$ vector

$$\begin{pmatrix} (-1)^0 \begin{pmatrix} N+1 \\ 0+1 \end{pmatrix} \\ \vdots \\ (-1)^k \begin{pmatrix} N+1 \\ k+1 \end{pmatrix} \\ \vdots \\ (-1)^N \begin{pmatrix} N+1 \\ N+1 \end{pmatrix} \end{pmatrix}, \tag{11.3.5}$$

and hence the first column of H_N^{-1} is equal to

$$\begin{pmatrix} \frac{(-1)^0}{0!} \begin{pmatrix} N+1 \\ 0+1 \end{pmatrix} \\ \vdots \\ \frac{(-1)^k}{k!} \begin{pmatrix} N+1 \\ k+1 \end{pmatrix} \\ \vdots \\ \frac{(-1)^N}{N!} \begin{pmatrix} N+1 \\ N+1 \end{pmatrix} \end{pmatrix}. \tag{11.3.6}$$

Remark 11.3.4. When $N = 1$ we have as a particular case that the function

$$F_1(z, w) = e^{z\overline{w}}(2 - |z - w|^2) \tag{11.3.7}$$

is positive definite in $\mathbb{C}$. A direct proof of this fact (without knowing the underlying inner product) is given in Exercise 7.2.5. The proof there is elementary, but does not provide insights on the associated reproducing kernel Hilbert space. On the other hand, specializing the previous analysis, it is not difficult to prove directly that (with $u = x + iy$)

$$e^{z\overline{w}}(2 - |z - w|^2) = \frac{1}{\pi} \iint_{\mathbb{C}} e^{u\overline{w}}(2 - |u - w|^2)e^{\overline{u}z}(2 - |\overline{u} - \overline{z}|^2)e^{-|u|^2}\, dxdy. \tag{11.3.8}$$

More precisely, (11.3.8) with $w = 0$ amounts to

$$2 - |z|^2 = \frac{1}{\pi} \iint_{\mathbb{C}} (2 - |u|^2)e^{\overline{u}z}(2 - |\overline{u} - \overline{z}|^2)e^{-|u|^2}\, dxdy$$
$$= \frac{1}{\pi} \iint_{\mathbb{C}} (2 - |u|^2)(2 - |z|^2 - |u|^2 + u\overline{z} + \overline{u}z)e^{\overline{u}z}e^{-|u|^2}\, dxdy.$$

The various integrals

$$\iint_{\mathbb{C}} |u|^4 e^{-|u|^2}\, dxdy, \dots$$

are given by (11.1.4) and the equality (11.3.8) readily follows when $w = 0$, that is,

$$2 - |z|^2 = \frac{1}{\pi} \iint_{\mathbb{C}} (2 - |u|^2) e^{\overline{u}z} (2 - |\overline{u} - \overline{z}|^2) e^{-|u|^2} dx dy.$$

The case of arbitrary w is obtained from the case $w = 0$ by replacing z by $z - w$ and then making the change of variable $u = t - w$ in the above integral. More precisely, with $t = a + ib$ we have:

$$2 - |z - w|^2 = \frac{1}{\pi} \iint_{\mathbb{C}} (2 - |u|^2) e^{\overline{u}(z-w)} (2 - |\overline{u} - \overline{z} + \overline{w}|^2) e^{-|u|^2} dx dy$$

$$= \frac{1}{\pi} \iint_{\mathbb{C}} (2 - |t - w|^2) e^{(\overline{t} - \overline{w})(z-w)} (2 - |\overline{t} - \overline{z}|^2) e^{-|t-w|^2} da db$$

$$= e^{-z\overline{w}} \frac{1}{\pi} \iint_{\mathbb{C}} (2 - |t - w|^2) e^{t\overline{w}} (2 - |\overline{t} - \overline{z}|^2) e^{\overline{t}z} e^{-|t|^2} da db,$$

since

$$e^{(\overline{t} - \overline{w})(z-w)} e^{-|t-w|^2} = e^{\overline{t}z + t\overline{w} - \overline{w}z - |t|^2}.$$

Bibliography

[1] S. Abou-Jaoudé and J. Chevalier. *Cahiers de mathématiques. Analyse I. Topologie.* O.C.D.L., 65 rue Claude Bernard, Paris 5, 1971.

[2] J. Agler and J. McCarthy. Complete Nevanlinna–Pick kernels. *J. Funct. Anal.*, 175:111–124, 2000.

[3] J. Agler and J. McCarthy. *Pick interpolation and Hilbert function spaces*, volume 44 of *Graduate Studies in Mathematics*. American Mathematical Society, Providence, RI, 2002.

[4] L. Ahlfors. *Complex analysis*. McGraw-Hill Book Co., third edition, 1978.

[5] N.I. Akhiezer. *The classical moment problem*. Hafner, New York, 1965.

[6] N.I. Akhiezer and I.M. Glazman. *Theory of linear operators. (Vol. I)*. Pitman Advanced Publishing Program, 1981.

[7] N.I. Akhiezer and I.M. Glazman. *Theory of linear operators. (Vol. II)*. Pitman Advanced Publishing Program, 1981.

[8] S.T. Ali, K. Górska, A. Horzela, and F.H. Szafraniec. Squeezed states and Hermite polynomials in a complex variable. *Journal of Mathematical Physics*, 55(1):012107, January 2014.

[9] D. Alpay. *The Schur algorithm, reproducing kernel spaces and system theory.* American Mathematical Society, Providence, RI, 2001. Translated from the 1998 French original by Stephen S. Wilson, Panoramas et Synthèses.

[10] D. Alpay. *A complex analysis problem book*. Birkhäuser/Springer Basel AG, Basel, 2011.

[11] D. Alpay, H. Attia, and D. Levanony. On the characteristics of a class of Gaussian processes within the white noise space setting. *Stochastic processes and applications*, 120:1074–1104, 2010.

[12] D. Alpay, T.Ya. Azizov, A. Dijksma, H. Langer, and G. Wanjala. The Schur algorithm for generalized Schur functions. IV. Unitary realizations. In *Current trends in operator theory and its applications*, volume 149 of *Oper. Theory Adv. Appl.*, pages 23–45. Birkhäuser, Basel, 2004.

[13] D. Alpay, V. Bolotnikov, and L. Rodman. Tangential interpolation with symmetries and two-point interpolation problem for matrix-valued H_2-functions. *Integral Equations Operator Theory*, 32(1):1–28, 1998.

[14] D. Alpay, F. Colombo, and I. Sabadini. Schur functions and their realizations in the slice hyperholomorphic setting. *Integral Equations and Operator Theory*, 72:253–289, 2012.

[15] D. Alpay, F. Colombo, and I. Sabadini. Schur analysis in the quaternionic setting: The Fueter regular and the slice regular case. Handbook of operator theory, 2015.

[16] D. Alpay, P. Dewilde, and H. Dym. Lossless inverse scattering and reproducing kernels for upper triangular operators. In I. Gohberg, editor, *Extension and interpolation of linear operators and matrix functions*, volume 47 of *Oper. Theory Adv. Appl.*, pages 61–135. Birkhäuser, Basel, 1990.

[17] D. Alpay, A. Dijksma, and H. Langer. The transformation of Issai Schur and related topics in an indefinite setting. In D. Alpay and V. Vinnikov, editors, *System theory, the Schur algorithm and multidimensional analysis*, volume 176 of *Oper. Theory Adv. Appl.*, pages 1–98. Birkhäuser, Basel, 2007.

[18] D. Alpay, A. Dijksma, H. Langer, and G. Wanjala. Basic boundary interpolation for generalized Schur functions and factorization of rational J-unitary matrix functions. In D. Alpay and I. Gohberg, editors, *Interpolation, Schur functions and moment problems*, volume 165 of *Oper. Theory Adv. Appl.*, pages 1–29. Birkhäuser Verlag, Basel, 2006.

[19] D. Alpay, A. Dijksma, J. Rovnyak, and H. de Snoo. *Schur functions, operator colligations, and reproducing kernel Pontryagin spaces*, volume 96 of *Operator theory: Advances and Applications*. Birkhäuser Verlag, Basel, 1997.

[20] D. Alpay, A. Dijksma, J. van der Ploeg, and H.S.V. de Snoo. *Holomorphic operators between Krein spaces and the number of squares of associated kernels*, volume 59 of *Operator Theory: Advances and Applications*, pages 11–29. Birkhäuser Verlag, Basel, 1992.

[21] D. Alpay, A. Dijksma, and D. Volok. Schur multipliers and de Branges–Rovnyak spaces: the multiscale case. *J. Operator Theory*, 61(1):87–118, 2009.

[22] D. Alpay and H. Dym. Hilbert spaces of analytic functions, inverse scattering and operator models, I. *Integral Equation and Operator Theory*, 7:589–641, 1984.

[23] D. Alpay and H. Dym. Hilbert spaces of analytic functions, inverse scattering and operator models, II. *Integral Equation and Operator Theory*, 8:145–180, 1985.

[24] D. Alpay and H. Dym. On a new class of reproducing kernel Hilbert spaces and a new generalization of the Iohvidov laws. *Linear Algebra Appl.*, 178:109–183, 1993.

[25] D. Alpay and H. Dym. On a new class of structured reproducing kernel Hilbert spaces. *J. Funct. Anal.*, 111:1–28, 1993.

[26] D. Alpay and H. Dym. On a new class of realization formulas and their applications. *Linear Algebra Appl.*, 241/243:3–84, 1996.

[27] D. Alpay, M.E. Luna Elizarraras, M. Shapiro, and D.C. Struppa. *Basics of functional analysis with bicomplex scalars, and bicomplex Schur analysis*. SpringerBriefs, 2014.

[28] D. Alpay and I. Gohberg. *On orthogonal matrix polynomials*, volume 34 of *Operator Theory: Advances and Applications*, pages 25–46. Birkhäuser Verlag, Basel, 1988.

[29] D. Alpay and I. Gohberg. Unitary rational matrix functions. In I. Gohberg, editor, *Topics in interpolation theory of rational matrix-valued functions*, volume 33 of *Operator Theory: Advances and Applications*, pages 175–222. Birkhäuser Verlag, Basel, 1988.

[30] D. Alpay and I. Gohberg. A trace formula for canonical differential expressions. *J. Funct. Anal.*, 197(2):489–525, 2003.

[31] D. Alpay and I. Gohberg. Pairs of selfadjoint operators and their invariants. *Algebra i Analiz*, 16(1):70–120, 2004.

[32] D. Alpay and I. Gohberg. Discrete analogs of canonical systems with pseudo-exponential potential. Definitions and formulas for the spectral matrix functions. In D. Alpay and I. Gohberg, editors, *The state space method. New results and new applications*, volume 161, pages 1–47. Birkhäuser Verlag, Basel, 2006.

[33] D. Alpay and P. Jorgensen. Reproducing kernel Hilbert spaces generated by the binomial coefficients. *Illinois Journal of Mathematics*, in press.

[34] D. Alpay, P. Jorgensen, and I. Lewkowicz. Extending wavelet filters: infinite dimensions, the nonrational case, and indefinite inner product spaces. In *Excursions in harmonic analysis. Volume 2*, Appl. Numer. Harmon. Anal., pages 69–111. Birkhäuser/Springer, New York, 2013.

[35] D. Alpay, P. Jorgensen, I. Lewkowicz, and I. Marziano. Representation formulas for Hardy space functions through the Cuntz relations and new interpolation problems. In Xiaoping Shen and Ahmed Zayed, editors, *Multiscale signal analysis and modeling*, pages 161–182. Springer, 2013.

[36] D. Alpay, P. Jorgensen, I. Lewkowicz, and D. Volok. Linear combination interpolation, Cuntz relations and infinite products. *ArXiv e-prints*, August 2014.

[37] D. Alpay, P. Jorgensen, R. Seager, and D. Volok. On discrete analytic functions: Products, rational functions and reproducing kernels. *Journal of Applied Mathematics and Computing*, 41:393–426, 2013.

[38] D. Alpay and T. Kaptanoğlu. Quaternionic Hilbert spaces and a von Neumann inequality. *Complex Var. Elliptic Equ.*, 57(6):667–675, 2012.

[39] D. Alpay, X. Mary, and D. Volok. Reproducing kernels: a survey. Unpublished manuscript.

[40] D. Alpay and T.M. Mills. A family of Hilbert spaces which are not reproducing kernel Hilbert spaces. *J. Anal. Appl.*, 1(2):107–111, 2003.

[41] D. Alpay and G. Salomon. New topological ℂ-algebras with applications in linear systems theory. *Infin. Dimens. Anal. Quantum Probab. Relat. Top.*, 15(2):1250011, 30, 2012.

[42] D. Alpay and G. Salomon. Non-commutative stochastic distributions and applications to linear systems theory. *Stochastic Process. Appl.*, 123(6):2303–2322, 2013.

[43] D. Alpay and G. Salomon. Topological convolution algebras. *J. Funct. Anal.*, 264(9):2224–2244, 2013.

[44] D. Alpay and M. Shapiro. Reproducing kernel quaternionic Pontryagin spaces. *Integral Equations and Operator Theory*, 50:431–476, 2004.

[45] D. Alpay, M. Shapiro, and D. Volok. Rational hyperholomorphic functions in R^4. *J. Funct. Anal.*, 221(1):122–149, 2005.

[46] D. Alpay and V. Vinnikov. Finite dimensional de Branges spaces on Riemann surfaces. *J. Funct. Anal.*, 189(2):283–324, 2002.

[47] V. Andreev and T. McNicholl. Computing conformal maps of finitely connected domains onto canonical slit domains. *Theory Comput. Syst.*, 50(2):354–369, 2012.

[48] T. Apostol. *Calculus. Volume I*. Xerox College Publishing, Waltham, Massachusetts, Second edition, 1969.

[49] T. Apostol. *Calculus. Volume II*. Xerox College Publishing, Waltham, Massachusetts, Second edition, 1969.

[50] J.-M. Arnaudies. *L'intégrale de Lebesgue sur la droite*. Vuibert, 1997.

[51] N. Aronszajn. La théorie générale des noyaux reproduisants et ses applications. *Math. Proc. Cambridge Phil. Soc.*, 39:133–153, 1944.

[52] N. Aronszajn. Theory of reproducing kernels. *Trans. Amer. Math. Soc.*, 68:337–404, 1950.

[53] N. Aronszajn and K.T. Smith. Theory of Bessel potentials. I. *Ann. Inst. Fourier (Grenoble)*, 11:385–475, 1961.

[54] S. Artstein-Avidan, Dan F., and V. Milman. Order isomorphisms on convex functions in windows. In *Geometric aspects of functional analysis*, volume 2050 of *Lecture Notes in Math.*, pages 61–122. Springer, Heidelberg, 2012.

[55] T.Ya. Azizov and I.S. Iohvidov. *Foundations of the theory of linear operators in spaces with indefinite metric.* Nauka, Moscow, 1986. (Russian). English translation: *Linear operators in spaces with an indefinite metric.* John Wiley, New York, 1989.

[56] J. Bak and D.J. Newman. *Complex analysis.* Undergraduate Texts in Mathematics. Springer-Verlag, New York, 1982.

[57] M. Balk. *Polyanalytic functions.* Akademie-Verlag, Berlin, 1991.

[58] J. Ball, I. Gohberg, and L. Rodman. *Interpolation of rational matrix functions,* volume 45 of *Operator Theory: Advances and Applications.* Birkhäuser Verlag, Basel, 1990.

[59] J. Ball, T. Trent, and V. Vinnikov. Interpolation and commutant lifting for multipliers on reproducing kernel Hilbert spaces. In *Proceedings of Conference in honor of the 60th birthday of M.A. Kaashoek,* volume 122 of *Operator Theory: Advances and Applications,* pages 89–138. Birkhäuser, 2001.

[60] V. Bargmann. On a Hilbert space of analytic functions and an associated integral transform. *Comm. Pure Appl. Math.,* 14:187–214, 1961.

[61] V. Bargmann. Remarks on a Hilbert space of analytic functions. *Proceedings of the National Academy of Arts,* 48:199–204, 1962.

[62] A. Baricz, J. Vesti, and M. Vuorinen. On Kaluza's sign criterion for reciprocal power series. *Ann. Univ. Mariae Curie-Skłodowska Sect. A,* 65(2):1–16, 2011.

[63] H. Bart, I. Gohberg, M.A. Kaashoek, and P. Van Dooren. Factorizations of transfer functions. *SIAM J. Control Optim.,* 18(6):675–696, 1980.

[64] H. Bart, I. Gohberg, and M.A. Kaashoek. *Minimal factorization of matrix and operator functions,* volume 1 of *Operator Theory: Advances and Applications.* Birkhäuser Verlag, Basel, 1979.

[65] H. Bart, I. Gohberg, and M.A. Kaashoek. Convolution equations and linear systems. *Integral Equations Operator Theory,* 5:283–340, 1982.

[66] J. Bass and P. Lévy. Propriétés des lois dont les fonctions caractéristiques sont $1/\mathrm{ch}\,z$, $z/\mathrm{sh}\,z$, $1/\mathrm{ch}^2 z$. *C. R. Acad. Sci. Paris,* 230:815–817, 1950.

[67] B. Beauzamy. *Introduction to operator theory and invariant subspaces.* North-Holland, 1988.

[68] C. Berenstein and R. Gay. *Complex variables,* volume 125 of *Graduate Texts in Mathematics.* Springer-Verlag, New York, 1991. An introduction.

[69] C. Berg, J. Christensen, and P. Ressel. *Harmonic analysis on semigroups,* volume 100 of *Graduate Texts in Mathematics.* Springer-Verlag, New York, 1984. Theory of positive definite and related functions.

[70] S. Bergman. *The kernel function and conformal mapping.* American Mathematical Society, 1950.

[71] A. Berlinet and C. Thomas-Agnan. *Reproducing kernel Hilbert spaces in probability and statistics.* Kluwer, 2004.

[72] F. Biagini, Y. Hu, B. Øksendal, and T. Zhang. *Stochastic calculus for fractional Brownian motion and applications.* Probability and its Applications (New York). Springer-Verlag London Ltd., London, 2008.

[73] Ph. Biane, J. Pitman, and M. Yor. Probability laws related to the Jacobi theta and Riemann zeta functions, and Brownian excursions. *Bull. Amer. Math. Soc. (N.S.)*, 38(4):435–465 (electronic), 2001.

[74] K. Bichteler. *Integration – a functional approach.* Modern Birkhäuser Classics. Birkhäuser Verlag, Basel, 2010. Reprint of the 1998 edition.

[75] A. Bloch. Les fonctions holomorphes et méromorphes dans le cercle-unité. *Mémorial des sciences mathématiques*, pages 1–61, 1926. Fascicule 20.

[76] J. Bognár. *Indefinite inner product spaces.* Ergebnisse der Mathematik und ihrer Grenzgebiete, Band 78. Springer-Verlag, Berlin, 1974.

[77] N. Bourbaki. *Éléments de mathématique. Topologie générale. Chapitres 1 à 4.* Hermann, Paris, 1971.

[78] N. Bourbaki. *Topologie générale. Chapitres 5 à 10.* Diffusion C.C.L.S., Paris, 1974.

[79] N. Bourbaki. *Espaces vectoriels topologiques.* Masson, 1981.

[80] M. Bożejko. Positive-definite kernels, length functions on groups and a non-commutative von Neumann inequality. *Studia Math.*, 95(2):107–118, 1989.

[81] M. Bożejko and T. Hasebe. On free infinite divisibility for classical Meixner distributions. *Probab. Math. Statist.*, 33(2):363–375, 2013.

[82] L. de Branges. Perturbations of self-adjoint transformations. *Amer. J. Math.*, 84:543–560, 1962.

[83] L. de Branges. Some Hilbert spaces of analytic functions I. *Trans. Amer. Math. Soc.*, 106:445–468, 1963.

[84] L. de Branges. *Espaces Hilbertiens de fonctions entières.* Masson, Paris, 1972.

[85] L. de Branges and J. Rovnyak. Canonical models in quantum scattering theory. In C. Wilcox, editor, *Perturbation theory and its applications in quantum mechanics*, pages 295–392. Wiley, New York, 1966.

[86] L. de Branges and J. Rovnyak. *Square summable power series.* Holt, Rinehart and Winston, New York, 1966.

[87] J. Bretagnolle, D. Dacunha-Castelle, and J.-L. Krivine. Lois stables et espaces L^p. *Ann. Inst. H. Poincaré Sect. B (N.S.)*, 2:231–259, 1965/1966.

[88] J. Bretagnolle, D. Dacunha-Castelle, and J.-L. Krivine. Lois stables et espaces L^p. In *Symposium on Probability Methods in Analysis (Loutraki, 1966)*, pages 48–54. Springer, Berlin, 1967.

[89] C. Brezinski. *Biorthogonality and its applications to numerical analysis*, volume 156 of *Monographs and Textbooks in Pure and Applied Mathematics*. Marcel Dekker Inc., New York, 1992.

[90] H. Brezis. *Analyse fonctionnelle*. Masson, Paris, 1987.

[91] R.B. Burckel. *An introduction to classical complex analysis, Vol. 1*. Birkhäuser, 1979.

[92] B. Calvo, J. Doyen, A. Calvo, and F. Boschet. *Exercices d'algèbre*. Armand Colin, 1970.

[93] Albert Camus. *The myth of Sisyphus and other essays*. Vintage Books, 1959. Reprinted by arrangements with Alfred A. Knopf, Inc.

[94] Albert Camus. *Le mythe de Sisyphe*. Gallimard, 1982. First published 1961.

[95] H. Cartan. *Théorie élémentaire des fonctions analytiques d'une ou plusieurs variables complexes*. Hermann, Paris, 1975.

[96] L.P. Castro, M.M. Rodrigues, and S. Saitoh. A fundamental theorem on initial value problems by using the theory of reproducing kernels. To appear in Complex Analysis and Operator Theory (2015).

[97] C. Chamfy. Fonctions méromorphes sur le cercle unité et leurs séries de Taylor. *Ann. Inst. Fourier*, 8:211–251, 1958.

[98] C.P. Chen and J. Choi. Asymptotic expansions for the constants of Landau and Lebesgue. *Adv. Math.*, 254:622–641, 2014.

[99] G. Choquet. *Topology*, volume XIX of *Pure and Applied Mathematics*. Academic Press, 1966.

[100] G. Choquet. *Cours d'analyse, Tome II: Topologie*. Masson, 120 bd Saint-Germain, Paris VI, 1973.

[101] S. Colombo. *Les transformations de Mellin et de Hankel: Applications à la physique mathématique*. Monographies du Centre d'Études Mathématiques en vue des Applications: B. – Méthodes de Calcul. Centre National de la Recherche Scientifique, Paris, 1959.

[102] E.H. Connell and P. Porcelli. An algorithm of J. Schur and the Taylor series. *Proc. Amer. Math. Soc.*, 13:232–235, 1962.

[103] C. de Boor and R.E. Lynch. On splines and their minimum properties. *J. Math. Mech.*, 15:953–969, 1966.

[104] L. de Branges. *Hilbert spaces of entire functions*. Prentice Hall Inc., Englewood Cliffs, N.J., 1968.

[105] G.F. de Montricher, R.A. Tapia, and J.R. Thompson. Nonparametric maximum likelihood estimation of probability densities by penalty function methods. *Ann. Statist.*, 3(6):1329–1348, 1975.

[106] G. de Rham. Sur les polygones générateurs de groupes fuchsiens. *Enseignement Math.*, 17:49–61, 1971.

[107] Ph. Delsarte, Y. Genin, and Y. Kamp. A generalization of the Levinson algorithm for Hermitian Toeplitz matrices with any rank profile. *IEEE Trans. Acoust. Speech Signal Process.*, 33(4):964–971, 1985.

[108] Ph. Delsarte, Y. Genin, and Y. Kamp. Pseudo-Carathéodory functions and Hermitian Toeplitz matrices. *Philips J. Res.*, 41(1):1–54, 1986.

[109] P. Dewilde and H. Dym. Lossless inverse scattering, digital filters, and estimation theory. *IEEE Trans. Inform. Theory*, 30(4):644–662, 1984.

[110] P. Dewilde and A.-J. van der Veen. *Time-varying systems and computations*. Kluwer Academic Publishers, Boston, MA, 1998.

[111] J. Dieudonné. *Eléments d'analyse, Volume 1: fondements de l'analyse moderne*. Gauthier-Villars, Paris, 1969.

[112] J. Dieudonné. *Eléments d'analyse, Tome 6: Chapitre XXII*. Gauthier-Villars, Paris, 1975.

[113] J. Dieudonné. *Eléments d'analyse, Tome 7: Chapitre XXIII*. Gauthier-Villars, Paris, 1978.

[114] J. Dieudonné. *Eléments d'analyse, Tome 2: Chapitres XII à XV*. Gauthier-Villars, Paris, 1982.

[115] J. Dieudonné and L. Schwarz. La dualité dans les espaces (F) et (LF). *Ann. Inst. Fourier*, 1:61–101, 1949.

[116] J. Dixmier. *Les algèbres d'opérateurs dans l'espace hilbertien (algèbres de von Neumann)*. Gauthier-Villars Éditeur, Paris, 1969. Deuxième édition, revue et augmentée, Cahiers Scientifiques, Fasc. XXV.

[117] W. Donoghue and P. Masani. A class of invalid assertions concerning function Hilbert spaces. *Bol. Soc. Mat. Mexicana* (2), 28(2):77–80, 1983.

[118] W.F. Donoghue. *Monotone matrix functions and analytic continuation*, volume 207 of *Die Grundlehren der mathematischen Wissenschaften*. Springer-Verlag, 1974.

[119] R.G. Douglas. *Banach algebra techniques in operator theory*. Academic Press, 1972.

[120] N. Dunford and J. Schwartz. *Linear operators*, volume 1. Interscience, 1957.

[121] P.L. Duren. *Theory of H^p spaces*. Academic Press, New York, 1970.

[122] H. Dym. *J-contractive matrix functions, reproducing kernel Hilbert spaces and interpolation*. Published for the Conference Board of the Mathematical Sciences, Washington, DC, 1989.

[123] H. Dym. *Shift, realizations and interpolation, Redux*, volume 73 of *Operator Theory: Advances and Applications*, pages 182–243. Birkhäuser Verlag, Basel, 1994.

[124] H. Dym. Linear fractional transformations, Riccati equations and bitangential interpolation, revisited. In D. Alpay, editor, *Reproducing kernel spaces and applications*, volume 143 of *Oper. Theory Adv. Appl.*, pages 171–212. Birkhäuser, Basel, 2003.

[125] H. Dym and A. Iacob. Positive definite extensions, canonical equations and inverse problems. In H. Dym and I. Gohberg, editors, *Proceedings of the workshop on applications of linear operator theory to systems and networks held at Rehovot, June* 13–16, 1983, volume 12 of *Operator Theory: Advances and Applications*, pages 141–240. Birkhäuser Verlag, Basel, 1984.

[126] H. Dym and H.P. McKean. *Gaussian processes, function theory and the inverse spectral problem*. Academic Press, 1976.

[127] K. Dzhaparidze and H. van Zanten. Kreĭn's spectral theory and the Paley–Wiener expansion for fractional Brownian motion. *The Annals of Probability*, 33(4):620–644, 2005.

[128] M.A. Evgravof, K. Béjanov, Y. Sidorov, M. Fédoruk, and M. Chabounine. *Recueil de problèmes sur la théorie des fonctions analytiques*. Éditions Mir, Moscou, 1974.

[129] J. Fay. *Theta functions on Riemann surfaces*. Springer-Verlag, Berlin, 1973. Lecture Notes in Mathematics, Vol. 352.

[130] A. Feintuch. *Robust control theory in Hilbert space*, volume 130 of *Applied Mathematical Sciences*. Springer-Verlag, New York, 1998.

[131] A. Feintuch and R. Saeks. Extended spaces and the resolution topology. *Internat. J. Control*, 33(2):347–354, 1981.

[132] A. Feintuch and R. Saeks. *System theory. A Hilbert space approach*, volume 102 of *Pure and Applied Mathematics*. Academic Press Inc. [Harcourt Brace Jovanovich Publishers], New York, 1982.

[133] P.A. Fillmore and J.P. Williams. On operator ranges. *Advances in Mathematics*, 7:254–281, 1971.

[134] V. Fock. Verallgemeinerung und Lösung der Diracschen statistischen Gleichung. *Z. f. Physik*, 49:339–357, 1928.

[135] L.R. Ford. *An introduction to the theory of automorphic functions*. Number 6 in Edinburgh Mathematical Tracts. Bell and Sons, 1915.

[136] L.R. Ford. *Automorphic functions*. McGraw-Hill Book company, 1929.

[137] E. Fornasini and G. Marchesini. State-space realization theory of two-dimensional filters. *IEEE Trans. Automatic Control*, AC–21(4):484–492, 1976.

[138] O. Forster. *Lectures on Riemann surfaces*, volume 81 of *Graduate Texts in Mathematics*. Springer-Verlag, 1981.

[139] S. Francinou, H. Gianella, and S. Nicolas. *Exercices de mathématiques. Oraux X-ENS. Analyse 1*. Cassini, 2003.

[140] E. Freitag and R. Busam. *Complex analysis*. Springer, 2005.

[141] E. Freitag and R. Busam. *Funktionentheorie 1*. Springer, 2006. 4. korrigierte und erweiterte Auflage.

[142] P.A. Fuhrmann. *Linear systems and operators in Hilbert space*. McGraw-Hill international book company, 1981.

[143] P.A. Fuhrmann. *A polynomial approach to linear algebra*. Universitext. Springer-Verlag, New York, 1996.

[144] W. Fulton. *Algebraic topology*, volume 153 of *Graduate Texts in Mathematics*. Springer-Verlag, New York, 1995. A first course.

[145] Th. Gamelin. *Complex analysis*. Undergraduate Texts in Mathematics. Springer-Verlag, New York, 2001.

[146] J. Garsoux. *Espaces vectoriels topologiques et distributions*. Avec la collaboration de Daniel Ribbens; Préface de Pierre Houzeau de Lehaie. Dunod, Paris, 1963.

[147] I.M. Gel'fand. A lemma in the theory of linear spaces. *Soobshch. Kharkov Mat. Obshch., Ser. 4*, 13:35–40, 1934.

[148] I.M. Gel'fand and N.Ya. Vilenkin. *Generalized functions. Vol. 4*. Academic Press [Harcourt Brace Jovanovich Publishers], New York, 1964 [1977]. Applications of harmonic analysis, Translated from the Russian by Amiel Feinstein.

[149] I.M. Gel'fand and A.M. Yaglom. Calculation of the amount of information about a random function contained in another such function. *Amer. Math. Soc. Transl. (2)*, 12:199–246, 1959.

[150] I.M. Gelfand and G.E. Shilov. *Generalized functions. Volume 2*. Academic Press, 1968.

[151] D. Givone and R. Roesser. Multidimensional linear iterative circuits – general properties. *IEEE Trans. Computers*, C-21:1067–1073, 1972.

[152] C. Godbillon. *Éléments de topologie algébrique*. Hermann, Paris, 1971.

[153] R. Godement. Les fonctions de type positif et la théorie des groupes. *Trans. Amer. Math. Soc.*, 63:1–84, 1948.

[154] R. Godement. *Cours d'algèbre*. Hermann, 1987.

[155] I. Gohberg, S. Goldberg, and M. Kaashoek. *Classes of linear operators. Vol. I*, volume 49 of *Operator Theory: Advances and Applications*. Birkhäuser Verlag, Basel, 1990.

[156] I. Gohberg, S. Goldberg, and M.A. Kaashoek. *Classes of linear operators. Vol. II*, volume 63 of *Operator Theory: Advances and Applications*. Birkhäuser Verlag, Basel, 1993.

[157] I. Gohberg, M.A. Kaashoek, and F. van Schagen. Szegö–Kac–Achiezer formulas in terms of realizations of the symbol. *J. Funct. Anal.*, 74:24–51, 1987.

[158] I. Gohberg and M.G. Kreĭn. *Introduction to the theory of linear nonselfadjoint operators*, volume 18 of *Translations of mathematical monographs*. American Mathematical Society, Rhode Island, 1969.

[159] M. Golomb and H.F. Weinberger. Optimal approximation and error bounds. In *On numerical approximation. Proceedings of a Symposium, Madison, April 21–23, 1958*, edited by R.E. Langer. Publication No. 1 of the Mathematics Research Center, U.S. Army, the University of Wisconsin, pages 117–190. The University of Wisconsin Press, Madison, Wis., 1959.

[160] J. Górniak. Remarks on positive definite operator valued functions in linear spaces. In *Probability theory on vector spaces (Proc. Conf., Trzebieszowice, 1977)*, volume 656 of *Lecture Notes in Math.*, pages 37–44. Springer, Berlin, 1978.

[161] J. Górniak. Locally convex spaces with factorization property. *Colloq. Math.*, 48(1):69–79, 1984.

[162] J. Górniak and A. Weron. Aronszajn–Kolmogorov type theorems for positive definite kernels in locally convex spaces. *Studia Math.*, 69(3):235–246, 1980/81.

[163] L. Grafakos. *Modern Fourier analysis*, volume 250 of *Graduate Texts in Mathematics*. Springer, New York, second edition, 2009.

[164] A. Grothendieck. Sur certains espaces de fonctions holomorphes. I. *J. Reine Angew. Math.*, 192:35–64, 1953.

[165] A. Grothendieck. Sur certains espaces de fonctions holomorphes. II. *J. Reine Angew. Math.*, 192:78–95, 1953.

[166] I.M. Guelfand and G.E. Shilov. *Les distributions. Tome 2*. Collection Universitaire de Mathématiques, No. 15. Dunod, Paris, 1964.

[167] I.M. Guelfand and N.Y. Vilenkin. *Les distributions. Tome 4: Applications de l'analyse harmonique.* Collection Universitaire de Mathématiques, No. 23. Dunod, Paris, 1967.

[168] R. Gunning. *Lectures on Riemann surfaces*, volume 2 of *Mathematical notes, Princeton University Press.* Springer-Verlag, Berlin, Heidelberg, New York, 1966.

[169] P. Halmos. *A Hilbert space problem book.* D. Van Nostrand Co., Inc., Princeton, N.J.-Toronto, Ont.-London, 1967.

[170] P.R. Halmos. *A Hilbert space problem book*, volume 19 of *Graduate Texts in Mathematics.* Springer-Verlag, New York, second edition, 1982. Encyclopedia of Mathematics and its Applications, 17.

[171] G.H. Hardy. *Divergent Series.* Oxford, at the Clarendon Press, 1949.

[172] R. Hartshorne. *Algebraic geometry*, volume 52 of *Graduate Texts in Mathematics.* Springer-Verlag, 1977.

[173] B. Hauchecorne. *Les contre-exemples en mathématiques.* Edition Marketing. 32 rue Bargue, 75015, Paris, 1988.

[174] M. Heins. *Selected topics in the classical theory of functions of a complex variable.* Athena Series: Selected Topics in Mathematics. Holt, Rinehart and Winston, New York, 1962.

[175] M. Heins. *Complex function theory.* Pure and Applied Mathematics, Vol. 28. Academic Press, New York, 1968.

[176] M. Heins. *Hardy classes on Riemann surfaces*, volume 98 of *Lecture Notes in Mathematics.* Springer-Verlag, 1969.

[177] E. Hille. A class of reciprocal functions. *Ann. of Math.* (2), 27(4):427–464, 1926.

[178] E. Hille. Contributions to the theory of Hermitian series. II. The representation problem. *Trans. Amer. Math. Soc.*, 47:80–94, 1940.

[179] E. Hille. Introduction to general theory of reproducing kernels. *Rocky Journal of Mathematics*, 2:321–368, 1972.

[180] K. Hoffman. *Banach spaces of analytic functions.* Dover Publications Inc., New York, 1988. Reprint of the 1962 original.

[181] Th. Hofmann, B. Schölkopf, and A.J. Smola. Kernel methods in machine learning. *Ann. Statist.*, 36(3):1171–1220, 2008.

[182] H. Holden, B. Øksendal, J. Ubøe, and T. Zhang. *Stochastic partial differential equations.* Probability and its Applications. Birkhäuser Boston Inc., Boston, MA, 1996.

[183] J. Horváth. *Topological vector spaces and distributions. Vol. I.* Addison-Wesley Publishing Co., Reading, Mass.-London-Don Mills, Ont., 1966.

[184] X. Hu, T. Ho, and H. Rabitz. The collocation method based on a generalized inverse multiquadric basis for bound-state problems. *Computer Physics Communications*, 113:168–179, 1998.

[185] X. Hu, T. Ho, and H. Rabitz. Variational reproducing kernel Hilbert space (RKHS) grid method for quantum mechanical bound-state problems. *Chem. Phys. Lett.*, 288:719–726, 1998.

[186] X. Hu, T. Ho, and H. Rabitz. Solving the bound-state Schrödinger equation by reproducing kernel interpolation. *Phys. Rev. E (3)*, 61(2):2074–2085, 2000.

[187] T. Huang. Some mapping properties of RC and RL driving-point impedance functions. *IEEE Transactions on Circuit Theory*, 12:257–259, 1965.

[188] Zh. Huang and J. Yan. *Introduction to infinite dimensional stochastic analysis*, volume 502 of *Mathematics and its Applications*. Kluwer Academic Publishers, Dordrecht, Chinese edition, 2000.

[189] I.S. Iohvidov and M.G. Kreĭn. Spectral theory of operators in spaces with indefinite metric. II. *Trudy Moskov. Mat. Obšč.*, 8:413–496, 1959.

[190] I.S. Iohvidov, M.G. Kreĭn, and H. Langer. *Introduction to the spectral theory of operators in spaces with an indefinite metric*. Akademie-Verlag, Berlin, 1982.

[191] J. von Neumann. Zur Algebra der Funktionaloperationen und Theorie der normalen Operatoren. *Math. Ann.*, 102(1):370–427, 1930.

[192] F. Jacobzon, S. Reich, and D. Shoikhet. Linear fractional mappings: invariant sets, semigroups and commutativity. *J. Fixed Point Theory Appl.*, 5(1):63–91, 2009.

[193] S. Janson. *Gaussian Hilbert spaces*, volume 129 of *Cambridge Tracts in Mathematics*. Cambridge University Press, Cambridge, 1997.

[194] Palle E.T. Jorgensen and Steen Pedersen. Dense analytic subspaces in fractal L^2-spaces. *J. Anal. Math.*, 75:185–228, 1998.

[195] G. Julia. *Leçons sur la représentation des domaines simplement connexes*, volume Fascicule 14 of *Cahiers scientifiques, publiés sous la direction de M. Gaston Julia*. Gauthier-Villars, 55, quai des Grands-Augustins, Paris, 1934. Leçons recueillies et rédigées par Georges Bourion et Jean Leray [Texte imprimé].

[196] G. Julia. *Exercices d'analyse. Tome II, Fascicule 1: Fonctions analytiques, développements en série, résidus, transformations analytiques, représentation conforme.* Deuxième édition, nouveau tirage. Gauthier-Villars Éditeur, Paris, 1969.

[197] R.E. Kalman. Linear stochastic filtering theory – reappraisal and outlook. In *Proc. Sympos. on System Theory (New York, 1965)*, pages 197–205. Polytechnic Press, Polytechnic Inst. Brooklyn, Brooklyn, N.Y., 1965.

[198] Th. Kaluza. Über die Koeffizienten reziproker Potenzreihen. *Math. Z.*, 28(1):161–170, 1928.

[199] Dmitry S. Kalyuzhnyi-Verbovetskiĭ and Victor Vinnikov. Non-commutative positive kernels and their matrix evaluations. *Proc. Amer. Math. Soc.*, 134(3):805–816 (electronic), 2006.

[200] Sh. Kantorovitz. *Topics in operator semigroups*, volume 281 of *Progress in Mathematics*. Birkhäuser Boston Inc., Boston, MA, 2010.

[201] D. Kendall. Renewal processes and their arithmetic. In *Symposium on Probability Methods in Analysis (Loutraki, 1966)*, volume 31 of *Lecture Notes in Mathematics*, pages 147–175. Springer, Berlin, 1967.

[202] S.M. Khaleelulla. *Counterexamples in topological vector spaces*, volume 936 of *Lecture Notes in Mathematics*. Springer-Verlag, Berlin, 1982.

[203] J. Kim and M.W. Wong. Schoenberg's theorem for positive definite functions on Heisenberg groups. In *Advances in analysis*, pages 265–273. World Sci. Publ., Hackensack, NJ, 2005.

[204] F. Kirwan. *Complex algebraic curves*, volume 23 of *London Mathematical Society Student Texts*. Cambridge University Press, 1993.

[205] H. Kober. A theorem on Banach spaces. *Compositio Math.*, 7:135–140, 1940.

[206] P. Koosis. *Introduction to H_p spaces*. Cambridge University Press, 1980.

[207] A.D. Košelev. The kernel function of a Hilbert space of functions that are polyanalytic in the disc. *Dokl. Akad. Nauk SSSR*, 232(2):277–279, 1977.

[208] M.G. Kreĭn and H. Langer. Über die verallgemeinerten Resolventen und die charakteristische Funktion eines isometrischen Operators im Raume Π_k. In *Hilbert space operators and operator algebras (Proc. Int. Conf. Tihany, 1970)*, pages 353–399. North-Holland, Amsterdam, 1972. Colloquia Math. Soc. János Bolyai.

[209] M.G. Kreĭn and A.A. Nudelman. *The Markov moment problem and extremal problems*, volume 50 of *Translations of mathematical monographs*. American Mathematical Society, Providence, Rhode Island, 1977.

[210] J. Lafontaine. *Introduction aux variétés différentielles*. Presse Universitaire de Grenoble, 1996.

[211] J. Lamperti. On the coefficients of reciprocal power series. *Amer. Math. Monthly*, 65:90–94, 1958.

[212] E. Landau. *Abschätzung der Koeffizientensumme einer Potenzreihe*. *Arch. Math. Phys.*, 21:250–255, 1913.

[213] H.J. Landau and R. Osserman. On analytic mappings of Riemann surfaces. *J. Analyse Math.*, 7:249–279, 1959/1960.

[214] S. Lang. *Algebra (third edition)*. Addison-Wesley, 1993.

[215] N.N. Lebedev. *Special functions and their applications*. Dover Publications Inc., New York, 1972. Revised edition, translated from the Russian and edited by Richard A. Silverman, Unabridged and corrected republication.

[216] D. Leborgne. *Calcul différentiel et géométrie*. Presses Universitaires de France, 1982.

[217] Y.W. Lee. *Statistical theory of communication*. John Wiley & Sons Inc., New York, 1960.

[218] D. Lehmann and C. Sacré. *Géométrie et topologie des surfaces*. Presses Universitaires de France, 1982.

[219] J. Lehner. *Discontinuous groups and automorphic functions*. Mathematical Surveys, No. VIII. American Mathematical Society, Providence, R.I., 1964.

[220] P. Lévy. *Calcul des probabilités*. Gauthier-Villars, Paris, 1925.

[221] D. Li and H. Queffélec. *Introduction à l'étude des espaces de Banach*, volume 12 of *Cours Spécialisés [Specialized Courses]*. Société Mathématique de France, Paris, 2004. Analyse et probabilités. [Analysis and probability theory].

[222] M.S. Livšic, N. Kravitski, A. Markus, and V. Vinnikov. *Commuting non-selfadjoint operators and their applications to system theory*. Kluwer, 1995.

[223] M. Loève. *Probability theory*. Third edition. D. Van Nostrand Co., Inc., Princeton, N.J.-Toronto, Ont.-London, 1963.

[224] E. Lukacs. *Characteristic functions*. Griffin's Statistical Monographs & Courses, No. 5. Hafner Publishing Co., New York, 1960.

[225] M.P. Malliavin. *Algèbre commutative*. Masson, Paris, 1990.

[226] L.E. Mansfield. On the optimal approximation of linear functionals in spaces of bivariate functions. *SIAM J. Numer. Anal.*, 8:115–126, 1971.

[227] M.J. Martín. Composition operators with linear fractional symbols and their adjoints. In *Proceedings of the First Advanced Course in Operator Theory and Complex Analysis*, pages 105–112. Univ. Sevilla Secr. Publ., Seville, 2006.

[228] C.R.F. Maunder. *Algebraic topology*. Dover Publications Inc., Mineola, NY, 1996. Reprint of the 1980 edition.

[229] S. Mazur and W. Orlicz. Grundlegende Eigenschaften der polynomischen Operationen. *Studia Math.*, pages 50–68, 1934.

[230] M. Mboup. On the structure of self-similar systems: a Hilbert space approach. In *Reproducing kernel spaces and applications*, volume 143 of *Oper. Theory Adv. Appl.*, pages 273–302. Birkhäuser, Basel, 2003.

[231] F.G. Mehler. Reihenentwicklungen nach Laplaceschen Funktionen höherer Ordnung. *Journal für Reine and Angewandte Mathematik*, 66:161–176, 1866.

[232] S. Mendelson. Learnability in Hilbert spaces with reproducing kernels. *J. Complexity*, 18(1):152–170, 2002.

[233] K. Menger. Die Metrik des Hilbertschen Raumes. *Anzeiger der Akademie der Wissenschaften in Wien, Mathematisch-Naturwissenschaftliche Klasse*, 65:159–160, 1928.

[234] H. Meschkovski. *Hilbertsche Räume mit Kernfunktion*. Springer-Verlag, 1962.

[235] Y. Meyer. *Ondelettes et opérateurs. I.* Actualités Mathématiques. [Current Mathematical Topics.] Hermann, Paris, 1990. Ondelettes. [Wavelets.]

[236] Y. Meyer and R.R. Coifman. *Ondelettes et opérateurs. III.* Actualités Mathématiques. [Current Mathematical Topics.] Hermann, Paris, 1991. Opérateurs multilinéaires. [Multilinear operators.]

[237] A.M. Molčanov. On conditions for discreteness of the spectrum of self-adjoint differential equations of the second order. *Trudy Moskov. Mat. Obšč.*, 2:169–199, 1953.

[238] Molière. *Le bourgeois gentilhomme.* Nouveaux Classiques Larousse. Librairie Larousse, 1965. First published 1670.

[239] P.S. Muhly and B. Solel. Progress in noncommutative function theory. *Sci. China Math.*, 54(11):2275–2294, 2011.

[240] B. Sz.-Nagy and C. Foias. *Harmonic analysis of operators on Hilbert spaces.* Akademia Kiado, Budapest, 1966.

[241] M.A. Naĭmark. *Linear differential operators. Part II: Linear differential operators in Hilbert space.* With additional material by the author, and a supplement by V.È. Ljance. Translated from the Russian by E.R. Dawson. English translation edited by W.N. Everitt. Frederick Ungar Publishing Co., New York, 1968.

[242] R. Narasimhan. *Compact Riemann surfaces.* Lectures in Mathematics, ETH Zürich. Birkhäuser Verlag, Basel, 1992.

[243] I.P. Natanson. *Theory of functions of a real variable. Volume I.* Frederick Ungar Publishing Co., New York, 1955. Translated by Leo F. Boron with the collaboration of Edwin Hewitt.

[244] Z. Nehari. *Conformal mapping.* McGraw-Hill Book Co., Inc., New York, Toronto, London, 1952.

[245] E. Nelson. *Topics in dynamics. I: Flows.* Mathematical Notes. Princeton University Press, Princeton, N.J., 1969.

[246] Y. Neretin. *Lectures on Gaussian integral operators and classical groups.* EMS Series of Lectures in Mathematics. European Mathematical Society (EMS), Zürich, 2011.

[247] J. Neveu. *Processus aléatoires gaussiens.* Number 34 in Séminaires de mathématiques supérieures. Les presses de l'université de Montréal, 1968.

[248] A. Nou. *Algèbre q-Gaussiennes.* Mémoire de DEA, Université de Franche-Comté, Juin 2000.

[249] A. Ogg. *Modular forms and Dirichlet series.* W.A. Benjamin, Inc., New York-Amsterdam, 1969.

[250] F. Pham. *Géométrie et calcul différentiel sur les variétés.* InterEditions, Paris, 1992. Cours, études et exercices pour la maîtrise de mathématiques. [Course, studies and exercises for the Masters in mathematics.]

[251] A. Pietsch. *Nuclear locally convex spaces.* Springer-Verlag, New York, 1972. Translated from the second German edition by William H. Ruckle, Ergebnisse der Mathematik und ihrer Grenzgebiete, Band 66.

[252] P. Porcelli and E.H. Connell. A proof of the power series expansion without Cauchy's formula. *Bull. Amer. Math. Soc.*, 67:177–181, 1961.

[253] V.P. Potapov. The multiplicative structure of J-contractive matrix-functions. *Trudy Moskow. Mat. Obs.*, 4:125–236, 1955. English translation in: American mathematical society translations (2), vol. 15, p. 131–243 (1960).

[254] P. Quiggin. For which reproducing kernel Hilbert spaces is Pick's theorem true? *Integral Equations Operator Theory*, 16:244–266, 1993.

[255] E. Ramis. *Exercices d'algèbre avec solutions développées.* Masson, Paris, 1970.

[256] M. Reed and B. Simon. *Methods of modern mathematical physics. I. Functional analysis.* Academic Press, New York, 1972.

[257] M. Reed and B. Simon. *Methods of modern mathematical physics. I.* Academic Press Inc. [Harcourt Brace Jovanovich Publishers], New York, second edition, 1980. Functional analysis.

[258] E. Reich. Elementary proof of a theorem on conformal rigidity. *Proc. Amer. Math. Soc.*, 17:644–645, 1966.

[259] F.M. Reza. On the schlicht behavior of certain impedance functions. *IRE Transactions on circuit theory*, 9:231–232, 1962.

[260] M. Rosenblum. *Generalized Hermite polynomials and the Bose-like oscillator calculus*, volume 73 of *Operator Theory: Advances and Applications*, pages 369–396. Birkhäuser Verlag, Basel, 1994.

[261] M. Rosenblum and J. Rovnyak. *Hardy classes and operator theory*. Birkhäuser Verlag, Basel, 1985.

[262] W.H. Ruckle. *Sequence spaces*, volume 49 of *Research Notes in Mathematics*. Pitman (Advanced Publishing Program), Boston, Mass.-London, 1981.

[263] W. Rudin. *Principles of mathematical analysis*. McGraw-Hill Book Co., New York, third edition, 1976. International Series in Pure and Applied Mathematics.

[264] W. Rudin. *Analyse réelle et complexe*. Masson, Paris, 1980.

[265] W. Rudin. *Real and complex analysis*. McGraw-Hill, 1982.

[266] W. Rudin. *Real and complex analysis*. McGraw-Hill Book Co., New York, third edition, 1987.

[267] S. Saitoh. *Theory of reproducing kernels and its applications*, volume 189. Longman scientific and technical, 1988.

[268] S. Saitoh. *Integral transforms, reproducing kernels and their applications*, volume 369 of *Pitman Research Notes in Mathematics Series*. Longman, Harlow, 1997.

[269] S. Saks and A. Zygmund. *Analytic functions*. Monografie Matematyczne, Tom XXVIII. Polskie Towarzystwo Matematyczne, Warszawa, 1952. Translated by E.J. Scott.

[270] G. Sansone. *Orthogonal functions*. Dover Publications, Inc., New York, 1991. Revised English Edition.

[271] G. Sansone and J. Gerretsen. *Lectures on the theory of functions of a complex variable. I. Holomorphic functions*. P. Noordhoff, Groningen, 1960.

[272] G. Sansone and J. Gerretsen. *Lectures on the theory of functions of a complex variable. II: Geometric theory*. Wolters-Noordhoff Publishing, Groningen, 1969.

[273] D. Sarason. Invariant subspaces. In C. Pearcy, editor, *Topics in operator theory*, pages 1–47. Math. Surveys, No. 13. Amer. Math. Soc., Providence, R.I., 1974.

[274] D. Sarason. *Sub-Hardy Hilbert spaces in the unit disk*, volume 10 of *University of Arkansas lecture notes in the mathematical sciences*. Wiley, New York, 1994.

[275] Z. Sasvári. The extension problem for positive definite functions. A short historical survey. In *Operator theory and indefinite inner product spaces*,

volume 163 of *Oper. Theory Adv. Appl.*, pages 365–379. Birkhäuser, Basel, 2006.

[276] R. Schaback and H. Wendland. Kernel techniques: from machine learning to meshless methods. *Acta Numer.*, 15:543–639, 2006.

[277] E. Schechter. *Handbook of analysis and its foundations*. Academic Press, Inc., San Diego, CA, 1997.

[278] I.J. Schoenberg. Metric spaces and completely monotone functions. *Ann. of Math.* (2), 39(4):811–841, 1938.

[279] I.J. Schoenberg. On certain metric spaces arising from Euclidean spaces by a change of metric and their imbedding in Hilbert space. *Ann. of Math.* (2), 38(4):787–793, 1937.

[280] I.J. Schoenberg. Metric spaces and positive definite functions. *Trans. Amer. Math. Soc.*, 44(3):522–536, 1938.

[281] I. Schur. Über die Potenzreihen, die im Innern des Einheitskreises beschränkt sind, I. *Journal für die Reine und Angewandte Mathematik*, 147:205–232, 1917. English translation in: I. Schur methods in operator theory and signal processing. (Operator theory: Advances and Applications OT 18 (1986), Birkhäuser Verlag), Basel.

[282] I. Schur. Über die Potenzreihen, die im Innern des Einheitskreises beschränkt sind, II. *Journal für die Reine und Angewandte Mathematik*, 148:122–145, 1918. English translation in: I. Schur methods in operator theory and signal processing. (Operator theory: Advances and Applications OT 18 (1986), Birkhäuser Verlag), Basel.

[283] L. Schwartz. Sous espaces hilbertiens d'espaces vectoriels topologiques et noyaux associés (noyaux reproduisants). *J. Analyse Math.*, 13:115–256, 1964.

[284] L. Schwartz. *Analyse*. Hermann, Paris, 1970. Deuxième partie: Topologie générale et analyse fonctionnelle, Collection Enseignement des Sciences, No. 11.

[285] L. Schwartz. *Analyse. Topologie générale et analyse fonctionnelle*, volume 11 of *Collection Enseignement des Sciences*. Hermann, 1970.

[286] S.L. Segal. *Nine introductions in complex analysis*, volume 53 of *North-Holland Mathematics Studies*. North-Holland Publishing Co., Amsterdam, 1981. Notas de Matemática [Mathematical Notes], 80.

[287] S. Shimorin. Complete Nevanlinna–Pick property of Dirichlet-type spaces. *J. Funct. Anal.*, 191(2):276–296, 2002.

[288] S. Shirali and H.L. Vasudeva. *Metric spaces*. Springer-Verlag London Ltd., London, 2006.

[289] C.L. Siegel. *Topics in complex function theory, Volume II*. Wiley Classics Library. Wiley Interscience, 1988.

[290] M. Sifi and F. Soltani. Generalized Fock spaces and Weyl relations for the Dunkl kernel on the real line. *J. Math. Anal. Appl.*, 270:92–106, 2002.

[291] A.J. Smola and B. Schölkopf. A tutorial on support vector regression. *Stat. Comput.*, 14(3):199–222, 2004.

[292] J. Snygg. Wave functions rotated in phase space. *Amer. J. Phys.*, 45(1):58–60, 1977.

[293] P. Sorjonen. Pontryagin Räume mit einem reproduzierenden Kern. *Ann. Acad. Fenn. Ser. A. I*, pages 1–30, 1973.

[294] E.H. Spanier. *Algebraic Topology*. Springer-Verlag, 1989. First published 1966.

[295] M. Spivak. *A comprehensive introduction to differential geometry. Vol. II*. Publish or Perish Inc., Wilmington, Del., second edition, 1979.

[296] T.A. Springer. *Linear algebraic groups*, volume 9 of *Progress in Mathematics*. Birkhäuser Boston, Mass., 1981.

[297] L. Steen and J.A. Seebach. *Counterexamples in topology*. Dover Publications Inc., Mineola, NY, 1995. Reprint of the second (1978) edition.

[298] S. Stoïlow. Sur les transformations continues et la topologie des fonctions analytiques. *Ann. Sci. École Norm. Sup. (3)*, 45:347–382, 1928.

[299] S. Stoïlow. *Leçons sur les principes topologiques de la théorie des fonctions analytiques. Deuxième édition, augmentée de notes sur les fonctions analytiques et leurs surfaces de Riemann*. Gauthier-Villars, Paris, 1938.

[300] B. Sz.-Nagy, C. Foias, H. Bercovici, and L. Kérchy. *Harmonic analysis of operators on Hilbert space*. Universitext. Springer, New York, second, enlarged edition, 2010.

[301] F. Szafraniec. The reproducing kernel property and its space: more or less standard examples and applications. Handbook of operator theory, 2015.

[302] F. Szafraniec. The reproducing kernel property and its space: the basics. Handbook of operator theory, 2015.

[303] T. Takagi. On an algebraic problem related to an analytic theorem of Carathéodory and Fejér and on an allied theorem of Landau. *Japanese journal of mathematics*, 1:83–93, 1924.

[304] T. Takagi. Remarks on an algebraic problem. *Japanese journal of mathematics*, 2:13–17, 1925.

[305] E.G.F. Thomas. A polarization identity for multilinear maps. *Indag. Math. (N.S.)*, 25(3):468–474, 2014. With an appendix by Tom H. Koornwinder.

[306] E. Toubiana and R. Sá Earp. *Introduction à la géométrie hyperbolique et aux surfaces de Riemann*. Diderot Éditeur, Arts et Sciences, 1997.

[307] F. Treves. *Topological vector spaces, distributions and kernels*. Academic Press, 1967.

[308] F. Tricomi. Les transformations de Fourier, Laplace, Gauss, et leurs applications au calcul des probabilités et à la statistique. *Ann. Inst. H. Poincaré*, 8(3):111–149, 1938.

[309] F. Tricomi. *Integral equations*. Dover Publications, Inc., New York, 1985. Reprint of the 1957 original.

[310] M. Tsuji. *Potential theory in modern function theory*. Maruzen, Tokyo, 1959.

[311] G. Våge. A general existence and uniqueness theorem for Wick-SDEs in $S^n_{-1,k}$. *Stochastic Sochastic Rep.*, 58:259–284, 1996.

[312] N.N. Vakhania. *Probability distributions on linear spaces*. North-Holland Publishing Co., New York, 1981. Translated from the Russian by I.I. Kotlarski, North-Holland Series in Probability and Applied Mathematics.

[313] A. van den Bos. Alternative interpretation of maximum entropy spectral analysis. *IEEE Trans. Inform. Theory*, 17:493–494, 1971.

[314] S.J.L. van Eijndhoven and J.L.H. Meyers. New orthogonality relations for the Hermite polynomials and related Hilbert spaces. *J. Math. Anal. Appl.*, 146(1):89–98, 1990.

[315] K. Venkatachaliengar. *Development of elliptic functions according to Ramanujan*, volume 6 of *Monographs in Number Theory*. World Scientific Publishing Co. Pte. Ltd., Hackensack, NJ, 2012. Edited, revised, and with a preface by Shaun Cooper.

[316] V. Vinnikov. Commuting operators and function theory on a Riemann surface. In *Holomorphic spaces (Berkeley, CA, 1995)*, pages 445–476. Cambridge Univ. Press, Cambridge, 1998.

[317] D.V. Voiculescu, K.J. Dykema, and A. Nica. *Free random variables*, volume 1 of *CRM Monograph Series*. American Mathematical Society, Providence, RI, 1992. A noncommutative probability approach to free products with applications to random matrices, operator algebras and harmonic analysis on free groups.

[318] G.F. Voronoi. Extension of the notion of the limit of the sum of terms of an infinite series. *Ann. of Math. (2)*, 33(3):422–428, 1932.

[319] G.N. Watson. Notes on Generating Functions of Polynomials: (2) Hermite Polynomials. *J. London Math. Soc.*, S1-8(3):194, 1933.

[320] G. Whyburn. *Topological analysis*. Second, revised edition. Princeton Mathematical Series, No. 23. Princeton University Press, Princeton, N.J., 1964.

[321] N. Wiener. Tauberian theorems. *Ann. of Math.*, 33:1–100, 1932.

[322] J.D. Maitland Wright. All operators on a Hilbert space are bounded. *Bull. Amer. Math. Soc.*, 79:1247–1250, 1973.

[323] B. Yazici and R.L. Kashyap. A class of second-order stationary self-similar processes for $1/f$ phenomena. *IEEE Trans. on Signal Processing*, 45:396–410, 1997.

[324] A. Yger. *Théorie et analyse du signal.* Mathématiques appliquées. Ellipses, Éditions Marketing S.A., 32 rue Bargue, Paris 15e, 1999.

[325] A.I. Zayed. Chromatic expansions and the Bargman transform. In Xiaoping Shen and Ahmed Zayed, editors, *Multiscale signal analysis and modeling*, pages 139–159. Springer, 2013.

[326] A.H. Zemanian. *Distribution theory and transform analysis.* Dover Publications Inc., New York, second edition, 1987. An introduction to generalized functions, with applications.

[327] A.H. Zemanian. *Realizability theory for continuous linear systems.* Dover Publications, Inc., New York, 1995.

[328] M. Zorn. Note on power series. *Bull. Amer. Math. Soc.*, 53:791–792, 1947.

Index

absolutely continuous function, 299
absorbing set, 200
additive regular function, 299
adjoint operator, 211
affine hyperplane, 203
algebra
  of sets, 299
  tensor algebra, 22
  Wiener, 37
algebraic function field, 24
algebraic supplement, 19
analytic structure, 145
annulus
  conformal equivalence
    of two annuli, 146
  definition, 64
arc-length, 85
argument principle, 75
atlas, 144
  differentiable, 144
  differential equivalent atlases, 144
automorphic function, 17
axiom of choice (and reproducing
  kernel Hilbert spaces), 356

backward shift operator, 445
backward shift realization, 35
Baire's theorem, 130, 152
balanced set, 200
Banach space, 203
Bargmann transform, 71, 476
  relation with Fourier transform,
    314
barrel, 200, 251

barreled space, 251
basic separation theorem, 20
basis of neighborhoods, 123
Bergman space, 204
  of polyanalytic functions, 464
  weighted, 462
Bernoulli numbers, 63
Bessel function, 297
  modified, of the second kind of
    order 0, 355
Blaschke factor
  disk case, 18
  right half-plane case, 370
Blaschke product, 370
Blaschke–Potapov factor, 41, 371
Bochner's theorem, 332, 341
  and characteristic functions of
    random variable, 347
Bolzano–Weierstrass property, 128
bosonic Fock space, 22
bounded set (in a topological vector
  space), 253
bounded variation function, 299

canonical projection, 13
Cantor set, 128
[CAPB], 6
Carathéodory function, 75, 87, 451
Carathéodory's theorem, 449
cartesian product, 136
Cauchy
  filter, 249
  sequence, 151
Cesàro mean, 61

characteristic function
    of a random variable, 347
characteristic operator function, 220
chart, 143
    at a point x, 143
Chu–Vandermonde formula, 29
closed graph therorem, 209
closed map, 137
closure of a set, 125
Colombo, F., 447
compact operator, 211
    example, 267
compact set, 127
compatible norms, 253
complementary space, 218
complementation, 218
complete metric space, 151
complete Nevanlinna–Pick kernel,
    61, 366
complex plane
    one point compactification, 129
complex projective plane
    as analytic manifold, 182
complex projective space, 143
    as analytic manifold, 145
composition operator, 363, 412
    example, 81
    in Bergmann space, 462
conditionally negative function, 347
conformal equivalence, 14
    of two annuli, 146
continuity
    criteria for sequential continuity,
    154
continuous function
    at a point, 132
    global definition, 132
continuous logarithm, 146
converging filter, 127
convex set
    internal point, 20
convolution
    tempered distributions, 260
convolution of sequences, 260

counterexample
    closure of a set larger than the
    set of limits of sequences, 125
    Hilbert space of functions which
    is not a reproducing kernel
    Hilbert space, 356
    topological vector space not
    locally convex, 251
covariance extension problem, 33
covariance function, 4
    definition, 347
    of the fractional Brownian
    motion, 5
Cuntz relations, 375, 413

de Branges' identity, 418
decimation operator, 81
degree
    of an extension field, 23
    trancendence, 24
derivative
    directional, 85
    normal, 85
diagonal process, 193, 281
diameter of a set, 196
diffeomorphism, 144
differentiable atlas, 144
differential structure, 144
dimension
    of a topological manifold, 142
directed set, 126
directional derivative, 85
disjoint union, 136
down-sampling operator, 81
Dunkl kernel, 292

equivalence relation, 13
    closed, 141
    open, 141
essential spectrum, 210
exactity relation, 105, 107, 400
example
    of an open map, 209
extended complex plane, 124
extended real line, 122

extended resolution space, 250

factor
 Blaschke (disk case), 18
 Blaschke (right half-plane case),
   370
 Blaschke–Potatov (first, second
   and third kind, 371
 Weierstrass, 68
factorization property, 254, 263, 352
field
 finite degree extension, 23
 finite generated extension, 23
 finite type extension, 23
filter, 126
 Cauchy, 249
 converging, 127
finite Blaschke product, 370
finite Blaschke product (open unit
   disk), 80
finite degree extension field, 23
finite generated extension field, 23
finite number of negative squares
   kernel, 332
finite type extension field, 23
first category set, 129
Fock space
 full, 22
formula
 Chu–Vandermonde, 29
 for the orthogonal projection on a
   closed sum, 206
 Herglotz representation, 87
 Mehler's, 291, 389
 Stieltjes–Perron inversion, 300
Fornasini–Marchesini realization, 40
Fourier rotation operator, 296
Fourier transform, 285
 an example of Bass and Lévy, 289
 Fourier rotation operator, 296
 inverse, 286
 of $\frac{1}{\cosh t}$, 289
 of a rational function, 288
 of the Hermite functions, 289

relation with Bargmann
   transform, 314
 spectrum, 296
Fréchet space, 255
 being barreled, 256
fractional Hardy space (half-plane),
   420
Fubini's theorem, 303
full Fock space, 22, 476
function
 absolutely continuous, 299
 automorphic, 17
 Carathéodory, 75, 87, 451
 Gamma, 74
 generalized Carathéodory, 75
 J-contractive, 221, 371
 Laguerre, 417
 lower semi-continuous, 133
 modified Bessel function, 355
 of bounded variation, 299
 positive real, 39
 Rademacher, 337
 subharmonic, 88
 uniformly continuous, 154
 Weierstrass $\wp$, 464

Gamma function, 74
gauge
 Minkowski, 252
generalized Carathéodory function,
   75
generalized Schur algorithm, 372
Givone–Roesser realization, 40
Gram matrix, 226
graph of an operator, 222
Green's formula, 64
Green's function, 342
 of the one-dimensional
   Schrödinger operator, 225
group
 action, 18
 separating, 143
 totally discontinuous, 143
 transitive, 18

automorphism, 18
  modular, 15
  special linear, 15

Hölder's inequality, 204
Hadamard product, 27
Hahn–Banach theorem, 19
Hamel basis, 19
Hankel matrix, 304, 473, 483
  example, 30, 474
Hankel transform, 297
Hardy space
  fractional Hardy space of the
    open unit disk, 409
  of the bidisk, 359
  of the open right half-plane, 416
  of the open unit disk, 406
    composition operator, 412, 413
Heine–Borel–Lebesgue property, 127
Herglotz representation formula, 87
Hermite functions, 289
  Fourier transform, 289
  Mehler's formula, 291
  normalized, 290
Hermite polynomials, 289, 304
  as orthogonal polynomials, 290
  definition, 67
  differential equation, 66
  formulas, 67
  generalized, 292
  reproducing kernel formula, 374
Hermitian form, 22
Hilbert–Schmidt operator, 212, 223,
  361
homeomorphic sets, 83
homeomorphism, 135
homogeneous polynomial, 143

indefinite inner product space, 225
index, 80
  of a subgroup, 15
induced topology, 123
inequality
  Hölder's, 204
  Minkowski, 204

infinite-dimensional analysis, 477
interior of a set, 125
internal point of a convex set, 20
inverse
  of the Fourier transform, 286
inverse Fourier transform, 286
inverse image, 12

J-contractive function, 221, 371
Jordan curve, 135
Jordan–Kronecker function, 267

Kaluza
  sequence, 62, 63
Kaluza's theorem, 61
Kreĭn space, 226

Laguerre function, 416, 417
Laguerre polynomial, 305, 417
Landau constants, 453
Laplace transform, 420
lifted norm, 377
linear relation, 54, 221
locally compact space, 130
locally convex space, 252
lower semi-continuous function, 133
Lyapunov equation, 369

MacDonald's function, 343, 355
manifold
  complex dimension, 145
  differential, 144
  dimension, 142
map
  closed, 137
  open, 137, 209
  proper, 138
matrix
  equation
    Lyapunov, 369
    Stein, 375
  Hadamard product, 27
  Halmos extension, 26
  Hankel, 304, 373
  matrix equations, 27
  nonnegative, 28

positive, 28
semi-simple, 24
signature matrix, 26
strictly positive, 28
sum of positive matrices, 32
Toeplitz, 33, 305
Mehler's formula, 291, 336, 389
meromorphic function
Mittag-Leffler expansion, 79
metric space
complete, 151
minimal realization, 36
Minkowski functional, 200, 252
Minkowski gauge, 252
modular group, 15
moment problem on an interval, 462
Montel space, 202
multiplication operator, 363
multiplier, 363

Nörlund means, 61
neighborhood, 123
neighborhood of infinity, 85
Nevanlinna–Pick interpolation
problem, 448
nonnegative matrix, 28
norm, 199
compatible norms, 155, 253
pairwise coordinated norms, 254
normal derivative, 85
normal family, 154

observability, 369, 411
observable pair of matrices, 411
one point compactification, 131
complex plane, 129
open ball in a metric space, 150
open connected simply connected
set, 83
operator
adjoint, 211
backward shift, 445
closed, 222
compact, 211
example, 267

composition and multiplication
(together), 363
Fourier rotation operator, 296
graph, 222
Hilbert–Schmidt, 212, 223, 361
positive, 214
second quantization of an
operator, 22
spectral radius, 210
outer normal, 85

pair of matrices
observable, 369, 411
parity of a permutation, 338
Parseval's equality, 206
partial fraction decomposition, 78
Poincaré Theta series, 17
point at infinity, 124
point spectrum, 210
polarization identity, 22
Polya's theorem, 295, 347
polyanalytic functions
Bergman space (disk), 464
polynomial
Laguerre, 417
polynomials
density, 463
Hermite, 67, 304
Laguerre, 305
Tchebycheff, of the second kind,
464
Pontryagin space, 226
positive definite
function, 332
kernel, 332
positive definite function
and infinite products, 69
and tensor product, 351
as sum of squares, 335
Polya's conditions, 295
positive matrix, 28
positive operator, 214
positive real function, 39
product topology, 137

projective curve, 143
projective space
　complex, 143
proper map, 138

quotient topology, 140
　universal property, 140

Rademacher functions, 337
rational function
　Fornasini–Marchesini realization,
　　40
　Givone–Roesser realization, 40
　J-inner, 371
　J-unitary on the real line, 370
　J-unitary on the unit circle, 375
　McMillan degree, 36
　minimal factorization, 36
　minimal product, 36
　partial fraction decomposition, 78
　realization centered at infinity, 35
　realization centered at the origin,
　　35
　schlicht behavior, 39
　unitary on the unit circle, 41
real topological manifold
　counterexamples, 142
　definition, 142
realization
　backward shift, 35
　Fornasini–Marchesini realization,
　　40
　Givone–Roesser, 40
　minimal, 36
relation, 13
　equivalence, 13
　linear, 221, 394
reproducing kernel, 352
　of complete Nevanlinna–Pick
　　type, 61, 366
　of the Bergman space of the
　　annulus, 266
　formula for the finite-dimensional
　　case, 367
　formula in the general case, 335

of the Bergman space of the
　annulus, 472
of the Fock space, 476
of the Fock space of polyanalytic
　functions, 5, 339
of the fractional Hardy space
　(open upper half-plane), 441
of the fractional Hardy space
　(right half-plane), 74
of the Hardy space (lower upper
　half-plane), 439
of the Hardy space (open upper
　half-plane), 439
of the Hardy space of the bidisk,
　359
reproducing kernel Hilbert space,
　335, 352
　and axiom of choice, 356
　composition operator, 363
　multiplication operator, 363
　sum of, 358
residual spectrum, 210
resolvent equation, 35, 210
resolvent identity, 35, 210
resolvent set, 209
Riemann sphere, 124
Riemann surface
　application of operator theory to,
　　214
Riemann's integral, 298
Riemann's lemma, 288
Riemann's mapping theorem, 14, 87
Riesz projection, 25, 211, 344
Rouché's theorem, 75

saturated set, 13
schlicht rational function, 39
Schrödinger operator, 225
Schur algorithm, 451
Schur complement, 32
Schur complement formula, 32
Schur function, 35
Schur's lemma, 31
Schur–Agler functions, 447

INDEX

Schwartz kernel theorem, 361
Schwartz space, 256
second category set, 129
second quantization, 22
second-order stochastic process, 347
semi-norm, 199, 200
semi-simple matrix, 24
separating group action, 143
sequence
    convolution of sequences, 260
    superexponential, 261
    with the Kaluza sign property,
        62, 74
sequential compactness, 154
sequential continuity, 154
set
    absorbing, 200
    balanced, 200
    barrel, 251
    boundary of a set, 126
    bounded (in a topological vector
        space), 253
    closure, 125
    compact, 127
    complement, 11
    diameter, 196
    first category, 129
    interior, 125
    meager, 129
    nowhere dense, 129
    open connected simply connected,
        83
    saturated, 13
    second category, 129
    simply connected, 1
    star-shaped, 82
    symmetric difference, 338
signature matrix, 26
simply connected set, 1
Sobolev space, 294, 353, 391, 392
space
    Kreĭn, 226
    Baire, 130
    Banach, 203

Bargmann, 476
Bargmann–Fock–Segal, 476
barreled, 134, 251
Bergman, 204
    of polyanalytic functions in the
        disk, 464
Fock, 476
Fréchet, 255
full Fock space, 22
Hardy fractional Hardy space
    (half-plane), 420
Hardy space of the open right
    half-plane, 416
Hardy space of the open unit
    disk, 406
indefinite inner product space,
    225
locally compact, 130
locally convex, 252
metric, 149
metric, totally bounded, 152
metrizable, 152
Montel, 202
path-connected, 134
Pontryagin, 226
regular, 124
Schwartz, 288
Schwartz (of smooth functions),
    256
Sobolev, 294, 353, 391, 392
symmetric Fock space, 22
topological, 122
topological vector, 249
ultra-metric, 149
weighted Bergman, 408
with reproducing kernel, 352
special linear group, 15
spectral radius, 210
spectrum, 209
    essential, 210
    point, 210
    residual, 210
squareroot
    of a positive operator

Hilbert space, 215
Stein equation, 375
stereographic
    metric, 156
    projection, 156
Stieltjes integral
    integration by parts formula, 300
Stieltjes–Perron inversion formula,
    300
strictly positive matrix, 28
strong algebras, 259
strong operator topology, 207
subharmonic function, 88
sum
    Cesàro, 61
    of positive matrices, 32
superexponential sequence, 261
surface
    topological, 142
symmetric Fock space, 22

tangent vector, 85
Taylor's formula with remainder, 3,
    388
tensor algebra, 22
tensor product, 21
    of positive definite functions, 351
theorem
    argument principle, 75
    Baire, 125, 130, 152
    Banach–Steinhaus, 208
    basic separation theorem, 20
    Beurling, 411, 444
    Brouwer's invariance of the
        domain theorem, 142
    Carathéodory (on the angular
        derivative), 449
    Casorati–Weierstrass, 84, 115
    closed graph, 209
    conformal equivalence
        of two annuli, 146
    Fubini, 303
    Green, 63

Hadamard three circles theorem,
    88
Hahn–Banach, 19
Hurwitz, 76
Kaluza, 61
Menger's imbedding theorem, 349
Mergelyan, 463
Mittag-Leffler expansion, 79
Molchanov's (on the discreteness
    of the spectrum of a
    differential expression), 224
on density of polynomials in a
    Lebesgue space, 463
on solvability of certain matrix
    equations, 27
on sum of positive kernels, 358
one point compactification, 131
open mapping theorem, 138
open mapping theorem (for
    operators between Banach
    spaces), 211
open mapping(analytic
    functions), 75
Picard's Big theorem, 84
Polya (on positive definite
    functions), 295, 347
Riemann's mapping theorem, 14
Riesz representation theorem for
    functionals in a Hilbert space,
    204
Rouché, 75
Schoenberg (on radial positive
    definite functions), 342
Schwartz' kernel theorem, 225,
    361, 389
spectral mapping, 217
Stoilow, 139
Tychonoff, 127
uniform boundedness, 208
Vitali's completeness theorem,
    303, 416
Wiener-Lévy, 37
Theta Fuchsian series, 17
Toeplitz matrix, 33, 305

INDEX

topological analysis, 136, 450
topological surface, 142
topological vector space, 249
    bounded set, 253
    factorization property, 254, 263
topology, 122
    coarser, 123
    finer, 123
    induced, 123
    product, 137
    quotient, 140
    strong operator, 207
    stronger, 123
    $\sigma(\mathcal{V}', \mathcal{V})$, 201
    weak operator, 207
    weak-$*$, 201
    weaker, 123
totally discontinuous group action, 143
transform
    Bargmann, 71
    Fourier, 285
    Fourier–Bessel, 297
    Hankel, 297
    Laplace, 420
    Mellin, 355
    Segal–Bargmann, 71
    z-transform, 407
transition map, 144
trigonometric moment problem, 451
type
    of an extension field, 23

Våge inequality, 262

weak operator topology, 207
weak-$*$ topology, 201
Weierstrass
    function $\wp$, 464
    factor, 68
Weierstrass function, 464
weighted composition operator, 363
    in the Hardy space of the disk, 413
    example in Fock space, 477

in the Hardy space of the open upper half-plane, 418
Wiener algebra, 37
winding number, 80

Zariski topology, 123, 158

Name Index

Agler, J., 36, 366
Aronszajn, N., 14, 206
Artstein-Avidan, S., 285
Azizov, Th., 372

Balk, M.B., 464
Ball, J., 447
Baricz, A., 62, 366
Bart, H., 287
Bass, J., 289, 309
Bercovici, H., 220
Bergman, S., 146
Berlinet, A., 335, 353
Bognár, J, 225
Bolotnikov, V., 18, 415, 447
Boschet, F., 12
Bożejko, M., 288, 352
Bretagnolle, J., 343
Brezis, H., 303
Brouwer, L.E.J., 142
Burckel, R., 1

Calvo, A., 12
Calvo, B., 12
Cartan, H., 281
Chamfy, C., 372
Chen, C.P., 453
Choi, J., 453
Connell, E.H., 136

Dacunha Castelle, D., 343
de Branges, L., 36, 220, 293, 297, 374, 397, 420
Delsarte, P., 372
Dieudonné, J., 20

Dijksma, A., 372
Donoghue, W., 356
Doyen, J., 12
Dym, H., 26, 397

Faifman, D., 285
Feintuch, A., 250
Foias, F., 220
Forster, O., 129
Fuhrmann, P.A., 35

Gelfand, I., 287
Genin, 372
Godement, R., 26
Gohberg, I., 220, 287
Gunning, R.C., 214
Górniak, J., 254, 263

Halmos, P, 14
Hardy, G.H., 61, 291
Hasebe, T., 288
Hille, E., 67, 290, 331

Iohvidov, I.S., 75

Jorgensen, P., 28
Julia, G., 68, 146

Kaashoek, R., 287
Kalman, R.E., 335
Kaluza,Th., 61, 366
Kalyuzhnyi-Verbovetskiĭ, D., 447
Kamp, Y., 372
Kashyap, R.L., 364
Kendall, D., 62
Kérchy, L., 220
Khaleelulla, S.M., 259

Kirwan, F., 143
Kober, H., 207
Kondratiev, Y., 259, 261
Košelev, A.D., 466
Kreĭn, M.G., 41, 75, 220, 332, 370, 371, 447
Krivine, J.-L., 343

Lamperti, J., 62, 366
Landau, E., 453
Langer, H., 41, 370–372, 447
Lebedev, N.N., 298
Leborgne,D., 142
Lee, T.W., 416
Lévy, P., 289, 292, 309
Li, D., 385
Loève, M., 347
Luna-Elizarrarás, M.E., 447

Masani, P., 356
Mazur, S., 23
Mboup, M., 420
McCarthy, J., 36
Mehler, F.G., 291, 336
Menger, K., 157, 349
Meschkowski, H., 336
Meyers, J., 258
Milman, V., 285
Molchanov, A.M., 224
Muhly, P., 36

Narasimhan, R., 250
Natanson, I.P., 300
Naĭmark, M. A., 223
Nehari, Z., 463
Nelson, E., 140
Neretin, Y., 475

Orlicz, W., 23

Polya, G., 295
Porcelli, P., 136
Potapov, V., 220, 371

Queffélec, H., 385

Rabitz, H., 225, 343

Ramis, E., 12
Rodman, L., 18, 415
Rosenblum, M., 292
Rovnyak, J., 36, 220, 374, 443
Ruckle, W., 279

Sabadini, I., 447
Saeks, R., 250
Saitoh, S., 331
Saks, S., 79, 99
Salomon, G., 259
Sarason, D., 307, 449
Schoenberg, I., 342, 350
Schur, I., 450
Schwartz, L., 361
Shannon, C.E., 341
Shapiro, M., 447
Siegel, C.L., 17
Sifi, M., 292
Snygg, J., 296
Solel, B., 36
Soltani, F., 292
Spanier, E.H., 142
Spivak, M., 142
Stoïlow, S., 121, 139
Struppa, D., 447
Sz.-Nagy, B., 220
Szafraniec, F., 331
Szpilrajn, E., 129

Thomas, E.G.F., 23
Thomas-Agnan, Ch., 335, 353
Treves, F., 250

Vakhania, N., 151
van den Bos, A., 33
van Eijndhoven, S., 258
Vesti, J., 62, 366
Vinnikov, V., 447
Volok, D., 447
Voronoi, G, 61
Vuorinen, M., 62, 366
Våge, G., 261

Wanjala, G., 372
Watson, G.N., 291

Weron, A., 254
Whyburn, G., 136

Yaglom, A.M., 287
Yazici, B., 364
Yger, A., 361

Zayed, A., 71
Zemanian, A., 361
Zorn, M., 129
Zygmund, A., 79, 99

Notation Index

1_A, characteristic function of the set A, 11

$B(0, R)$, the open disk centered at the origin and with radius R, 67

$B(x, r)$, open ball with center x and radius r, 150

$B_c(x, r)$, closed ball with center x and radius r, 150

C_R, the circle centered at the origin and of radius R, 79, 438

$\check{f}$, inverse Fourier transform, 286

ℓ_2, space of square summable sequences of complex numbers indexed by $\mathbb{N}$, 153

$\mathbb{C}_+$, open upper half-plane, 17

$\mathbb{C}_-$, open lower half-plane, 439

$\mathbb{C}_r$, open right half-plane, 3, 39

$\mathbb{D}$, the open unit disk, 3

$\mathbb{P}^n$, complex projective space, 143

$\mathbb{R}_+$, the half-line $[0, \infty)$, 3

$\mathbb{T}$, unit circle, 37

$\mathbf{L}(\mathcal{V})$, equal to $\mathbf{L}(\mathcal{V}, \mathcal{V})$, 207

$\mathbf{L}(\mathcal{V}_1, \mathcal{V}_2)$, the space of linear continuous operators between normed spaces, 207

$\mathbf{L}_2(\mathbb{R})$, Lebesgue space, 6

$\mathcal{B}_p(\Omega)$, Bergman space, 204

$\mathcal{P}(X)$, set of all subsets of X, 11

$\mathscr{S}$, Schwartz space of smooth functions, 256

$\mathscr{S}$, real-valued Schwartz functions, 256

$\mathscr{S}'_{\mathbb{R}}$, real tempered distributions, 256

$\mathscr{S}_{\mathbb{R}}$, space of real-valued Schwartz functions, 345

$\widehat{f}$, Fourier transform, 285

$\xi_n(z)$, Hermite function, 289

$b_a(z)$, Blaschke factor (open unit disk), 18

$d\mu_g = \frac{1}{\sqrt{\pi}} e^{-x^2} dx$, the Gaussian probability measure, 303

$k_s(z, w)$, kernel equal to $\frac{1 - s(z)\overline{s(w)}}{1 - z\overline{w}}$, 36

$\mathrm{Dom}\,(T)$, domain of the operator T, 211

$\mathrm{ran}\,M$, range of the matrix (or of the operator) A, 31

nted in the United States
Bookmasters